Engineering Materials

This series provides topical information on innovative, structural and functional materials and composites with applications in optical, electrical, mechanical, civil, aeronautical, medical, bio- and nano-engineering. The individual volumes are complete, comprehensive monographs covering the structure, properties, manufacturing process and applications of these materials. This multidisciplinary series is devoted to professionals, students and all those interested in the latest developments in the Materials Science field, that look for a carefully selected collection of high quality review articles on their respective field of expertise.

Indexed at Compendex (2021) and Scopus (2022)

Neetu Talreja · Divya Chauhan ·
Mohammad Ashfaq
Editors

Two-dimensional Hybrid Composites

Synthesis, Properties and Applications

Springer

Editors
Neetu Talreja
Alliance University Bangalore
Bangalore, India

Divya Chauhan
Ernst & Young Global LLP
New Delhi, Delhi, India

Mohammad Ashfaq
Chandigarh University
Mohali, India

ISSN 1612-1317 ISSN 1868-1212 (electronic)
Engineering Materials
ISBN 978-981-99-8009-3 ISBN 978-981-99-8010-9 (eBook)
https://doi.org/10.1007/978-981-99-8010-9

This Springer imprint is published by the registered company Springer Nature Singapore Pte Ltd.
The registered company address is: 152 Beach Road, #21-01/04 Gateway East, Singapore 189721, Singapore

Paper in this product is recyclable.

Preface

This book "Two-dimensional (2D) based Hybrid Composites and Their Applications," explores the fascinating world of hybrid composites that incorporate 2D materials, exploring their synthesis, characterization, properties, and diverse applications. As researchers, academicians, and engineers seek novel ways to address complex challenges across industries ranging from environments to medicine, this book provides an insightful overview of the potential solutions that 2D-based hybrid composites offer.

The chapters begins by introducing the basic concepts of 2D materials, highlighting their unique structures, properties, and potential applications. Introduction of 2D material also covers the synthesis of 2D materials with precise control over their dimensions and properties. After synthesis, characterization techniques, including microscopy, spectroscopy, and diffraction, are explored to provide a comprehensive understanding of the structure and properties of these materials.

In this book, we majorly delve into various applications of two-dimensional nanomaterials such as solar cell, environment, biomedical, sensing, and agriculture that enable the applicability of 2D materials in diverse fields. The journey continues by delving into the properties that make 2D materials stand out as building blocks for hybrid composites. Their exceptional mechanical strength, electrical conductivity, and thermal stability make them ideal candidates for reinforcing traditional materials. The synergistic effects of combining 2D materials with other components, such as polymers, metals, and ceramics, create composites with enhanced mechanical, electrical, and thermal properties. This section explores the mechanisms behind these improvements and presents case studies showcasing the diversity of composite properties achievable through smart material selection and design.

The book's main focus lies in exploring the myriad applications of 2D-based hybrid composites and their applications in environment, energy, agriculture, and medicine. From lightweight and high-strength to flexible and conductive electronics, these composites are transforming industries by overcoming the limitations of conventional materials. The chapters highlight cutting-edge research and real-world applications, emphasizing how hybrid composites form a shape.

No exploration of a groundbreaking field would be complete without addressing the challenges that lie ahead. These chapters examine the barriers of researchers and engineers must overcome to fully harness the potential of 2D-based hybrid composites. Whether it's scalability of synthesis methods, understanding the interfacial interactions, or addressing environmental concerns, these obstacles provide valuable insights for the next generation of materials scientists.

In conclusion, "Two-dimensional (2D) based Hybrid Composites and Their Applications" is an enlightening journey into the world of materials science that blends the exceptional properties of 2D materials with the versatility of composite engineering. This book serves as a guide for researchers, students, and professionals interested in understanding the synthesis, properties, and applications of these extraordinary materials. It encapsulates the spirit of innovation that drives materials science forward and opens the door to a new era of advanced materials with limitless possibilities.

Bangalore, India — Neetu Talreja
New Delhi, India — Divya Chauhan
Mohali, India — Mohammad Ashfaq

Contents

Boron Nitride and Its Hybrids: Synthesis, Properties and Potential Applications

Kulwinder Singh, Sawini, Anup Thakur, and Akshay Kumar

Abstract Two-dimensional (2D) materials and their hybrid structures have garnered immense attention over the past two decades. The distinct properties exhibited by these materials offer potential across various domains, from electronics to energy applications. As our understanding of these materials deepens, and synthesis methods improve, the potential applications appear limitless. From reshaping electronics to revolutionizing energy storage, 2D materials, and their hybrids are undeniably at the forefront of the next wave of technological advancements. The 2D materials, such as graphene, transition metal dichalcogenides (TMDs), and hexagonal boron nitride (h-BN), have garnered immense attention over the past decade due to their unique properties, which differ significantly from their bulk counterparts. This chapter overviews the properties, applications, and motivation towards using boron nitride (BN) and its hybrid-based materials. Detailed properties of BN and its nanocomposites are elaborated. Different synthesis methods are also included. It also includes the systematic literature review of BN and its composites with various materials for gas sensing, photodetection, and other applications perspective.

K. Singh (✉)
University Centre for Research and Development, Chandigarh University, Mohali 140 413, Punjab, India
e-mail: ksingh.nano@gmail.com

Sawini
Department of Physics, Chandigarh University, Mohali 140 413, Punjab, India

A. Thakur
Advanced Materials Research Lab., Department of Basic and Applied Sciences, Punjabi University, Patiala 147 002, Punjab, India

A. Kumar
Department of Physics, Sardar Patel University, Mandi 175 001, Himachal Pradesh, India

N. Talreja et al. (eds.), *Two-dimensional Hybrid Composites*, Engineering Materials,
https://doi.org/10.1007/978-981-99-8010-9_1

1 Introduction

Materials with at least one dimension in the nano-range are considered nanomaterials. Nanomaterials can be divided into four categories based on dimensionality, as shown in Fig. 1. (a) Zero-dimensional, (b) One-dimensional, (c) Two-dimensional, and (d) Three-dimensional nanomaterials, respectively.

The story of 2D materials starts with graphene. Before its isolation in 2004 by researchers Andre Geim and Konstantin Novoselov at the University of Manchester, graphene was theoretically known but had never been isolated as a standalone material. This discovery, made using a simple technique involving Scotch tape and graphite, led to a Nobel Prize in Physics in 2010. Graphene exhibited remarkable properties: extraordinary electrical conductivity, mechanical strength, and thermal conductivity, setting the stage for subsequent research in the field. After discovering graphene, the scientific community began considering other materials that could exist in a 2D form. These efforts bore fruit, leading to the identification of several 2D materials, including:

i. Transition metal dichalcogenides (TMDs): Unlike graphene, which is a semimetal, TMDs like molybdenum di-sulfide (MoS_2), tungsten disulfide (WS_2), and molybdenum di-selenide ($MoSe_2$) are semiconductors. These exhibit properties make them suitable for optoelectronic applications, such as photodetectors and light-emitting diodes.
ii. Hexagonal boron nitride (h-BN): Often referred to as "white graphene," h-BN is a large band gap semiconductor, making it a valuable substrate or dielectric layer for electronic devices and useful in other applications.
iii. Phosphorene: This is a monolayer form of black phosphorus. With a direct bandgap and high charge carrier mobility, it has potential in electronics and optoelectronics.
iv. MXenes: A family of 2D transition metal carbides, nitrides, and carbonitrides with promising applications in energy storage and catalysis.

In the last two decades, functional materials such as large bandgap semiconductor materials have gained the attention of scientists due to their potential usage

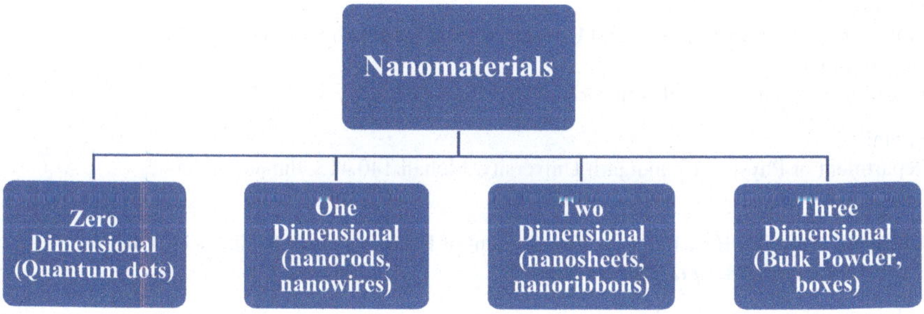

Fig. 1 Classification of nanomaterials

in nanoelectronics, photonics, and optoelectronics. These materials show tunable properties at nanoscale than their bulk form, making them suitable for the aforementioned applications [1]. Numerous studies have been devoted and reported to investigate semiconductor properties at nanoscale regimes [2–5]. The 2D nanostructures such as graphene, boron nitride (BN), and MoS_2 have gained the intensive attention of scientists and industrialists due to their fascinating thermal, electrical, and mechanical properties [6–9]. In 2004, graphene has been formed and isolated for the first time [7, 10]. This invention opens a new research field related to 2D nanomaterials, which is growing exponentially. A single atomic thick carbon layer with a honeycomb lattice is known as graphene, one of the glorified materials in the first decade of the twenty-first century. Graphene shows noteworthy thermal, mechanical, and electrical properties than other materials [11, 12]. High surface area and high conductivity make it a promising candidate useful for sensing purposes. The selectivity of graphene towards different molecules makes it useful in membrane applications [13, 14]. These properties of graphene also triggered the application research field towards shielding materials [15]. The physicochemical properties of graphene lead to the attention of the academic community as well as chemical and materials companies depending upon the usage of graphene-based products. This started the research and observation related to thin sheets of different layered nanomaterials like BN [16], TMDs such as iron disulfide (FeS_2) [17], MoS_2 [18], tungsten di-selenide (WSe_2) [19], WS_2 [20], tin sulfide (SnS_2) [21] etc. and MXenes, such as phosphorene [22], carbonitrides [23], and other transition metal carbides.

The structural and chemical properties of 2D nanomaterials are related to their dimensionality [24] and offer enormous prospects for different fields, from fundamental to applied research. Nitrides-based on boron possess excellent properties which show their potentiality for sensing, biomedical, optical, and electrical applications. Tunable bandgap of BN and its hybrids make them a suitable candidates for gas sensing and photodetection applications.

2 Structure and Properties Perspectives of BN and Its Hybrids

BN is one of the nitride-based on boron, possessing an equal number of boron and nitrogen atoms [25]. In 1842, BN was synthesized by Balman for the first time by a chemical reaction. Boric acid was processed with potassium cyanide, which forms BN, which was not stable in nature [26]. With the development of technology and synthesis methods, BN has become an inexpensive material. For the first time, boron nitride nanotubes (BNNTs) were obtained by Chopra et al. in 1995 [27]. BNNTs possess properties independent of nanotube chirality, diameter, and number of tube walls [28]. The 2D BN nanostructures were first synthesized in 2004 [29]. BN nanosheets, also known as white graphene, have a honeycomb structure similar to that of graphene, with alternating boron and nitrogen atoms consisting of strong

sp^2 covalent in-plane bonding and weak van der Waals forces between layers [30–32]. Allotropes of BN are structurally isoelectronic to carbon allotropes and show similar properties. It exhibits various crystalline structures such as hexagonal (h-BN), wurtzite (w-BN), cubic (c-BN), and rhombohedral (r-BN) [33]. Compared to its analogous material, carbon, the cubic phase of BN is the most stable among the other phases. Cubic BN has a crystal structure analogous to diamond with sp^3-bonded atomic arrangements with ABCABC layered stacking. This material is a promising candidate for the replacement of diamonds. BN also possesses a wurtzite crystalline form with sp^3-hybridization. The crystal structure of c-BN, rhombohedral, and wurtzite BN is shown in Fig. 2.

The r-BN possesses ABCABC threefold layered stacking. On the other hand, h-BN is equivalent to graphite [34] with sp^2-bonded atomic arrangements like graphite structure, making it softer than cubic phase [16, 35]. Nanosheets of BN have similar layered structure (AA'A) and properties to graphene [36]. Both materials are composed of layers of hexagonal crystal lattices. Graphite has carbon atoms

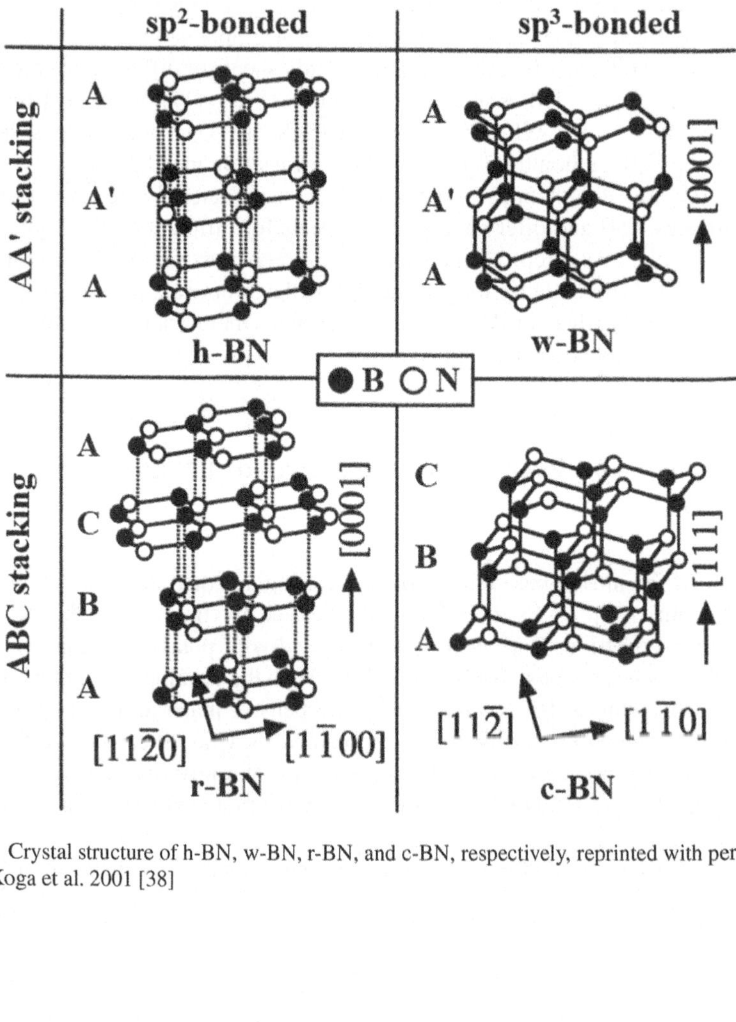

Fig. 2 Crystal structure of h-BN, w-BN, r-BN, and c-BN, respectively, reprinted with permission from Koga et al. 2001 [38]

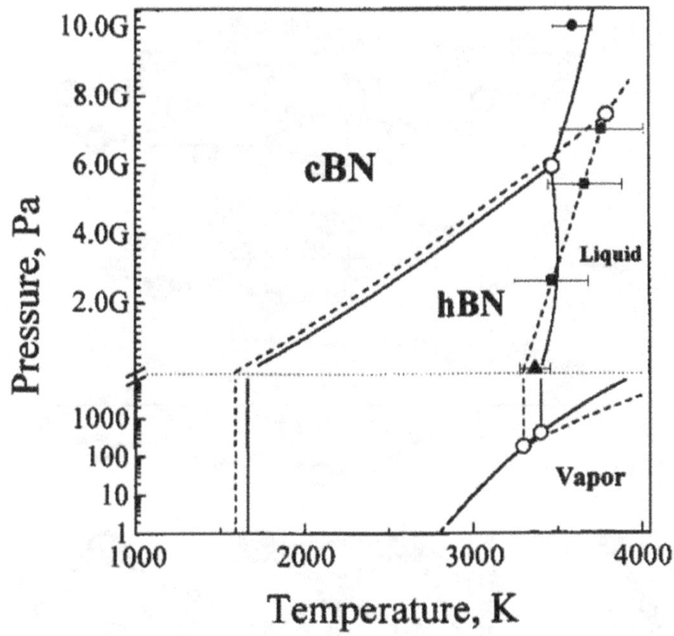

Fig. 3 Phase diagram of BN reproduced with permission from Solozhenko et al. [41]

at all lattice points, while h-BN is composed of alternating boron and nitrogen atoms [37]. The relative stability of different forms of BN is a concern for materials scientists. Depending upon temperature and pressure conditions, the relative stability of h-BN and c-BN varies at a particular set of conditions than other forms, as shown in Fig. 3. Phase transformation mainly depends upon the material's purity, grain size, and concentration of defects present in the material [39–41].

BN shows high chemical stability, corrosion resistance, thermal stability, and tunable solubility in water and other solvents [25, 35]. Due to high chemical stability, BN does not react with iron-based metals below high temperatures such as 1500 °C. The graphitic-like crystalline form of BN, i.e., hexagonal phase, possesses lubricating properties at low and high temperatures. Hexagonal-BN (h-BN) is one of the polymorphs of BN synthesized by chemical process. Like graphene structure, h-BN has a 2D honeycomb-like covalently bonded atomic structure. The atomic representation in the crystal structure of h-BN is shown in Fig. 4. Van der Waals forces bond together different layers and strong covalent bonds in the plane. This polymorph of BN exhibits large bandgap energy in the range of 3.6 to 6.5 eV [42, 43], extending its potential for optical applications. Table 1 shows the different properties of h-BN.

Structural properties of h-BN can be modified by varying synthesis conditions and by forming hybrids with other semiconductor materials. Composites/hybrids of BN with monoclinic zirconia are useful in steel production for varying the thermal shock of the material and casting. BN/titanium diboride (TiB_2) and BN/aluminum nitride

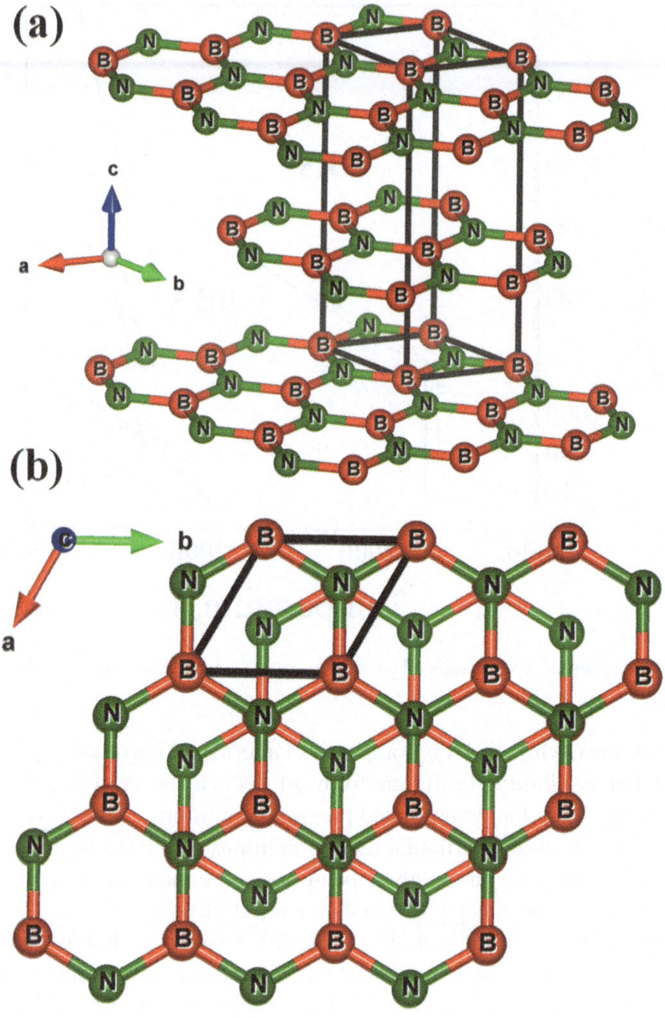

Fig. 4 Crystal structure of h-BN; **a** Side view and **b** Top view, respectively

(AlN) composites are useful in the metalizing industry for evaporation boats [45]. Composites of BN with titanium dioxide (TiO_2) [46], tin dioxide (SnO_2) [47], silver phosphate (Ag_3PO_4) [48], and zinc oxide (ZnO) [49] are useful in catalytic applications. Hybrids of BN with NiO show the tunability of different properties, potentially required for gas sensing and photodetection properties [50, 51]. BN's potential as a reinforcing agent is significant. When embedded in polymers or ceramics, it enhances the material's overall strength, thermal stability, and, in some cases, electrical properties. Such BN-reinforced composites have applications in aerospace field, where materials need to be strong, lightweight, and thermally stable. Similarly, BN composites can strengthen the automotive sector without adding significant weight, leading to

Table 1 General characteristics of h-BN, reproduced with permission from Jiang et al. [44]

Property	Value and units
Bond length	0.144 nm
Bond energy	4 eV
Melting point	3150–3400 K
Interlayer spacing	0.333 nm
Bulk modulus	12.5–65 GPa
Young's modulus	0.81–1.3 TPa
Tensile strength	27–83.3 MPa
Hardness	400 kg/mm
Thermal conductivity	400 W/mK
Bandgap energy	3.5–6.5 eV
Breakdown voltage	~ 7 MV/cm
Dielectric constant	3.9–4.3
Appearance color	White/semi-transparent

more fuel-efficient vehicles. BN/polymers and carbon fiber-reinforced BN composites are useful in reinforcement [52, 53], the biocompatibility of BN and its composites makes it a potential candidate for biomedical applications [35, 54], BN/nickel oxide (NiO_x) and BN/nickel (Ni) composites [55], h-BN/perovskite/h-BN structure [56] and BN/MoS_2/WSe_2 heterojunction are useful in Photovoltaic/solar cells applications [57]. Integrated BN nanostructures (BNNSs) in silicon carbon nitride (SiCN) and BN nanotubes (BNNTs) loaded silicon oxy-carbide (SiOC) are useful in electrochemical applications [58]. BNNS/poly vinyl alcohol (PVA) nanocomposites are convenient in membranes [59]. BNNS@cupper sulfide (CuS) composites are suitable for sensing applications [60].

Hexagonal-BN is an old metallurgical material with outstanding electrical and thermal properties [61], good resistance to corrosion, low density, higher melting point, and excellent chemical stability [62]. Research on h-BN-based materials has grown recently [25, 63]. This is not only due to the high thermal and chemical stability of h-BN but also the possibility of forming exactly one atomic layer of h-BN on metal surfaces [29] used in combination with graphene in electronic devices. Moreover, much attention has been paid to h-BN due to its unique electronic and optical properties, such as a tunable large direct bandgap and a negative electron affinity [64].

Nowadays, composites-based on BN have attracted the attention of material scientists due to the superior properties than that of the individual components. BN-based nanocomposites possess structural tunability, high chemical stability, high thermal conductivity, high mechanical strength, and improved adsorption and absorption properties [65]. Depending upon the potential of BN-based composites, possible research fields for these materials are environment, energy, gas sensing, optoelectrical, and health. Despite advancements in the preparation of BN-based nanocomposites, practical applications face challenges that must be overcome to realize these materials in device-based applications. Devices-based on BN composites show their potential for sensing, photodetection, and water purification applications. This whole will be possible with the realization of BN composites for these applications by studying the mechanisms and properties of these composite materials. Therefore, this work mainly focuses on two applications, i.e., gas sensing and photodetection.

3 BN and Its Hybrids from a Chemical Gas Sensor Perspective

A gas sensor is a device that determines the presence and concentration of gases in different environments. In other words, a gas sensor is a device with different components that selectively respond to the variation in the surrounding medium and provide an equivalent electrical signal [66]. The schematic of a gas sensor is shown in Fig. 5.

Depending upon different factors such as types of gases, concentration, and other factors; the sensing device gives a respective behavior either in potential difference with a variation of resistance or current of the sensing materials [67]. This respective variation provides the output of a sensor. Detection and determination of the nature of gas moieties depends on the material used for sensing. The basic principle of a chemical gas sensor is discussed below:

When the gas molecules are adsorbed on the sensing material, electrons get transferred or withdrawn from the adsorbent material, which leads to a change in the current or resistance of the material corresponding to the adsorption of gas molecules. In other words, the active layer of the sensing material provides the resistance or current change with the interaction of gaseous molecules and forms the output corresponding to respective property variation [68]. Gas sensors engrossed the vast attention of scientists. These are widely used in various fields such as personal and military safety [69], mining monitoring, environmental monitoring, emission control [70], medical diagnostics [71], and controlling the production process in industry as well as in the agriculture field [72]. Metal oxides are widely investigated for

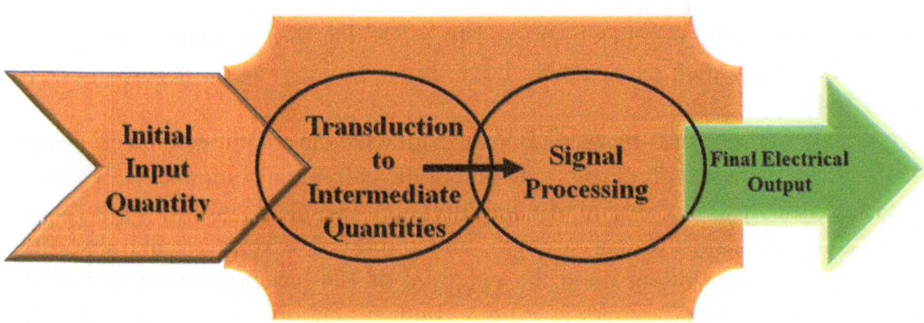

Fig. 5 Schematic of sensor principle

gas-sensing applications due to their high sensitivity and low cost. However, metal oxide-based sensors have drawbacks, including large energy consumption, which ultimately increases the cost factor, high operation temperature, and poor selectivity [73]. Thus, numerous studies have been reported to develop new sensors with lower operating temperatures based on conducting polymers and carbon nanotubes [74, 75]. These sensors lack long response timing, large recovery times, degradation, and poor stability with difficult processing processes. Various factors such as sensitivity range, temperature range, humidity effects, corrosion, size, over-range protection, cost, availability, lifetime, power consumption, self-test capability, stability, repeatability, linearity, error, response time, and frequency range must be considered before choosing a material for gas sensing applications. On the other hand, the unique properties of 2D nanomaterials show their potential for sensing applications, which can advance the sensing field by improving sensitivity, selectivity, operating temperature, and stability [76]. BN and its hybrid materials show their candidature towards their usage in sensing applications due to large bandgap energy, tunable structural properties, and selective adsorption of gases on different sites, which can enhance the sensitivity, stability, and performance of the sensor.

4 BN and Its Hybrids from a Photodetector Perspective

A device that converts the incident light to electrical output, either in the form of current or voltage, is called a photodetector. In other words, the photodetector is an optical receiver device converting the incoming light signal into an electrical response. Also, photodetectors are one type of converter that converts the optical input to electrical output [77]. On the other hand, the conversion of light input to electrical signals by harvesting the incident light, separation of exciton, and facilitation of charge carrier transport towards electrodes is known as photodetection. Figure 6 shows the schematic of a photodetection process of a photodetector.

To determine the performance of a photodetector, numerous figures of merits such as responsivity, specific detectivity, photoconductive gain, external quantum efficiency, and linear dynamic range must be considered [78]. Photodetectors are classified into two types: (a) Thermal photodetectors, which detect the incident light with rising temperature after light absorption, and (b) Photon detectors, which detect the light via creation of electron–hole pairs after the absorption of light. The intensity of incident light determines the carrier concentration. Therefore, higher intensity leads to the formation of a high density of charge carriers, improving the response of a photodetector.

When the incident light possesses sufficient energy and strikes the surface of a detector, the electron gets excited to the excited state and generates a hole in the ground state. This process leads to the generation of free electron–hole pairs. If the absorption of incident light takes place in the depletion region, then the generated electrons and holes are swept from the depletion region to electrodes, giving the photocurrent. Photodetector devices are dependent on the wavelength of the incident

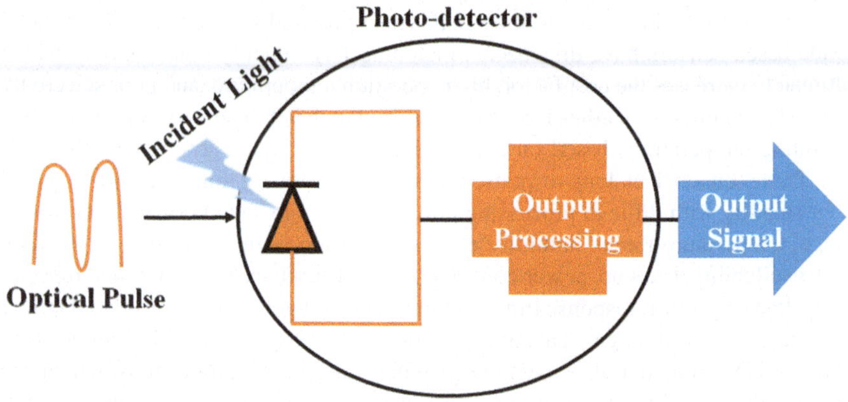

Fig. 6 Schematic of photodetector

radiations. Therefore, for choosing a material for photodetectors, some points must be considered, such as working wavelength region, sensitivity, photon conversion, absorption coefficient, fast response, low noise, and low cost. The absorption coefficient of a semiconductor material determines the radiation penetration depth and is further responsible for electron–hole pair generation. Detection of charge carriers is related to the lifetime of a photo-detecting material and the distance charged carriers travel before collection in electrodes. Reducing the device's size can minimize the distance traveled by charge carriers. Doping and forming composites also reduces the depletion region and improves the collection of generated electron–hole pairs. BN-based composites showed that their structural and optical properties can be tuned by varying synthesis conditions, which is the basic requirement for choosing a material for photodetectors. The bandgap tunability of BN and its hybrid materials and good electrical properties make them useful for photodetection and photo-sensing applications.

5 Synthesis of BN and Its Hybrid Nanostructures

This section provides an overview of the synthesis methods of BN nanostructures and their modifications like doping, functionalization, and hybrid/composite formation. Furthermore, this part also gives a detailed summary of generalized applications and the usefulness of BN-based compounds for gas sensing and photodetection applications. In spite of the extraordinary properties of nanostructures of BN, carbon nanostructures are much more popular from a research perspective than BN nanostructures due to easy synthesis methods. Research activities on nanostructures of carbon and BN were started in the same year. Still, the limitations of synthesis conditions and techniques leave the BN research far behind as compared to carbon [79]. The techniques used for the synthesis of BN include plasma sputtering [80], chemical vapor deposition (CVD) [81], chemical exfoliation [82], chemical blowing [83], ball milling [84], solvothermal method [35] and liquid exfoliation of a bulk h-BN [31]. Some of the synthesis methods are listed in Fig. 7. Singh et al. reported synthesizing boron nitride nanostructures using the solvothermal method. Structural analysis revealed that the synthesized material is preferentially grown, which deforms the structure and leads to sheets wrapping to form nanotubes [35]. Singh et al. have also demonstrated the synthesis procedure for preparing boron nitride composites with TiO_2 and SnO_2. The authors employed the chemical method for synthesizing nanocomposites and studied their structural, morphological, and compositional properties [46, 47]. Sajjad et al. have demonstrated the synthesis of BNNSs by pulsed laser deposition (PLD) technique and studied the structural properties of synthesized nanostructures [85]. Borek et al. have produced boron nitride materials with high surface area by vacuum pyrolysis. Authors have found that the pore structure of BN might be modified to yield a high surface area and primarily micro-porous solid [86]. Janik et al. have produced the boron nitride material using the polymeric precursor method. The results suggested that significant adsorption selectivity for molecules of similar diameter might be acquired by changing the boron nitride polymer precursor/ processing conditions [87].

Fig. 7 Synthesis methods of BN nanostructures

Yang et al. have reported a sonochemical method for preparing nanocomposites of gold nanoparticles/boron nitride sheets and studied the structural and morphological properties [88]. Xiong et al. synthesized carbon-doped BN (C-BN) using a soft template and carbon source. Authors have shown that exposure of atoms along the edges of the pores leads to stronger Lewis acid–base interactions between the analyte and C-BN. These interactions cause excellent adsorption performance of porous C-BN [89]. Kimura et al. have reported the deposition of polycrystalline BN films on Si substrates [90]. Also, the authors have investigated the electrical properties of polycrystalline BN films. Results revealed that the current becomes space-charge limited due to the absorption of hydronium ions in the BN films [91]. Li et al. have reported a facile fabrication route for embedding organic dyes into activated BN. Embedding of dye molecules into BN makes the composite a phosphor. Results revealed that the synthesized composite showed high thermal stability and enhancement in Förster resonance energy transfer (FRET), which can emit in the broad visible region (500–650 nm) under blue light excitation. The authors applied the formed phosphor along with a light-emitting diode (LED) chip in a transparent epoxy resin to achieve high performance, excellent current stability, and thermal conductivity [92]. Wu et al. reported BN-based polymer nanocomposites for high thermal stability, flexibility, and excellent thermal conductance prepared using 1D nanofiber and 2D BN nanosheets. Results showed that the polymer chains possessed a well-crystalline structure due to low coiling and less entanglement, rapidly enhancing the axial thermal conductivity. The obtained results suggested that composite films can be useful in thermal management of microelectrodes used at high temperatures [93]. Tay et al. have investigated layered BN, 3D interconnected networks, and BN/polydimethylsiloxane composite foams for multifunctional capacitive sensors. Results revealed that BN and polydimethylsiloxane's synergic effects lead to high compressibility, good mechanical resilience, cyclic performance, and high elasticity. Furthermore, due to the non-conductive nature of BN/polydimethylsiloxane, foams were used in capacitive sensors as a dielectric layer. The demonstrated device is useful in multiple sensing functions [94].

6 Applications of BN Nanostructures and Their Composites/Hybrids

BN nanostructures have a wide band gap possessing unique characteristics such as high thermal stability [95], emission in deep ultraviolet regions [96], unique wettability properties [97], and tunable electrical conductivity [98]. Due to these properties, BN and its composites are useful in various applications. The usage of BN has been limited to some applications in recent decades. With the advancement in processing techniques and device technology, scientists and industrialists have given attention to the usefulness of this material and its other compounds. Comparable properties of

BN to carbon counter parts open fields for its applications in photocatalytic, biological, sensor, photodetectors, lubricants, and defense fields. Different applications of BN and its composites/hybrid nanostructures are shown in Fig. 8.

Recent studies showed that BN nanotubes and other nanostructures are non-toxic in nature and further useful for biomedical applications such as diagnostic by boron neutron capture therapy (BNCT), drug delivery, etc. [35, 54, 99, 100]. For usage of material in biomedical applications, it must be biocompatible, selective, and stable. Kaur et al. have reported the boron neutron capture therapy (BNCT) results of ^{10}BN nanostructures [54]. Singh et al. have also represented the biocompatibility and potential of BNCT applications of BN nanotubes [35]. Ciofani et al. have demonstrated that BN does not affect the viability of normal cell lines or the functioning of metabolism but adversely affects the tumor cell lines [101]. Bhattacharya et al. have investigated the design and properties of a biosensor device-based on boron-nitrogen substituted graphene nanoribbon. Results showed the potential of the material for biosensing

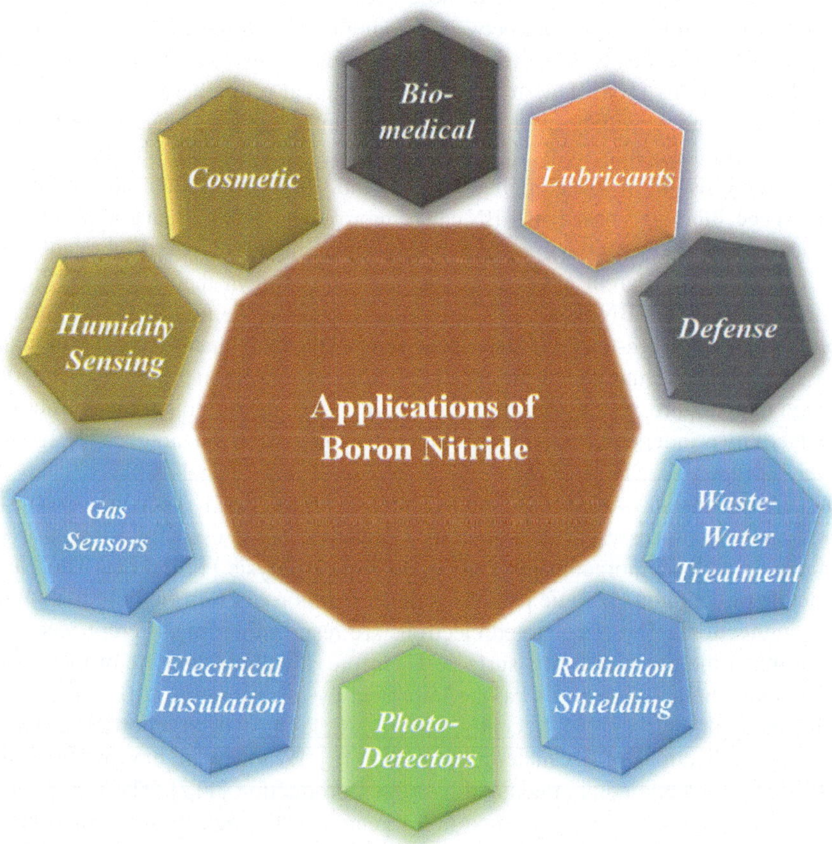

Fig. 8 Applications of BN and its hybrid nanostructures

applications [102]. As BN exhibits high thermal conductivity and chemical stability, spacecraft, packaging, and radiation protection gears can be formed using this material [103–105]. Also, the BNNSs possess superior thermal and mechanical properties, which can be used to form composite materials. Reinforcing with the polymeric matrices can improve the thermal conductivity of polymeric compounds [106]. BN nanotubes also act as a matrix composites reinforcing agent [107]. Boron content in h-BN possesses a high adsorption cross section, making h-BN useful in radiation shielding applications. Compared to other hydrogen and lead-based materials, nanostructured h-BN takes less weight and volume to protect the targeted material. By modifying the BN coating thickness, desired shielding properties can be obtained [104]. Due to mechanical and thermal properties, BN is referenced as an advanced engineered material. Therefore, high thermal conductivity and strength make this material useful in defense and security applications. Due to the lightweight and high strength of BN composites, these materials are used for the production of bullet proof products for army personnel. Along-with these products, transparent vehicle armors are also made from these ceramic-based compounds. A layer of ceramic material is sandwiched between two high-strength materials to form bullet proof protection gears. Boron nitride-based propellants have been used for military-based applications [108]. Due to superior mechanical properties, h-BN possesses low friction and is widely used as a dry lubricant. High resistance to oxidation, high-temperature conductivity, light weight matrix, and water hydrophobicity make this material useful in space applications. As the water droplets are not present in outer space, lubricants based on h-BN can be more advantageous than graphite-based lubricants [109]. Due to the low friction coefficient, the h-BN-based lubricants can be inserted into different alloy materials and are useful in their solid and liquid state. Reinforcing h-BN into resins and rubbers leads to the production of various solid lubricants. High-temperature conductivity makes this material useful in high-pressure and temperature applications [110]. BN acts as a green lubrication material at high temperatures and is useful in engine oil [111]. At low temperatures, thermal conductivity is independent of the chirality and diameter of BN nanotubes [112]. Due to these properties, BN nanostructures can be useful in flat panel displays. Theoretical calculations showed that BN and nanoparticle-decorated sheets are useful in photocatalytic processes. Due to the large bandgap, BN is used for adsorption of organic as well as inorganic pollutant adsorption catalysis [113, 114]. The basic principle of the photocatalytic process is the generation of electron–hole pairs with light interactions. These generated electrons are responsible for different reactions occurring at the photocatalyst's surface. Therefore, the photocatalytic properties of other semiconductor materials can be improved by incorporating h-BN within the semiconductor matrix [115]. Recently, various articles have been reported related to catalytic applications of BN materials [16, 17, 116]. Postole et al. have reported the oxidation of organic compounds and hydrogen production [117]. The h-BN facilitates the separation of electrons and holes generated through light interactions to avoid recombination in the photocatalytic reaction. As the optical properties of h-BN can be altered in various ways, such as doping, intercalation, functionalization, and composite formation, the absorption edge can be modified as per requirements. This modification in the adsorption edge suppresses the recombination process and leads to the improvement of photocatalytic properties. Singh et al. have designed BN-SnO_2 nanostructures and reported their photocatalytic applications [47]. The results showed that composite materials possess higher catalytic properties than organic dyes. Singh el al. have reported the synthesis of BN/TiO_2 nanostructures and their photocatalytic applications towards organic and inorganic pollutants [46]. BN-based composites's partial polar nature is responsible for the movements of holes toward the semiconductor surface, facilitating photocatalytic performance [118]. Wang et al. have synthesized the sub-micro-sized boxes of BN using a solvothermal method and studied their photocatalytic properties. Authors also formed the composites of BN with SnO_2 and showed photocatalytic performance towards methyl orange dye [119]. Furthermore, BN nanostructures are the potential candidate materials for the realization of lasers due to their wide bandgap energy [120]. In recent years, material scientists have been working on the synthesis techniques of nanostructured BN so that its properties can be explored for industrial applications such as gas sensing and photodetection.

6.1 BN and Its Hybrids/Nanocomposites in Gas Sensing Applications

The tunable structural and surface properties of h-BN make it a useful material for sensing applications. Sajjad et al. have studied the gas-sensing properties of h-BN towards reducing gases. The results revealed that the BNNSs-based gas sensor possesses good sensitivity and performance [85]. Kimura et al. have reported the usage of polycrystalline BN towards humidity sensing properties of h-BN and showed the switching characteristics [90]. Yu et al. have reported the partial polar character of h-BN and studied the sensing properties of pristine and Au-decorated BNNTS toward humidity. The results showed that the decoration of BN with gold nanoparticles enhances the sensing ability towards humidity at room temperature [121]. Sun et al. have performed the first principal density-functional theory (DFT) and dispersion-DFT calculations to study the absorption properties of BN nanomaterials. The results revealed that negatively charged BN nano-sorbents have high selectivity for separating carbon dioxide (CO_2) from its mixtures with methane (CH_4) and/or H_2 [122]. Liu et al. have developed BN film-based sensors for detecting mercuric ion (Hg^{2+}) ions. Authors demonstrated that dansyl-chloride-functionalized BN films possess fluorescence quenching upon adding metal ions in an ethanol solution and high selectivity and concentration-dependent characteristics to Hg^{2+} ions [123]. Zeng et al. have performed theoretical calculations using DFT and studied the adsorption of hydrogen cyanide on pure and metal-doped BNNTs. Results revealed that metal-doping leads to charge distribution variation and thus, adsorption properties get changed [124]. Yoo et al. have performed theoretical as well as experimental studies for hydrogen sensing by BNNTs decorated alloys. Results revealed that the fabricated device possesses good sensor response with low power consumption [125]. Weber et al. have used the vapor–liquid-solid growth and atomic layer deposition (ALD) method to prepare BN-coated ZnO nanowires and studied their sensing properties towards hydrogen gas. Results revealed that the synergic effects of BN and platinum (Pt) decoration lead to sensor response improvement [126]. Vessally et al. have studied the adsorption properties of cyanogen chloride (ClCN) on different structures of BN. The authors also studied the effect of Al-doping on the adsorption properties and Al-doping helped to detect ClCN [127]. Shan et al. have investigated the reactivity of BN nanocones towards toluene gas using theoretical calculations. Adsorption properties revealed the potential of BN nanocones for detection of toluene gas [128]. Rakhshi et al. have performed the theoretical calculations on BNNTs with and without an electric field. The authors also studied the adsorption properties of ammonia gas molecules. Results revealed that geometric as well as electronic structure variations affect the ammonia adsorption properties, which shows the potential for gas sensing applications [129]. Mohsennia et al. have performed DFT calculations on BNNTs towards the adsorption of ammonia gas. The authors also studied the effect of Ni and Pt decoration on the adsorption properties of BNNTs in the absence and the presence of the electric field. Results revealed that the decoration leads to decreased symmetry, which improved the chemical activity towards ammonia molecules. Results suggested that these compounds can be experimentally explored for ammonia gas sensing applications [130]. He et al. have investigated the electronic and sensing properties of BNNTs towards C_2H_2. The authors also demonstrated the effect of modification of BNNTs with metal oxides such as Fe_2O_3, NiO, and TiO_2 on sensing properties. Results revealed that the decoration of BNNTs with metal oxides generated vacant states near the valance band, improving electrical conductivity. This study shows the potential of metal oxide/BN decorations for gas-sensing applications [131]. Authors have also performed the DFT calculations on CuO-modified BNNTs and studied the adsorption properties towards C_2H_2, H_2, and CH_4. Results revealed that the modification of BNNTs with CuO leads to improvement in the reactivity and sensitivity towards gases in the oil. This study suggested the potential of BN-based composites for sensing applications [132]. He et al. reported synthesizing and modifying BNNS with sulfate using chemical exfoliation. Experimental as well as theoretical results showed that the modification enhances the sensing properties of BNNS towards NO_2 gas [133]. Ammar et al. have employed DFT calculations for studying the effect of Mn and Fe doping on the adsorption properties of BN nanocages towards various gases such as CO, NO, and NH_3. Results revealed that doping with the transition metal ions improves the adsorption properties of BN nanocages towards gaseous molecules [134]. Badran et al. have studied the adsorption of ammonia gas molecules on BN nanocages using theoretical calculations under different environments. The authors also studied the effect of electric field on adsorption properties of BN nanocages. This study provides insights into the potential usage of BN structures for ammonia gas sensing [135]. Xu et al. reported an enzymatic procedure based on a catalase biosensor for detecting forchlorfenuron (CPPU) using BN. Results revealed that CPPU can be determined with the detection limit in the micro-range [136]. Liu et al. have studied the reactivity and properties of Si-doped BN sheets using the

DFT approach. The results revealed that Si-doping enhanced the chemical reactivity of the h-BN sheet toward CO, NH_3, O_2, and NO in various ways, suggesting its potential application for developing gas sensors or metal-free catalysts [137]. Huang et al. have designed a sub-micrometer-sized pH sensor based on biotin-fluorescein-functionalized multi-walled BNNTs decorated with Ag nanoparticles. Results have also revealed hybrid BNNTs can detect pH values in a sub-micrometer regime [138]. Wang et al. have performed theoretical calculations to study the reactivity of pristine and Si-doped single-walled BNNTs towards the hydrogen cyanide (HCN) molecule. Based on calculated results, authors have suggested that the Si-doped BNNT is expected to be a potential resource for detecting the presence of toxic HCN [139]. Mukhopadhyay et al. have employed a plane-wave pseudo-potential approach within the local density approximation (LDA) of DFT to study the sensitivity of BN towards biomolecules. Authors suggested that BNNTs be a better substrate for protein immobilization than carbon nanotubes (CNTs) due to their higher sensitivity toward amino acid polarity [140]. Choi et al. have performed the ab-initio calculations to investigate the adsorption properties of BNNTs. The result implies that boron-rich BNNT could capture CO_2 effectively at ambient conditions [141]. Guo et al. have performed DFT calculations with dispersion correction. Results showed that under an external electric field, a h-BN sheet can become an effective sorbent for CO_2 capture [142]. Sun et al. have implemented a DFT approach to investigate the adsorption of N_2 and CH_4 on different charge states of BN nanocage fullerene ($B_{36}N_{36}$). Their results indicated that $B_{36}N_{36}$ possesses high selectivity in separating N_2 from CH_4 [143]. Zhao et al. have theoretically investigated the electronic properties of BNNTs using DFT calculations. The effect of different gas molecule adsorptions on BNNTs, as well as interactions, are also elaborated. Results showed the potential for gas sensing application [144]. Paura et al. have studied the absorption properties of armchairs and zigzag single-walled BNNT using DFT. The effect of vacancy defects on the adsorption of CO_2 is also elaborated in this study. Results revealed that the adsorption of CO_2 takes place through the chemical adsorption process on the top of the defect site [145]. Rimola et al. have performed the DFT calculations to study the adsorption properties of various molecules (H_2O, NH_3, and HCOOH as polar molecules and C_6H_6, CH_4 as non-polar) on zig-zag single-walled BNNTs and BN monolayer [146]. Barbary et al. have performed theoretical calculations using DFT to investigate the adsorption properties of single-walled BNNTs. Authors have found that the best BNNT for adsorbing the CO, CO_2, NO, and NO_2 gas molecules is (5, 0) BNNT [147]. Feng et al. have studied the healing effect/doping of oxygen in the h-BN sheet through DFT calculations. Authors have also suggested that nitrogen vacancies can be healed or transformed to yield an O-doped BN sheet via NO_2 adsorption [148]. Neklyudov et al. have studied the characteristics of BN in relation to the trapping and de-trapping of ion-implanted deuterium, as well as the radiation-induced desorption of gases from its surface under deuterium ion bombardment [149]. Zhang et al. have investigated the adsorption of alkaline-earth metal (Mg/Ca) on carbon-doped h-BN sheets using first-principles calculations. Results revealed that NO, H_2O and O_2 are chemisorbed, while CO, H_2, and CO_2 are physisorbed on h-BN sheets [150]. Choudhuri et al. have performed theoretical estimations using DFT to study the gas (CO,

CO_2, NO, and NO_2) sensing behavior/mechanism of pure and doped (B@, N@, and B-N@) graphene surfaces. Authors have also found that graphene with B-N co-doping can be highly sensitive and selective for semiconductor-based gas sensors [151]. BN nanotubes have semiconducting properties, making them very useful in electronics compared to CNTs with a bandgap energy dependent upon parameters [152]. In contrast to traditional adsorbents, BN nanostructures have small bending moduli [153, 154] and may experience conformational changes due to the physisorption of gas molecules. Such behavior should improve their surface adsorption and hence have implications for developing novel applications such as gas sensors [155]. Recently, Ataca et al. calculated the effect of ad-atoms adsorbed on h-BN sheets and of substituting foreign atoms for B and N atoms in the honeycomb structure [156]. It was concluded that such functionalization leads to dramatic electronic structure modifications. However, there is a lack of straight forward experimental method to overcome the difficulty of detecting morphological changes at the atomic level after coverage of adsorbates [157]. Singh et al. have demonstrated the usefulness of BN/NiO hybrids for the detection of hydrogen at room temperature as well as ammonia gas at different temperatures, respectively [158, 159].

6.2 BN and Its Hybrids/Nanocomposites in Photodetection Applications

The h-BN shows intense emission in the ultraviolet region of the solar spectrum, and the optical properties can be tunable in various ways, either by doping, functionalization or hybrid/composite formation. This modification shifts the absorption edge, which is useful in photodetector applications. Several reports have expressed the usefulness of BN-based compounds, such as c-BN, h-BN nanosheets, h-BN epilayers, and graphene-hBN hybrids, for photodetection applications [7, 160–168]. Zhou et al. reported the fabrication of metal-semiconductor-metal (MSM) heterostructure based on h-BN for photodetection applications. The results revealed the structure is stable under light radiation up-to 100 °C [30]. Liao et al. synthesized the c-BN films by sputtering a h-BN sintered target. The films obtained were characterized by scanning electron microscopy, infrared absorption, and UV–visible transmission spectra. The photodetector devices of c-BN films were fabricated by photolithographic techniques [160]. Li et al. have synthesized h-BN epilayers by metal–organic CVD and studied their dielectric strength, optical absorption, and potential use in deep-ultraviolet detector material [161]. Sajjad et al. reported the synthesis of a few atomic-layer BN nanosheets and studied their photodetection properties. Results show potential of nanosheets for photodetection applications [162]. Ju et al. reported photoinduced doping in Van der Waals heterostructures consisting of graphene and BN layers. It enables flexible and repeatable writing and erasing of charge doping in graphene with visible light [163]. Li et al. have formed a diode by introducing 2D h-BN into graphene/gallium arsenide heterostructure to suppress the static charge transfer for solar cells as well as photodetectors. The authors studied the photodetection properties of fabricated heterostructure in different light regions [164]. Li et al. have prepared the BN films by plasma-enhanced CVD and described the effect of sulfur doping on the optical properties of deposited films. The results revealed that the fabricated structure of S-doped BN is appropriate for solar blind detection applications [165]. Lu et al. have demonstrated a ZnO quantum dot photo-doped graphene/h-BN/gallium nitride (GaN)-hetero-structured photodetector. The results revealed that the barrier height at the graphene/GaN heterojunction interface increases with an increase in the h-BN layer, which leads to a decrease in dark current and further improves the on/off current ratio of the device, consequently, the device's responsivity increases [166]. BenMoussa et al. have reported photodetectors based on diamond, c-BN, and wurtzite AlN. Results revealed that Diamond, c-BN, AlN, and MSM photodetectors are sensitive and stable under UV irradiation [169]. Ahmad et al. deposited the thin films of BN using a sputtering technique from a BN target. The authors revealed that the UV sensitivity of the BN films follows the temperature change and is fairly constant in a temperature range of –40 °C to 80 °C [170]. Zhang et al. have reported the synthesis of h-BN nanocrystals by flux growth route and mechanically exfoliated the h-BN to form thin layers. Authors have studied the photodetection properties in two configurations, i.e., top contact and bottom contact. Results revealed that h-BN is promising for detection applications [171]. Moradinasab et al. studied the optical properties of 1D super-lattices formed by armchair graphene nanoribbons embedded in h-BN super-lattices. A set of tight binding parameters is proposed, resulting in excellent agreement with first-principal calculations [172]. Soltani et al. have fabricated deep-ultraviolet solar-blind photodetectors based on high-quality BN films with a MSM configuration and studied the photodetection properties [173]. Shiue et al. demonstrated an on-chip ultrafast photodetector based on a 2D heterostructure consisting of high-quality graphene encapsulated in h-BN and studied the photodetection behavior in different regions. Results revealed that the hot-electron effects are responsible for the good response of the fabricated device [174]. Rivera et al. have reported the preparation of 2D BN nanosheets using the PLD method and studied their photodetection properties. Results revealed that the fabricated devices exhibit good detection properties to weaker light intensities at high temperatures [175]. Furthermore, authors have found the potential usage of fabricated devices up-to 400 °C for UV detection applications [176]. Lin et al. have fabricated the photodetector using 2D BN nanosheets with 1D cellulose nano-fibres. Prepared detectors possess good photodetection properties. Results inferred that these detectors can be useful up-to 200 °C [177]. Chu et al. reported the bottom configuration based on h-BN coupled SnS_2 for photodetection applications. Results revealed that the device possesses good detection properties. Results also suggested that the fabricated device structure can be transferred to flexible substrates for future applications [178]. Lu et al. have fabricated the photodetector device based on black phosphorus/h-BN/graphene junctions. Results revealed that BN acts like a tunneling barrier, and the device performs well in the mid-infra-red region [179]. Tan et al. have deposited the h-BN films by ambient pressure CVD method and studied the morphological and elemental properties. The authors also demonstrated the photodetection properties of deposited h-BN films.

Results revealed that deposited films possess low dark current as well as good detectivity. This study shows the potential of h-BN for photodetection applications [180]. Thompson et al. have demonstrated the vertical configuration of graphene/h-BN and studied their photodetection properties. Results inferred that the fabricated device is responsive to different light intensities [181]. Wang et al. have prepared C-doped h-BN using the sputtering technique and studied their photodetection properties. Results revealed that C-doping after a certain amount leads to a decrease in crystallinity, which affects the detection properties. It was inferred that carbon-doping improved the detection properties of h-BN [182]. Kaushik et al. have prepared the h-BN flakes using mechanical exfoliation method. The detector device was fabricated using electron beam lithography. Results revealed that the fabricated device possesses good photodetection properties in the UV region [183]. Singh et al. reported the photodetection properties of BN/NiO hybrids and studied the impact of loading of gold nanoparticles on photodetection properties of BN/NiO hybrids [184]. The aforementioned work done by various researchers shows the potential usefulness of BN and BN-based hybrids/nanocomposites in various applications.

7 Conclusions

This work covers the crystal structure, different phases, properties, and synthesis methods of BN and its hybrids/nanocomposites. Literature shows that the BN and BN-based hybrids have potential in various applications such as multi-function composites, photocatalysis, electronics, photodetectors, biological applications, etc. In all the allotropes of boron nitride, graphene like 2D h-BN, has the potential to be explored in sensing and optoelectronic applications. It also shows that this material's electronic properties are relatively superior to gapless graphene due to the tunable band gap. Theoretical studies on BN and its nanocomposites have been done extensively, but tough synthesis conditions were the main reason for this material to be less explored. These above conditions form the basis of motivation to choose h-BN and its nanocomposites for photodetection and sensing applications.

References

1. Shelke, N.T., Karle, S.C., Karche, B.R.: Photoresponse properties of CdSe thin film photodetector. J. Mater. Sci.: Mater. Electron. **31**, 15061–15069 (2020)
2. Lieber, C.M., Wang, Z.L.: Functional nanowires. Mater. Res. Bull. **32**, 99–108 (2007)
3. Pan, Z.W., Dai, Z.R., Wang, Z.L.: Nanobelts of semiconducting oxides. Science **291**, 1947–1949 (2001)
4. Law, M., Goldberger, J., Yang, P.: Semiconductor nanowires and nanotubes. Annu. Rev. Mater. Res. **34**, 83–122 (2004)
5. Xia, Y., Yang, P., Sun, Y., Wu, Y., Mayers, B., Gates, B., Yin, Y., Kim, F., Yan, H.: One-dimensional nanostructures: synthesis, characterization, and applications. Adv. Mater. **15**, 353–389 (2003)
6. Balandin, A.A., Ghosh, S., Bao, W., Calizo, I., Teweldebrhan, D., Miao, F., Lau, C.N.: Superior thermal conductivity of single-layer graphene. Nano Lett. **8**, 902–907 (2008)
7. Novoselov, K.S., Geim, A.K., Morozov, S.V., Jiang, D., Katsnelson, M.I., Grigorieva, I.V., Dubonos, S., Firsov, A.A.: Two-dimensional gas of massless Dirac fermions in graphene. Nature **438**, 197–200 (2005)
8. Lee, C., Wei, X., Kysar, J.W., Hone, J.: Measurement of the elastic properties and intrinsic strength of monolayer graphene. Science **321**, 385–388 (2008)
9. Anichini, C., Czepa, W., Pakulski, D., Aliprandi, A., Ciesielski, A., Samorì, P.: Chemical sensing with 2D materials. Chem. Soc. Rev. **47**, 4860–4908 (2018)
10. Randviir, E.P., Brownson, D.A., Banks, C.E.: A decade of graphene research: production, applications and outlook. Mater. Today **17**, 426–432 (2014)
11. Papageorgiou, D.G., Kinloch, I.A., Young, R.J.: Mechanical properties of graphene and graphene-based nanocomposites. Prog. Mater. Sci. **90**, 75–127 (2017)
12. Li, X., Tao, L., Chen, Z., Fang, H., Li, X., Wang, X., Xu, J.B., Zhu, H.: Graphene and related two-dimensional materials: structure-property relationships for electronics and optoelectronics. Appl. Phys. Rev. **4**, 021306 (2017)
13. Berry, V.: Impermeability of graphene and its applications. Carbon **62**, 1–10 (2013)
14. Guo, F., Silverberg, G., Bowers, S., Kim, S.P., Datta, D., Shenoy, V., Hurt, R.H.: Graphene-based environmental barriers. Environ. Sci. Technol. **46**, 7717–7724 (2012)
15. Prasai, D., Tuberquia, J.C., Harl, R.R., Jennings, G.K., Bolotin, K.I.: Graphene: corrosion-inhibiting coating. ACS Nano **6**, 1102–1108 (2012)
16. Wang, Q.H., Kalantar-Zadeh, K., Kis, A., Coleman, J.N., Strano, M.S.: Electronics and optoelectronics of two-dimensional transition metal dichalcogenides. Nat. Nanotechnol. **7**, 699–712 (2012)
17. Kaur, G., Singh, B., Singh, P., Singh, K., Thakur, A., Kumar, M., Bala, R., Kumar, A.: Iron disulfide (FeS_2): a promising material for removal of industrial pollutants. ChemistrySelect **2**, 2166–2173 (2017)
18. Krishnan, U., Kaur, M., Singh, K., Kumar, M., Kumar, A.: A synoptic review of MoS_2: synthesis to applications. Superlattice Microst. **128**, 274–297 (2019)
19. Kaur, M., Singh, K., Chauhan, I., Singh, H., Sharma, R.K., Vij, A., Thakur, A., Kumar, A.: Low temperature carrier transport mechanism and photo-conductivity of WSe_2. J. Alloys Compd. **869**, 159369 (2021)
20. Eftekhari, A.: Tungsten dichalcogenides (WS_2, WSe_2, and WTe_2): materials chemistry and applications. J. Mater. Chem A **5**, 18299–18325 (2017)
21. Li, Y., Tang, L., Li, R., Xiang, J., Teng, K.S., Lau, S.P.: SnS_2 quantum dots: facile synthesis, properties, and applications in ultraviolet photodetector. Chinese Phys. B **28**, 037801 (2019)
22. Liu, H., Neal, A.T., Zhu, Z., Luo, Z., Xu, X., Tománek, D., Ye, P.D.: Phosphorene: an unexplored 2D semiconductor with a high hole mobility. ACS Nano **8**, 4033–4041 (2014)
23. Barsoum, M.W.: The $M_{N+1}AX_N$ phases: a new class of solids: thermodynamically stable nanolaminates. Prog. Solid State Chem. **28**, 201–281 (2000)
24. Gupta, A., Sakthivel, T., Seal, S.: Recent development in 2D materials beyond grapheme. Prog. Mater. Sci. **73**, 44–126 (2015)

25. Zhang, K., Feng, Y., Wang, F., Yang, Z., Wang, J.: Two dimensional hexagonal boron nitride (2D-hBN): synthesis, properties and applications. J. Mater. Chem. C **5**, 11992–12022 (2017)
26. Balmain, W.H.: Bemerkungen über die Bildung von Verbindungen des Bors und Siliciums mit Stickstoff und gewissen Metallen (Remarks on the formation of compounds of boron and silicon with nitrogen and certain metals). J. Prakt. Chem. **27**, 422–430 (1842)
27. Chopra, N.G., Luyken, R.J., Cherrey, K., Crespi, V.H., Cohen, M.L., Louie, S.G., Zettl, A.: Boron nitride nanotubes. Science **269**, 966–967 (1995)
28. Loiseau, A., Willaime, F., Demoncy, N., Hug, G., Pascard, H.: Boron nitride nanotubes with reduced numbers of layers synthesized by arc discharge. Phys. Rev. Lett. **76**, 4737 (1996)
29. Corso, M., Auwarter, W., Muntwiler, M., Tamai, A., Greber, T., Osterwalder, J.: Boron nitride nanomesh. Science **303**, 217–220 (2004)
30. Zhou, A.F., Aldalbahi, A., Feng, P.: Vertical metal-semiconductor-metal deep UV photodetectors based on hexagonal boron nitride nanosheets prepared by laser plasma deposition. Opt. Mater. Express **6**, 3286–3292 (2016)
31. Zhi, C., Bando, Y., Tang, C., Kuwahara, H., Golberg, D.: Large-scale fabrication of boron nitride nanosheets and their utilization in polymeric composites with improved thermal and mechanical properties. Adv. Mater. **21**, 2889–2893 (2009)
32. Mateti, S., Wong, C.S., Liu, Z., Yang, W., Li, Y., Li, L.H., Chen, Y.: Biocompatibility of boron nitride nanosheets. Nano Res. **11**, 334–342 (2017)
33. Pakdel, A., Zhi, C., Bando, Y., Golberg, D.: Low-dimensional boron nitride nanomaterials. Mater. Today **15**, 256–265 (2012)
34. Samantaray, C.B., Singh, R.N.: Review of synthesis and properties of cubic boron nitride (c-BN) thin films. Int. Mater. Rev. **50**, 313–344 (2005)
35. Singh, B., Kaur, G., Singh, P., Singh, K., Kumar, B., Vij, A., Kumar, M., Bala, R., Meena, R., Singh, A., Thakur, A.: Nanostructured boron nitride with high water dispersibility for boron neutron capture therapy. Sci. Rep. **6**, 35535 (2016)
36. Boldrin, L., Scarpa, F., Chowdhury, R., Adhikari, S.: Effective mechanical properties of hexagonal boron nitride nanosheets. Nanotechnology **22**, 505702 (2011)
37. Ishigami, M., Aloni, S., Zettl, A.: Properties of boron nitride nanotubes. AIP Conf. Proc. **696**, 94–99 (2003)
38. Koga, H., Nakamura, Y., Watanabe, M., Yoshida, T.: Molecular dynamics study of deposition mechanism of cubic boron nitride. Sci. Technol. Adv. Mater. **2**, 349–356 (2001)
39. Xu, Y.N., Ching, W.Y.: Calculation of ground-state and optical properties of boron nitrides in the hexagonal, cubic, and wurtzite structures. Phys. Rev. B **44**, 7787 (1991)
40. Sokołowska, A., Wronikowski, M.: The phase diagram (p, T, E) of boron nitride. J. Cryst. Growth **76**, 511–513 (1986)
41. Solozhenko, V.L., Turkevich, V.Z., Holzapfel, W.B.: Refined phase diagram of boron nitride. J. Phys. Chem. B **103**, 2903–2905 (1999)
42. Solozhenko, V.L., Lazarenko, A.G., Petitet, J.P., Kanaev, A.V.: Bandgap energy of graphite-like hexagonal boron nitride. J. Phys. Chem. Solids **62**, 1331–1334 (2001)
43. Watanabe, K., Taniguchi, T., Kanda, H.: Direct-bandgap properties and evidence for ultraviolet lasing of hexagonal boron nitride single crystal. Nat. Mater. **3**, 404–409 (2004)
44. Jiang, X.F., Weng, Q., Wang, X.B., Li, X., Zhang, J., Golberg, D., Bando, Y.: Recent progress on fabrications and applications of boron nitride nanomaterials: a review. J. Mater. Sci. Technol. **31**, 589–598 (2015)
45. Eichler, J., Lesniak, C.: Boron nitride (BN) and BN composites for high-temperature applications. J. Eur. Ceram. Soc. **28**, 1105–1109 (2008)
46. Singh, B., Singh, P., Singh, K., Sharma, J., Kumar, M., Bala, R., Meena, R., Sharma, S.K., Kumar, A.: Nanostructured BN-TiO_2 composite with ultra-high photocatalytic activity. New J. Chem. **41**, 11640–11646 (2017)
47. Singh, B., Singh, K., Kumar, M., Thakur, S., Kumar, A.: Insights of preferred growth, elemental and morphological properties of BN/SnO_2 composite for photocatalytic applications towards organic pollutants. Chem. Phys. **531**, 110659 (2020)

48. Zhu, C., Zheng, J., Fang, L., Hu, P., Liu, Y., Cao, X., Wu, M.: Advanced visible-light driven photocatalyst with enhanced charge separation fabricated by facile deposition of Ag_3PO_4 nanoparticles on graphene-like h-BN nanosheets. J. Mol. Catal. A: Chem. **424**, 135–144 (2016)
49. Fu, X., Hu, Y., Zhang, T., Chen, S.: The role of ball milled h-BN in the enhanced photocatalytic activity: a study based on the model of ZnO. Appl. Surf. Sci. **280**, 828–835 (2013)
50. Singh, K., Kaur, M., Chauhan, I., Meena, R., Singh, J., Thakur, A., Kumar, A.: Tailoring of structural, morphological and optical properties of boron nitride/nickel oxide ($BN_{100\text{-}x}/NiO_x$) nanocomposites. J. Clust. Sci. **32**, 865–873 (2021)
51. Singh, K., Thakur, A., Awasthi, A., Kumar, A.: Structural, morphological and temperature-dependent electrical properties of BN/NiO nanocomposites. J. Mater. Sci. Mater. Electron. **31**, 13158–13166 (2020)
52. Wang, Z., Li, Q., Chen, Z., Liu, J., Liu, T., Li, H., Zheng, S.: Enhanced comprehensive properties of nylon-6 nanocomposites by aniline-modified boron nitride nanosheets. Ind. Eng. Chem. Res. **57**, 11005–11013 (2018)
53. Zhong, W., Wang, S., Li, J., Bechelany, M.C., Ghisleni, R., Rossignol, F., Balan, C., Chartier, T., Bernard, S., Miele, P.: Design of carbon fiber reinforced boron nitride matrix composites by vacuum-assisted polyborazylene transfer molding and pyrolysis. J. Eur. Ceram. Soc. **33**, 2979–2992 (2013)
54. Kaur, M., Singh, P., Singh, K., Gaharwar, U.S., Meena, R., Kumar, M., Nakagawa, F., Wu, S., Suzuki, M., Nakamura, H., Kumar, A.: Boron Nitride (^{10}BN) a prospective material for treatment of cancer by boron neutron capture therapy (BNCT). Mater. Lett. **259**, 126832 (2020)
55. Mann, D.S., Patil, P., Kim, D.H., Kwon, S.N.: Na, SI: Boron nitride-incorporated NiO_x as a hole transport material for high-performance pin planner pervoskite solar cells. J. Power. Sources **477**, 228738 (2020)
56. Seitz, M., Gant, P., Castellanos-Gomez, A., Prins, F.: Long-term stabilization of two-dimensional perovskites by encapsulation with hexagonal boron nitride. Nanomaterials **9**, 1120 (2019)
57. Cho, A.J., Kwon, J.Y.: Hexagonal boron nitride for surface passivation of two-dimensional van der Waals heterojunction solar cells. ACS Appl. Mater. Interfaces **11**, 39765–39771 (2019)
58. Abass, M.A., Syed, A.A., Gervais, C., Singh, G.: Synthesis and electrochemical performance of a polymer-derived silicon oxycarbide/boron nitride nanotube composite. RSC Adv. **7**, 21576–21584 (2017)
59. Gonzalez-Ortiz, D., Pochat-Bohatier, C., Gassara, S., Cambedouzou, J., Bechelany, M., Miele, P.: Development of novel h-BNNS/PVA porous membranes via Pickering emulsion templating. Green Chem. **20**, 4319–4329 (2018)
60. Zhang, Y., Wang, Y.N., Sun, X.T., Chen, L., Xu, Z.R.: Boron nitride nanosheet/CuS nanocomposites as mimetic peroxidase for sensitive colorimetric detection of cholesterol. Sens. Actuator. B Chem. **246**, 118–126 (2017)
61. Liu, D., Lei, W., Qin, S., Klika, K.D., Chen, Y.: Superior adsorption of pharmaceutical molecules by highly porous BN nanosheets. Phys. Chem. Chem. Phys. **18**, 84–88 (2016)
62. Liu, D., He, L., Lei, W., Klika, K.D., Kong, L., Chen, Y.: Multifunctional polymer/porous boron nitride nanosheet membranes for superior trapping emulsified oils and organic molecules. Adv. Mater. Interfaces **2**, 1500228 (2015)
63. McMillan, P.F.: New materials from high-pressure experiments. Nat. Mater. **1**, 19–25 (2002)
64. Simonov, K.A., Vinogradov, N.A., Ng, M.L., Vinogradov, A.S., Mårtensson, N., Preobrajenski, A.B.: Controllable oxidation of h-BN monolayer on Ir (111) studied by core-level spectroscopies. Surf. Sci. **606**, 564–570 (2012)
65. Gonzalez-Ortiz, D., Salameh, C., Bechelany, M., Miele, P.: Nanostructured boron nitride–based materials: synthesis and applications. Mater. Today Adv. **8**, 100107 (2020)
66. Yang, S., Jiang, C., Wei, S.H.: Gas sensing in 2D materials. Appl. Phys. Rev. **4**, 21304 (2017)
67. Janata, J.: Introduction: modern topics in chemical sensing. Chem. Rev. **108**, 327–328 (2008)

68. Kim, Y.S., Ha, S.C., Kim, K., Yang, H., Choi, S.Y., Kim, Y.T., Park, J.T., Lee, C.H., Choi, J., Paek, J., Lee, K.: Room-temperature semiconductor gas sensor based on nonstoichiometric tungsten oxide nanorod film. Appl. Phys. Lett. **86**, 213105 (2005)
69. Pak, Y., Kim, S.M., Jeong, H., Kang, C.G., Park, J.S., Song, H., Lee, R., Myoung, N., Lee, B.H., Seo, S., Kim, J.T.: Palladium-decorated hydrogen-gas sensors using periodically aligned graphene nanoribbons. ACS Appl. Mater. Interfaces **6**, 13293–13298 (2014)
70. Fine, G.F., Cavanagh, L.M., Afonja, A., Binions, R.: Metal oxide semiconductor gas sensors in environmental monitoring. Sensors **10**, 5469–5502 (2010)
71. Li, W., Geng, X., Guo, Y., Rong, J., Gong, Y., Wu, L., Zhang, X., Li, P., Xu, J., Cheng, G., Sun, M.: Reduced graphene oxide electrically contacted graphene sensor for highly sensitive nitric oxide detection. ACS Nano **5**, 6955–6961 (2011)
72. Zhang, J., Liu, X., Neri, G., Pinna, N.: Nanostructured materials for room-temperature gas sensors. Adv. Mater. **28**, 795–831 (2016)
73. Wang, C., Yin, L., Zhang, L., Xiang, D., Gao, R.: Metal oxide gas sensors: sensitivity and influencing factors. Sensors **10**, 2088–2106 (2010)
74. Waghuley, S.A., Yenorkar, S.M., Yawale, S.S., Yawale, S.P.: Application of chemically synthesized conducting polymer-polypyrrole as a carbon dioxide gas sensor. Sens. Actuator. B **128**, 366–373 (2008)
75. Li, J., Lu, Y., Ye, Q., Cinke, M., Han, J., Meyyappan, M.: Carbon nanotube sensors for gas and organic vapor detection. Nano Lett. **3**, 929–933 (2003)
76. Bag, A., Lee, N.E.: Gas sensing heterostructures based on two-dimensional nanostructures materials: a review. Nanoscale **7**, 13367–13383 (2019)
77. Hui, R.: Chapter 4 – Photodetectors, Editor(s): Rongqing Hui, Introduction to fiber-optic communications. Academic Press, pp. 125 (2020)
78. Nasiri, N., Tricoli, A.: Chapter 5—Nanomaterials-based UV photodetectors. pp. 123–149. Editor(s): Thomas, S., Grohens, Y., Pottathara, Y.B.: In micro and nano technologies, Industrial applications of nanomaterials. Elsevier (2019)
79. Angizi, S., Khalaj, M., Alem, S.A.A., Pakdel, A., Willander, M., Hatamie, A., Simchi, A.: Review—towards the two-dimensional hexagonal boron nitride (2D h-BN) electrochemical sensing platforms. J. Electrochem. Soc. **167**, 126513 (2020)
80. Anzai, A., Nishiyama, F., Yamanaka, S., Inumaru, K.: Thin film growth of boron nitride on α-Al_2O_3 (001) substrates by reactive sputtering. Mater. Res. Bull. **46**, 2230–2234 (2011)
81. Lin, Y., Connell, J.W.: Advances in 2D boron nitride nanostructures: nanosheets, nanoribbons, nanomeshes, and hybrids with graphene. Nanoscale **4**, 6908–6939 (2012)
82. Wang, Y., Shi, Z., Yin, J.: Boron nitride nanosheets: large-scale exfoliation in methanesulfonic acid and their composites with polybenzimidazole. J. Mater. Chem. **21**, 11371–11377 (2011)
83. Wang, X., Zhi, C., Li, L., Zeng, H., Li, C., Mitome, M., Golberg, D., Bando, Y.: "Chemical blowing" of thin-walled bubbles: high-throughput fabrication of large-area, few-layered BN and C_x-BN nanosheets. Adv. Mater. **23**, 4072–4076 (2011)
84. Deepika, Li, L.H., Glushenkov, A.M., Hait, S.K., Hodgson, P., Chen, Y.: High-efficient production of boron nitride nanosheets via an optimized ball milling process for lubrication in oil. Sci. Rep. **4**, 7288 (2014)
85. Sajjad, M., Feng, P.: Study the gas sensing properties of boron nitride nanosheets. Mater. Res. Bull. **49**, 35–38 (2014)
86. Borek, T.T., Ackerman, W., Hua, D.W., Paine, R.T., Smith, D.M.: Highly microporous boron nitride for gas adsorption. Langmuir **7**, 2844–2846 (1991)
87. Janik, J.F., Ackerman, W.C., Paine, R.T., Hua, D.W., Maskara, A., Smith, D.M.: Boron nitride as a selective gas adsorbent. Langmuir **10**, 514–518 (1994)
88. Yang, G.H., Abulizi, A., Zhu, J.J.: Sonochemical fabrication of gold nanoparticles–boron nitride sheets nanocomposites for enzymeless hydrogen peroxide detection. Ultrason. Sonochem. **21**, 1958–1963 (2014)
89. Xiong, J., Zhu, W., Li, H., Yang, L., Chao, Y., Wu, P., Xun, S., Jiang, W., Zhang, M., Li, H.: Carbon-doped porous boron nitride: metal-free adsorbents for sulfur removal from fuels. J. Mater. Chem. A **3**, 12738–12747 (2015)

90. Kimura, T., Yamamoto, K., Shimizu, T., Yugo, S.: Humidity-sensitive electrical properties and switching characteristics of BN films. Thin Solid Films **70**, 351–362 (1980)
91. Kimura, T., Yamamoto, K., Yugo, S.: Humidity sensitive electrical conduction of polycrystalline boron nitride films. Jpn. J. Appl. Phys. **17**, 1871 (1978)
92. Li, J., Lin, J., Huang, Y., Xu, X., Liu, Z., Xue, Y., Ding, X., Luo, H., Jin, P., Zhang, J., Zou, J.: Organic fluorescent dyes supported on activated boron nitride: a promising blue light excited phosphors for high-performance white light-emitting diodes. Sci. Rep. **5**, 8492 (2015)
93. Wu, K., Wang, J., Liu, D., Lei, C., Liu, D., Lei, W., Fu, Q.: Highly thermoconductive, thermostable, and super-flexible film by engineering 1D rigid rod-like aramid nanofiber/2D boron nitride nanosheets. Adv. Mater. **32**, 1906939 (2020)
94. Tay, R.Y., Li, H., Lin, J., Wang, H., Lim, J.S.K., Chen, S., Leong, W.L., Tsang, S.H., Teo, E.H.T.: Lightweight, Superelastic boron nitride/ polydimethylsiloxane foam as air dielectric substitute for multifunctional capacitive sensor applications. Adv. Funct. Mater. **30**, 1909604 (2020)
95. Li, L.H., Cervenka, J., Watanabe, K., Taniguchi, T., Chen, Y.: Strong oxidation resistance of atomically thin boron nitride nanosheets. ACS Nano **8**, 1457–1462 (2014)
96. Li, L., Li, L.H., Chen, Y., Dai, X.J., Lamb, P.R., Cheng, B.M., Lin, M.Y., Liu, X.: High-quality boron nitride nanoribbons: unzipping during nanotube synthesis. Angew. Chem. Int. Ed. **125**, 4306–4310 (2013)
97. Zeng, H., Zhi, C., Zhang, Z., Wei, X., Wang, X., Guo, W., Bando, Y., Golberg, D.: "White graphenes": boron nitride nanoribbons via boron nitride nanotube unwrapping. Nano Lett. **10**, 5049–5055 (2010)
98. Taha-Tijerina, J., Narayanan, T.N., Gao, G., Rohde, M., Tsentalovich, D.A., Pasquali, M., Ajayan, P.M.: Electrically insulating thermal nano-oils using 2D fillers. ACS Nano **6**, 1214–1220 (2012)
99. Lee, S.H., Kim, M.J., Ahn, S., Koh, B.: Purification of boron nitride nanotubes enhances biological application properties. Int. J. Mol. Sci. **21**, 1529 (2020)
100. Domi, B., Bhorkar, K., Rumbo, C., Sygellou, L., Yannopoulos, S.N., Barros, R., Quesada, R., Tamayo-Ramos, J.A.: Assessment of physico-chemical and toxicological properties of commercial 2D boron nitride nanopowder and nanoplatelets. Int. J. Mol. Sci. **22**, 567 (2021)
101. Ciofani, G., Ricotti, L., Danti, S., Moscato, S., Nesti, C., D'Alessandro, D., Dinucci, D., Chiellini, F., Pietrabissa, A., Petrini, M., Menciassi, A.: Investigation of interactions between poly-L-lysine-coated boron nitride nanotubes and C_2C_{12} cells: up-take, cytocompatibility, and differentiation. Int. J. Nanomedicine **5**, 285–298 (2010)
102. Bhattacharya, B., Singh, N.B., Sarkar, U.: Interaction of boron-nitrogen substitued graphene nanoribbon with nucleobases: the idea of biosensor. Soft Nanoscience Letters **3**, 43–45 (2013)
103. Kang, J.H., Park, C., Sauti, G., Smith, M.W., Jordan, K.C., Lowther, S.E., Bryant, R.G.: High kinetic energy penetrator shielding and high wear resistance materials fabricated with boron nitride nanotubes (BNNTS) and BNNT polymer composites. Patent no. US 2012/0186742 A1 (2012)
104. Harrison, C., Weaver, S., Bertelsen, C., Burgett, E., Hertel, N., Grulke, E.: Polyethylene/boron nitride composites for space radiation shielding. J. Appl. Polym. Sci. **109**, 2529–2538 (2008)
105. Ashrafi, B., Jakubinek, M.B., Martinez-Rubi, Y., Rahmat, M., Djokic, D., Laqua, K., Park, D., Kim, K.S., Simard, B., Yousefpour, A.: Multifunctional fiber reinforced polymer composites using carbon and boron nitride nanotubes. Acta Astronaut. **141**, 57–63 (2017)
106. Zhi, C., Bando, Y., Tang, C., Golberg, D.: Boron nitride nanotubes. Mater. Sci. Eng. R. Rep. **70**, 92–111 (2010)
107. Tang, D.M., Ren, C.L., Wei, X., Wang, M.S., Liu, C., Bando, Y., Golberg, D.: Mechanical properties of bamboo-like boron nitride nanotubes by in situ TEM and MD simulations: strengthening effect of interlocked joint interfaces. ACS Nano **5**, 7362–7368 (2011)
108. Manning, T., Field, R., Klingaman, K., Fair, M., Bolognini, J., Crownover, R., Adam, C.P., Panchal, V., Rozumov, E., Grau, H., Matter, P.: Innovative boron nitride-doped propellants. Def. Technol. **12**, 69–80 (2016)

109. Çelik, O.N., Ay, N., Göncü, Y.: Effect of nano hexagonal boron nitride lubricant additives on the friction and wear properties of AISI 4140 steel. Particul. Sci. Technol. **31**, 501–506 (2013)
110. Saito, T., Honda, F.: Chemical contribution to friction behavior of sintered hexagonal boron nitride in water. Wear **237**, 253–260 (2000)
111. Pakdel, A., Zhi, C., Bando, Y., Nakayama, T., Golberg, D.: Boron nitride nanosheet coatings with controllable water repellency. ACS Nano **5**, 6507–6515 (2011)
112. Bando, Y., Ogawa, K., Golberg, D.: Insulating nanocables': Invar Fe–Ni alloy nanorods inside BN nanotubes. Chem. Phys. Lett. **347**, 349–354 (2001)
113. Chen, X., Jia, S., Ding, N., Shi, J., Wang, Z.: Capture of aromatic organic pollutants by hexagonal boron nitride nanosheets: density functional theoretical and molecular dynamic investigation. Environ. Sci. Nano **3**, 1493–1503 (2016)
114. Li, J., Xiao, X., Xu, X., Lin, J., Huang, Y., Xue, Y., Jin, P., Zou, J., Tang, C.: Activated boron nitride as an effective adsorbent for metal ions and organic pollutants. Sci. Rep. **3**, 3208 (2013)
115. Li, M., Wang, Y., Tang, P., Xie, N., Zhao, Y., Liu, X., Hu, G., Xie, J., Zhao, Y., Tang, J., Zhang, T.: Graphene with atomic-level in-plane decoration of h-BN domains for efficient photocatalysis. Chem. Mater. **29**, 2769–2776 (2017)
116. Gautam, C., Tiwary, C.S., Jose, S., Brunetto, G., Ozden, S., Vinod, S., Raghavan, P., Biradar, S., Galvao, D.S., Ajayan, P.M.: Synthesis of low-density, carbon-doped, porous hexagonal boron nitride solids. ACS Nano **9**, 12088–12095 (2015)
117. Postole, G., Gervasini, A., Caldararu, M., Bonnetot, B., Auroux, A.: Is BN an appropriate support for metal oxide catalysts? Appl. Catal. A **325**, 227 (2007)
118. Song, Y., Xu, H., Wang, C., Chen, J., Yan, J., Xu, Y., Li, Y., Liu, C., Li, H., Lei, Y.: Graphene-analogue boron nitride/Ag_3PO_4 composite for efficient visible-light-driven photocatalysis. RSC Adv. **4**, 56853–56862 (2014)
119. Wang, M., Li, M., Xu, L., Wang, L., Ju, Z., Li, G., Qian, Y.: High yield synthesis of novel boron nitride submicro-boxes and their photocatalytic application under visible light irradiation. Catal. Sci. Technol. **1**, 1159–1165 (2011)
120. Taniguchi, T., Watanabe, K., Kubota, Y., Tsuda, O.: Production of a hexagonal boron nitride crystal body capable of emitting out ultraviolet radiation. US Patent **7**, 811–909 (2010)
121. Yu, Y., Chen, H., Liu, Y., Li, L.H., Chen, Y.: Humidity sensing properties of single Au-decorated boron nitride nanotubes. Electrochem. Commun. **30**, 29–33 (2013)
122. Sun, Q., Li, Z., Searles, D.J., Chen, Y., Lu, G., Du, A.: Charge-controlled switchable CO_2 capture on boron nitride nanomaterials. J. Am. Chem. Soc. **135**, 8246–8253 (2013)
123. Liu, W.M., Zhao, W.W., Zhang, H.Y., Wang, P.F., Chong, Y.M., Ye, Q., Zou, Y.S., Zhang, W.J., Zapien, J.A., Bello, I., Lee, S.T.: A cubic boron nitride film-based fluorescent sensor for detecting Hg^{2+}. App. Phys. Lett. **94**, 183105 (2009)
124. Zeng, B., Li, W., Chen, H., Li, G.: First-principles study on the adsorption of hydrogen cyanide on the metal doped (8, 0) boron nitride nanotubes. Chem. Phys. Lett. **730**, 513–520 (2019)
125. Yoo, S.M., Sharma, B., Kim, J.S.: Boron nitride nanotubes (BNNTs) decorated Pd-ternary alloy ($Pd_{63.2}Ni_{34.3}Co_{2.5}$) for H_2 sensing. Int. J. Hydrog. Energy **46**, 12263–12270 (2020)
126. Weber, M., Kim, J.Y., Lee, J.H., Kim, J.H., Iatsunskyi, I., Coy, E., Miele, P., Bechelany, M., Kim, S.S.: Highly efficient hydrogen sensors based on Pd nanoparticles supported on boron nitride coated ZnO nanowires. J. Mater. Chem. A **7**, 8107–8116 (2019)
127. Vessally, E., Moladoust, R., Mousavi-Khoshdel, S.M., Esrafili, M.D., Hosseinian, A., Edjlali, L.: The ClCN adsorption on the pristine and Al-doped boron nitride nanosheet, nanocage, and nanocone: density functional studies. Thin Solid Films **645**, 363–369 (2020)
128. Shan, S., Junhong, W., Wu, L.: The pristine BN nanocone as a toluene gas sensor: a computational study. Sustain. Chem. Pharm. **17**, 100277 (2020)
129. Rakhshi, M., Mohsennia, M., Rasa, H., Sameti, M.R.: First-principle study of ammonia molecules adsorption on boron nitride nanotubes in presence and absence of static electric field and ion field. Vacuum **155**, 456–464 (2018)
130. Mohsennia, M., Rakhshi, M., Rasa, H.: A computational study on interactions of Ni- and Pt-doped boron nitride nano tubes with NH_3 in presence and absence of electric fields. Comput. Theor. Chem. **1136**, 1–9 (2018)

131. He, X., Gui, Y., Liu, K., Xu, L.: Comparison of sensing and electronic properties of C_2H_2 on different transition metal oxide nanoparticles (Fe_2O_3, NiO, TiO_2) modified BNNT (10, 0). Appl. Surf. Sci. **521**, 146463 (2020)
132. He, X., Gui, Y., Xie, J., Liu, X., Wang, Q., Tang, C.: A DFT study of dissolved gas (C_2H_2, H_2, CH_4) detection in oil on CuO modified BNNT. Appl. Surf. Sci. **500**, 144030 (2020)
133. He, Y., Li, D., Gao, W., Yin, H., Chen, F., Sun, Y.: High-performance NO_2 sensors based on spontaneously functionalized hexagonal boron nitride nanosheets via chemical exfoliation. Nanoscale **11**, 21909–21916 (2019)
134. Ammar, H.Y., Badran, H.M., Eid, K.M.: TM-doped $B_{12}N_{12}$ nnanocage(TM = Mn, Fe) as a sensor for CO, NO, and NH_3 gases: A DFT and TD-DFT study. Mater. Today Commun. **25**, 101681 (2020)
135. Badran, H.M., Eid, K.M., Ammar, H.Y.: A DFT study on the effect of the external electric field on ammonia interaction with boron nitride nnanocage. J. Phys. Chem. Solids **141**, 109399 (2020)
136. Xu, Q., Cai, L., Zhao, H., Tang, J., Shen, Y., Hu, X., Zeng, H.: Forchlorfenuron detection based on its inhibitory effect towards catalase immobilized on boron nitride substrate. Biosens. Bioelectron. **63**, 294–300 (2015)
137. Liu, Y.J., Gao, B., Xu, D., Wang, H.M., Zhao, J.X.: Theoretical study on Si-doped hexagonal boron nitride (h-BN) sheet: electronic, magnetic properties, and reactivity. Phys. Lett. A **378**, 2989–2994 (2014)
138. Huang, Q., Bando, Y., Zhao, L., Zhi, C.Y., Golberg, D.: PH sensor based on boron nitride nanotubes. Nanotechnology **20**, 415501 (2009)
139. Wang, R., Zhang, D., Liu, Y., Liu, C.: A theoretical study of silicon-doped boron nitride nanotubes serving as a potential chemical sensor for hydrogen cyanide. Nanotechnology **20**, 505704 (2009)
140. Mukhopadhyay, S., Scheicher, R.H., Pandey, R., Karna, S.P.: Sensitivity of boron nitride nanotubes toward biomolecules of different polarities. J. Phys. Chem. Lett. **2**, 2442–2447 (2011)
141. Choi, H., Park, Y.C., Kim, Y.H., Lee, Y.S.: Ambient carbon dioxide capture by boron-rich boron nitride nanotube. J. Am. Chem. Soc. **133**, 2084–2087 (2011)
142. Guo, H., Zhang, W., Lu, N., Zhuo, Z., Zeng, X.C., Wu, X., Yang, J.: CO_2 capture on h-bn sheet with high selectivity controlled by external electric field. J. Phys. Chem. C **119**, 6912–6917 (2015)
143. Sun, Q., Sun, C., Du, A., Li, Z.: Charged-controlled separation of nitrogen from natural gas using boron nitride fullerene. J. Phys. Chem. C **118**, 30006–30012 (2014)
144. Zhao, J.X., Ding, Y.H.: Theoretical studies of the interaction of an open-ended boron nitride nanotube (BNNT) with gas molecules. J. Phys. Chem. C **112**, 20206–20211 (2008)
145. Paura, E.N.C., Cunha, W.F.da, Roncaratti, L.F., Martins, J.B.L., Silva, G.M.e, Gargano, R.: CO_2 adsorption on single-walled boron nitride nanotubes containing vacancy defects. RSC Adv. **5**, 27412–27420 (2015)
146. Rimola, A., Sodupe, M.: Physisorption vs. chemisorption of probe molecules on boron nitride nanomaterials: the effect of surface curvature. Phys. Chem. **15**, 13190–13198 (2013)
147. El-Barbary, A.A., Eid, K.M., Kamel, M.A., Taha, H.O., Ismail, G.H.: Adsorption of CO, CO_2, NO and NO_2 on boron nitride nanotubes: DFT study. J. Surf. Eng. Mater. Adv. Technol. **5**, 154 (2015)
148. Feng, J.W., Liu, Y.J., Zhao, J.X.: Theoretical prediction of the mechanisms for defect healing or oxygen doping in a hexagonal boron nitride (h-BN) sheet with nitrogen vacancies by NO_2 molecules. J. Mol. Model. **20**, 2307 (2014)
149. Neklyudov, I.M., Rybalko, V.F., Morozov, A.N., Pistryak, S.V., Kulish, V.G., Buzhinsky, O.I., Opimakh, I.V.: Implantation, radiation-induced and thermoactivated desorption from boron nitride irradiated with deuterium ions. Plasma Devices Oper. **3**, 151–156 (1994)
150. Zhang, N.C., Ren, J., Peng, X.J.: Mg/Ca decorated on carbon-doped boron nitride sheet: application for gas adsorption. Fuller. Nanotub. Car. N. **24**, 298–304 (2016)

151. Choudhuri, I., Patra, N., Mahata, A., Ahuja, R., Pathak, B.: B-N@ graphene: highly sensitive and selective gas sensor. J. Phys. Chem. C **119**, 24827–24836 (2015)
152. Rubio, A., Corkill, J.L., Cohen, M.L.: Theory of graphitic boron nitride nanotubes. Phys. Rev. B **49**, 5081 (1994)
153. Xu, Z., Buehler, M.J.: Geometry controls conformation of graphene sheets: membranes, ribbons, and scrolls. ACS Nano **4**, 3869–3876 (2010)
154. Zhang, D.B., Akatyeva, E., Dumitrică, T.: Bending ultrathin graphene at the margins of continuum mechanics. Phys. Rev. Lett. **106**, 255503 (2011)
155. Koltonow, A.R., Huang, J.: Two-dimensional nanofluidics. Science **351**, 1395–1396 (2016)
156. Ataca, C., Ciraci, S.: Functionalization of BN honeycomb structure by adsorption and substitution of foreign atoms. Phys. Rev. B **82**, 165402 (2010)
157. Svatek, S.A., Scott, O.R., Rivett, J.P., Wright, K., Baldoni, M., Bichoutskaia, E., Taniguchi, T., Watanabe, K., Marsden, A.J., Wilson, N.R.: Beton, PH: Adsorbate-induced curvature and stiffening of graphene. Nano Lett. **15**, 159–164 (2015)
158. Singh, K., Kaur, M., Chauhan, I., Awasthi, A., Kumar, M., Thakur, A., Kumar, A.: BN/NiO Nanocomposites: Structural, defect chemistry and electrical properties in hydrogen gas atmosphere. Ceram. Int. **46**, 26233–26237 (2020)
159. Singh, K., Kaur, M., Chauhan, I., Singh, H., Awasthi, A., Kumar, M., Thakur, A., Kumar, A.: Effect of ammonia gas on electrical properties of boron nitride/nickel oxide (BN_{80}/NiO_{20}) nanocomposite. J. Mater. Sci. Mater. Electron. **32**, 5556–5566 (2021)
160. Liao, K.J., Wang, W.L., Cai, C.Z., Wang, X.S.: Kong, CY: UV Photodetectors of c-BN Films. Int. J. Mod. Phys. B **16**, 1115–1119 (2002)
161. Li, J., Majety, S., Dahal, R., Zhao, W.P., Lin, J.Y., Jiang, H.X.: Dielectric strength, optical absorption, and deep ultraviolet detectors of hexagonal boron nitride epilayers. Appl. Phys. Lett. **101**, 171112 (2012)
162. Sajjad, M., Jadwisienczak, W.M., Feng, P.: Nanoscale structure study of boron nitride nanosheets and development of a deep-UV photodetector. Nanoscale **6**, 4577–4582 (2014)
163. Ju, L., Velasco, J., Jr., Huang, E., Kahn, S., Nosiglia, C., Tsai, H.Z., Yang, W., Taniguchi, T., Watanabe, K., Zhang, Y., Zhang, G.: Photoinduced doping in heterostructures of graphene and boron nitride. Nat. Nanotechnol. **9**, 348–352 (2014)
164. Li, X., Lin, S., Lin, X., Xu, Z., Wang, P., Zhang, S., Zhong, H., Xu, W., Wu, Z., Fang, W.: Graphene/h-BN/GaAs sandwich diode as solar cell and photodetector. Opt. Express **24**, 134–145 (2016)
165. Li, Y.B., Zhong, J.W., Zhou, L.M., Sun, C.L., Wang, X., Yang, H.S., Milne, W.: Deep ultraviolet photodetector based on sulphur-doped cubic boron nitride thin film. Mater. Sci. Forum **879**, 1117–1122 (2016)
166. Lu, Y., Wu, Z., Xu, W., Lin, S.: ZnO quantum dot-doped graphene/h-BN/GaN-heterostructure ultraviolet photodetector with extremely high responsivity. Nanotechnology **27**, 48LT03 (2016)
167. Dean, C.R., Young, A.F., Meric, I., Lee, C., Wang, L., Sorgenfrei, S., Watanabe, K., Taniguchi, T., Kim, P., Shepard, K.L., Hone, J.: Boron nitride substrates for high-quality graphene electronics. Nat. Nanotechnol. **5**, 722–726 (2010)
168. Zhi, C., Bando, Y., Tang, C., Golberg, D., Xie, R., Sekigushi, T.: Phonon characteristics and cathodoluminescence of boron nitride nanotubes. Appl. Phys. Lett. **86**, 213110 (2005)
169. BenMoussa, A., Soltani, A., Schühle, U., Haenen, K., Chong, Y.M., Zhang, W.J., Dahal, R., Lin, J.Y., Jiang, H.X., Barkad, H.A., BenMoussa, B.: Recent developments of wide-bandgap semiconductor based UV sensors. Diamond Relat. Mater. **18**, 860–864 (2009)
170. Ahmad, N., Lichtman, D.: Study of boron nitride thin films for ultraviolet-sensor applications. Sens. Actuators **18**, 397–405 (1989)
171. Zhang, N., Song, Y., Zhou, K., Wang, W.: Enhanced performance of solar-blind photodetector of hexagonal boron nitride with bottom-contact electrodes. AIP Adv. **10**, 085013 (2020)
172. Moradinasab, M., Pourfath, M., Fathipour, M., Kosina, H.: Numerical study of graphene superlattice-based photodetectors. IEEE Trans. Electron Dev. **62**, 593–600 (2015)

173. Soltani, A., Barkad, H.A., Mattalah, M., Benbakhti, B., Jaeger, J.C.D., Chong, Y.M., Zou, Y.S., Zhang, W.J., Lee, S.T., BenMoussa, A., Giordanengo, B., Hochedez, J.F.: 193 nm deep-ultraviolet solar-blind cubic boron nitride based photodetectors. Appl. Phys. Lett. **92**, 053501 (2008)
174. Shiue, R.J., Gao, Y., Wang, Y., Peng, C., Robertson, A.D., Efetov, D.K., Assefa, S., Koppens, F.H., Hone, J., Englund, D.: High-responsivity graphene–boron nitride photodetector and autocorrelator in a silicon photonic integrated circuit. Nano lett. **15**, 7288–7293 (2015)
175. Rivera, M., Velázquez, R., Aldalbahi, A., Zhou, A.F., Feng, P.: High operating temperature and low power consumption boron nitride nanosheets based broadband UV photodetector. Sci. Rep. **7**, 42973 (2017)
176. Rivera, M., Velázquez, R., Aldalbahi, A., Zhou, A.F., Feng, P.X.: UV photodetector based on energy bandgap shifted hexagonal boron nitride nanosheets for high-temperature environments. J. Phys. D Appl. Phys. **51**, 045102 (2018)
177. Lin, C.H., Fu, H.C., Cheng, B., Tsai, M.L., Luo, W., Zhou, L., Jang, S.H., Hu, L., He, JH: A flexible solar-blind 2D boron nitride nanopaper-based photodetector with high thermal resistance. npj 2D Mater. Appl. **2**, 23 (2018)
178. Chu, D., Pak, S.W., Kim, E.K.: Locally gated SnS_2/hBN thin film transistors with a broadband photoresponse. Sci. Rep. **8**, 10585 (2018)
179. Lu, Q., Yu, L., Liu, Y., Zhang, J., Han, G., Hao, Y.: Low-noise mid-infrared photodetection in BP/h-BN/Graphene van der Waals heterojunctions. Materials **12**, 2532 (2019)
180. Tan, B., Yang, H., Hu, Y., Gao, F., Wang, L., Dai, M., Zhang, S., Shang, H., Chen, H., Hu, P.: Synthesis of high-quality multilayer hexagonal boron nitride films on Au foils for ultrahigh rejection ratio solar-blind photodetection. ACS Appl. Mater. Interfaces **12**, 28351–28359 (2020)
181. Thompson, J.E., Smalley, D., Ishigami, M.: Solar-blind ultraviolet photodetectors based on vertical graphene-hexagonal boron nitride heterostructures. MRS Adv. **5**, 1993–2002 (2020)
182. Wang, Y., Meng, J., Tian, Y., Chen, Y., Wang, G., Yin, Z., Jin, P., You, J., Wu, J., Zhang, X.: Deep ultraviolet photodetectors based on carbon-doped two-dimensional hexagonal boron nitride. ACS Appl. Mater. Interfaces **12**, 27361 (2020)
183. Kaushik, S., Sorifi, S., Singh, R.: Study of temperature dependent behavior of h-BN nanoflakes based deep UV photodetector. Photonics Nanostructures: Fundam. Appl. **43**, 100887 (2020)
184. Singh, K., Kaur, G., Kaur, M., Chauhan, I., Kumar, M., Thakur, A., Kumar, A.: Photoconductivity of gold nanoparticles loaded boron nitride/nickel oxide nanocomposites. Chem. Phys. Lett. **762**, 138153 (2021)

Two-Dimensional (2D) Materials Incorporated PMMA Polymeric Nanocomposites: Synthesis and Applications

Sonika, Sushil Kumar Verma, Varatharajan Prasannavenkadesan, and Shailendra Rajput

Abstract This chapter discusses the two-dimensional (2D) based PMMA polymer nanocomposites. Due to their unique structural and electrochemical properties, recently, 2D materials have attracted remarkable attention for various energy applications, including energy harvesting and energy storage devices. Over the past 20 years, polymer nanocomposites have been increasingly significant in the field of material research. While this innovative method is still in its early stages of research, it has the potential for success in a variety of applications. These 2D-material-based composites could expand the application domains of conventional 2D-based materials and devices. The application of composites has been expanded by incorporating polymers and metal structures with 2D materials. Some technical issues, including energy storage for future applications, water purification, low-energy electronics, bio-inspired computing, biomedical applications, and effective catalysis, are seen to be suitable for the composite material-oriented technique. Polymer nanocomposites present a lot of potential, which has made them an increasingly common subject of study in recent years. The extent to which we can overcome the obstacles will determine how nanocomposites develop and are used.

Keywords Nanocomposite · PMMA · Energy harvesting · Clay · Polymerization

---

Sonika
Department of Physics, Rajiv Gandhi University, Itanagar 791112, India

S. K. Verma
Department of Chemical Engineering, Indian Institute of Technology Guwahati, Guwahati 781039, India

V. Prasannavenkadesan
Department of Chemical Engineering, COE SusPol, Indian Institute of Technology Guwahati, Guwahati 781039, India

S. Rajput (✉)
College of Engineering, Xi'an International University, Xi'an 710077, China
e-mail: shailendra@xaiu.edu.cn

N. Talreja et al. (eds.), *Two-dimensional Hybrid Composites*, Engineering Materials,
https://doi.org/10.1007/978-981-99-8010-9_2

1 Introduction

Recently, it has been easy to synthesize nanoscale materials like carbon nanotubes, graphene sheets, etc., to manufacture novel materials with better properties [1–4]. Polymer nanocomposites are newly developing materials containing nanoscale fillers with a characteristic length of less than 100 nm evenly disseminated in the matrix. Recently, these materials have shown significant improvements in mechanical properties and aided the development of novel multifunctional attributes (like electrical and thermal) that are not attainable with microfillers [5–14].

Two-dimensional (2D) materials are constructed from numerous thin layers with weak van der Waal connections. A layer that is only one atom thick may be in a few nanometers. In the 2D plane, the electrons are not constrained; however, their third-dimensional movement has constraints due to quantum mechanics. Materials like quantum dots, nano-ribbons, tubes, and wires are classified as zero-dimensional (0D). Finally, three-dimensional (3D) materials, such as nanoballs and nanocones, are other examples. Materials that come into the 2D material category consist of graphene, transition metal dichalcogenides (TMDs), transition metal oxides (TMOs), graphitic C_3N_4, 2D clay materials, hexagonal boron nitride (hBN), black phosphorus (BP), and germanene. These materials are often only a single atom thick. The 2D-material composite is fabricated with a combination of these materials. These novel materials outperform conventional materials and composites, such as metals, wood, reinforced plastics, fiber, fiberglass, Kevlar, carbon fiber, and polymers. These materials are considered prospective candidates for leading-edge devices because of their excellent mechanical strength, flexibility, stiffness, and enhanced electrical and optical properties [15–35].

In particular, new developments in developing methods facilitate high-homogeneousness and large-area 2D materials, reducing manufacturing costs and improving the standard [36, 37]. To achieve the best material properties, engineered 2D composite materials necessitate organized and interactional techniques (Fig. 1). Recently, polymer nanocomposites have drawn much attention to obtain better mechanical, thermal, optical, and barrier characteristics than pure polymers. Synthetic polymer materials have replaced conventional inorganic and natural polymeric materials. As these innovative materials are flammable, they should be modified by adding flame-retardant (FR) chemicals to reduce their flammability. The use of several halogenated FR additions has recently been banned by environmental legislation, sparking a hunt for substitute FR additives. As they may concurrently enhance the polymer nanocomposite's physical and flammability qualities, nanoparticle fillers are very appealing for this application.

Researchers have recently shown that organic–inorganic nanoscale composites are particularly interesting because they frequently show hybrid properties. One example is a hybrid of organic polymers and inorganic clay minerals, which adhere to the broad family of 2:1 layered silicates [39]. It has been identified that the dispersed phase has an essential effect on the polymer composite properties. The dimensions may typically be in one or more ultrafine sizes between 1 and 10 nm. Due to their distinct

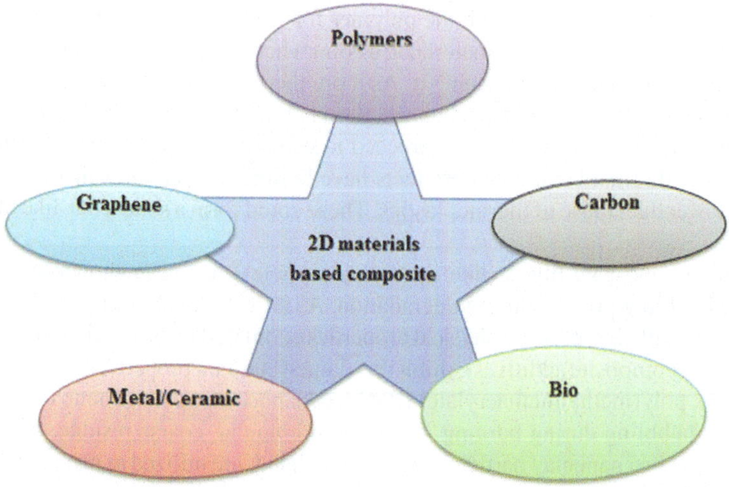

Fig. 1 Classification of 2D-materials based composites [38]

nanoscale structure, polymer–clay nanocomposites have distinctive characteristics. The advantages have been achieved in polymer–clay nanocomposites' thermal and mechanical properties [40–42].

Poly(methyl methacrylate, or PMMA), an optically transparent polymer of industrial significance, has excellent dimensional stability, strong scratch resistance, weather ability, and light resistance. Furthermore, PMMA nanocomposites could have reduced gas penetrability and improved thermal and mechanical properties despite losing optical clarity. PMMA isn't thermally steady and will decay at around 220 °C, giving a lot of monomers. PMMA has a limiting oxygen index of 17, making it easy to burn and produce toxic gases, smoke, and heat. As a result, the goal is to improve its mechanical strength and excellent physical properties while maintaining its excellent thermal stability. Since PMMA can be easily polymerized using various techniques, it has frequently been employed as a standard thermoplastic polymer structure. The PMMA shows a high modulus and high warm strength. However, the nanocomposite structure of the polymer can offer more thorough mechanical and thermal attributes improvements. The polymer is the guest in nanocomposite applications, while the nanoscale clay is the host. These nanocomposites may have improved mechanical and thermal properties substantially due to favorable guest–host interactions.

Toyota researchers were the first to propose this idea when they discovered that organophilic clay and polyamide-6 could be used to make a nanocomposite. Because clays are used to reinforce polymeric materials, clay/polymer nanocomposites have recently garnered much interest [43–45]. The mechanical strength [46], thermal stability [47], flammability [48], and gas permeability [49] of clay/polymer nanocomposites often differ significantly from those of the polymer alone. Well-dispersed nanocomposites are only possible if the polymer and clay work together [47]. The clay requires compatibility with the polymer for well-dispersed nanocomposites to form [47]. Heating a polymer material to a point where its thermal degradation occurs is usually the initial step in burning it. As more thermal breakdown products diffuse from the adjacent molten plastic, bubbles form below the heated polymer surface and develop [50]. As fuel vapors, they proceed to develop into the gas phase. The bulk of polymer thermal breakdown products have boiling temperatures that are significantly lower than those of thermoplastics. These could form a solid, char-like thermal barrier.

Adding nanoscale fillers (like nano clay particles) has been shown to stop the intense bubbling process during degradation. As an additional incentive, these filler particles usually improve the physical properties compared to the matrix [40, 51]. But, not all nanocomposite additives exhibit this flame-retardant property. As with unfilled materials, poly(methylmethacrylate) (PMMA)/nanocomposites of nano clay exhibit vigorous bubbling during burning, resulting in granular, coarse residue. Nanocomposites based on nanoclay particles, on the other hand, are utilized as protective solids [52–55].

The protective layer performs an essential role in lowering flammability. These clay-particle coatings usually experience significant lateral surface splitting and continuous bubbling. Although it is evident that these extended nanoparticles have potential as flame retardants, more research is required to comprehend the physical mechanisms underlying this behavior and improve the filler's efficiency. PMMA nanocomposites have also been processed using the solution, melt, and other methods [56–58]. The melt technique exceeds environmental standards by not using solvents and is compatible with conventional polymer extrusion and blending techniques. Regarding the production of PCNs, melt intercalation generally has advantages in terms of economics and the environment. Indeed, an intercalated structure is produced when the polymer chain of polar polymers like PMMA quickly enters the galleries of the clay layer. The entropy increase from the contrasting interaction between the clay and polymer layers effectively makes up for the polymer chain's entropy loss.

Polymer layered silicate nanocomposites stand out to scientists because of their extraordinary behavior. Many different polymer/clay hybrids or polymer/clay nanocomposites have been reported, such as nylon 6/clay hybrid [59–61], epoxy polymer/clay nanocomposite [40, 62], acrylic polymer/clay hybrid [63], polystyrene (PS)/clay nanocomposite [64, 65], and so on. The addition of only a minimal amount of clay, typically less than 5% inorganic, to a polymeric matrix has polymers and organoclay containing functional groups, such as amide or imide in nylon, oxirane ring in epoxy polymer, unsaturated (double) bonds or hydroxyl group in clay modified organic chain, and so on, could only be exfoliated and dispersed homogeneously in a small number of instances. The chemical interaction between the intercalating monomer or polymer and the silicate layers is required, such as:

$$\textbf{MMT-N}^{+}\textbf{(Me)}_3\textbf{-R-CH} = \textbf{CH}_2 + \textbf{nCH} = \textbf{CH-R'} \rightarrow \textbf{MMT-N}^{+}\textbf{(Me)}_3\textbf{-R-(CH}_2\textbf{-CH-R')n-}$$

Natural materials such as clay typically consist of fine-grained minerals. Due to the changing water content, it can exhibit fluidity and harden when dried or burnt. Most clay deposits comprise clay minerals, adding flexibility and hardening when dried or burnt. Various quantities of water are trapped in the mineral structure due to the polar attraction. In addition, non-plastic organic compounds can be identified in clay deposits [66–68].

1.1 Clay

The growing marketable interest in nanocomposites could indicate a continuing fascination with applying nano clays to modify polymeric materials for various applications [69, 70]. Since centuries ago, clay has been utilized in construction, industry, and agriculture. Clay-based materials like ceramics, earthenware, drainage pipes, and floor and wall tiles have been applied by men for a long time. Clay's curious property allows it to grow and form when saturated while maintaining the container's shape after drying. This distinctive behavior has been used to develop several traditional clay products. Although there are many varieties of minerals, clay is one among them. In chemistry, all clay minerals are referred to as hydrous silicates. With respect to the origin, clay minerals can be classified into: The surface weathering of rock or shale, a black rock composed of layers of condensed clay, silt, or mud, produces residual clays in a variety of methods. The additional names for residual clays are transported clay and sedimentary clay. Rocks containing silica and alumina and mixing rocks can all deliver the remaining dirt. The second form of clay, transported clay, is eroded from the origin and accumulated at a distant location.

Clay minerals often have a fine-grained, sheet-like transform in their native state. Phyllosilicates are the usual name for hydrous silicates with a sheet structure [71–74]. The diameter of natural clay particles is less than 0.004 mm. Aluminum, iron, mica, quartz, and feldspar oxides can all range in diameter from 0.002 to 0.001 mm [75]. Colloidal clay particles of smaller dimensions (0.001 mm) are in layered silicates.

The four significant classifications of clay minerals are based primarily on differences in the layered structure, as shown in Table 1. Examples of these consist of the illite group, the chlorite group, the montmorillonite/smectite group, and the kaolinite group. The chemical formula of the three components that constitute the kaolinite group, kaolinite, dickite, and nacrite, is $Al_2Si_2O_5(OH)_4$. When members of this group have distinct structures but the same formula, they are considered polymorphs. Silicate sheets (Si_2O_5) connected to the aluminum oxide/hydroxide layers constitute every component. Strong bonds connect the two different kinds of layers combined. These clays produce plastic, rubber, paper, paint, ceramics, and fillers.

One or two constituents of the significant smectite clay category are montmorillonite, powder, pyrophyllite, saponite, and nontronite. The chemical formula for this group is (Ca, Na, H) $(Al, Mg, Fe, Zn)_2$ $(Si, Al)_4O_{10}(OH)_2XH_2O$. The main distinction between the individuals in this category can be found in their chemical compositions. In the layer structure, silicate layers are sandwiched between an aluminum oxide/hydroxide layer [$Al_2(OH)_4$]. These materials are applied in paints, rubbers, drilling muds, plasticizers in molding sands, and porcelain that are resistant to acid, heat, and electricity.

Table 1 Major Clay mineral categories, determined by variations in layered structure)

Group name	Member minerals
Kaolinite	Kaolinite, dickite, nacrite
Montmorillonite-smectite	Montmorillonite, Pyrophyllite, talc
Illite	Vermicullite, sauconite
Chlorite	Amesite, chamosite, cookeite, nimite

Clays are used in various scientific domains due to their natural abundance and ability to be chemically and physically changed to satisfy practical technological demands. Industry and academics have recently drawn particular attention to polymer/layered silicate nanocomposites. This book chapter emphasizes the modification of montmorillonite clay (MMT). MMT is a common nanofiller considered in the manufacture of polymer nanocomposites. The structure and need for modification are discussed in this section (Fig. 2).

In several organizational contrasts, clays are distinguished from fine-grained soils. The particle sizes of silts, fine-grained soils, are typically larger than those of clays. Various naturally occurring deposits contain both silts and clays. But, there is considerable similarity in size and properties. The separation typically takes place at a particle size of 2 m for geologists and soil scientists (clays are finer than silts), 4–5 m for sedimentologists, and 1 m for colloid chemists. Based on the soil's plasticity, geotechnical engineers differentiate between clays and silts using ISO 14688 grades: Silts are relatively larger than clay particles (0.063 mm).

Depending on the academic source, clays can be divided into three or four major groups: kaolinite, montmorillonite-smectite, illite, and chlorite. These categories contain approximately thirty distinct kinds of 'pure' clay, but most 'natural' clay is a mixture of these unique kinds and other weathered minerals.

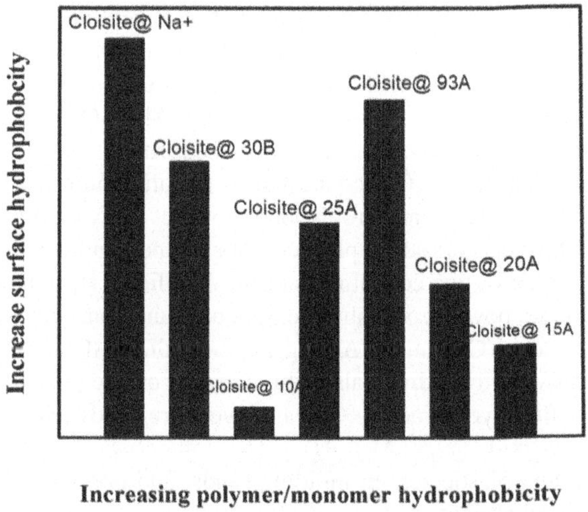

Fig. 2 Cloisite@ selection chart based on Polymer chemistry

1.2 Clay Minerals

Clay minerals contain different proportions of alkali metals, earths, and cations. Clays produce flat hexagonal sheets because they share structural characteristics with micas. It is typical to find the consequences of low-temperature hydrothermal alteration and weathering, such as feldspar and clay mineral weathering.

1.3 Groups of Clay Minerals

Clay minerals are divided into the following groups:

- Kaolin group: Kaolinite, dickite, halloysite, and nacrite.
- Smectite group: Dioctahedral smectites like nontronite; trioctahedral smectites like saponite.
- Illites: The group of clay micas and widely spread minerals.
- The chlorite group contains numerous minerals that seem like one another but vary chemically.

The clay MMT is of several tens of stacked nanolayers (tactoids) (Fig. 3). Thickness of layers is 1 nm, and interlayer spacing is 1 nm. Clay minerals are categorized by:

1. Tetrahedra of two dimensions composed of a corner-shared combination of SiO_4 and AlO_4. This two-dimensional hexagonal array is illustrated in Fig. 3. These

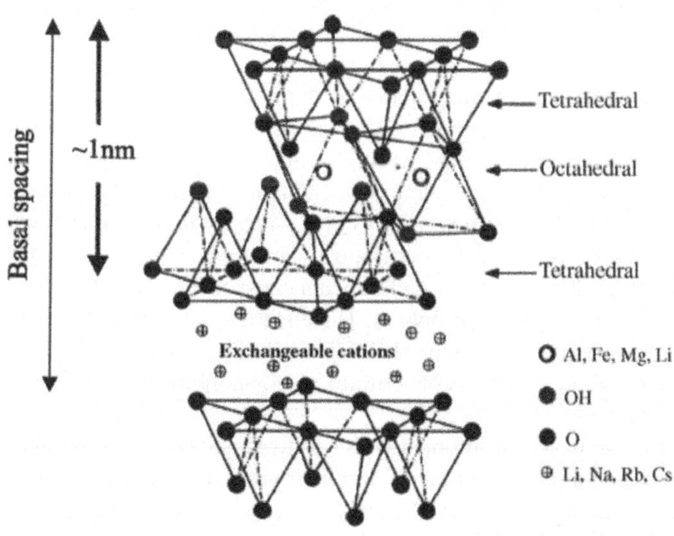

Fig. 3 Schematic diagram of clay [76]

tetrahedral sheets with the chemical formula $(Al\text{-}Si)_3O_4$ share three oxygen atoms with other tetrahedra at its vertex. The fourth vertex does not belong to any other tetrahedron, and the tetrahedra all "point" in the same direction.

2. As shown in Fig. 3, tetrahedral sheets in clay are constantly connected to octahedral sheets comprised of small cations like aluminum or magnesium and coordinated by six oxygen atoms.
3. The unknown vertex from the tetrahedral sheet similarly contributes to one side of the octahedral sheet. Still, an additional oxygen atom is at the center of the six tetrahedra above the gap in the tetrahedral sheet. This oxygen atom produces an OH group with a hydrogen atom in the clay structure.
4. The Classification of clays is based on how octahedral and tetrahedral sheets are divided into layers. In the unlikely circumstance that each layer contained just one octahedral bunch and one tetrahedral bunch, the clay is considered 1:1 earth. The alternative, described as 2:1 clay, consists of two tetrahedral sheets, one of which has an unshared vertex pointing in the same direction as the reverse sheets of octahedron vertex.
5. This composition determines whether the layer is impartial or has a net negative charge. If the layers are charged, interlayer cations like Na^+ or K^+ balance this charge. Water may also be present in the interlayer in any case. A stack of layers separated by an interlayer forms the crystal structure.

Because they are minerals that are found in nature, common clays have natural variations in their composition. The purity of the clay may influence the final properties. Numerous substrates are aluminosilicates, which consist of silica SiO_4 tetrahedra attached in different ways to alumina AlO_6 octahedra and have a sheet-like (layered) structure. The 2:1 division of the tetrahedra by the octahedra results in the formation of smectite clays. Montmorillonite is the variety of smectite clay that is most frequent. Other metals, such as magnesium, could replace aluminum in the crystal structure. The counterions are mainly present in the interlayer spacing of the clay to balance the charge that the sheets carry on their edges and surfaces. The chemical composition of the clay specifically determines this charge. The Thickness of the layers is about 1 nm, and their aspect ratios range from 100 to 1500.

The clays' surface areas per gram can frequently exceed hundreds of m^2. The different ion exchange abilities of the clays, such as those for cations, assist in distinguishing groups. The cation-exchange capacity of soil refers to how much it can absorb and exchange cations. A significant consequence of clays' charged nature is that they are often highly hydrophilic. The effective development of polymer–clay nanocomposites requires changing the clay's polarity to make it "organophilic." A generally hydrophilic clay can convert into an organophilic clay with ion exchange with an organic cation.

Polymeric composites reinforced with additives gained research interests due to the commercial significance of modifying their properties: layered inorganic fillers, fibers, and particulates [77, 78]. Recently, nanometer-thin layered aluminosilicate inorganic fillers like clay, talc, and mica have received much attention and intense research [59, 79]. Although the ability of polymers to interact with artificial clays

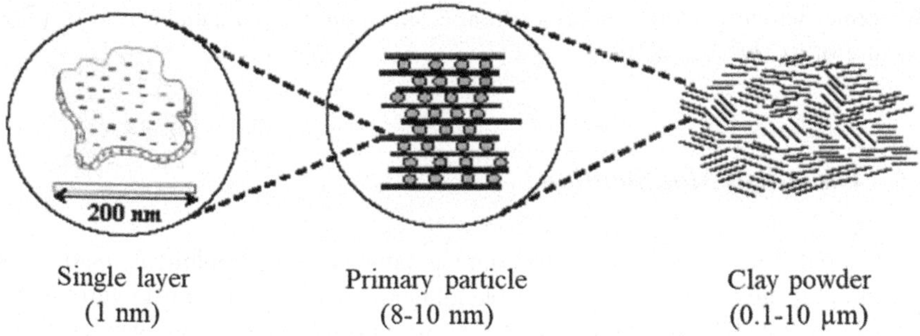

Fig. 4 Structural concept of clay [81]

and appropriately modified clay minerals is explored [80, 81], the study of polymer/clay nanocomposites has only attracted significant interest.

1.4 Montmorillonite Clay (MMT)

The phyllosilicate mineral MMT is often soft and crystallizes into minute clay particles. It is named after the French town of Montmorillon. Figure 4 illustrates the 2:1 clay montmorillonite, a smectite family member with plate-shaped particles and an average diameter of 1 μm. It is an essential constituent in the volcanic ash weathering product bentonite.

Depending on the quantity of water it contains, montmorillonite increases enormously in volume. The chemical name is hydrated sodium calcium aluminum magnesium silicate hydroxide, and represented as $Na,Ca,_{0.33}(Al,Mg)_2(Si_4O_{10})(OH)_2.nH_2O)$. The precise ratio of cations varies depending on the source, but frequent alternatives include potassium, iron, and other cations. It frequently coexists with illite, muscovite, chlorite, cookeite, and kaolinite.

1.5 Uses of Clay

Montmorillonite expands when water is added. Yet, due to water entering the molecular gaps between the layers and subsequent adsorption, it expands significantly more than other clays. The sort of exchangeable cation in the sample significantly impacts how much expansion occurs. The clay can grow to many times its original volume if sodium is the dominating exchangeable cation present. As a result, sodium montmorillonite is now a key ingredient in non-explosive agents used to fracture rocks and destroy concrete buildings when employing explosive charges is inappropriate.

Bentonite, which contains montmorillonite, is suitable as an annular seal or plug because of its swelling ability.

1.6 Chemical Modification

The layered silicate clays montmorillonite and others are hydrophilic by nature. As a result, they are not well adapted for blending and reacting with most hydrophobic polymer matrices. Moreover, electrostatic forces firmly hold the clay platelet stacks together. Figure 5 illustrates that counterions attract to the clay platelets' net negative charge. Platelet stacks firmly bound together can be formed when two nearby platelets share counterions.

These factors need treating the clay before using it to produce a nanocomposite. These clay platelet stacks are bigger in all dimensions than one nanometer. Developing a composite from untreated clay is not recommended, as most of the clay would be confined and incapable of interacting with the matrix. The clay surface can be modified using the popular simple technique of ion swapping to make it more responsive to an organic matrix. Small molecule cations can replace the cations since they are not closely bound to the clay surface.

For instance, the grey cations in the Fig. 6 are sodium ions. Several of them have been substituted with different cations. It is companionable with an organic matrix if the black cations were quaternary ammonium ions with long alkyl chains. The compatibility of montmorillonite clay with a wide range of matrix polymers may be achieved by exchanging it with different organic cations. The clay platelets are also separated during this process, making it easier for them to intercalate and exfoliate.

More than 60 years have passed since the discovery of a pure layered silicate [82–86]. They possess 2:1 phyllosilicate structural characteristics with the well-known. They pertain to the group of clay minerals, which are more common. Two tetrahedral silica sheets and an edge-shared octahedral sheet of either alumina or magnesia are fused to produce their two-dimensional layers. In arranging these layers, van der Waals gaps or galleries can be made.

The charge shortfall caused by isomorphous substitution (such as tetrahedral Si^{4+} by Al^{3+} or octahedral Al^{3+} by Mg^{2+}) is balanced by cations, often Na^{+} and/or Ca^{2+}, in the galleries (sometimes referred to as interlayers). These cations are exchangeable since they are not structural and can quickly be altered by highly

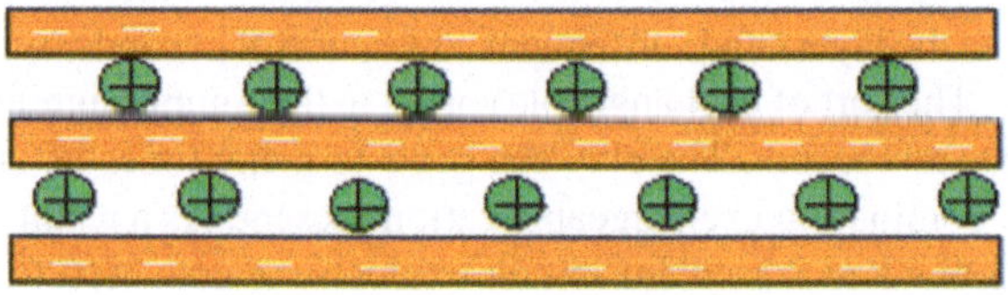

Fig. 5 Role of Cations and Anions in between clay layers [82]

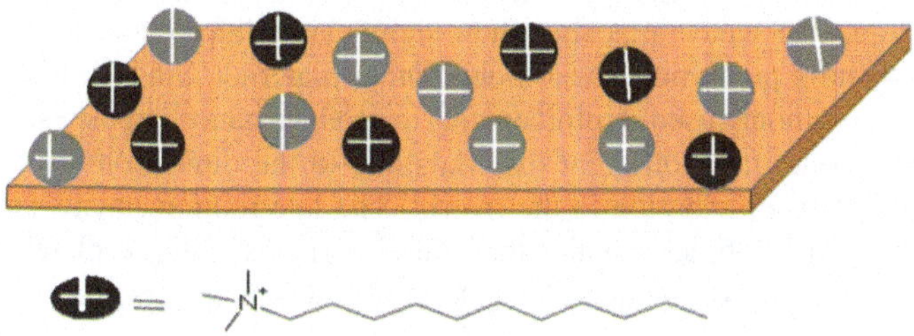

Fig. 6 Chemical interaction of modifier with clay [82]

charged atoms. The organic modifiers in the galleries allow the hydrophilic surface to become organophilic. To maximize their compatibility with a specific polymer, the organic modifiers' functionality, packing density, and length may all be considered when developing organically modified layered silicates. Cations in the underlayer layer or, more typically, replacements inside the lattice can neutralize and connect montmorillonite, the most common smectite clay.

The montmorillonite's structural characteristics determine the degree of substitution. While substitution can only go so far in the tetrahedral plane, roughly 15% of it can go to completion there. When organic cations are added to clay to change its surface, the clay materials' thermal behavior will be dramatically affected. Why do we develop nanocomposites?

Diminutive size of the filles

- The ratio of surface/volume is high
- Close spacing between fillers
- Mechanical properties;
- Greater ductility without an associated loss of strength
- Resistance to scratch
- Optical Properties;
- Light transmission properties depend on particle size

1.7 2D-Materials Nanocomposites

Nanocomposites are Composite materials with at least one scattered particle dimension in the nanoscale range that integrate one or more distinct components to improve performance characteristics. Nanocomposites are comprised of inorganic particles in ultrafine size that are uniformly dispersed throughout the polymer matrix and usually range in size from 1 to 1000 nm. Government, academia, and industry researchers are currently fascinated with these materials due to their remaining properties. The polymer layered silicate (PLS) nanocomposites attain the required properties with a significantly smaller ceramic material than equivalent glass- or inorganic-reinforced polymers.

In addition, PLS nanocomposites show significantly improved heat stability and self-extinguishing properties. Even though other materials can be utilized, natural silicates are commonly used reinforcements to produce nanocomposites. Nanocomposites are distinct from different materials due to the dimensions of the silicate particles and their distribution in the matrix. The talc reinforcements currently in application are 1400 times smaller than these particles. Silicate platelets usually have surfaces that are 1000 nm broad and only a few nanometers thick.

1.8 Categories of Nanocomposites

Silicate layers usually have a high aspect ratio and a thickness of about 1 nm. When appropriately dispersed throughout the polymer matrix, a few wt% of layered silicates can result in significantly more surface area than conventional composite materials. Three distinct PLS nanocomposites are thermodynamically attainable based on the interfacial interaction strength. The nanocomposites may be categorized into three groups based on how the polymer matrix is distributed between the clay layers:

(1) Phase-separated/Aggregated
(2) Intercalated
(3) Exfoliated

(a) Phase Separated

The characteristics of a phase-separated composite are consistent with those of conventional micro-composites and develop when the polymer cannot penetrate the silicate sheets. The tactoids matrix still contains the organoclay layers together. The mixing processes have little impact on the separation between the silicate layers. Microcomposites are formed due to phase separation but do not provide the intended advantages. The micro-composite is the least preferred and most likely of these morphologies. In this instance, the clay acts like a dispersed, "inactive" component that barely interacts with the surrounding polymer matrix. As the composite cannot take advantage of the material advantages fundamental to the clay's nano-dimensions, no enhancements in the polymer's attributes are expected.

(b) Intercalated Nanocomposite

Despite the quantity of clay compared to polymer, a polymer matrix is inserted into the layered silicate. Typically, intercalated nanocomposites are produced by layering a few molecular layers of polymer. The composites' properties frequently resemble those of ceramic materials [88]. A multilayered structure with alternating polymers and silicate polymer diffusion into clay lamellae arises from intercalation, which is the condition in which extended polymer chains lie between the clay layers. Nonetheless, the original clay structure is still visible. Silicate layers are separated by less than 2–3 nm due to intercalations of polymer chains. While not significant, these nanocomposites demonstrate improved characteristics (Fig. 7).

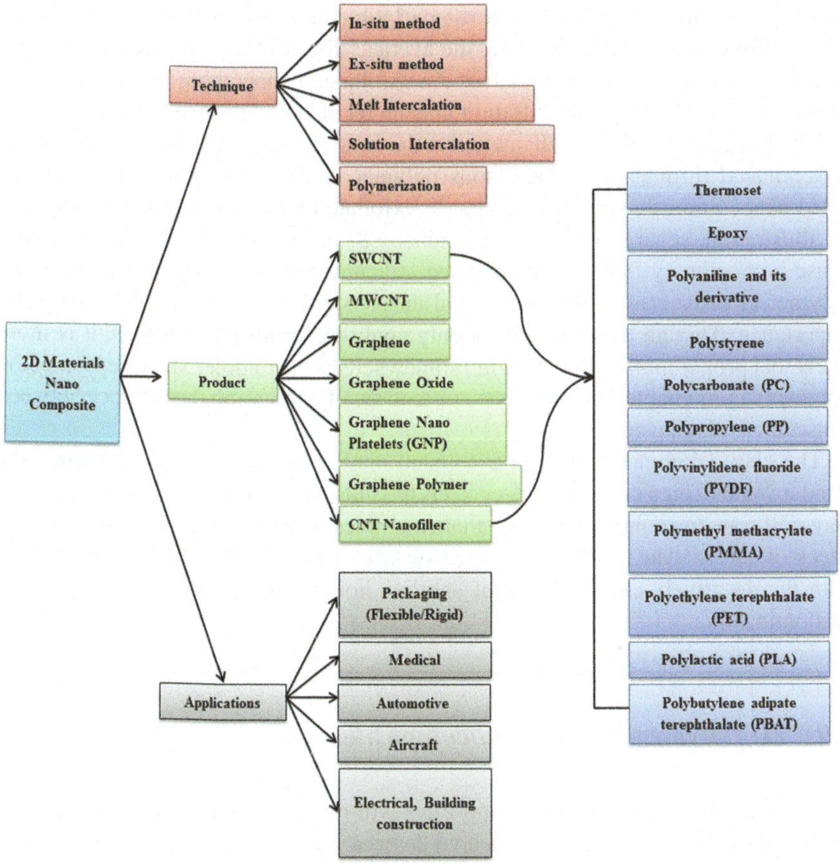

Fig. 7 2D materials-based nanocomposite possibilities [87]

In PLS nanocomposites, the two elevated forms (or delaminated) of microstructures are normally intercalated and exfoliated. Extended polymer chains can be visible between the host silicate layers of an intercalated hybrid, which usually differ by 1–4 nm. Intercalation leads to properly ordered structures with alternative polymer and inorganic layers in several layers. The silicate layers in an exfoliated hybrid are in a random arrangement within the matrix, and the spacing is roughly equal to the polymer's radius of gyration. A hierarchy of morphologies in real PLS nanocomposites lies between the idealized microstructures.

(c) Exfoliated Nanocomposite

The nanocomposite's structure in the exfoliated or delaminated stage, when the periodic arrangement has vanished, it or not, however, is well illuminated by WAXD. Weak peak intensity, bias towards the surface area, and poor peak resolution are all possible challenges with WAXD, particularly in composites with low clay concentration. Depending on clay loading, the average distance between the various clay layers in a continuous polymer matrix. An exfoliated nanocomposite typically has a substantially lower clay concentration than an intercalated nanocomposite.

When silicate layers with diameters in the nano range are evenly and thoroughly disseminated throughout a continuous polymer matrix, the exfoliated or delaminated nanocomposite structure results. In exfoliated nanocomposites, the polymer, for illustration, separates the clay platelets by 8–10 nm or more. It exhibits higher phase homogeneity than the intercalated structure because it maximizes with the polymer, effectively affecting mechanical and physical properties, which is of specific importance. Though these assertions only apply under ideal conditions, it is more likely that proper morphology will fall somewhere in the middle. The insertion of the polymer chains enables the interlayer space to swell, increasing the distance between consecutive clay layers.

The clay delaminates, and an exfoliated nanocomposite is produced when the Vander Waals forces can no longer keep the clay layers together. As a result, the clay will transform into a structure that resembles a "house of cards" rather than a "deck of cards." The positively charged surface of one clay layer will then connect with the negatively charged edge of another, holding the exfoliated clay together predominantly.

1.9 2D Polymer-Based Nanocomposites

In the twenty-first century, nanocomposites will be one of the most advanced technological substitutes for lead-engineered devices. The dispersion phase of polymer-based nanocomposites has a size range of 1 to 100 nm. The ultrafine powder will be incorporated with matrices to produce polymer-based nanocomposites. Most composite materials are made from inorganic powders/polymers through melting, solvent mixing, co-deposition assembly, and simple blending. In nanocomposite materials, molecular dispersion levels extend from microscopic composites of ultrafine dispersions to the molecular scale. They split into polymer/polymer and polymer/fill structures. The categories are used to comprehend how nanocomposites behave. Polymer-based nanocomposites have unique benefits regarding their characteristics and capabilities, as mentioned below;

(1) **Property development:** The materials exhibit improved mechanical, crystalline, optical, and acoustic functions and enhanced heat distortion temperature and crystalline rate.
(2) **Properties economizing composites:** The essential raw materials, such as organic polymers and inorganic MMT, are easily accessible.
(3) **Sustaining novel technology:** The vast majority of the time, conventional techniques and tools are used for producing these nanomaterials. This vital aspect could promote the industrialization of the manufacture of these nanocomposites.

The Classification separates systems into polymer/polymer, polymer/oligomer, and polymer/filler systems. Metal particles, non-metal powder, semiconductor material, or conductor powder have been added to organic/inorganic polymer/filler systems such as PET/layered silicate systems.

1.10 Synthesis Techniques

The sol–gel, chemical vapor deposition (CVD), and ball-milling techniques are frequently used to process nanoparticles. One of the most critical issues that must be addressed is particle dispersion in nanocomposite materials. The following section will discuss the dispersion and any associated preparations and synthesis. A particle surface treatment can be required to make the particles compatible with the polymer matrix to distribute the particles. Surface treatment has a strong correlation with nanoparticle stability. To deposit or cover layers on the exterior of the particles, dispersion processes for nanoparticles dispersed in polymers include gas vapor deposition (GVD) and CVD.

The substance in the deposited layers can be either organic or inorganic. Surface treatments, such as polymer films, oligomers, and surfactants, are organic. Ionization, degradation, or mechanical milling methods are also performed to establish polymer powders more compatible with superfine inorganic particles. Surface treatments make nanoparticles so they can be evenly distributed in nanocomposite matrixes. They encompass several facets, briefly discussed here and thoroughly explained in the following chapters.

(a) Sol–gel Technique

Hydrating the treatment agent with the nanoparticle precursors produces a sol using this technique. After recovering the gel, the sol coats the particle in a surface layer. Nanocomposites are made in a single pot by incorporating a few monomers of polymers that dissolve in a solvent (such as polyvinylpyrrolidone (PVP) or water, such as polyvinyl alcohol (PVA)). This technique provides a smaller selection area for organic polymers and nanomaterials because the polymer is soluble in the medium but is especially well-liked for nanocomposites.

(b) Nanoprecursor Method

Nanoprecursor methods can be used when the base materials and various layers are provided. Reducing almost all oxide compounds or their parts into ultrafine powder is possible. Some are challenging to produce into composite nanoparticles or nanocomposites due to their convoluted compositions. These components are converted into a precursor, an intermediate form that allows for the production of homogenous raw materials with very high purity.

(c) Intercalation Method

Treatment reagents are intercalated into the layer space for layered compounds or silicates. These intercalates, which are nanomaterial precursors with regulated layer shapes, are exfoliated during the polymerization technique. Three conflicting effects may be used to justify the postulated driving force of this mechanism, namely:

i. The conformational entropy of the flexible polymer chains is decreased when confined between the silicate layers.
ii. To accommodate the polymers, the organic modifiers' conformational entropy increased.
iii. A net energy gain resulting from the appearance of a second, more beneficial contact between the polymer and modified layered silicates than the organic interaction with the silicate, which is not intercalated.

This method mixes the layered silicate while still molten inside the polymer matrix without a solvent. Using standard techniques like extrusion and injection molding, a thermoplastic polymer is physically combined with organophilic clay at a high temperature. Next, to produce nanocomposites, the polymer chains are intercalated or exfoliated. This approach is frequently used to produce thermoplastic nanocomposites. This method of melt intercalation allows for the processing of polymers unsuitable for adsorption or in-situ polymerization; layered clays are combined directly with the polymer matrix while still molten. Nanocomposites may be produced using the melt intercalation process in about ten minutes. Environmentally friendly method: No solvent is necessary. Using standard plastic extrusion and molding techniques, nanocomposites may be treated. Factors affecting the formation of PNC via melt intercalation:

- Molecular architecture of the surface modifier
- Compatibiliser concentration
- Processing shear and temperature
- Polymer type and molecular weight
- Presence of additives

(d) Blending Methods

Nanoparticles are modified by blending them with materials having lower molecular weights. By blending or mixing, particles with beneficial substances on them are produced. A core (inorganic)-shell (polymer cover) structure is frequently produced when the treatment agent is a polymer. The nanoparticles are combined with an intermediary substance rather than directly with the polymers. For instance, to make the intermediate pellets, which have a silica concentration of up to 10–30% (by wt), maleic anhydride-treated nano-silica is combined with polypropylene. Subsequently, dispersion in the polymer matrix reduces this intermediate.

(e) Filling Techniques

It is possible to develop hybrid nanoparticles or nanocomposites by mechanically mixing powder, dissolving liquid, or melting the polymers into the pores/surfaces/interfaces of inorganic particles. This flexible technique can be used in various ways to synthesize or process nanocomposites. One such method involves a dry mixing process comparable to that used for producing polymer alloys, like blends of solid particles and other polymers.

(f) In-Situ Polymerization

In-situ, polymerization was the first approach utilized to produce polymer/clay nanocomposites. Three continuous phases make up the in-situ polymerization process. The nanoscale additives are first given the proper surface pretreatments, which are then distributed into a monomer and polymerized. In-situ nanocomposites are produced during polymerization. When monomers are intercalated into layered clays, they can be polymerized by heat, radiation, pre-intercalated initiators, or catalysts. This technique is known as in-situ polymerization. A suitable initiator or curing agent can initiate the polymerization reaction.

To produce delaminated nanocomposites, the intra-gallery polymerization reaction kinetics must be greater than the extra-gallery polymerization reaction kinetics. The driving forces are the strength of the interaction between the monomer and the clay surface and the enthalpy variation during interlayer polymerization. In-situ polymerization has several advantages of being relatively quick, easier to handle, and potentially resulting in better final products. Ideal for low- or non-soluble polymers, a usual thermoset nanocomposite manufacturing technique.

(g) Melt Blending

Another intriguing technique developed for creating organic–inorganic nanocomposites by a twin screw extruder is melt intercalation, a type of melt mixing. This method entails annealing a polymer and clay combination statically or under shear above the glass transition temperature (Tg). The polymer chains may enter the clay galleries that have already been treated during this procedure, producing either intercalation- or exfoliation-type nanocomposites. Melt mixing, which can fully use well-built polymer processing machinery such as extruders or injectors, is the fastest way to commercialize novel nanocomposites.

1.11 Characterization of Polymer–Clay Nanocomposites

Traditionally, polymer layered silicate nanocomposite structures have been explained using X-ray diffraction (XRD), Fourier transform infrared spectroscopy (FTIR), transmission electron microscopy (TEM), and atomic force microscopy (AFM). Because of the periodic structure, XRD determines the interlayer spacing. The most used method for analyzing the structure and sporadically for researching the kinetics of the manufacturing process of polymer nanocomposites is XRD, particularly wide-angle XRD. If a nanostructure is intercalated or exfoliated, it can be determined by observing its position, shape, and strength of the basal reflections of the materials' XRD patterns. In contrast, in an intercalated nanocomposite, the finite layer expansion leads to the emergence of a new basal reflection that corresponds to the increased gallery height. Using XRD, it is simple to uncover the interlayer spacing.

However, such a technique cannot reveal the geographic distribution of the clay layers. Systematic analysis of systems with a spreading peak and waning intensity is also challenging. From this vantage point, the information provided by XRD patterns is insufficient to shed light on the process by which nanocomposites are formed and their eventual structure. Unfortunately, to offer quantitative information in an XRD "silent" nanocomposite, XRD does not give clear information on the structure of the nanocomposites in the absence of a registry or a disordered nanocomposite.

By contrast, TEM may offer qualitative data on the structure, morphology, distribution, and structural defects. TEM is a very effective method. It offers a way to determine the uniformity of the mixing process and descriptions of the spatial correlation of the layered silicates. Individual crystallites of several nm layers are also seen in TEM images. The combined use of wide-angle XRD and small-angle X-ray scattering (SAXS) can produce a quantitative analysis of the structure in polymer nanocomposites. The structural development of polymer nanocomposites may be understood using Fourier transform infrared (FTIR) and Raman spectroscopy.

2 Conclusions

Polymer nanocomposites constitute an interesting and capable alternative to conventional materials because of the dispersion of nanoscale clay platelets and their enormously enhanced mechanical, thermal, barrier, optical, electrical, and other physical and chemical performance characteristics. As a result, numerous organizations have invested heavily in developing nano clays and polymer nanocomposites. More and more commercial goods are now readily available. Moreover, applications for clay-based polymer nanocomposites in the biological and bioengineering disciplines have shown promise. We outlined a model 2D-material-based composite, their characterizations, classifications, and advantages in the context of physics and mechanics associated with recent studies. We additionally emphasized other aspects that formed new nanocomposites based on polymers and metals and the most recent developments of recently synthesized 2D materials enabling the manufacturing of new nanocomposite species.

References

1. Alexandre, M., Dubois, P.: Polymer-layered silicate nanocomposites: preparation, properties and uses of a new class of materials. Mater. Sci. Eng. R. Rep. **28**(1–2), 1–63 (2000)
2. Lu, X., Yu, M., Huang, H., Ruoff, R.S.: Tailoring graphite with the goal of achieving single sheets. Nanotechnology **10**(3), 269 (1999)
3. Rao, A.M., Richter, E., Bandow, S., Chase, B., Eklund, P.C., Williams, K.A., Dresselhaus, M.S.: Diameter-selective Raman scattering from vibrational modes in carbon nanotubes. Science **275**(5297), 187–191 (1997)
4. Wu, T.M., Hsu, S.F., Wu, J.Y.: Effect of the premelting temperature and sample thickness on the polymorphic behavior of syndiotactic polystyrene/clay nanocomposites. J. Polym. Sci. B Polym. Phys. **41**(14), 1730–1738 (2003)

5. Sumita, M., Tsukihi, H., Miyasaka, K., Ishikawa, K.: Dynamic mechanical properties of polypropylene composites filled with ultrafine particles. J. Appl. Polym. Sci. **29**(5), 1523–1530 (1984)
6. Gonsalves, K.E., Chen, X., Baraton, M.I.: Mechanistic investigation of the preparation of polymer/ceramic nanocomposites. Nanostructured Mater. **9**(1–8), 181–184 (1997)
7. Ash, B.J., Stone, J., Rogers, D.F., Schadler, L.S., Siegel, R.W., Benicewicz, B.C., Apple, T.: Investigation into the thermal and mechanical behavior of PMMA/alumina nanocomposites. MRS Online Proc. Library **661**, KK2–1 (2000)
8. Becker, C., Krug, H., Schmidt, H.: Tailoring of thermomechanical properties of thermoplastic nanocomposites by surface modification of nanoscale silica particles. MRS Online Proc. Library (OPL) **435**, 237 (1996)
9. Ramanathan, T., Liu, H., Brinson, L.C.: Functionalized SWNT/polymer nanocomposites for dramatic property improvement. J. Polym. Sci. B Polym. Phys. **43**(17), 2269–2279 (2005)
10. LeBaron, P.C., Wang, Z., Pinnavaia, T.J.: Polymer-layered silicate nanocomposites: an overview. Appl. Clay Sci. **15**(1–2), 11–29 (1999)
11. Zheng, W., Lu, X., Wong, S.C.: Electrical and mechanical properties of expanded graphite-reinforced high-density polyethylene. J. Appl. Polym. Sci. **91**(5), 2781–2788 (2004)
12. Xu, J., Hu, Y., Song, L., Wang, Q., Fan, W., Chen, Z.: Preparation and characterization of poly (vinyl alcohol)/graphite oxide nanocomposite. Carbon **3**(40), 450–451 (2002)
13. Pan, Y.X., Yu, Z.Z., Ou, Y.C., Hu, G.H.: A new process of fabricating electrically conducting nylon 6/graphite nanocomposites via intercalation polymerization. J. Polym. Sci. B Polym. Phys. **38**(12), 1626–1633 (2000)
14. Chen, G., Weng, W., Wu, D., Wu, C.: PMMA/graphite nanosheets composite and its conducting properties. Eur. Polym. **39**(12), 2329–2335 (2003)
15. Le, C.T., Clark, D.J., Ullah, F., Jang, J.I., Senthilkumar, V., Sim, Y., Kim, Y.S.: Impact of selenium doping on resonant second-harmonic generation in monolayer MoS_2. ACS Photonics **4**(1), 38–44 (2017)
16. Akashi, R., Ochi, M., Bordács, S., Suzuki, R., Tokura, Y., Iwasa, Y., Arita, R: Two-dimensional valley electrons and excitons in Noncentrosymmetric 3 R– MoS 2. Phys. Rev. Appl. **4**(1), 014002 (2015)
17. Pham, P.V.: Hexagon flower quantum dot-like Cu pattern formation during low-pressure chemical vapor deposited graphene growth on a liquid Cu/W substrate. ACS Omega **3**(7), 8036–8041 (2018)
18. Pham, V.P., Nguyen, M.T., Park, J.W., Kwak, S.S., Nguyen, D.H.T., Mun, M.K., Yeom, G.Y.: Chlorine-trapped CVD bilayer graphene for resistive pressure sensor with high detection limit and high sensitivity. 2D Mater. **4**(2), 025049 (2017)
19. Pham, V.P., Mishra, A., Yeom, G.Y.: The enhancement of Hall mobility and conductivity of CVD graphene through radical doping and vacuum annealing. RSC Adv. **7**(26), 16104–16108 (2017)
20. Pham, V.P., Kim, K.H., Jeon, M.H., Lee, S.H., Kim, K.N., Yeom, G.Y.: Low damage pre-doping on CVD graphene/Cu using a chlorine inductively coupled plasma. Carbon **95**, 664–671 (2015)
21. Kim, K.N., Pham, V.P., Yeom, G.Y.: Chlorine radical doping of a few layer graphene with low damage. ECS J. Solid State Sci. Techno. **4**(6), N5095 (2015)
22. Pham, V.P., Kim, K.N., Jeon, M.H., Kim, K.S., Yeom, G.Y.: Cyclic chlorine trap-doping for transparent, conductive, thermally stable and damage-free graphene. Nanoscale **6**(24), 15301–15308 (2014)
23. Pham, V.P., Kim, D.S., Kim, K.S., Park, J.W., Yang, K.C., Lee, S.H., Kim, K.N.: Low energy BCl3 plasma doping of few-layer graphene. Sci. Adv. Mater. **8**(4), 884–890 (2016)
24. Pham, P.V.: Cleaning of graphene surfaces by low-pressure air plasma. R. Soc. Open Sci. **5**(5), 172395 (2018)
25. Pham, P.V.: A library of doped-graphene images via transmission electron microscopy. C. **4**(2), 34 (2018)
26. Pham, V.P.: Graphene etching: how could it be etched. Curr. Graphene Sci. **2**(1), 15–20 (2018)

27. Pham, V.P., Yeom, G.Y.: Recent advances in doping of molybdenum disulfide: industrial applications and future prospects. Adv. Mater. **28**(41), 9024–9059 (2016)
28. Pham, V.P., Jang, H.S., Whang, D., Choi, J.Y.: Direct growth of graphene on rigid and flexible substrates: Progress, applications, and challenges. Chem. Soc. Rev. **46**(20), 6276–6300 (2017)
29. Min-Dianey, K.A.A., Le, T.K., Choi, J.R., Pham, P.V.: Advanced optical detection through the use of a deformably transferred nanofilm. Nanomaterials **11**(3), 816 (2021)
30. Zhang, B., Luo, J., Fang, Y.: Access to functionalized alkynylcyclopropanes via reductive radical-polar crossover-based reactions of 1, 3-enynes with alkyl radicals. Org. Biomol. Chem. **21**(4), 732–737 (2023)
31. Jeevanandham, G., Jerome, R., Murugan, N., Preethika, M., Vediappan, K., Sundramoorthy, A. K.: Nickel oxide decorated MoS 2 nanosheet-based non-enzymatic sensor for the selective detection of glucose. RSC Adv. **10**(2), 643–654
32. Murugan, N., Jerome, R., Preethika, M., Sundaramurthy, A., Sundramoorthy, A.K.: 2D-titanium carbide (MXene) based selective electrochemical sensor for simultaneous detection of ascorbic acid, dopamine and uric acid. J. Mater. Sci. Techno. **72**, 122–131 (2021)
33. Rajendran, J., Reshetilov, A.N., Sundramoorthy, A.K.: An electrochemically exfoliated graphene/poly (3, 4-ethylenedioxythiophene) nanocomposite-based electrochemical sensor for the detection of nicotine. Mater. Adv. **2**(10), 3336–3345 (2021)
34. Jerome, R., Sundramoorthy, A.K.: Preparation of hexagonal boron nitride doped graphene film modified sensor for selective electrochemical detection of nicotine in tobacco sample. Anal. Chim. Acta **1132**, 110–120 (2020)
35. Jerome, R., Keerthivasan, P.V., Murugan, N., Devi, N.R., Sundramoorthy, A.K.: Preparation of stable CuO/Boron nitride nanocomposite modified electrode for selective electrochemical detection of L-Cysteine. ChemistrySelect **5**(29), 9111–9118 (2020)
36. Senthilkumar, V., Tam, L.C., Kim, Y.S., Sim, Y., Seong, M.J., Jang, J.I.: Direct vapor phase growth process and robust photoluminescence properties of large area MoS 2 layers. Nano Res. **7**, 1759–1768 (2014)
37. Nguyen, A.D., Nguyen, T.K., Le, C.T., Kim, S., Ullah, F., Lee, Y., Kim, Y.S.: Nitrogen-plasma-treated continuous monolayer MoS2 for improving hydrogen evolution reaction. ACS Omega **4**(25), 21509–21515 (2019)
38. Qadir, A., Le, T.K., Malik, M., Min-Dianey, K.A.A., Saeed, I., Yu, Y., Pham, P.V.: Representative 2D-material-based nanocomposites and their emerging applications: a review. RSC Adv. **11**(39), 23860–23880 (2021)
39. Pinnavaia, T.J.: Intercalated clay catalysts. Science **220**(4595), 365–371 (1983)
40. Giannelis, E.P.: Polymer layered silicate nanocomposites. Adv. Mater. **8**(1), 29–35 (1996)
41. Kojima, Y., Usuki, A., Kawasumi, M., Okada, A., Fukushima, Y., Kurauchi, T., Kamigaito, O.: Mechanical properties of nylon 6-clay hybrid. J. Mater. Res. **8**(5), 1185–1189 (1993)
42. Lan, T., Pinnavaia, T.J.: Clay-reinforced epoxy nanocomposites. Chem. Mater. **6**(12), 2216–2219 (1994)
43. Zeng, C., Lee, L.J.: Poly (methyl methacrylate) and polystyrene/clay nanocomposites prepared by in-situ polymerization. Macromolecules **34**(12), 4098–4103 (2001)
44. Manias, E., Touny, A., Wu, L., Strawhecker, K., Lu, B., Chung, T.C.: Polypropylene/montmorillonite nanocomposites. Review of the synthetic routes and materials properties. Chem. Mater. **13**(10), 3516–3523 (2001)
45. Kim, Y.K., Choi, Y.S., Wang, K.H., Chung, I.J.: Synthesis of exfoliated PS/Na− MMT nanocomposites via emulsion polymerization. Chem. Mater. **14**(12), 4990–4995 (2002)
46. Wang, D., Zhu, J., Yao, Q., Wilkie, C.A.: A comparison of various methods for the preparation of polystyrene and poly (methyl methacrylate) clay nanocomposites. Chem. Mater. **14**(9), 3837–3843 (2002)
47. Uhl, F.M., Wilkie, C.A.: Polystyrene/graphite nanocomposites: effect on thermal stability. Polym. Degrad. Stab. **76**(1), 111–122 (2002)
48. Su, S., Wilkie, C.A.: Exfoliated poly (methyl methacrylate) and polystyrene nanocomposites occur when the clay cation contains a vinyl monomer. J. Polym. Sci. A Polym. Chem. **41**(8), 1124–1135 (2003)

49. Kashiwagi, T.: Polymer combustion and flammability—Role of the condensed phase. In Symposium (International) on Combustion **25**, 1423–1437 (1994).
50. Clift, R., Grace, J.R., Bubbles, M.W.: Drops and Particles. Academic Press, New York (1978)
51. Wang, Z., Pinnavaia, T.J.: Hybrid organic-inorganic nanocomposites: exfoliation of magadiite nanolayers in an elastomeric epoxy polymer. Chem. Mater. **10**, 1820–1826 (1998)
52. Gilman, J.W., Kashiwagi, T.: Nanocomposites: A revolutionary new flame retardant approach. SAMPE J. **33**, 40–46 (1997)
53. Gilman, J.W., Jackson, C.L., Morgan, A.B., Harris, R., Manias, E., Giannelis, E.P., Phillips, S.H.: Flammability properties of polymer-layered-silicate nanocomposites. Polypropylene and polystyrene nanocomposites. Chem. Mater. **12**, 1866–1873 (2000)
54. Zhu, J., Morgan, A.B., Lamelas, F.J., Wilkie, C.A.: Fire properties of polystyrene-clay nanocomposites. Chem. Mater. **13**, 3774–3780 (2001)
55. Zanetti, M., Kashiwagi, T., Falqui, L., Camino, G.: Cone calorimeter combustion and gasification studies of polymer layered silicate nanocomposites. Chem. Mater. **14**, 881–887 (2002)
56. Kumar, S., Jog, J.P., Natarajan, U.: Preparation and characterization of poly (methyl methacrylate)–clay nanocomposites via melt intercalation: the effect of organoclay on the structure and thermal properties. J. Appl. Polym. Sci. **89**(5), 1186–1194 (2003)
57. Tiwari, R.R., Natarajan, U.: Influence of organic modification on mechanical properties of melt processed intercalated poly(methyl methacrylate)–organoclay nanocomposites. J. Appl. Polym. Sci. **105**(5), 2433–2443 (2007)
58. Liaw, J.H., Hsueh, T.Y., Tan, T.S., Wang, Y., Chiao, S.M.: Twin-screw compounding of poly(methylmethacrylate)/clay nanocomposites: effects of compounding temperature and matrix molecular weight. Polym. Int. **56**, 1045–1052 (2007)
59. Giannelis, E., Krishnamoorti, R., Manias, E.: Polymer-silicate nanocomposites: model systems for confined polymers and polymer brushes. Adv. Polym. Sci. **138**, 107–148 (1999)
60. Kojima, Y., Usuki, A., Kawasmi, M., Okada, A., Kurauchi, T., Kamigaito, O.: Sorption of water in nylon 6-clay hybrid. J. Appl. Poly. Sci. **49**(7), 1259–1264 (1993)
61. Kojima, Y., Fukumori, K., Usuki, A., Okada, A., Kurauchi, T.: Gas permeabilities in rubber-clay hybrid. J. Mater. Sci. Lett. **12**, 889–890 (1993)
62. Vaia, R.A., Ishii, H., Giannelis, E.P.: Synthesis and properties of two-dimensional nanostructures by direct intercalation of polymer melts in layered silicates. Chem. Mater. **5**(12), 1694–1696 (1993)
63. Sikka, M., Cerini, L.N., Ghosh, S.S., Winey, K.I.: Melt intercalation of polystyrene in layered silicates. J. Poly. Sci. B Poly. Phy. **34**(8), 1443–1449 (1996)
64. Lee, D.C., Jang, L.W.: Preparation and characterization of PMMA–Clay hybrid composite by emulsion polymerization. J. Appl. Poly. Sci. **61**(7), 1117–1122 (1996)
65. Biasci, A., Aglietto, M., Ruggeri, G., Ciardelli, F.: Functionalization of montmorillonite by methyl methacrylate polymers containing side-chain ammonium cations. Polymer **35**(15), 3296–3304 (1994)
66. Kawasumi, M., Hasegawa, N., Kato, M., Usuki, A., Okada, A.: Preparation and mechanical properties of polypropylene−clay hybrids. Macromolecules **30**(20), 6333–6338 (1997)
67. Kelly, P., Akelah, A., Qutubuddin, S., Moet, A.: Reduction of residual stress in montmorillonite/epoxy compounds. J. Mater. Sci. **29**, 2274–2280 (1994)
68. Sherman, L M.: Nanocomposites, Plastic Technology, Online Article, Home Page: www.ptonline.com/articles/200705fa.html, pp. 1–2
69. McWilliams, A.: Nanocomposites, Nanoparticles, Nanoclays, and Nanotubes, June 2006, Research Report No. GB- NANO21C, Home Page: www.electronics.ca, Accessed June 15, 2007
70. Siddiqui, M.A., Ahmed, Z.: Arab. J. Sci. Eng. **30**(2A), 196 (2005)
71. Clay—Columbia Encyclopedia Article about Clay, Home Page: www.columbia.thefreedictionary.com, Accessed July 10 2007
72. Clay Minerals—Sci-Tech Encyclopedia, McGraw Hill, Home Page: www.answers.com, Accessed July 9, 2007

73. Mineral Gallery—The Phyllosilicate Subclass, Home Page: www.mineral.galleries.com, accessed July 10, 2007.
74. Montmorillonite, Home Page: www.en.wikipedia.org, Accessed July 9, 2007.
75. Anon.: Estimating Clay Content, US Department of Transportation, Federal Highway
76. Xia, Y., Rubino, M., Auras, R.: Chapter Seven—Interaction of nanoclay-reinforced packaging nanocomposites with food simulants and compost environment. Adv. Food Nutr. Res. **88**, 275–298 (2019)
77. Malaika, S.A., Golovoy, A., Wilkie, C.A. (eds.): Chemistry and Technology of Polymer Additives (1st Edition). Blackwell Sci, Oxford (1999)
78. Solomon, D.H., Hawthorne, D.G. (eds.): Chemistry of Pigments and Fillers. Krieger, Malabar, FL (1991)
79. Alexandre, M., Dubois, P.: Polymer-layered silicate nanocomposites: preparation, properties and uses of a new class of materials. Mat. Sci. Eng. **28**, 1–63 (2000)
80. Theng, B.K.G.: Formation and properties of clay-polymer complexes. Develop. Soil Sci. **9**, 3–362 (1979)
81. Theng, B.K.G.: The chemistry of clay-organic reactions. Clays Clay Miner. **23**, 336 (1975)
82. Gieseking, J.E.: The Mechanism of cation exchange in the montmorillonite-beidellite-nontronite type of clay minerals. Soil Sci. **47**(1), 1–14 (1939)
83. Richards, L.A.: A pressure—membrane extraction apparatus for soil solution. Soil Sci. **51**(5), 125–132 (1941)
84. Earnest, C.M.: Thermal Analysis in Geosciences, pp. 288–312. Springer (1991)
85. Newman, A.C.D.: Chemistry of Clay and Clay Minerals (1987)
86. Earnest, C. M. : Thermal Analysis Application Study (1980)
87. Qadir, A., Le, T.K., Malik, M., et al.: Representative 2D-material-based nanocomposites and their emerging applications: a review. RSC Adv. **11**, 23860 (2021)
88. Muduli, S.P., Parida, S., Behura, S.K., Rajput, S., Rout, S.K., Sareen, S.: Synergistic effect of graphene on dielectric and piezoelectric characteristic of PVDF-(BZT-BCT) composite for energy harvesting applications. Polym. Adv. Technol. **33**(10), 3628–3642 (2022)

Two-Dimensional Transition Metal Oxides (TMOs) for Solar Cell Applications

Ramakrishna Madaka, Dinesh kumar, Beauty Pandey, Narendra Bandaru, Jagadeeswara Rao Dasari, Lalhriat Zuala, and Jatindra Kumar Rath

Abstract Recent research has shown that two-dimensional transition metal oxides (TMOs) are potential materials for improving the stability and efficiency of solar cells. The brief overview of TMOs' potential for use in solar cell applications in this abstract emphasizes their special qualities, synthesis techniques, and most recent developments. Due to their various chemical compositions and electrical structures, transition metal oxides have been thoroughly researched for their potential in renewable energy technologies, particularly in photovoltaics. Due to their fascinating electrical characteristics and simplicity of integration into solar cell topologies, two-dimensional TMOs like vanadium pentoxide (V_2O_5), tungsten diselenide (WSe_2), TiO_2 (titanium dioxide), zinc oxide (ZnO), tin dioxide (SnO_2) has drawn a lot of attention. TMOs' distinctive electrical band structures make effective charge separation and light absorption possible, both of which are essential for the performance of solar cells. TMOs have variable bandgaps that enable them to absorb various sun-spectrum wavelengths. TMOs are a good choice for single-junction and tandem solar cells because of their superior charge transfer capabilities and tunability. TMOs' adaptability in terms of synthesis techniques is one of their main benefits. High-quality TMO thin films have been created using various methods, including chemical vapor deposition (CVD), liquid-phase exfoliation, and atomic layer deposition

R. Madaka (✉) · B. Pandey
School of Sciences, Woxsen University, Hyderabad 502345, Telangana, India
e-mail: ramakrishna.madaka@woxsen.edu.in

D. kumar
Functional Materials and Device Laboratory, School of Energy Science and Engineering, Indian Institute of Technology Kharagpur, Kharagpur 721302, West Bengal, India

R. Madaka · N. Bandaru · J. K. Rath (✉)
Department of Physics, Indian Institute of Technology Madras, Chennai 600036, Taminadu, India
e-mail: jkr@iitm.ac.in

J. R. Dasari
Department of Physics, S.V Government Degree College, Parvathipuram 535501, Andhra Pradesh, India

L. Zuala
Physical Sciences Research Centre (PSRC), Pachhunga University College, Mizoram University, Aizawl 796001, Mizoram, India

N. Talreja et al. (eds.), *Two-dimensional Hybrid Composites*, Engineering Materials,
https://doi.org/10.1007/978-981-99-8010-9_3

(ALD). These techniques enable the optimization of TMO features for particular solar cell applications by providing fine control over thickness, shape, and composition. The development of TMO-based solar cells has been impressive recently. Particularly, tandem solar cells profit from the superior optical qualities of TMOs. Researchers have significantly increased power conversion efficiency by mixing TMOs with other semiconducting substances like silicon or perovskites. TMOs have been included in many solar cell topologies, such as thin-film, organic, and quantum dot solar cells, proving their versatility. The long-term survival of solar cell technologies depends on TMOs' exceptional stability and durability in adverse environmental conditions. They are desirable options for outdoor applications and have longer operational lifetimes because of their resistance to corrosion and photo-induced degradation.

1 Introduction

Energy crisis and atmospheric contamination are the two most common challenges infesting the world universally [1–6]. These seemingly inevitable global issues are the direct byproduct of large-scale industrialization and rapid population expansion. Therefore, developing high-efficiency, environment-friendly, and sustainable technologies is imperative to generate clean energy and restore the environment. Sunlight has appeared as a promising remedy due to its natural abundance and ease of availability and its ability to drive environmentally safe reactions that can convert solar energy to chemical or electrical energy or solar fuels. Moreover, solar energy is seen as the most cost-effective replacement for fossil fuels, whose continuous usage as the prevalent energy source has increased pollution and contributed to climate change. Owing to its pivotal role in the solar-to-electricity transformation process, solar cells have been a hot topic for research for decades. Thus, solar cells have been regarded as one of the fascinating green energy technologies that have the potential to convert this enormous amount of solar energy into electricity. The solar cell technology can be broadly categorized into three generations [6–9].

The first generation comprised solar cells made of monocrystalline silicon and polycrystalline silicon. Second-generation basically consisted of thin-film solar cells that were synthesized from CIGS (copper indium gallium selenide), CIS (copper indium diselenide), CdS (cadmium sulfide), CdTe (cadmium telluride), the amorphous silicon, etc. And the third-generation solar cells include PSCs (perovskite solar cells), DSSCs (dye-sensitized solar cells), quantum dot QDSCs sensitized (solar cells), polymer solar cells, and all such ground-breaking photo-voltaic (PV) technologies. Due to its narrow bandgap of 1.12 eV, availability, affordability, nontoxicity, better purity, suitable minority carrier lifetime, extremely low grain boundary defects, easily tunable resistivity, and well-established fabrication procedures, silicon (Si), which makes up more than 90% of the photovoltaic market today, is used extensively in solar-cell design [2, 10–12]. The solar-to-electricity conversion efficiency and the cost per watt of the power generated are arguably the two determining factors by which the success of Si-based PV technology can be evaluated. Are likely to be.

With record power conversion efficiencies of about 25–26%, silicon solar cells have dominated the renewable energy industry over the last 50 years.

Although crystalline silicon-based solar cells are an established technology with comparatively reasonable costs in several markets, the photovoltaic community strives to achieve improved efficiencies, lesser costs, and lower environmental damages [1, 10, 13, 14]. This involves developing different technologies to minimize the price per watt and increase the watt-per-gram utilization of Si-based solar cells. A culmination of various efforts has thus opened many avenues for investigating different revolutionary approaches in photovoltaics. Though there is unprecedented progress, there are still many bottlenecks involved. The challenges that need reform include degradation against water, oxygen, and radiation, lack of scalability, and stability of device structure, which severely limits the extent of commercialization. Additionally, the long-term instability of the hole transport layer and electron transport layer, as well as their high cost, is another important factor affecting the solar cells' effectiveness [13–16]. Owing to this, investigations of different strategies that could upgrade the overall performance and stability of the device are constantly underway. The last few decades have seen an extensive volume of effort dedicated to developing a variety of semiconductors either in the form of 2D (two-dimensional) thin films, 1D (one-dimensional) nanostructures or 0D (zero-dimensional) nanoparticles as the potential candidates in the photovoltaic industry [17–19]. In recent years, there has been a huge paradigm change in the global photovoltaics arena where the demand has shifted more on wearable, portable, and flexible solar-powered gadgets, making the production of supple, light, and efficient power production solar resources necessary. Therefore, 2D materials with all their exotic properties are believed to play critical roles in the fabrication and performance of the next generation of solar cells [10, 20].

Recent times witnessed sporadic interest in 2D materials characterized by high aspect ratio, thick atomic layers, possession-plane covalent bonding, and contrasting properties compared to their bulk counterparts [20–24]. These materials were not unknown, but the exploration of their 2D structure augmented in the last few years with the segregation of graphene. These compounds, like graphite, are available in layered structures, making it easy to exfoliate them into smaller thicknesses with single or multilayers. These 2D nanomaterials include various sheets and thin films of conducting polymers, carbon nanotubes (CNTs), metal nanowires, metal nanotubes, conducting oxides, transition metal dichalcogenides (TMDs), and transition metal oxides (TMOs) that have been inspected in search for a low-cost, mechanically stable and high-performance photovoltaic material [6, 25–27]. Moreover, features like exceptional transparency, excellent flexibility, high conductivity, outstanding scalability, good modularity, and ease of roll-to-roll production make 2D materials stand out further. Fine-quality monoatomic thin layers of 2D materials enable effective control of the electrostatic conductivity. Due to the decreased number of dangling bonds, the charge carrier scattering is much lesser in 2D materials. Owing to this wonderful amalgamation of extraordinary properties, the research inclination towards producing and exfoliating different 2D materials has renewed extensively. Consequently, the library of fascinating 2D compounds is expanding with each passing year and attributes a surplus of about 150 exotic materials that can easily be cleaved into sub-nanometre 2D monolayers [21, 28–31].

By far, TMOs (transition metal oxides), due to their interesting physio-chemical properties and promising potential, are one of the most popular categories of 2D materials that are significantly used for different solar cell applications [32–35]. TMOs are compounds of oxygen and multivalent transition metals. Due to the variable oxidation states of the transition metals, TMOs exist in different forms like monoxide, dioxide, trioxide, etc. TMOs offer a unique behavior in terms of their preferred conductivity for one type of charge carrier (electrons/holes) while inhibiting the flow of the other kind, assisting in the efficient separation of photogenerated charge carriers [36–38]. The highly desired TMOs for solar cell applications include TiO_2 (titanium dioxide), zinc oxide (ZnO), tin dioxide (SnO_2), molybdenum trioxide (MoO_3), vanadium pentoxide (V_2O_5), tungsten trioxide (WO_3), etc. As compared to their bulk versions, these 2D nanomaterials exhibit greater surface area and customizable chemical properties [39–45]. Their excellent electronic properties can be attributed to the lack of an interlayer in their single-layer nanosheets. They are usually highly transparent wide-bandgap semiconductors with low contact resistivities. Furthermore, they display outstanding optical and electronic features that can be tailored easily by maneuvering their thickness. As the 2D TMOs can be engineered easily from their parent layered compounds by a variety of exfoliation approaches and temperatures processing techniques, the thickness of the exfoliated 2D TMOs matches with the thickness of the individual layer of TMO, which in the majority of cases is less than 5 nm. Since their surface-to-volume ratio is practically ~1, the 2D TMO structures expose almost all their components to the surface, making it convenient to be tuned for desired applications.

All the unique features of 2D TMOs thus make it a highly in-demand candidate for present-day solar cell applications and the photovoltaic industry. The present chapter provides an exciting review of different solar cell applications of various two-dimensional transition metal oxides. The article evolved from the urgent need to develop straightforward, low-cost technology to create solar cells that can achieve high efficiencies. The work is significantly focused on 2D TMOs' role in such solar cell applications. A comprehensive investigation of the overall performance of 2D TMO-based solar cells and their different photovoltaic parameters, such as open circuit voltage, short current density, fill factor, and efficiency under various conditions, are reported. The study discusses the concepts of flexible, ultrathin, and tandem solar cells based on 2D TMOs. The chapter concludes with a brief overview of persisting limitations and the future outlook on the technical challenges and research necessities that are essential for meeting the requirements of next-generation photovoltaics driven by 2D TMOs.

Recent years have seen notable breakthroughs in solar cell technology as a result of the ongoing research for sustainable and clean energy sources. For a variety of applications, including solar cells, two-dimensional transition metal oxides (2D TMOs) have shown promise [46–54]. Due to their atomically thin structure, which enables customized electrical and optoelectronic capabilities, these materials have special features [55, 56]. Due to its unique characteristics, including high carrier mobility, configurable band gaps, and wide surface areas, 2D TMOs have drawn a lot of attention in research and development [57–61]. They are ideally suited for next-generation optoelectronic devices because of these characteristics, especially solar cells, where effective light absorption and charge transfer are essential. Vanadium pentoxide (V_2O_5) has become a well-known choice for solar cell applications among the numerous 2D TMOs [62, 63]. V_2O_5 has a layered structure made up of stacked vanadium oxide sheets, which can be exfoliated to create ultrathin layers with special characteristics that can be included in solar cell topologies to improve performance and efficiency [64–66].

There are a number of reasons why V_2O_5 is used in solar cell applications. First, V2O5 is a material well suited for enabling charge transport in solar cell devices because of its good electrical conductivity. Second, V_2O_5 has suitable bandgap and optical properties that enable efficient light absorption across a broad range of the solar spectrum [67, 68]. This high carrier mobility of V_2O_5 enables efficient extraction of photogenerated charges, reducing recombination losses and enhancing overall device performance. V_2O_5 is particularly appealing for solar cell applications because it can absorb light in the visible and near-infrared spectrum, which maximizes light gathering. Additionally, the bandgap engineering capabilities of V_2O_5 by doping and alloying enable fine-tuning of its electrical and optoelectronic characteristics [69–71]. This adaptability creates opportunities for improving solar cells' absorption, charge separation, and charge transport characteristics based on V_2O_5. The use of V_2O_5 in solar cell applications is also consistent with the current research emphasis on sustainable and ecologically friendly materials. The advantages of V_2O_5 over traditional solar cell materials are its abundance, affordability, and low toxicity profile [72].

We intend to thoroughly review the use of V_2O_5 as a 2D TMO material for solar cell applications in this book chapter. We will examine $V_2O_{5's}$ structural and electrical characteristics as well as the synthesis and manufacturing methods utilized to create it as a 2D material. We will also go over how it is integrated into different solar cell layouts and look at the optoelectronic characteristics and device performance that V_2O_5 inclusion yields. We will also talk about the stability issues and environmental effects of V_2O_5-based solar cells. This chapter's overall goal is to demonstrate the enormous potential of V_2O_5 as a 2D TMO material for improving solar cell technology. We can open the door to the creation of extremely effective, scalable, and environmentally friendly solar cell technology by comprehending the special qualities and possibilities provided by V_2O_5.

2 Techniques for Synthesis and Fabrication

The methods used to create and produce V_2O_5 as a two-dimensional material are examined in the synthesis and manufacturing procedures section. It explores several fabrication processes for V_2O_5-based solar cell devices, such as thin film deposition, nanostructure development, and material integration. The section also goes through the methods used to characterize the structural and optical characteristics of V_2O_5 films, shedding light on the production procedures and methods used to create V_2O_5-based solar cells [73].

2.1 *Techniques for Synthesizing and Acquiring V2O5 as a Two-Dimensional Material*

A variety of procedures that allow for the controlled development and construction of atomically thin layers are used in the synthesis and manufacture of V_2O_5 as a two-dimensional (2D) material. Researchers can manipulate V_2O_5 characteristics using these techniques to make it more or less suitable for use in solar cells. Mechanical exfoliation is one method that is frequently used to create 2D V_2O_5 [74–78]. This process entails mechanically separating thick layers of bulk V_2O_5 crystals using sticky tape or other tools. Researchers can isolate single-layer or few-layer V_2O_5 sheets with perfect control over their thickness by repeatedly peeling off small flakes from the bulk material. High-quality 2D V_2O_5 samples may be produced by mechanical exfoliation and used for additional research and device development.

Another popular technique for creating V_2O_5 as a 2D material is chemical vapor deposition (CVD). This method involves adding an oxidizing agent and a vanadium precursor gas to a high-temperature reaction chamber [79, 80]. The precursor gas undergoes chemical reactions in controlled environments that result in the synthesis of V_2O_5. As a consequence, atomically thin layers of V_2O_5 may be placed onto a substrate. Scalability is a benefit of CVD that makes it possible to produce large-area V_2O_5 films appropriate for solar cell applications. Liquid-phase exfoliation is an additional method for producing 2D V_2O_5. This technique disperses bulk V_2O_5 in a suitable solvent, and then thin layers of the material are created by ultra-sonication or other mechanical techniques [52, 81, 82]. This method makes it possible to create V_2O_5 suspensions of various thicknesses, which may then be processed further to create devices. Another way to make 2D V_2O_5 is by using hydrothermal synthesis. A precursor solution containing vanadium ions is put through this process at high temperatures and pressures [83, 84]. In these circumstances, vanadium ions react with an oxidizing agent, creating V_2O_5 nanostructures with precise dimensions. The hydrothermal synthesis procedure allows the V_2O_5 nanostructures to be customized in terms of size, shape, and crystallinity, allowing for the creation of novel solar cell technologies.

Other techniques for the production of V_2O_5 as a 2D material have been examined, including molecular beam epitaxy (MBE), atomic layer deposition (ALD), and electrodeposition [85–89]. These approaches enable precision layer thickness control and the production of high-quality V2O5 films with specific required characteristics. Finally, the various synthesis procedures mentioned above provide researchers with a number of options for making 2D V_2O_5 materials. Using the correct approach, researchers may adjust the characteristics of the V_2O_5 layers to meet the special needs of solar cell applications, such as optimization of electrical conductivity, bandgap engineering, and charge transport characteristics.

2.2 Device Fabrication Methods for V_2O_5-based Solar Cells

Well-established manufacturing procedures that provide the exact placement and controlled inclusion of V_2O_5 layers within the device structure are necessary to effectively integrate V_2O_5 into solar cell topologies. These methods are essential for determining how well and efficiently V_2O_5-based solar cells function. The heterojunction method is one manufacturing method frequently used for V_2O_5-based solar cells [90–96]. The fabrication process typically entails depositing the V_2O_5 layer onto a suitable substrate using methods such as spin coating, spray coating, or vapor deposition [71]. In this method, V_2O_5 is used as the active layer in combination with other complementary materials, such as organic semiconductors or inorganic nanomaterials, to form a heterojunction interface. The complementary material is then placed on top of the V2O5 layer to create the heterojunction. The compatibility of the materials, the required layer thickness, and the device's design all influence the chosen deposition process.

Utilizing V_2O_5 in tandem or multilayered solar cell designs is an alternative strategy [97]. This method involves stacking V_2O_5 layers with other substances, such as perovskites or other 2D TMOs, to produce a series of absorbers with various bandgaps. This maximizes light harvesting by enabling the effective absorption of a wider spectrum of sunlight. Tandem structures are commonly created by progressively depositing the various layers using techniques like CVD, thermal evaporation, or solution-based approaches, followed by proper encapsulation and electrode deposition [98–101]. For the mass production of V_2O_5-based solar cells at low cost, solution-based processes like spin coating and inkjet printing are especially well suited. These techniques entail depositing V_2O_5 precursor solutions onto the substrate, which are then subjected to thermal or chemical processing in order to transform the precursors into the appropriate V_2O_5 phase. While inkjet printing allows for fine patterning of V_2O_5 layers and offers freedom in device design, the spin coating offers strong control over film thickness and homogeneity [102, 103].

Lithography methods such as photolithography and electron beam lithography are used for the exact patterning and defining of V2O5 structures in solar cell devices [69]. With the use of electron beams or masks, these methods selectively expose and develop the V_2O_5 layers to produce complex patterns or electrode architectures. Making high-efficiency V_2O_5-based solar cells with improved light trapping and charge collecting capabilities makes use of lithography methods. Post-processing techniques like thermal annealing or surface treatments may be used to improve the film quality, increase crystallinity, or change the surface characteristics of the V2O5 layers after the manufacture of V2O5-based solar cell devices [104, 105]. These post-processing procedures can improve charge transmission, lower faults, and increase general device stability, all of which substantially influence the device's performance.

In this manufacturing process, V_2O_5 layers are deposited using processes like spin coating, vapor deposition, or inkjet printing, and then complimentary materials are added to create heterojunction interfaces or tandem structures. The performance and functionality of V_2O_5-based solar cells may be further improved via lithography methods and post-processing procedures. The preferred device architecture, scalability, compatibility with other materials, and cost considerations all play a role in the manufacturing technology selection.

2.3 *Techniques for Characterizing V_2O_5 to Examine Its Structural and Optical Characteristics*

It is essential to use characterization techniques to examine the structural and optical characteristics of V_2O_5-based solar cells in order to comprehend and improve their performance. There are several methods available for this. Transmission electron microscopy (TEM) and scanning electron microscopy (SEM) are often employed techniques for the analysis of the size, thickness, and crystalline structure of V_2O_5 layers. The use of these techniques enables the attainment of high-resolution imaging, which has the potential to uncover defects or grain boundaries that might have an impact on the performance of materials [52, 106, 107].

The technique of X-ray diffraction (XRD) is utilized to examine the crystal structure and phase purity of V_2O_5 [108, 109]. By analyzing diffraction patterns, researchers can determine the crystallographic phases present within a material. As proposed by Urea-Begara and Shvets, Raman spectroscopy has been identified as an additional valuable technique for the characterization of V_2O_5 [110, 111]. The provided data includes the crystal symmetry and vibrational modes of the material, which is useful for detecting doping or strain-induced modifications and confirming the structural integrity of the substance. Optical characterization techniques such as UV–Vis spectroscopy and photoluminescence spectroscopy are used to determine the optical properties of V_2O_5, including its light absorption and emission behavior [112, 113]. These methodologies are able to provide exhaustive data regarding the bandgap, absorption rates, and excitonic properties of a given material.

Synthesis and manufacturing techniques employed in the production of V_2O_5 as a two-dimensional material and its integration into solar cell topologies significantly impact the device's efficacy. In order to improve understanding of the behavior of V_2O_5 and optimize the design of solar cells, researchers can assess its structural and optical properties using a variety of characterization techniques.

3 Structural and Electronic Characteristics of V_2O_5

3.1 *Crystal Structure and Morphology of V_2O_5*

Understanding V_2O_5 characteristics and behavior in solar cell applications requires understanding its crystal structure and morphology. The layered, crystal structure of V_2O_5 significantly influences the material's electrical and optoelectronic characteristics. Vanadium oxide sheets are stacked to form the V_2O_5 crystal structure. Each sheet is made up of warped VO6 octahedral that are linked together by common oxygen atoms to form a two-dimensional network [114–116]. Vanadium oxide can be easily exfoliated into atomically thin 2D layers because these vanadium oxide layers are only bound together by weak van der Waals interactions. V_2O_5 has distinct features due to its layered structure. A network of V–O bonds is formed within the layers by the vanadium atoms' deformed octahedral coordination [117]. The electrical characteristics of V_2O_5 may be modulated by this structure, which has both metallic and insulating areas. Depending on the synthesis technique and processing circumstances, V_2O_5 shape can change. V_2O_5 generally appears as aggregates or crystalline particles in its bulk state. Variables like precursor concentration, reaction temperature, and growth kinetics during synthesis affect the V_2O_5 morphology. V_2O_5 can, however, exhibit a range of morphologies when exfoliated into thin layers, including nanosheets, nanowires, and nanoribbons [118, 119]. The surface area, surface chemistry, and interaction with light of V_2O_5 are significantly influenced by its shape [120, 121]. For instance, high aspect ratio nanoscale V_2O_5 structures can offer more surface area, enhancing light absorption and charge carrier formation. The energy bandgap, charge transport, and other features that are dependent on morphology are likewise influenced by the morphology of V_2O_5 (Fig. 1).

Fig. 1 Perspective view of two layers of V2O5. V atoms are represented as grey balls and O atoms as red balls. Weak van der Waals bonds are omitted for clarity [71]

Various characterization techniques are often employed to investigate the morphology and crystal structure of V_2O_5. The high-resolution imaging capabilities provided by scanning electron microscopy (SEM) and transmission electron microscopy (TEM) enable the observation of the surface morphology, particle size, and shape of V_2O_5. The aforementioned approaches provide the capability to detect structural flaws or impurities that might potentially influence the operational efficiency of V_2O_5-based solar cells. In addition, they shed light on the relationships between the structural composition and the resulting properties. Utilizing X-ray diffraction (XRD) analysis to examine the crystal structure of V_2O_5 is a highly effective technique. V_2O_5 crystallographic phases, lattice parameters, and layer orientation can be determined through the use of diffraction patterns resulting from X-ray interactions with the crystal. Examining crystallinity, phase purity, and structural modifications induced by doping or post-processing interventions is facilitated by X-ray diffraction (XRD). However, the crystal structure and geometry of V_2O_5 significantly affect its electrical, optical, and catalytic properties. Due to its lamellar structure and capacity to form atomically thin layers, V_2O_5 is a flexible material for solar cell applications. Using characterization techniques such as SEM, TEM, and XRD, researchers can analyze the crystal structure and morphology of V_2O_5, which sheds light on the material's behavior and facilitates the development of efficient V_2O_5-based solar cells.

3.2 Electronic Band Configuration and Energy Levels of V_2O_5

The optoelectronic characteristics of V_2O_5 and its applicability for solar cell applications are strongly influenced by its electronic band structure and energy levels. Researchers can learn more about the optical absorption, charge transport, and photoconversion processes of V_2O_5 by comprehending its energy band structure [122–126].

The arrangement of electrons in the valence and conduction bands, which specify the energy levels accessible for electronic transitions, gives birth to the electronic band structure of v2o5. According to Le, T. K., Beke, S., and Yalagala, V_2O_5 is typically thought of as a broad bandgap semiconductor, with its bandgap lying between 1.6 and 2.4 eV depending on parameters including crystal structure, stoichiometry, and doping [71, 127, 128]. The oxygen 2p orbitals dominate the valence band of V_2O_5, but the vanadium 3d and oxygen 2p orbitals combine to create the conduction band. The lowest energy needed for electrons to move from the valence band to the conduction band, enabling the absorption of light with particular wavelengths, depends on the energy bandgap between the valence and conduction bands. Alloying, doping, strain, and defect engineering are a few of the variables that can change the electrical band structure of V_2O_5. For instance, the location of the energy bands may be changed by adding dopant atoms or changing stoichiometry, which effectively tunes the bandgap and affects the material's optoelectronic capabilities [129]. This opens up possibilities for bandgap engineering and modifying the properties of V_2O_5's light absorption for solar cell applications. The charge transport characteristics of V_2O_5 are also strongly influenced by the energy levels inside its band structure. In its crystal structure, V_2O_5 displays both metallic and insulating areas, resulting in mixed electrical conductivity [116, 130, 131]. Electrons are localized in insulating materials, while they are delocalized and free to flow in metallic materials. V_2O_5 has a range of electrical conductivity due to this special property, which may be further modified by doping or altering the crystal structure.

The energy levels of defect states, such as oxygen vacancies or vanadium interstitials, may also be seen in V_2O_5 [132, 133]. These defect states can affect the dynamics of charge carriers, recombination procedures, and overall device performance. For V_2O_5-based solar cells to work as well as possible, it is essential to comprehend and manage the energy levels associated with faults. Several spectroscopy methods are frequently used to experimentally explore the electronic band structure and energy levels of V_2O_5 [134, 135]. Photoemission spectroscopy (PES) measures the kinetic energy and intensity of the emitted electrons in response to photon excitation, disclosing information about the electronic structure and energy levels [136, 137]. This method establishes the location of the Fermi level, the distribution of energy bands, and the presence of defect states. Other spectroscopic techniques, such as optical absorption, photoluminescence, and impedance spectroscopy, can be used to examine the energy levels and charge transport properties of V_2O_5-based materials. The material's electronic band structure and energy levels influence the optoelectronic properties of V_2O_5, including light absorption, charge transport, and photoconversion efficacy. V_2O_5 is a prospective material for use in solar cell applications due to its large bandgap and ability to manipulate energy levels through alloying, doping, and defect engineering techniques. Characterization techniques, such as photoemission spectroscopy (PES) and optical spectroscopy, play a crucial role in developing and improving V_2O_5-based solar cells by revealing the electrical band structure and energy levels of V_2O_5.

3.3 Doping and Defect Effects on the Characteristics of V_2O_5

Doping and the presence of defects can substantially affect the electrical, optical, and structural properties of V_2O_5. By purposefully injecting dopants or changing the defect concentration, researchers may modify the characteristics of V_2O_5 for certain applications, such as solar cell devices [129, 138]. The conductivity and electrical band structure of V_2O_5 can change as a result of doping or the introduction of foreign atoms into the material's crystal lattice. For example, doping V_2O_5 with transition metal ions like titanium (Ti), niobium (Nb), or tungsten (W) can change the bandgap's size and shape as well as the concentration of charge carriers [139–141]. These changes in the electronic structure can enhance light absorption, improve charge transport, and optimize the overall performance of V_2O_5-based solar cells.

Additionally, doping may create new energy levels in the bandgap of V_2O_5. These dopant-induced energy levels may impact the mobility, recombination rates, and overall device efficiency of charge carriers, which act as trap states [142–144]. Researchers are able to modify the defect levels, reduce charge trapping, and increase the charge carrier lifetimes in V_2O_5-based solar cells by carefully choosing the dopants and managing their concentration. The characteristics of V_2O_5 can be severely impacted by defects such as oxygen vacancies, vanadium interstitials, or impurities [85, 142, 143]. These flaws introduce localized energy levels inside the bandgap, which affects optical characteristics, electronic conductivity, and charge carrier dynamics. For instance, oxygen vacancies can serve as electron donors, raising the electron concentration and changing the way that V_2O_5 transports charges. Vanadium interstitials, on the other hand, can serve as electron acceptors, altering the concentration and transit of holes.

Defects can potentially compromise V_2O_5's structural integrity, causing strain or lattice deformation. Defect-induced strain can affect charge carrier mobility, carrier lifespan, and the performance of V_2O_5-based solar cells by changing the electronic band structure, affecting the energy bandgap and band alignment [144, 145]. To investigate the impact of doping and defects on the characteristics of V_2O_5, characterization methods such as X-ray photoelectron spectroscopy (XPS), Raman spectroscopy, and electron paramagnetic resonance (EPR) spectroscopy are frequently used [146, 147]. With these methods, it is possible to understand how the electrical structure, chemical makeup, and defect concentration vary in doped or defective V_2O_5 materials. Researchers can deliberately add dopants or manage the concentrations of defects to modify the characteristics of V_2O_5 to improve its functionality in solar cell applications. Light absorption, charge transfer, and overall efficiency of V_2O_5-based solar cells may all be increased by optimizing the kinds, concentrations, and defect engineering techniques. This provides opportunities for adjusting the electrical and optical characteristics of V_2O_5 by doping and the existence of faults in V_2O_5. It is possible to alter the energy levels, bandgap, charge carrier concentrations, and transport parameters by carefully managing doping and defect engineering. Researchers can improve V_2O_5's performance for solar cell applications by using characterization methods, which offer insightful information on the impact of doping and flaws.

4 Solar Cell Architectures Using V_2O_5

An overview of the various solar cell architectures and their working principles.

Each of the several designs used to construct solar cells has its own set of operating principles and features. Exploring the potential of two-dimensional transition metal oxides, such as V_2O_5, in solar cell applications necessitates an understanding of the various topologies of solar cells. This section provides an overview of a few typical solar cell configurations and explains how they work.

Silicon solar cells with a single junction

Single-junction silicon solar cells, which operate on the p–n junction principle, are one of the most popular solar cell layouts [148–151]. Photons from the sun induce the silicon material to form electron–hole pairs in the depletion zone of the p–n junction. The internal electric field of the junction separates the charges, resulting in a voltage differential and an electric current. This design can gain from the addition of V_2O_5 as a passivation layer (Shown in Fig. 2) on the silicon surface in the context of V_2O_5 [71, 152, 153]. The addition of a thin coating of V_2O_5 can increase charge carrier lifespan, decrease surface recombination, and boost total solar cell efficiency.

Sunlight is absorbed by tiny layers of semiconductor materials in thin-film solar cells. Cadmium telluride (CdTe), copper indium gallium selenide (CIGS), and amorphous silicon (a-Si) are a few examples of materials used in thin-film solar cells. These thin layers increase production flexibility, use less material, and provide affordable solutions. The particular material used determines the thin-film solar cell's operating principle. For instance, incoming photons in CdTe solar cells produce electron–hole pairs in the CdTe absorber layer. The resultant charges are then gathered at the electrodes to produce electricity [154–156]. In the case of V_2O_5, it can function as a transparent conducting layer or a buffer layer in thin-film solar cells. Due to V_2O_5's excellent electrical conductivity and optical transparency make effective charge transfer and light transmission possible. Additionally, the use of V_2O_5 as

Fig. 2 Schematic of Vox-based device

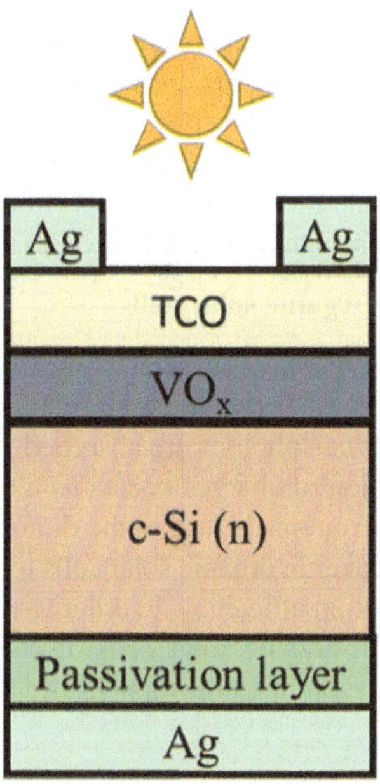

a buffer layer can promote charge extraction, improve energy level alignment, and minimize losses at the absorber-electrode interface [157, 158].

Dye-sensitive solar cells (DSSCs)

The mesoporous titanium dioxide (TiO2) sheet is covered with a light-absorbing dye in dye-sensitized solar cells, also referred to as Grätzel cells [159–161]. The dye absorbs photons from sunshine and releases excited electrons as a result. An electric current is produced when these electrons are introduced into the TiO_2 layer and subsequently collected at a conductive substrate. By accepting the electrons from the counter electrode, the redox couple—typically made up of an electrolyte containing iodide/triiodide ions—enables the regeneration of the dye. With regard to V2O5, it can play a variety of roles in DSSCs. V2O5 can be used as the counter electrode's substance in place of more expensive traditional metals like platinum (Pt) [162, 163]. In order to facilitate effective dye regeneration, it can also function as a catalyst, assisting in the reduction of triiodide ions (I3-) to iodide ions (I-) at the counter electrode. To improve the stability and longevity of DSSCs, V_2O_5 can also be used as a passivation material or protective layer.

Perovskite solar cells

Due to their promise of high efficiency and simplicity of production, perovskite solar cells have attracted a lot of attention. They are built using perovskite materials made of organometallic halides, such as methyl ammonium lead iodide ($CH_3NH_3PbI_3$). In order for perovskite solar cells to function, photons must be absorbed. This causes excitons (electron–hole pairs) to form within the perovskite layer [164, 165]. Following the separation of the excitons, the electrons and holes go to the electron transport layer and the hole transport layer, respectively, which eventually contributes to the electric current. V_2O_5 can function as a hole transport layer (HTL) or a hole-blocking layer in perovskite solar cells. V_2O_5 can function as a hole-blocking layer, preventing charge recombination at the interface between the perovskite absorber and the electron transport layer, and as an HTL, enables effective charge extraction from the perovskite layer to the electrode, enhancing device performance [166, 167].

Organic solar cells

Organic semiconductors, usually conjugated polymers or tiny molecules, are used as the active layer in organic solar cells. Excitons (electron–hole pairs) are created when photons are absorbed within the organic material [106, 168, 169]. The separation of charges occurs as a result of exciton dissociation, which occurs when these excitons diffuse to the donor–acceptor contact. V_2O_5 can function as an interfacial layer in organic solar cells to improve charge extraction and increase exciton dissociation efficiency. In order to maximize charge-collecting efficiency and reduce losses in organic solar cells, it can also function as a transparent electrode or a charge recombination suppressor [170, 171].

Multi-junction or Tandem Solar Cells

Solar cells having numerous bandgap layers stacked on top of one another are known as tandem or multi-junction solar cells. A wider variety of photon energies may be effectively used since each layer efficiently absorbs a particular portion of the sun spectrum [87, 172]. In order to increase total voltage and improve device efficiency, the operating concept requires cascading the absorbed photon energy over the layers. Charge extraction, effective current matching, and overall device optimization are made possible by the use of V_2O_5 as a selective contact layer or a passivation layer between various sub-cells [173, 174].

To explore the potential of 2D transition metal oxide materials like V_2O_5 in boosting their performance, it is essential to comprehend the operating principles and features of these various solar cell layouts. Researchers can examine the integration and optimization of V_2O_5 inside various solar cell topologies to increase the efficiency and general performance of the devices by using its special qualities, such as its electrical conductivity, bandgap engineering potential, and surface reactivity.

5 Optoelectronic Properties and Performance of Devices

The optoelectronic characteristics of this two-dimensional transition metal oxide greatly influence the device performance of V2O5-based solar cells. To fully use the potential of V_2O_5 in solar cell applications, it is crucial to comprehend its optical characteristics, charge transport dynamics, and efficiency improvement techniques [74, 126, 175–177]. The optoelectronic characteristics of V_2O_5 are discussed in this part, along with how they affect device performance. Topics covered include light absorption processes, charge transport, recombination kinetics, and methods for increasing efficiency.

5.1 Optical Characteristics of V_2O_5's and Light-Absorbing Mechanisms

Unique optical characteristics that V_2O_5 demonstrates are crucial for applications in solar cells. Its capacity to capture solar energy depends on its bandgap, energy levels, and light-absorption processes. By modifying the material's stoichiometry and structure, the bandgap of V_2O_5 may be changed, allowing for the optimization of certain solar spectra. In V_2O_5, light absorption can take place either directly between energy states or indirectly through defect levels [127, 147]. It is important to comprehend these light absorption mechanisms for V2O5-based solar cells to capture light as efficiently as possible.

5.2 Charge Transport and Recombination Processes in Solar Cells Based on V_2O_5

In solar cells based on V_2O_5, efficient charge transfer and recombination suppression are essential for obtaining high device performance. The charge extraction and collection effectiveness are directly impacted by the charge transport characteristics of V_2O_5, such as carrier mobility and diffusion length. To maximize device performance, it is essential to comprehend the charge transport processes and the variables influencing carrier mobility and diffusion in V_2O_5 films [178]. The efficiency of the whole device as well as the lifetimes of the charge carriers, can be considerably impacted by recombination mechanisms, including radiative and non-radiative recombination. Designing solar cells using V_2O_5 must prioritize reducing recombination losses and increasing charge carrier lifetime.

5.3 Efficiency-Improving Techniques and Difficulties for Solar Cells Based on V_2O_5

The efficiency of V_2O_5-based solar cells may be increased using a variety of techniques. One method is to minimize energy level mismatches and interfacial recombination by optimizing the interface between V_2O_5 and other materials in the device stack. The efficiency of charge extraction and collection can be increased by designing heterojunctions, interlayers, or buffer layers. Additionally, adding nanostructures like nanowires or nanosheets can improve the capabilities of charge transport and light absorption capabilities. Conductivity and bandgap engineering may also be enhanced by doping V_2O_5 with other elements or including dopants in the device construction. However, several difficulties must be overcome, including concerns about stability, scalability of manufacturing methods, and financial constraints. It remains a considerable difficulty to produce stable and effective V_2O_5-based solar cells on a wide scale, necessitating more study and development.

To maximize their efficiency and realize their promise in practical applications, V_2O_5-based solar cells must have their optoelectronic characteristics and device performance understood. Researchers may learn more about the charge transport processes, recombination kinetics, and light absorption characteristics of V_2O_5 by using sophisticated characterization techniques and theoretical modeling. This information is crucial for directing the development of innovative device designs and putting efficiency-improving techniques into practice. The performance of V_2O_5-based solar cells is heavily influenced by the optoelectronic features of V_2O_5, including its optical properties, charge transport kinetics, and recombination processes. Researchers may create techniques to improve device efficiency and handle problems with stability, scalability, and cost by thoroughly understanding these features. Realizing the full potential of two-dimensional transition metal oxides in developing photovoltaics requires continued study and investigation of V_2O_5 in solar cell applications.

6 Stability and Environmental Factors

The viability of V_2O_5-based solar cell technologies depends critically on their stability and environmental effects. For long-term performance to be improved, it is essential to comprehend the stability problems and degradation mechanisms of V_2O_5 [179, 180]. Vanadium hydroxides or vanadium forms can be produced as a result of moisture-induced reactions, which lower electrical conductivity and charge transfer. Thermal instability and oxidation can potentially change an object's electrical and optoelectronic characteristics. An environmental impact assessment is necessary to determine if V2O5-based solar cell technologies are sustainable [181]. Considerations are made for things such as raw material extraction, manufacturing energy use, and end-of-life management. The production of V_2O_5 uses hazardous chemicals, which impact the environment. Assessing resource depletion, energy use, greenhouse gas emissions, and toxicological consequences helps to identify problem areas and promotes ethical behavior.

Strategies for environmental protection and stability improvement can be used to improve V_2O_5-based solar cell technology. Techniques for surface passivation, protective coatings, and encapsulating layers stop chemical deterioration and moisture intrusion. The deterioration and unintended reaction products are reduced via interface engineering. Novel synthesis techniques improve the thermal characteristics and stability of V2O5 films. Advanced characterization techniques make monitoring device deterioration and creating more stable V2O5-based solar cells possible. To address environmental problems, researchers might investigate environmentally friendly synthesis pathways and renewable energy sources. Minimizing the environmental effect requires increasing energy efficiency and lowering the use of harmful chemicals. Solar cells made of V_2O_5 guarantee responsible material management and support the circular economy when they are properly recycled and disposed of. V_2O_5-based solar cell technologies can advance towards greater sustainability by implementing stability enhancement and environmental responsibility initiatives. For V_2O_5-based solar cell technologies to advance while reducing their environmental impact, ongoing research on stability enhancement, environmentally friendly procedures, and recycling is essential.

7 Future Perspectives

The stability and environmental effects of V_2O_5-based solar cell technologies are key factors in their adoption. Long-term effectiveness depends on an understanding of V_2O_5's deterioration processes. Vanadium hydroxides or vanadates can occur as a result of moisture-induced reactions, impeding charge transmission. While heat instability impacts optoelectronic behavior, oxidation may change its electrical characteristics. An environmental impact evaluation is necessary to evaluate the viability of solar cells based on V2O5. Considerations are made for things like raw material extraction, energy use, and end-of-life management. Hazardous substances influence the environmental impact of V2O5 production. Responsible behavior is encouraged through assessing energy usage, emissions, and toxicological impacts.

Technologies based on V_2O_5 can be improved by enhancing stability and being environmentally conscious. Interface engineering and surface passivation stop moisture from entering and causing damage. New synthesis techniques that result in V2O5 films without flaws improve stability. Device deterioration may be tracked using sophisticated characterization methods. Researchers can investigate environmentally friendly synthesis pathways and renewable energy sources. Environmental effect is minimized through increasing energy efficiency and lowering the use of harmful chemicals. Responsible materials management is ensured through appropriate recycling and disposal. V_2O_5-based solar cell technologies can advance sustainably by implementing stability enhancement and environmental responsibility measures. To progress while reducing the negative effects on the environment, ongoing study is essential on stability enhancement, environmentally friendly procedures, and recycling.

8 Application of Two-Dimensional Transition Metal Oxide and Chalcogenide-Based Material in Solar Cells

8.1 Application of 2-Dimensional (2D) TiS_2, TiO_2, ZnS, ZnSe, ZnO in Solar Cells

2D materials refer to materials with at least one dimension thickness of only a few atomic layers, typically less than 100 nm. Transition metal oxides (TMO), transition metal chalcogenides (TMC), and transition metal dichalcogenides (TMDs) are composed of transition metal cations and chalcogen (sulfur, selenium or tellurium) anions and exhibit a wide range of optoelectronic properties [182]. Thus, Two-dimensional TMC, TMO, and TMD have proven helpful as electron transport layers (ETL) and hole transport layers (HTL) in solar cells. ETL are n-type semiconductors with high electron mobility, typically electron affinity higher than the absorber layer, and unfavorable transport properties for the photogenerated hole and vice versa for the HTL [183]. In this section, we will explore the application of oxide-based 2D structures or nanosheets based on TiS_2, TiO_2, ZnS, ZnSe, and ZnO as charge transport layers in various types of solar cells, including DSSC, perovskite, organic, and inorganic solar cells. Additionally, we will discuss the findings of several research studies in this field.

8.2 TiS_2 and TiO_2 2D Material

Huckaba et al. [184] investigated TiS_2 as a cost-effective and room temperature processed alternative to the commonly used HTL spiro-OMeTAD. The transmittance and absorbance spectrum of the 2D TiS_2 film, deposited on the fluorine-doped tin oxide (FTO) substrate using ultraviolet (UV)-visible absorption spectroscopy, was studied. The UV–Visible spectrum of material provides information about the energy required to excite electrons from the valence band to the conduction band, which is related to the bandgap energy. The bandgap of the semiconductor sample can be determined using the Tauc plot and plotting (αhν)2 as a function of the photon energy (hν). The plot usually shows a straight line in the high-energy region, corresponding to the material's direct bandgap absorption [185]. The energy corresponding to the intersection of the extrapolated straight line with the x-axis gives the value of the optical bandgap of the material. The semiconductor bandgap of the 2D layer was observed at 1.8 eV. The X-ray diffraction (XRD) spectrum of the TiS_2 thin film and powder sample was observed to have an amorphous nature. X-ray photoelectron spectroscopy (XPS) shows the core electron level spectrum of the TiS_2 film. The low-binding-energy contributions for both the S 2p (S $2p_{3/2}$ at 161.3 eV and S $2p_{1/2}$ at 162.5 eV) and Ti 2p (Ti $2p_{3/2}$ at 457.1 eV and Ti $2p_{1/2}$ at 463.8 eV) correspond to the expected Ti:S ratio of 1:2. The XPS study also revealed the features of TiO_2 film due to the partial oxidation of TiS_2. In XPS, an X-ray beam is directed at a sample surface, and the electrons emitted from the surface are collected and analyzed. XPS can identify the chemical composition of surfaces and thin films with high sensitivity and accuracy. XPS can also provide information about surface atoms' oxidation state and electronic structure [186]. The UPS spectrum confirms the work function of the TiS_2 layer at 4.3 eV and the valence band maximum onset at 1.53 eV. The band structure alignment of the TiS_2 layer concerning the mixed perovskite absorber layer confirms the favorable energy landscape for photogenerated hole transport towards HTL and metal contacts. In UPS, a sample is irradiated with ultraviolet (UV) light of known energy. Electrons in the valence band absorb the energy of the photons, and some of them are ejected from the sample surface. The ejected electrons are then analyzed to determine their kinetic and binding energy relative to the vacuum level [187].

The investigated perovskite solar cell with the device structure FTO/Compact TiO_2/meso-TiO_2/Perovskite absorber/TiS_2 (Spiro-OMeTAD, no HTL) with different HTL material was observed and explored. The results show that TiS_2 and Spiro-OMeTAD efficiently extract holes from the absorber layer, as evidenced by the diminished PL intensity. Photoluminescence (PL) is the emission of light by a material when it absorbs photons of light. When a semiconductor absorbs light, an electron in the valence band can be excited to the conduction band, leaving behind a hole in the valence band. The electron and hole can recombine, emitting a photon of light in the process. The semiconductor material having direct bandgap and maximum radiative recombination will yield maximum PL intensity. At the same time, the efficient extraction of electrons or holes by the charge transport layer will diminish the PL intensity signal [188]. Furthermore, the measured carrier lifetime using the RC time constant was higher than that of the doped Spiro-OMeTAD, ranging from 1–10 μs.

Alias et al. [189] examined the effect of a TiS_2/PEDOT: PSS composite transport layer as a counter electrode in a DSSC-based solar cell. They observed that the PEDOT: PSS and TiS_2 composite provided uniform coverage of the FTO substrate compared to a bare TiS_2 thin film, which exhibited large particle agglomeration. They also noted that the PEDOT: PSS and TiS_2 composite led to a rougher surface, allowing lower charge transfer resistance. The maximum power conversion efficiency (PCE) was observed for 10 wt% of TiO_2/PEDOT: PSS composite. Additionally, TiS_2 and PEDOT: PSS composite improved electrocatalytic activity, as confirmed by scanning electron microscopy (SEM), atomic force microscopy (AFM), and capacitance–voltage (CV) spectroscopic techniques. The authors observed the highest charge transfer resistance from the Tafel polarization plot for bare TiS_2 due to poor surface coverage and linkage with the FTO substrate. A Tafel plot shows the relationship between the electrochemical reaction rate and the applied potential. It is used to study the kinetics of electrochemical reactions and to determine the electrochemical activity and mechanism of various materials. The Tafel slope can be used to determine the rate constant for the electrochemical reaction, as well as the mechanism of the reaction. A steep slope indicates that the rate-determining step is electron transfer, while a shallow slope indicates that the rate-determining step is a chemical reaction [190]. Huang et al. [191] used a solution and room temperature processed electron transport layer in a planar n-i-p-based perovskite solar cell. They observed a high PCE of 18.79% after the UV ozone (UVO) treatment of a 2D TiS_2 layer, as shown in the *JV* response of Fig. 3d. XRD (Fig. 3a), AFM, and transmission electron microscopy (TEM) were used to confirm the formation of nanosheets of TiS_2 material. The authors observed that the UVO treatment modified the TiS_2 layer's work function from 4.79 eV to 4.64 eV, as shown in Fig. 3b. They also found that structural defects at the TiS_2 surface, such as S vacancies, were passivated by partial oxidation, confirmed via the XPS technique, as shown in Fig. 3c. The optimized UVO time was observed at 15 min. Yin et al. used TiS_2 as a low-temperature and solution-processed ETL in perovskite-based solar cells. They observed a TiS_2 bandgap at 1.7 eV, a valence band position at 5.72 eV, and a conduction band at 4.02 eV using the UPS. They also measured the carrier lifetime at 5.9 ns compared to bare perovskite at 57.9 ns using a Time-resolved photoluminescence (TRPL) study. TRPL is a technique used to study the dynamics of photogenerated charge carriers in a material. It involves measuring the luminescence signal emitted from a sample after it has been excited by a short light pulse. TRPL experiments monitor the luminescence signal as a function of time after the excitation pulse. The decay of the luminescence signal over time provides information about the lifetime of the photogenerated charge carriers in the material [192].

The optimized device performance was observed at 1.5 mg/ml concentration of TiS_2 material and a rapid spin speed of 5000–6000 rpm. Lickederer et al. [12] demonstrated the potential of transforming TiO_2 film into TiS_2 nanosheets by high-temperature treatment in H_2S. With a 2-h H_2S treated sample, the best device-performing parameters were observed. The conversion of TiO_2 into 2D TiS_2 was

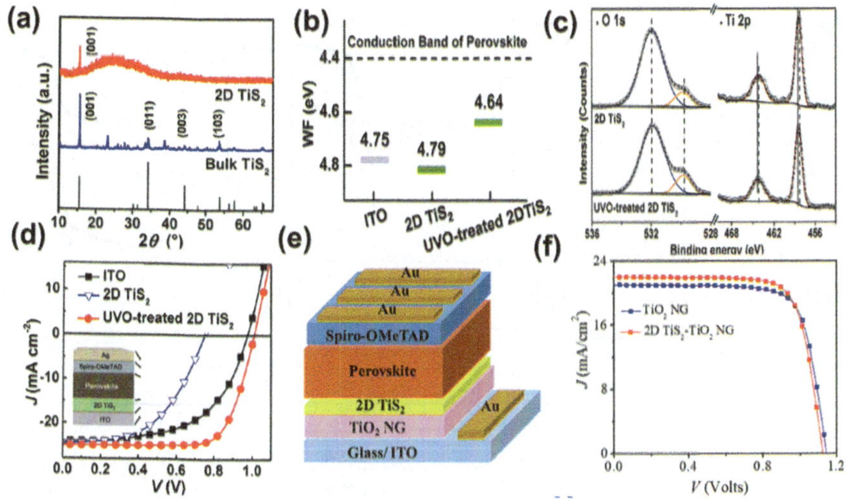

Fig. 3 **a** The X-ray diffraction study of bulk vs 2D TiS_2 sample. **b** UV-Ozone dependent work function of 2D TiS_2 layer **c** X-ray photoelectron spectroscopy (XPS) of 2D TiS_2 layer confirms various elements' presence. **d** JV response of perovskite solar cell w/o and w UVO treatment of TiS_2 2D layer as the electron transport layer. **e** Schematic of device design of the perovskite solar cell with a dual layer of 2D TiS_2 and TiO_2 nano grass (NG) as ETL. **f** The JV response of the fabricated device confirms the surface passivation impact of the 2D TiS_2 layer. Fig. (a–d) [188] Fig. (e–f) [189]

confirmed using SEM, XRD, XPS, energy dispersive x-ray spectroscopy (EDX), and cyclic voltammetry. Figure 3e shows the device schematic of a perovskite-based solar cell by passivating the TiO_2 NG surface using TiS_2 2D atomically thin layer. Figure 3f confirms the passivation impact by the 2D TiS_2 layer. [189]. Figure 4a shows the XRD spectrum of the TiO_2 NG layer with a thin 2D TiS_2 layer. Authors have confirmed a smaller leakage current in a modified device with a smaller diode quality factor ($n = 0.53$ ~ indicating domination of trimolecular recombination rate) compared to $n = 2.65$ (due to high non-radiative recombination centers), indicating a strong presence of surface defects at TiO_2 NG and perovskite interface as shown in Fig. 4b. Furthermore, the lower trap density and higher charge carrier mobility (Fig. 4c), smaller charge transfer resistance (R_{ct}) and higher charge recombination resistance (R_{rec}) (Fig. 4d) were observed in case of a modified device with 2D TiS_2 and TiO_2 NG layers in comparison to the pristine device with only TiO_2 NG layer as ETL.

The 3D TiO_2 layer is widely explored in perovskite-based solar cells [195, 196]. Researchers have also explored various 2D nanosheets based on TiO_2 for use in solar cells. For instance, Xu et al. [197] modified the surface of TiO_2 nanosheets by coating g-C_3N_4 to create the TiO_2@C_3N_4 heterostructure (Fig. 4e) in DSSC-based solar cells. This modification led to an improvement in the morphology as confirmed by the XRD spectrum of various g-C_3N_4-modified TiO_2 at different urea to TiO_2 weight ratios (%), and the best *JV* response was observed with CTS-8 as shown in

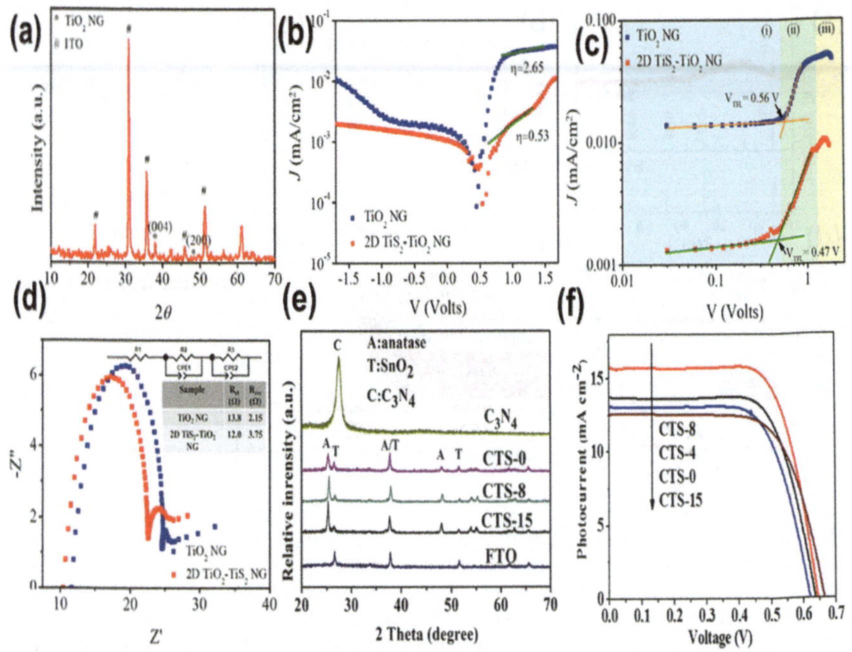

Fig. 4 **a** The XRD spectrum of 2D TiS_2-TiO_2 NG with various diffraction peaks. **b** The dark JV response of only TiO_2NG-based device and TiO_2 NG and TiS_2 stack layers. **c** Trap density and charge carrier mobility are determined using double log scale dark JV response of TiO_2 NG-based device and TiO_2 NG and TiS_2 stack layers. **d** The Nyquist plot of TiO_2 NG-based device and TiO_2 NG and TiS_2 stack layers. **e** XRD spectrum of TiO2 nanosheets with Urea incorporation. **f** The response of the overall device with various organic modifications. Figure a, d [193] Fig. e, f [194]

Fig. 4f. Etgar et al. [198] synthesized 30 nm TiO_2 nanosheets using the hydrothermal route in PbS QDs-based solar cells, demonstrating decent device performance.

8.3 ZnS, ZnSe, and ZnO 2D Material

Similarly, researchers have also explored ZnO, ZnS, and ZnSe nanosheets for use in various solar cell architectures. Ohtake et al. [199] used a ZnSe buffer layer with a thickness of less than 100 nm in Cu(InGa) Se_2 absorber layer-based thin-film solar cells. The buffer layer is used to engineer the interface between the active layer and the contact material to reduce losses due to charge recombination. Buffer layers can be used to modify the work function of the electrode or to passivate the surface of the active layer to improve the collection efficiency of charge carriers [197, 200]. The broader bandgap nature of the ZnSe buffer layer improved spectral response compared to the widely used CdS buffer layer in these devices. The spectral response of a solar cell refers to how well the cell converts light of different wavelengths (colors) into electrical energy [195]. Li et al. [201] studied the solution and low-temperature processed ZnSe layer as an ETL in a planar perovskite solar cell. The XRD pattern of the ZnSe film deposited on the FTO substrate with characteristic peaks matches the zinc blende structure with an interplanar spacing of 0.33 nm, as shown in Fig. 5a. The schematic band diagram of the ZnSe layer sandwiched between TiO_2 and the perovskite layer, as confirmed using the UPS spectra, indicated a favorable energetic landscape for electron transfers towards the metal contact (Fig. 5b). The *JV* response of the overall device with various reaction times of keeping the synthesis pot at elevated temperature for the ZnSe layer showed that a 2-h reaction time resulted in the best device performance (Fig. 5c). The morphology of the ZnSe thin film deposited on the FTO substrate also changed with the change in reaction time. Moreover, the ZnSe layer exhibited higher conductivity than the TiO_2 deposited layer sandwiched between the gold electrode and FTO substrate, as shown in Fig. 5d; such a bi-ETL for effective bandgap engineering can be highly crucial for single-junction perovskite solar cells [197] and perovskite-perovskite multijunction solar cells [200].

Cardenas et al. [199] used ZnS as a buffer layer in CdTe solar cells using the chemical bath deposition technique. Zinc sulfate and thiourea were precursors in the ammonium hydroxide complexing agent. The thermal treatment of ZnS film was carried out in the presence of oxygen, argon, and $CdCl_2$. $CdCl_2$, argon, and oxygen

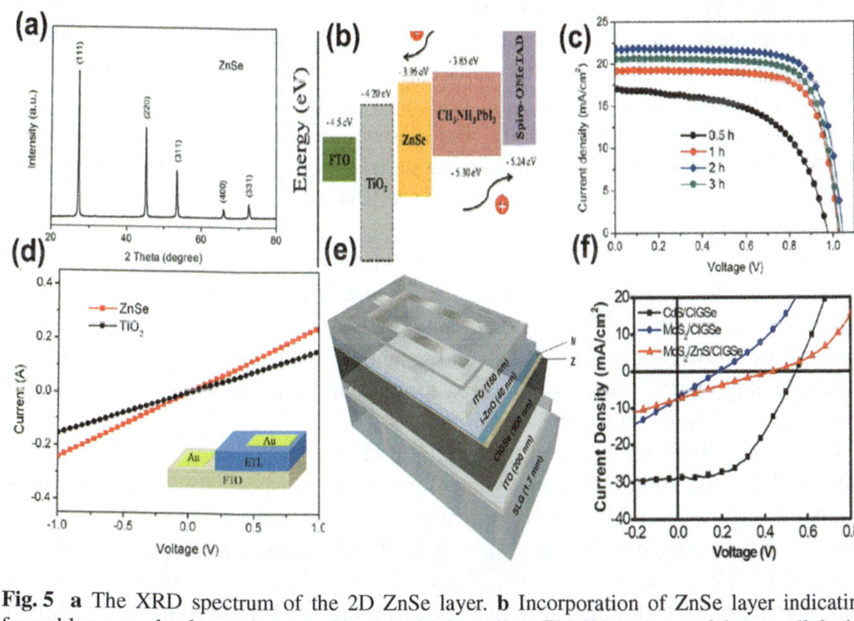

Fig. 5 **a** The XRD spectrum of the 2D ZnSe layer. **b** Incorporation of ZnSe layer indicating favorable energy landscape in perovskite-based solar cells. **c** The JV response of the overall device with various annealing times. d The JV response of ZnSe and TiO2 layer coated on FTO substrate. **e** Device schematic of CIGSe-based solar cell incorporating 2D MoS_2-ZnS buffer layers. **f** The JV response of the devices with various ETL layer modifications. Figure a, d [196] e, f [198]

improved the visible and near-infrared transmittance spectrum by up to 80–90% in ZnS-based thin films, resulting in the best device performance when employed as a buffer layer. To serve the purpose of surface passivation, Park et al. [202] sandwiched the ZnS layer in-between MoS_2 and the absorber CIGSe layer. With the introduction of the ZnS layer in the device stack Fig. 5e, the authors observed an improvement in the device shunt resistance, small series resistance, ideality factor, and reverse saturation current and better *JV* response compared to pristine solar cell as shown in Fig. 5f. In addition to ZnS, ZnO nanosheets have also been explored for application in solar cells. Bi et al. [203] studied 2D ZnO as an ETL synthesized using sol–gel, atomic layer deposition, and ZnO nanorods in organic solar cells with P3HT-PCBM as the absorber layer. 2D ZnO deposited by the sol–gel method provided the best device performance due to the better crystalline structure facilitated by the sol–gel method. Finally, ZnO nanostructures have also been explored in quantum dot-sensitized solar cells [201, 204, 205] and DSSC-based solar cells [202].

9 Conclusion

In conclusion, the utilization of V_2O_5 in solar cells holds significant promise for the advancement of photovoltaic technology. V_2O_5-based solar cells offer a pathway towards improved efficiency, enhanced stability, and increased scalability, given the ongoing research and development efforts in this field. With further exploration and refinement, V2O5-based solar cells can emerge as a vital contributor to the future of renewable energy production, effectively addressing critical challenges and opening up new avenues for research and innovation. Additionally, the application of TiS_2, TiO_2, ZnS, ZnSe, and ZnO as charge transport layers in various solar cell architectures, including dye-sensitized solar cells (DSSC), perovskite, organic, and inorganic solar cells, have demonstrated their effectiveness. By revisiting the roles of these materials as electron transport layers (ETL) and hole transport layers (HTL), we have provided insights into their electronic properties, energy level spectra, and their positioning within the energy band diagram among different solar cell absorber layers. This knowledge enhances our understanding of their critical roles in facilitating efficient charge transport and separation, thereby contributing to the overall performance of these solar cells.

In summary, both V_2O_5-based solar cells and the application of TiS_2, TiO_2, ZnS, ZnSe, and ZnO as charge transport layers offer exciting prospects for the advancement of solar energy technology. These innovations not only enhance the efficiency and stability of solar cells but also pave the way for the development of more sustainable and scalable photovoltaic solutions, ultimately bolstering the future of renewable energy generation. Continued research and exploration of these materials and their applications are crucial steps in realizing the full potential of solar energy as a clean and sustainable power source.

References

1. Zhao, Y., Yu, L., Sun, M.: Recent progress in emerging 2D layered materials for organic solar cells. Solar Energy **218,** 621–638 (2021). https://doi.org/10.1016/j.solener.2021.02.066
2. Yu, C., Xu, S., Yao, J., Han, S.: Recent advances in and new perspectives on crystalline silicon solar cells with carrier-selective passivation contacts. Crystals **8,** 430 (2018). https://doi.org/10.3390/cryst8110430
3. Pandey, B., Roy, S.C.: Carbon-based Nanomaterials as Photocatalysts and their performance attributes. In: Carbon Nanostructures: Fundamentals to Applications, pp. 5–1. AIP Publishing LLC, Melville, New York (2021)
4. Pandey, B., Shankar, D.: Recent trends in metallic nanocomposites for sensing and electrochemical devices. Nanoparticles Reinforced Metal Nanocomposites: Mechanical Performance and Durability, pp. 237–271 (2023)
5. Soonmin, H.: A review of metal oxide thin films in solar cell applications. Int. J. Thin. Film. Sci. Tec. **11**(1), 37–45 (2022)
6. Agyei-Tuffour, B., Mensah-Darkwa, K., NframahAmpong, D., AdzoAddae, E., SelasieGbadam, G., Darko, C.N.S., Akyaw1, A.O., Adjah, J., Asare, J., Neill, G.: Nanomaterials in 2-dimensions for flexible solar cell applications—a review. J. Goosen, Cogent Eng. **9**, 2143034 (2022). https://doi.org/10.1080/23311916.2022.2143034
7. Wu, C., Wang, D., Zhang, Y., Gu, F., Liu, G., Zhu, N., Luo, W., Han, D., Guo, X., Qu, B., Wang, S., Bian, Z., Chen, Z., Xiao, L.: FAPbI3Flexible solar cells with a record efficiency of 19.38% fabricated in air via ligand and additive synergetic process. Adv. Funct. Mater. **29**(34), 1902974 (2019). https://doi.org/10.1002/adfm.201902974
8. Koo, D., Jung, S., Seo, J., Jeong, G., Choi, Y., Lee, J., Lee, S.M.: Flexible organic solar cells over 15% efficiency with polyimide-integrated graphene electrodes. Joule **4**(5), 1021–1034 (2020). https://doi.org/10.1016/j.joule.2020.02.012
9. Nassiri Nazif, K., Daus, A., Hong, J., Lee, N., Vaziri, S., Kumar, A., Nitta, F., Chen, M.E., Kanaian, S., Islam, R., Kim, K.-H., Park, J.-H., Poon, A.S., Brongersma, M.L., Pop, E., Saraswat, K.C.: High-specific-power flexible transition metal dichalcogenide solar cells. Nat. Commun. **12**(1), 7034 (2021). https://doi.org/10.1038/s41467-021-27195-7
10. Das, S., Pandey, D., Thomas, J., Roy, T.: The role of graphene and other 2D materials in solar photovoltaics. Adv. Mater. 1802722 (2018). https://doi.org/10.1002/adma.201802722
11. Taguchi, M., Yano, A., Tohoda, S., Matsuyama, K., Nakamura, Y., Nishiwaki, T., Fujita, K., Maruyama, E.: IEEE J. Photovoltaics **4**, 96 (2014)
12. Zheng, C., Kammen, D.M.: An innovation-focused roadmap for a sustainable global photovoltaic industry. Energy Policy **67**, 159–169 (2014)
13. Brus, V.V., Lee, J., Luginbuhl, B.R., Ko, S.-J., Bazan, G.C., Nguyen, T.-Q.: Solution-processed semitransparent organic photovoltaics: from molecular design to device performance. Adv. Mater. **31**, 1900904 (2019)
14. Yu, P., Zhen, Y., Dong, H., Hu, W.: Crystal engineering of organic optoelectronic materials. Chem **5**, 2814–2853 (2019)
15. Schulz, P., Cowan, S.R., Guan, Z.-L., Garcia, A., Olson, D.C., Kahn, A.: NiOX/ MoO3Bi-Layers as efficient hole extraction contacts in organic solar cells. Adv. Funct. Mater. **24**, 701–706 (2014)
16. Ruscello, M., Sarkar, T., Levitsky, A., Matrone, G.M., Droseros, N., Schlisske, S., Sachs, E., Reiser, P., Mankel, E., Kowalsky, W., Banerji, N., Stingelin, N., Frey, G.L., Hernandez-Sosa, G.: Nanocomposite of nickel oxide nanoparticles and polyethylene oxide as printable hole transport layer for organic solar cells. Sustain. Energy Fuels **3**, 1418–1426 (2019)
17. Kumbhakar, P., Gowda, C.C., Mahapatra, P.L., Mukherjee, M., Malviya, K.D., Chaker, M., Chandra, A., Lahiri, B., Ajayan, P.M., Jariwala, D., Singh, A., Sekhar Tiwary, C.: Emerging 2D metal oxides and their applications. Materials Today **45**, 142–168 (2021)
18. Comini, E., Baratto, C., Faglia, G., Ferroni, M., Vomiero, A., Sberveglieri, G.: Quasi-one dimensional metal oxide semiconductors: Preparation, characterization and application as

chemical sensors. Progress Mater. Sci. **54**, 1–67 (2009), https://doi.org/10.1016/j.pmatsci.2008.06.003
19. Melo de Lima, L.R., Martins, F.P., Lagarinhos, J.N., Santos, L., Lima, P., Torcato, R., Marques, P.A.A.P., Rodrigues, D.L., Melo, S., Grilo, J., Oliveira, J.M.: Characterization of commercial graphene-based materials for application in thermoplastic nanocomposites. Mater. Today Proc. **20**(Part 3), 383–390 (2020)
20. Akinwande, D., Petrone, N., Hone, J.: Nat. Commun. **5**, 5678 (2014)
21. Nawz, T., Safdar, A., Hussain, M., Lee, D.S., Siyar, M.: Graphene to advanced MoS2: a review of structure, synthesis, and optoelectronic device application. Crystals 10, 902 (2020). https://doi.org/10.3390/cryst10100902
22. Geim, A.K., Novoselov, K.S., Jiang, D., Schedin, F., Booth, T.J., Khotkevich, V.V., Morozov, S.V.: Two-dimensional atomic crystals. Proc. Natl. Acad. Sci. U.S.A. **102**, 10451–10453 (2005)
23. Novoselov, K.S., Mishchenko, A., Carvalho, A., Castro Neto, A.H.: 2D materials and van der Waals heterostructures. Science **353**, 6298 (2016)
24. Balendhran, S., Walia, S., Nili, H., Zhen Ou, J., Zhuiykov, S., Kaner, R.B., Sriram, S., Bhaskaran, M., Kalantar-zadeh, K.: Two-dimensional molybdenum trioxide and dichalcogenides. Adv. Funct. Mater. (2013). https://doi.org/10.1002/adfm.201300125
25. Hou, C.-H., Shyue, J.-J., Su, W.-F., Tsai, F.-Y.: Catalytic metal-induced crystallization of sol–gel metal oxides for high-efficiency flexible perovskite solar cells. J. Mater. Chem. A **6**(34), 16450–16457 (2018). https://doi.org/10.1039/C8TA05973J
26. Li, C., Cong, S., Tian, Z., Song, Y., Yu, L., Lu, C., Shao, Y., Li, J., Zou, G., Rümmeli, M.H., Dou, S., Sun, J., Liu, Z.: Flexible perovskite solar cell-drivenphoto-rechargeable lithium-ion capacitor for self-powered wearable strain sensors. Nano Energy **60**, 247–256 (2019). https://doi.org/10.1016/j.nanoen.2019.03.061
27. Wang, Y., Chen, Q., Zhang, G., Xiao, C., Wei, Y., Li, W.: Ultrathin flexible transparent composite Agyei-Tuffour et al., Cogent Eng. **9**, 2143034 (2022). https://doi.org/10.1080/23311916.2022.2143034 Page 29 of 31 electrode via semi-embedding silver nanowires in a colorless polyimide for high-performance ultraflexible organic solar cells. Mater. Sci. ACS Appl. Mater. Interfaces **14**, 5699–5708. https://doi.org/10.1021/acsami.1c18866
28. Butler, S.Z., Hollen, S.M., Cao, L., Cui, Y., Gupta, J.A., Gutiérrez, H.R., Heinz, T.F., Hong, S.S., Huang, J., Ismach, A.F., et al.: Progress, challenges, and opportunities in two-dimensional materials beyond graphene. ACS Nano **7**, 2898–2926 (2013)
29. Lauhon, L.J., Marks, T.J., Hersam, M.C., Sangwan, V.K., Jariwala, D.: Emerging device applications for semiconducting two-dimensional transition metal dichalcogenides. ACS Nano **8**, 1102–1120 (2014)
30. Duan, X., Wang, C., Pan, A., Yu, R., Duan, X.: Two-dimensional transition metal dichalcogenides as atomically thin semiconductors: opportunities and challenges. Chem. Soc. Rev. **44**, 8859–8876 (2015)
31. Zhang, H.: Introduction: 2D materials chemistry. Chem. Rev. **118**, 6089–6090 (2018)
32. Chander, S., Kant Tripathi, S.: Recent advancement in efficient metal oxide-based flexible perovskite solar cells: a short review. Mater. Adv. **3**, 7198–7211 (2022). https://doi.org/10.1039/D2MA00700B
33. Ogale, S.B., Venkatesan, T.V., Blamire, M.: Functional Metal Oxides: New Science and Novel Applications. Wiley-VCH, Singapore (2013)
34. Gerling, L.G., Mahato, S., Voz, C., Alcubilla, R., Puigdollers, J: Characterization of transition metal oxide/silicon heterojunctions for solar cell applications. Appl. Sci. **5**(4), 695–705 (2015). https://doi.org/10.3390/app5040695
35. Würfel, P.: Physics of Solar Cells, 1st ed.; Wiley-VCH: Weinheim, Germany, pp. 113–114 (2005)
36. Radisavljevic, B., Radenovic, A., Brivio, J., Giacometti, V., Kis, A.: Nat. Nanotechnol. **6**, 147 (2011)
37. Kumar, A., Ahluwalia, P.K.: Eur. Phys. J. B. **85**, 186 (2012)
38. Mak, K.F., Lee, C., Hone, J., Shan, J., Heinz, T.F.: Phys. Rev. Lett. **105**, 136805 (2010)

39. Haque, F., Daeneke, T., Kalantar-zadeh, K., Ou, J.Z.: Two-dimensional transition metal oxide and chalcogenide based photocatalysts. Nano-Micro Lett. **10**, 23 (2018). https://doi.org/10.1007/s40820-017-0176-y
40. Kim, S.S., et al.: Chem. Mater. **27**, 4222–4228 (2015). https://doi.org/10.1021/cm504802j
41. Geng, F., et al.: J. Am. Chem. Soc. **136**, 5491–5500 (2014). https://doi.org/10.1021/ja501587y
42. Toth, P.S.: From two-dimensional materials to their heterostructures: an electrochemist's perspective. Appl. Mater. Today **8**, 68–103 (2017)
43. Pumera, M., Sofer, Z., Ambrosi, A.: Layered transition metal dichalcogenides for electrochemical energygeneration and storage. J. Mater. Chem. A **2**, 8981–8987 (2014)
44. Choi, H.S., et al.: Appl. Phys. Lett. **113**, 243103 (2018). https://doi.org/10.1063/1.5060627
45. L. Gao et al.: 2D Mater. **4**, 025031 (2017). https://doi.org/10.1088/2053-1583/aa5b1b
46. Haque, F., Daeneke, T., Kalantar-Zadeh, K., Ou, J.Z.: Two-dimensional transition metal oxide and chalcogenide-based photocatalysts. Nano-micro Lett. **10**(2), 1–27 (2018)
47. Battaglia, C., Yin, X., Zheng, M., Sharp, I.D., Chen, T., McDonnell, S., Azcatl, A., Carraro, C., Ma, B., Maboudian, R., Wallace, R.M.: Hole selective MoO x contact for silicon solar cells. Nano Lett. **14**(2), 967–971 (2014)
48. Guan, M., Wang, Q., Zhang, X., Bao, J., Gong, X., Liu, Y.: Two-dimensional transition metal oxide and hydroxide-based hierarchical architectures for advanced supercapacitor materials. Front. Chem. **8**, 390 (2020)
49. Bellani, S., Bartolotta, A., Agresti, A., Calogero, G., Grancini, G., Di Carlo, A., Kymakis, E., Bonaccorso, F.: Solution-processed two-dimensional materials for next-generation photovoltaics. Chem. Soc. Rev. **50**(21), 11870–11965 (2021)
50. Ming, X., Guo, A., Wang, G., Wang, X.: Two-dimensional defective tungsten oxide nanosheets as high performance photo-absorbers for efficient solar steam generation. Sol. Energy Mater. Sol. Cells **185**, 333–341 (2018)
51. Choudhary, N., Islam, M.A., Kim, J.H., Ko, T.J., Schropp, A., Hurtado, L., Weitzman, D., Zhai, L., Jung, Y.: Two-dimensional transition metal dichalcogenide hybrid materials for energy applications. Nano Today **19**, 16–40 (2018)
52. Zhang, C.J., Park, S.H., O'Brien, S.E., Seral-Ascaso, A., Liang, M., Hanlon, D., Krishnan, D., Crossley, A., McEvoy, N., Coleman, J.N., Nicolosi, V.: Liquid exfoliation of interlayer spacing-tunable 2D vanadium oxide nanosheets: High capacity and rate handling Li-ion battery cathodes. Nano Energy **39**, 151–161 (2017)
53. Wei, Z.: Functionalization of two-dimensional transition metal oxides for the sensing applications (Doctoral dissertation, Ghent University) (2021)
54. Liu, M., Su, B., Tang, Y., Jiang, X., Yu, A.: Recent advances in nanostructured vanadium oxides and composites for energy conversion. Adv. Energy Mater. **7**(23), 1700885 (2017)
55. Van Gog, H., Li, W.F., Fang, C., Koster, R.S., Dijkstra, M., van Huis, M.: Thermal stability and electronic and magnetic properties of atomically thin 2D transition metal oxides. npj 2D Mater. Appl. **3**(1), 18 (2019)
56. Mahmood, N., De Castro, I.A., Pramoda, K., Khoshmanesh, K., Bhargava, S.K., Kalantar-Zadeh, K.: Atomically thin two-dimensional metal oxide nanosheets and their heterostructures for energy storage. Energy Storage Mater. **16**, 455–480 (2019)
57. Fang, J., Zhou, Z., Xiao, M., Lou, Z., Wei, Z., Shen, G.: Recent advances in low-dimensional semiconductor nanomaterials and their applications in high-performance photodetectors. InfoMat **2**(2), 291–317 (2020)
58. Mannhart, J., Schlom, D.G.: Oxide interfaces—an opportunity for electronics. Science **327**(5973), 1607–1611 (2010)
59. Goniakowski, J., Finocchi, F., Noguera, C.: Polarity of oxide surfaces and nanostructures. Rep. Prog. Phys. **71**(1), 016501 (2007)
60. Osada, M., Sasaki, T.: Exfoliated oxide nanosheets: new solution to nanoelectronics. J. Mater. Chem. **19**(17), 2503–2511 (2009)
61. Sasaki, T., Watanabe, M.: Semiconductor nanosheet crystallites of quasi-TiO_2 and their optical properties. J. Phys. Chem. B **101**(49), 10159–10161 (1997)

62. Dahiya, H., Suthar, R., Singh, M.K., Singhal, R., Karak, S., Sharma, G.D.: Facile approach for efficient non-fullerene-based binary and ternary organic solar cells using hydrated vanadium pentoxide as a hole transport layer. ACS Appl. Energy Mater. **6**(6), 3442–3451 (2023)
63. Cheng, J., Ren, X., Zhu, H.L., Mao, J., Liang, C., Zhuang, J., Roy, V.A., Choy, W.C.: Pre-and post-treatments free nanocomposite based hole transport layer for high performance organic solar cells with considerably enhanced reproducibility. Nano Energy **34**, 76–85 (2017)
64. Zhang, W., Wang, Q., Chen, Y., Wang, Z., Wee, A.T.: Van der waals stacked 2D layered materials for optoelectronics. 2D Mater. **3**(2), 022001 (2016)
65. Zhu, J., Cao, L., Wu, Y., Gong, Y., Liu, Z., Hoster, H.E., Zhang, Y., Zhang, S., Yang, S., Yan, Q., Ajayan, P.M.: Building 3D structures of vanadium pentoxide nanosheets and application as electrodes in supercapacitors. Nano Lett. **13**(11), 5408–5413 (2013)
66. Filonenko, V.P., Sundberg, M., Werner, P.E., Zibrov, I.P.: Structure of a high-pressure phase of vanadium pentoxide, β-V2O5. Acta Crystallogr. B **60**(4), 375–381 (2004)
67. Bo, Z., Qing, S., De-Yan, H.: First-principles calculations on the electronic and vibrational properties of β-V2O5. Chin. Phys. B **18**(11), 4988 (2009)
68. Rahman, M.F., Mahmud, N., Alam, I., Ali, M.H., Moon, M.M.A., Kuddus, A., Toki, G.F., Rubel, M.H.K., Al Asad, M.A., Hossain, M.K.: Design and numerical analysis of CIGS-based solar cell with V2O5 as the BSF layer to enhance photovoltaic performance. AIP Adv. **13**(4) (2023)
69. Tom, T., Ros, E., López-Pintó, N., Miguel Asensi, J., Andreu, J., Bertomeu, J., Puigdollers, J., Voz, C.: Influence of co-sputtered Ag: Al ultra-thin layers in transparent V2O5/Ag: Al/AZO hole-selective electrodes for silicon solar cells. Materials **13**(21), 4905 (2020)
70. Zhao, J., Ma, F.J., Ho-Baillie, A., Bremner, S.: June. Spin coated nickel oxide and vanadium oxide layers on silicon for a carrier selective contact solar cell. In: 2017 IEEE 44th Photovoltaic Specialist Conference (PVSC), pp. 1845–1849. IEEE (2017)
71. Beke, S.: A review of the growth of V2O5 films from 1885 to 2010. Thin Solid Films **519**(6), 1761–1771 (2011)
72. Bensalah, N., De Luna, Y.: Recent progress in layered manganese and vanadium oxide cathodes for Zn-Ion batteries. Energ. Technol. **9**(5), 2100011 (2021)
73. Xie, H., Li, Z., Cheng, L., Haidry, A.A., Tao, J., Xu, Y., Xu, K. and Ou, J.Z.: Recent advances in the fabrication of 2D metal oxides. Iscience **25**(1) (2022)
74. Sucharitakul, S., Ye, G., Lambrecht, W.R., Bhandari, C., Gross, A., He, R., Poelman, H., Gao, X.P.: V2o5: A 2d van der Waals oxide with strong in-plane electrical and optical anisotropy. ACS Appl. Mater. Interfaces. **9**(28), 23949–23956 (2017)
75. Qi, Y., Qin, K., Jian, Z., Yang, X., Jin, W., Tan, Y., Zou, Y., Chen, W.: Ag-functionalized exfoliated V 2 O 5 nanosheets: a flexible and binder-free cathode for lithium-ion batteries. J. Mater. Sci. **54**, 12713–12722 (2019)
76. Wang, C., Yi, Y., Li, H., Wu, P., Li, M., Jiang, W., Chen, Z., Li, H., Zhu, W., Dai, S.: Rapid gas-assisted exfoliation promises V2O5 nanosheets for high performance lithium-sulfur batteries. Nano Energy **67**, 104253 (2020)
77. Zhou, K., Shang, G., Hsu, H.H., Han, S.T., Roy, V.A., Zhou, Y.: Emerging 2D Metal Oxides: From Synthesis to Device Integration. Advanced Materials, p. 2207774 (2023)
78. Lira-Cantu, M. (ed.): The Future of Semiconductor Oxides in Next-Generation Solar Cells. Elsevier (2017)
79. Raiford, J.A., Oyakhire, S.T., Bent, S.F.: Applications of atomic layer deposition and chemical vapor deposition for perovskite solar cells. Energy Environ. Sci. **13**(7), 1997–2023 (2020)
80. Chu, S., Zhao, R., Liu, R., Gao, Y., Wang, X., Liu, C., Chen, J., Zhou, H.: Atomic-layer-deposited ultra thin VOx film as a hole transport layer for perovskite solar cells. Semicond Sci. Technol. **33**(11), 115016 (2018)
81. Zhang, R., Averianov, T., Andris, R., Zachman, M.J., Pomerantseva, E.: Liquid phase exfoliation of chemically prelithiated bilayered vanadium oxide in aqueous media for Li-Ion batteries. J. Phys. Chem. C **127**(2), 919–929 (2023)
82. Rui, X., Lu, Z., Yu, H., Yang, D., Hng, H.H., Lim, T.M., Yan, Q.: Ultrathin V 2 O 5 nanosheet cathodes: realizing ultrafast reversible lithium storage. Nanoscale **5**(2), 556–560 (2013)

83. Mu, J., Wang, J., Hao, J., Cao, P., Zhao, S., Zeng, W., Miao, B., Xu, S.: Hydrothermal synthesis and electrochemical properties of V2O5 nanomaterials with different dimensions. Ceram. Int. **41**(10), 12626–12632 (2015)
84. Gnanasekar, S., Nirmala Grace, A.: V2O5 nanosheets as an efficient, low-cost pt-free alternate counter electrode for dye-sensitized solar cells. ChemNanoMat **8**(2), e202100382 (2022)
85. Fan, L.L., Chen, S., Luo, Z.L., Liu, Q.H., Wu, Y.F., Song, L., Ji, D.X., Wang, P., Chu, W.S., Gao, C., Zou, C.W.: Strain dynamics of ultrathin VO2 film grown on TiO2 (001) and the associated phase transition modulation. Nano Lett. **14**(7), 4036–4043 (2014)
86. Prasadam, V.P., Bahlawane, N., Mattelaer, F., Rampelberg, G., Detavernier, C., Fang, L., Jiang, Y., Martens, K., Parkin, I.P., Papakonstantinou, I.: Atomic layer deposition of vanadium oxides: process and application review. Mater. Today Chem. **12**, 396–423 (2019)
87. Raiford, J.A., Belisle, R.A., Bush, K.A., Prasanna, R., Palmstrom, A.F., McGehee, M.D., Bent, S.F.: Atomic layer deposition of vanadium oxide to reduce parasitic absorption and improve stability in n–i–p perovskite solar cells for tandems. Sustain. Energy Fuels **3**(6), 1517–1525 (2019)
88. Hasannezhad, F., Naji, L., Arvand, M.: Influencing parameters on the electrochemical growth of V2O5 nanorods on ITO as interfacial layer in bulk heterojunction polymer solar cells. Mater. Sci. Semicond. Process. **139**, 106333 (2022)
89. Liu, X., Zeng, J., Yang, H., Zhou, K., Pan, D.: V 2 O 5-Based nanomaterials: synthesis and their applications. RSC Adv. **8**(8), 4014–4031 (2018)
90. Alkarsifi, R., Ackermann, J., Margeat, O.: Hole transport layers in organic solar cells: a review. J. Metals Mater. Minerals **32**(4), 1–22 (2022)
91. Martín, I., López, G., Garin, M., Ortega, P., Voz, C., Jia, G., Gawlik, A., Plentz, J.: Thin c-Si solar cells based on VO x Heterojunctions with texturized rear surface. IEEE J. Photovoltaics **11**(6), 1358–1362 (2021)
92. Narushima, S., Mizoguchi, H., Shimizu, K.I., Ueda, K., Ohta, H., Hirano, M., Kamiya, T., Hosono, H.: A p-type amorphous oxide semiconductor and room temperature fabrication of amorphous oxide p–n heterojunction diodes. Adv. Mater. **15**(17), 1409–1413 (2003)
93. Zhang, L., Jiang, C., Wu, C., Ju, H., Jiang, G., Liu, W., Zhu, C., Chen, T.: V2O5 as hole transporting material for efficient all inorganic Sb2S3 solar cells. ACS Appl. Mater. Interfaces. **10**(32), 27098–27105 (2018)
94. Ahmmed, S., Aktar, A., Ismail, A.B.M.: Role of a solution-processed V2O5 hole extracting layer on the performance of CuO–ZnO-based solar cells. ACS Omega **6**(19), 12631–12639 (2021)
95. Shuvo, N., Al Askary, M.A.H., Hai, T.B., Alam, S.U.: August. Effect of thin tunnel layers MnO3 and V2O5 on the enhancement of single-layer organic solar cell efficiency. In: Journal of Physics: Conference Series (Vol. 2312, No. 1, p. 012052). IOP Publishing (2022)
96. Su, S.H., Lin, W.K., Chiang, W.T., Lin, Y.F., Yokoyama, M.: Application of inorganic/organic stacked hole transporting layer in organic solar cells. Japanese J. Appl. Phys. **51**(2S), 02BK03 (2012)
97. Almora, O., Gerling, L.G., Voz, C., Alcubilla, R., Puigdollers, J., Garcia-Belmonte, G.: Superior performance of V2O5 as hole selective contact over other transition metal oxides in silicon heterojunction solar cells. Sol. Energy Mater. Sol. Cells **168**, 221–226 (2017)
98. Ameri, T., Dennler, G., Lungenschmied, C., Brabec, C.J.: Organic tandem solar cells: a review. Energy Environ. Sci. **2**(4), 347–363 (2009)
99. Cheng, Y., Ding, L.: Perovskite/Si tandem solar cells: fundamentals, advances, challenges, and novel applications. SusMat **1**(3), 324–344 (2021)
100. Keppner, H., Meier, J., Torres, P., Fischer, D., Shah, A.: Microcrystalline silicon and micromorph tandem solar cells. Appl. Phys. A **69**, 169–177 (1999)
101. Yu, Z.J., Leilaeioun, M., Holman, Z.: Selecting tandem partners for silicon solar cells. Nat. Energy **1**(11), 1–4 (2016)
102. Terán-Escobar, G., Pampel, J., Caicedo, J.M., Lira-Cantú, M.: Low-temperature, solution-processed, layered V 2 O 5 hydrate as the hole-transport layer for stable organic solar cells. Energy Environ. Sci. **6**(10), 3088–3098 (2013)

103. Sun, H., Hou, X., Wei, Q., Liu, H., Yang, K., Wang, W., An, Q., Rong, Y.: Low-temperature solution-processed p-type vanadium oxide for perovskite solar cells. Chem. Commun. **52**(52), 8099–8102 (2016)
104. Gerling, L.G., Masmitja, G., Voz, C., Ortega, P., Puigdollers, J., Alcubilla, R.: Back junction n-type silicon heterojunction solar cells with V2O5 hole-selective contact. Energy Procedia **92**, 633–637 (2016)
105. Zilberberg, K., Trost, S., Meyer, J., Kahn, A., Behrendt, A., Lützenkirchen-Hecht, D., Frahm, R., Riedl, T.: Inverted organic solar cells with sol–gel processed high work-function vanadium oxide hole-extraction layers. Adv. Func. Mater. **21**(24), 4776–4783 (2011)
106. Wang, H.Q., Li, N., Guldal, N.S., Brabec, C.J.: Nanocrystal V2O5 thin film as hole-extraction layer in normal architecture organic solar cells. Org. Electron. **13**(12), 3014–3021 (2012)
107. Corr, S.A., Grossman, M., Furman, J.D., Melot, B.C., Cheetham, A.K., Heier, K.R., Seshadri, R.: Controlled reduction of vanadium oxide nanoscrolls: crystal structure, morphology, and electrical properties. Chem. Mater. **20**(20), 6396–6404 (2008)
108. Javed, M.S., Lei, H., Wang, Z., Liu, B.T., Cai, X., Mai, W.: 2D V2O5 nanosheets as a binder-free high-energy cathode for ultrafast aqueous and flexible Zn-ion batteries. Nano Energy **70**, 104573 (2020)
109. Xu, Y., Dunwell, M., Fei, L., Fu, E., Lin, Q., Patterson, B., Yuan, B., Deng, S., Andersen, P., Luo, H., Zou, G.: Two-dimensional V2O5 sheet network as electrode for lithium-ion batteries. ACS Appl. Mater. Interfaces. **6**(22), 20408–20413 (2014)
110. Ureña-Begara, F., Crunteanu, A., Raskin, J.P.: Raman and XPS characterization of vanadium oxide thin films with temperature. Appl. Surf. Sci. **403**, 717–727 (2017)
111. Shvets, P., Dikaya, O., Maksimova, K., Goikhman, A.: A review of Raman spectroscopy of vanadium oxides. J. Raman Spectrosc. **50**(8), 1226–1244 (2019)
112. Wang, S., Wu, L., Zhang, H., Wang, Z., Qin, Q., Wang, X., Lu, Y., Li, L., Li, M.: Facile synthesis of two dimensional (2D) V2O5 nanosheets film towards photodetectors. Materials **15**(23), 8313 (2022)
113. Dadigala, R., Bandi, R., Gangapuram, B.R., Dasari, A., Belay, H.H., Guttena, V.: Fabrication of novel 1D/2D V2O5/g-C3N4 composites as Z-scheme photocatalysts for CR degradation and Cr (VI) reduction under sunlight irradiation. J. Environ. Chem. Eng. **7**(1), 102822 (2019)
114. Yao, J., Li, Y., Massé, R.C., Uchaker, E., Cao, G.: Revitalized interest in vanadium pentoxide as cathode material for lithium-ion batteries and beyond. Energy Storage Mater. **11**, 205–259 (2018)
115. Wang, Y., Cao, G.: Synthesis and enhanced intercalation properties of nanostructured vanadium oxides. Chem. Mater. **18**(12), 2787–2804 (2006)
116. Hu, P., Hu, P., Vu, T.D., Li, M., Wang, S., Ke, Y., Zeng, X., Mai, L., Long, Y.: Vanadium oxide: phase diagrams, structures, synthesis, and applications. Chem. Rev. **123**(8), 4353–4415 (2023)
117. Sykora, R.E., Ok, K.M., Halasyamani, P.S., Wells, D.M., Albrecht-Schmitt, T.E.: New one-dimensional vanadyl iodates: hydrothermal preparation, structures, and NLO properties of A [VO2 (IO3) 2](A= K, Rb) and A [(VO) 2 (IO3) 3O2](A= NH4, Rb, Cs). Chem. Mater. **14**(6), 2741–2749 (2002)
118. Suh, J.Y., Lopez, R., Feldman, L.C., Haglund, R.F., Jr.: Semiconductor to metal phase transition in the nucleation and growth of VO 2 nanoparticles and thin films. J. Appl. Phys. **96**(2), 1209–1213 (2004)
119. Bergerud, A.J.: Phase Stability and Transformations in Vanadium Oxide Nanocrystals. University of California, Berkeley (2016)
120. Ibrahim, I., Belessiotis, G.V., Arfanis, M.K., Athanasekou, C., Philippopoulos, A.I., Mitsopoulou, C.A., Romanos, G.E., Falaras, P.: Surfactant effects on the synthesis of redox bifunctional v2o5 photocatalysts. Materials **13**(20), 4665 (2020)
121. Rita Kakkar, N.: Recent advancements on warfare agents/metal oxides surface chemistry and their simulation study. Adv. Mater. Lett. **4**(7), 508–521 (2013)
122. Le, T.K., Pham, P.V., Dong, C.L., Bahlawane, N., Vernardou, D., Mjejri, I., Rougier, A., Kim, S.W.: Recent advances in vanadium pentoxide (V 2 O 5) towards related applications

in chromogenics and beyond: fundamentals, progress, and perspectives. J. Mater. Chem. C **10**(11), 4019–4071 (2022)
123. Bullett, D.W.: The energy band structure of V2O5: a simpler theoretical approach. J. Phys. C: Solid State Phys. **13**(23), L595 (1980)
124. Lambrecht, W., Djafari-Rouhani, B., Lannoo, M. and Vennik, J.: The energy band structure of V2O5. I. Theoretical approach and band calculations. J. Phys. C Solid State Phys. **13**(13), 2485 (1980)
125. Parker, J.C., Geiser, U.W., Lam, D.J., Xu, Y., Ching, W.Y.: Optical properties of the vanadium oxides VO2 and V2O5. J. Am. Ceram. Soc. **73**(11), 3206–3208 (1990)
126. Schneider, K.: Optical properties and electronic structure of V2O5, V2O3 and VO2. J. Mater. Sci. Mater. Electron. **31**(13), 10478–10488 (2020)
127. Le, T.K., Kang, M., Kim, S.W.: A review on the optical characterization of V2O5 micro-nanostructures. Ceram. Int. **45**(13), 15781–15798 (2019)
128. Yalagala, B.P., Sahatiya, P., Kolli, C.S.R., Khandelwal, S., Mattela, V., Badhulika, S.: V2O5 nanosheets for flexible memristors and broadband photodetectors. ACS Appl. Nano Mater. **2**(2), 937–947 (2019)
129. Srilakshmi, P., Maheswari, A.U., Sajeev, V., Sivakumar, M.: Tuning the optical bandgap of V2O5 nanoparticles by doping transition metal ions. Mater. Today Proc. **18**, 1375–1379 (2019)
130. Kang, M., Kim, I., Kim, S.W., Ryu, J.W., Park, H.Y.: Metal-insulator transition without structural phase transition in V2O5 film. Appl. Phys. Lett. **98**(13) (2011)
131. Surnev, S., Ramsey, M.G., Netzer, F.P.: Vanadium oxide surface studies. Prog. Surf. Sci. **73**(4–8), 117–165 (2003)
132. Ngamwongwan, L., Fongkaew, I., Jungthawan, S., Hirunsit, P., Limpijumnong, S., Suthirakun, S.: Electronic and thermodynamic properties of native point defects in V 2 O 5: A first-principles study. Phys. Chem. Chem. Phys. **23**(19), 11374–11387 (2021)
133. Cui, Y., Liu, B., Chen, L., Luo, H., Gao, Y.: Formation energies of intrinsic point defects in monoclinic VO2 studied by first-principles calculations. AIP Adv. **6**(10) (2016)
134. Ramana, C.V., Hussain, O.M., Naidu, B.S., Reddy, P.J.: Spectroscopic characterization of electron beam evaporated V2O5 thin films. Thin Solid Films **305**(1–2), 219–226 (1997)
135. Chakrabarti, A., Hermann, K., Druzinic, R., Witko, M., Wagner, F., Petersen, M.: Geometric and electronic structure of vanadium pentoxide: a density functional bulk and surface study. Phys. Rev. B **59**(16), 10583 (1999)
136. Vuurman, M.A., Wachs, I.E., Hirt, A.M.: Structural determination of supported V2O5-WO3/TiO2 catalysts by in situ Raman spectroscopy and X-ray photoelectron spectroscopy. J. Phys. Chem. **95**(24), 9928–9931 (1991)
137. Wu, Q.H., Thißen, A., Jaegermann, W.: Photoelectron spectroscopic study of Na intercalation into V2O5 thin films. Solid State Ionics **167**(1–2), 155–163 (2004)
138. Phatyenchuen, S., Pongthawornsakun, B., Panpranot, J., Praserthdam, P.: Effect of transition metal dopants (M= Nb, La, Zr, and Y) on the M-TiO2 supported V2O5 catalysts in the selective oxidation of H2S to elemental sulfur. J. Environ. Chem. Eng. **6**(5), 5655–5661 (2018)
139. Wang, B., Huang, Y., Han, Y., Zhao, S., Ding, W., Zhang, W., Li, R., Wu, X., Jiang, Q., Li, Y., Gao, D.: Single tungsten atoms modified porous V2O5 nanoflowers with enhanced electrochromic performance. Cell Reports Phys. Sci. **4**(5) (2023)
140. Devan, R.S., Patil, R.A., Lin, J.H., Ma, Y.R.: One-dimensional metal-oxide nanostructures: recent developments in synthesis, characterization, and applications. Adv. Func. Mater. **22**(16), 3326–3370 (2012)
141. Warwick, M.E., Binions, R.: Advances in thermochromic vanadium dioxide films. J. Mater. Chem. A **2**(10), 3275–3292 (2014)
142. Lambrecht, W., Djafari-Rouhani, B., Vennik, J.: Theoretical study of the vanadyl-oxygen vacancy in V2O5: tight-binding Green function calculation, optical properties and neutral vacancy ground-state splitting. J. Phys. C: Solid State Phys. **19**(3), 369 (1986)
143. Shen, K., Yang, R., Wang, D., Jeng, M., Chaudhary, S., Ho, K., Wang, D.: Stable CdTe solar cell with V2O5 as a back contact buffer layer. Sol. Energy Mater. Sol. Cells **144**, 500–508 (2016)

144. Sivakumar, A., Dhas, S.S.J., Sivaprakash, P., Raj, A.D., Kumar, R.S., Arumugam, S., Prabhu, S., Ramesh, R., Chakraborty, S., Dhas, S.M.B.: Shock wave recovery experiments on α-V2O5 nano-crystalline materials: a potential material for energy storage applications. J. Alloy. Compd. **929**, 167180 (2022)
145. Basyooni, M.A., Al-Dossari, M., Zaki, S.E., Eker, Y.R., Yilmaz, M., Shaban, M.: Tuning the metal–insulator transition properties of VO2 thin films with the synergetic combination of oxygen vacancies, strain engineering, and tungsten doping. Nanomaterials **12**(9), 1470 (2022)
146. Anquillare, E.L.: X-ray spectroscopy of electronic band structure in vanadium oxide nanoparticles (Doctoral dissertation, Boston University) (2021)
147. Wang, Q., Brier, M., Joshi, S., Puntambekar, A., Chakrapani, V.: Defect-induced Burstein-Moss shift in reduced V 2 O 5 nanostructures. Phys. Rev. B **94**(24), 245305 (2016)
148. Liu, C., Zhang, L., Yu, G., Wang, T., Wu, X., Xu, L., Lin, P., Cui, C., Yu, X., Wang, P.: Tuning oxygen vacancies in vanadium-doped molybdenum oxide for silicon solar cells with hole selective contact. Mater. Sci. Semicond. Process. **146**, 106687 (2022)
149. Chen, C.Y., Wei, T.C., Hsiao, P.H., Hung, C.H.: Vanadium oxide as transparent carrier-selective layer in silicon hybrid solar cells promoting photovoltaic performances. ACS Appl. Energy Mater. **2**(7), 4873–4881 (2019)
150. He, Z., Xiao, B., Liu, F., Wu, H., Yang, Y., Xiao, S., Wang, C., Russell, T.P., Cao, Y.: Single-junction polymer solar cells with high efficiency and photovoltage. Nat. Photonics **9**(3), 174–179 (2015)
151. Kayes, B.M., Nie, H., Twist, R., Spruytte, S.G., Reinhardt, F., Kizilyalli, I.C., Higashi, G.S.: June. 27.6% conversion efficiency, a new record for single-junction solar cells under 1 sun illumination. In: 2011 37th IEEE Photovoltaic Specialists Conference, pp. 000004–000008. IEEE (2011)
152. Costals, E.R., Masmitjà, G., Almache, E., Pusay, B., Tiwari, K., Saucedo, E., Raj, C.J., Kim, B.C., Puigdollers, J., Martin, I., Voz, C.: Atomic layer deposition of vanadium oxide films for crystalline silicon solar cells. Mater. Adv. **3**(1), 337–345 (2022)
153. Bandaru, N., Kanakala, R., Madaka, R., Dsouza, N., Maurya, R., Rath, J.K.: The structural, optical, and electrical properties of thermal evaporation-deposited V2OX films for use in silicon heterojunction solar cells. J. Mater. Sci. Mater. Electron. **34**(12), 1015 (2023)
154. Chopra, K.L., Paulson, P.D., Dutta, V.: Thin-film solar cells: an overview. Prog. Photovoltaics Res. Appl. **12**(2–3), 69–92 (2004)
155. Green, M.A.: Thin-film solar cells: review of materials, technologies and commercial status. J. Mater. Sci.: Mater. Electron. **18**, 15–19 (2007)
156. Shah, A., Torres, P., Tscharner, R., Wyrsch, N., Keppner, H.: Photovoltaic technology: the case for thin-film solar cells. Science **285**(5428), 692–698 (1999)
157. Jin, S., Jung, B.J., Song, C.K., Kwak, J.: Room-temperature and solution-processed vanadium oxide buffer layer for efficient charge injection in bottom-contact organic field-effect transistors. Curr. Appl. Phys. **14**(12), 1809–1812 (2014)
158. Meyer, J., Hamwi, S., Kröger, M., Kowalsky, W., Riedl, T., Kahn, A.: Transition metal oxides for organic electronics: energetics, device physics and applications. Adv. Mater. **24**(40), 5408–5427 (2012)
159. Gnanasekar, S., Sonar, P., Jain, S.M., Jeong, S.K., Grace, A.N.: Performance evaluation of a low-cost, novel vanadium nitride xerogel (VNXG) as a platinum-free electrocatalyst for dye-sensitized solar cells. RSC Adv. **10**(67), 41177–41186 (2020)
160. Mutta, G.R., Popuri, S.R., Maciejczyk, M., Robertson, N., Vasundhara, M., Wilson, J.I., Bennett, N.S.: V2O5 as an inexpensive counter electrode for dye sensitized solar cells. Mater. Res. Express **3**(3), 035501 (2016)
161. Elbohy, H., Thapa, A., Poudel, P., Adhikary, N., Venkatesan, S., Qiao, Q.: Vanadium oxide as new charge recombination blocking layer for high efficiency dye-sensitized solar cells. Nano Energy **13**, 368–375 (2015)
162. Gao, C., Han, Q., Wu, M.: Review on transition metal compounds based counter electrode for dye-sensitized solar cells. J. Energy Chem. **27**(3), 703–712 (2018)

163. Wang, G., Hou, S., Yan, C., Lin, Y., Liu, S.: Three-dimensional porous vanadium nitride nanoribbon aerogels as Pt-free counter electrode for high-performance dye-sensitized solar cells. Chem. Eng. J. **322**, 611–617 (2017)
164. Wang, D., Elumalai, N.K., Mahmud, M.A., Wright, M., Upama, M.B., Chan, K.H., Xu, C., Haque, F., Conibeer, G., Uddin, A.: V2O5-PEDOT: PSS bilayer as hole transport layer for highly efficient and stable perovskite solar cells. Org. Electron. **53**, 66–73 (2018)
165. Guo, Q., Wang, C., Li, J., Bai, Y., Wang, F., Liu, L., Zhang, B., Hayat, T., Alsaedi, A., Tan, Z.A.: Low-temperature solution-processed vanadium oxide as hole transport layer for efficient and stable perovskite solar cells. Phys. Chem. Chem. Phys. **20**(33), 21746–21754 (2018)
166. Duan, C., Zhao, M., Zhao, C., Wang, Y., Li, J., Han, W., Hu, Q., Yao, L., Jian, H., Lu, F., Jiu, T.: Inverted CH3NH3PbI3 perovskite solar cells based on solution-processed V2O5 film combined with P3CT salt as hole transport layer. Mater. Today Energy **9**, 487–495 (2018)
167. Tepliakova, M.M., Mikheeva, A.N., Frolova, L.A., Boldyreva, A.G., Elakshar, A., Novikov, A.V., Tsarev, S.A., Ustinova, M.I., Yamilova, O.R., Nasibulin, A.G., Aldoshin, S.M.: Incorporation of vanadium (V) oxide in hybrid hole transport layer enables long-term operational stability of perovskite solar cells. J. Phys. Chem. Lett. **11**(14), 5563–5568 (2020)
168. Beliatis, M.J., Helgesen, M., García-Valverde, R., Corazza, M., Roth, B., Carlé, J.E., Jørgensen, M., Krebs, F.C., Gevorgyan, S.A.: Slot-Die-coated V2O5 as hole transport layer for flexible organic solar cells and optoelectronic devices. Adv. Eng. Mater. **18**(8), 1494–1503 (2016)
169. Alhummiany, H., Rafique, S., Sulaiman, K.: XPS analysis of the improved operational stability of organic solar cells using a V2O5 and PEDOT: PSS composite layer: effect of varied atmospheric conditions. J. Phys. Chem. C **121**(14), 7649–7658 (2017)
170. Rafique, S., Abdullah, S.M., Mahmoud, W.E., Al-Ghamdi, A.A., Sulaiman, K.: Stability enhancement in organic solar cells by incorporating V 2 O 5 nanoparticles in the hole transport layer. RSC Adv. **6**(55), 50043–50052 (2016)
171. Zhang, F., Xu, X., Tang, W., Zhang, J., Zhuo, Z., Wang, J., Wang, J., Xu, Z., Wang, Y.: Recent development of the inverted configuration organic solar cells. Sol. Energy Mater. Sol. Cells **95**(7), 1785–1799 (2011)
172. Yuan, Y., Huang, J., Li, G.: Intermediate layers in tandem organic solar cells (2011)
173. Yin, Z., Wei, J., Zheng, Q.: Interfacial materials for organic solar cells: recent advances and perspectives. Adv. Sci. **3**(8), 1500362 (2016)
174. Wang, Y., Wenisch, R., Schlatmann, R., Lauermann, I.: Inorganic materials as hole selective contacts and intermediate tunnel junction layer for monolithic Perovskite-CIGSe tandem solar cells. Adv. Energy Mater. **8**(30), 1801692 (2018)
175. David, S.P., Soosaimanickam, A., Sakthivel, T., Sambandam, B., Sivaramalingam, A.: Thin film metal oxides for displays and other optoelectronic applications. Metal Metal Oxides Energy Electron. **55**, 185 (2020)
176. Das, T., Tosoni, S., Pacchioni, G.: Structural and electronic properties of bulk and ultrathin layers of V2O5 and MoO3. Comput. Mater. Sci. **163**, 230–240 (2019)
177. Wang, Y., Li, Z., Sheng, X., Zhang, Z.: Synthesis and optical properties of V2O5 nanorods. J. Chem. Phys. **126**(16) (2007)
178. Ma, H., Yip, H.L., Huang, F., Jen, A.K.Y.: Interface engineering for organic electronics. Adv. Func. Mater. **20**(9), 1371–1388 (2010)
179. Vattikuti, S.P., Police, A.K.R., Shim, J., Byon, C.: In situ fabrication of the Bi2O3–V2O5 hybrid embedded with graphitic carbon nitride nanosheets: oxygen vacancies mediated enhanced visible-light–driven photocatalytic degradation of organic pollutants and hydrogen evolution. Appl. Surf. Sci. **447**, 740–756 (2018)
180. Duan, L., Uddin, A.: Defects and stability of perovskite solar cells: a critical analysis. Mater. Chem. Front. **6**(4), 400–417 (2022)
181. Zafar, M., Yun, J.Y., Kim, D.H.: Highly stable inverted organic photovoltaic cells with a V 2 O 5 hole transport layer. Korean J. Chem. Eng. **34**, 1504–1508 (2017)
182. Choudhary, N., Islam, M.A., Kim, J.H., Ko, T.-J., Schropp, A., Hurtado, L., Weitzman, D., Zhai, L., Jung, Y.: Nano Today **19**, 16–40 (2018)

183. Jena, A.K., Kulkarni, A., Miyasaka, T.: Chem. Rev. **119**, 3036–3103 (2019)
184. Huckaba, A.J., Gharibzadeh, S., Ralaiarisoa, M., Roldán-Carmona, C., Mohammadian, N., Grancini, G., Lee, Y., Amsalem, P., Plichta, E.J., Koch, N., Moshaii, A., Nazeeruddin, M.K.: Small Methods, 1 (2017)
185. Kulkarni, A., Singh, T., Jena, A.K., Pinpithak, P., Ikegami, M., Miyasaka, T.: ACS Appl. Mater. Interfaces. **10**, 9547–9554 (2018)
186. Boro, B., Porwal, S., Kumar, D., Mishra, S., Ghosh, S., Kansal, S., Chandra, A., Singh, T.: Catalysis Res. **02**, 033 (2022)
187. Whitten, J.E.: Appl. Surface Sci. Adv. **13**, 100384 (2023)
188. Huang, P., Yuan, L., Zhang, K., Chen, Q., Zhou, Y., Song, B., Li, Y.: ACS Appl. Mater. Interfaces **10**, 14796–14802 (2018)
189. Alias, N., Ali Umar, A., Malek, N.A.A., Liu, K., Li, X., Abdullah, N.A., Rosli, M.M., Abd Rahman, M.Y., Shi, Z., Zhang, X., Zhang, H., Liu, F., Wang, J., Zhan, Y.: ACS Appl. Mater Interfaces **13**, 3051–3061 (2021)
190. Kumar, D., Porwal, S., Singh, T.: Emergent Mater. **5**, 987–1020 (2022)
191. Li, C.-T., Lee, C.-P., Li, Y.-Y., Yeh, M.-H., Ho, K.-C.: J. Mater. Chem. A, 1 (2013)
192. Singh, T., Udagawa, Y., Ikegami, M., Kunugita, H., Ema, K., Miyasaka, T.: APL Mater., 5 (2017)
193. Singh, T.: T. Miyasaka **8**, 1700677 (2018)
194. Xu, J., Wang, G., Fan, J., Liu, B., Cao, S., Yu, J.: J. Power Sources **274**, 77–84 (2015)
195. Mishra, S., Ghosh, S., Boro, B., Kumar, D., Porwal, S., Paul, M., Dixit, H., Singh, T.: Energy Adv. **1**, 761–792 (2022)
196. Li, X., Yang, J., Jiang, Q., Lai, H., Li, S., Xin, J., Chu, W., Hou, J.: ACS Nano **12**, 5605–5614 (2018)
197. Bansal, N.K., Porwal, S., Dixit, H., Kumar, D., Singh, T.: Energ. Technol. **11**, 2201395 (2023)
198. Park, J.H., Kim, D., Shin, S.S., Jo, Y., Cho, J.-S., Park, J., Kim, T.: Appl. Surface Sci., 558 (2021)
199. Ortega-Cardenas, J.A., Albor-Aguilera, M.L., Hernandez-Vasquez, C., Flores-Marquez, J.M., Rueda-Morales, G., Rangel-Kuoppa, V.T., González-Trujillo, M.Á., Yee-Madeira, H.: Mater. Res. Express, 6 (2019)
200. Kumar, D., Bansal, N.K., Dixit, H., Kulkarni, A., Singh, T.: Adv. Theory Simulations **6**, 2200800 (2023)
201. Mati Ur, R., Xie, F., Li, Y., Wei, M.: J. Electroanalyt. Chem. **840**, 160–164 (2019)
202. Suresh, S., Pandikumar, A., Murugesan, S., Ramaraj, R., Raj, S.P.: Sol. Energy **85**, 1787–1793 (2011)
203. Bi, S., Li, Y., Liu, Y., Ouyang, Z., Jiang, C.: J. Sci. Adv. Mater. Dev. **3**, 428–432 (2018)
204. Kim, H., Yong, K.: Phys. Chem. Chem. Phys. **15**, 2109–2116 (2013)
205. Marimuthu, T., Anandhan, N., Thangamuthu, R., Mummoorthi, M., Rajendran, S., Ravi, G.: Mater. Res. Express, 2 (2015)

MXene-Based Two-Dimensional (2D) Hybrid Materials and Their Applications Towards an Environment

Upasana Bagri, Vijaykumar Patel, Ajit Kumar Sharma, Tim Tim Mashangva, Mukesh Kumar, Ashima Thakur, and Harmanjit Singh

Abstract MXenes have obtained noticeable interest cause of their unique features (physical and chemical features). MXenes are considered auspicious applicants considering the resolution of ecological and energy issues because of their distinctive stacked nanostructure, multiple functionalities at the surface, the excess of these compounds on earth, and their appealing optical, electrical, and thermal properties. Due to its large area, flexible chemical composition, and readily modifiable compositions of elements, MXenes need to become a viable choice to enhance photocatalytic efficiency in renewable energy and ecological treatment applications. Cause of their layered nanostructure with an abundance of functionality, they are with outstanding adjustable performance and are simple to mix with other materials, like metallic oxides, polymers, organic hybrids, and carbonaceous materials, to satisfy the demands of high-performance applications. MXenes are excellent catalysts because of their multiple interlayer groups, surface group activities, and adaptable layer spacing. The MXenes family contains more than 30 distinct members, all of which have been investigated and effectively used as catalysts. The fabrication, mechanism at the surface, and uses of MXenes with associated nanocomposites are covered in this chapter. We also discuss MXenes principles and their respective manufacturing methods, such as exfoliation delamination, HF etching, hydrothermal, polymerization, etc., to better understand. MXenes have excelled as photocatalysts for photochemical degradation, carbon dioxide reduction, hydrogen evolution, and nitrogen fixation. Moreover, surface flaws of MXenes offer lots of CO_2 adsorption sites. Also, these materials' superior 2D-nanomaterial structure and fast electron transport pathways contribute to their extremely effective oxidation reaction activity. The effectiveness of heterostructures based on MXene and their nanocomposite photocatalysts for removing organic pollutants is also thoroughly analyzed in

U. Bagri · V. Patel · A. K. Sharma (✉) · T. T. Mashangva · M. Kumar · A. Thakur · H. Singh (✉)
School of Chemical Engineering and Physical Sciences, Lovely Professional University, Jalandhar, Punjab, India
e-mail: ajitsharma2003@gmail.com

H. Singh
e-mail: harmanjit.singh@lpu.co.in

N. Talreja et al. (eds.), *Two-dimensional Hybrid Composites*, Engineering Materials,
https://doi.org/10.1007/978-981-99-8010-9_4

this chapter. Lastly, a future direction for energy and ecological sciences research is suggested.

Keywords Photocatalyst · Photodegradation · MXene · Adsorption · Organic pollutants

1 Introduction

Organic contaminants are the ones that mostly harm water resources. Water bodies contain a variety of organic pollutants, including industrial solvents, medicines, pesticides, plasticizers, phenolic compounds, and many more [1–3]. Across the world, water pollution is a problem that negatively influences aquatic ecosystems and public health. Heavy metals, pesticides, and other toxic and dangerous contaminants are established in Egypt's rivers and interchangeable conditions arise inside India's rivers [4, 5]. In China, the effects of water pollution are so severe that numerous diseases caused by water toxins have caused millions of people to suffer and even die [6]. Pharmaceuticals are globally used to control illness and upgrade the well-being of human beings and other ecological communities.

Pharmaceutical drugs are extremely ingested through several pharmaceutics because of their powerful effects despite hazardous pathogens and microbes [7, 8]. In Fig. 1, there is research data based on research regarding the percentage of world water consumption and waste water output [382]. Many nanomaterials exist that have been employed to purify water, whereas MXene nanocomposites have been studied to be used as a water disinfectant and water purifier. While they are soluble in water, nanoparticles are not regarded as contaminants. Therefore, it is essential to understand their potential hazards, and getting rid of them can be difficult. By photocatalysis, photon energy is converted into potential (chemical) energy (hydrogen) [9]. Researchers have extensively studied two-dimensional (2D) nanomaterials for their innovative catalytic and medicinal uses [10–14].

MXene-based composites have gained the most interest due to their interesting physicochemical features [16–18]. MXenes ought to be broadly fabricated utilizing several different kinds of top-down and bottom-up techniques, with a common formula of $Mn^{+1}XnTx$ [19–22], comprising ball-milling, ultrasonic synthesis, pyrolysis, chemical vapor deposition, molten salts etching, electrochemical etching, the hydrothermal process, and solvothermal procedures [23–28]. However, several issues relating to the use of lacking or rarely toxic substances, the best environmentally friendly circumstances, an advancement toward structural or stability defects, delamination techniques, MXenes oxidation, and the eradication of aluminum layer etch-out residues required to be more advancely investigated [29–33]. For the photocatalytic degradation of pollutants, nanostructures that are based on MXenes having manageable chemical features for surface, large surface area, distinctive optical/thermal features, regular planar structures, outstanding metal conductivity, hydrophilicity

Fig. 1 Percentage of world water consumption and waste water output [15]

and affluence of derivatives have received substantial research [34–38]. In addition to their abilities for catalytic removal and photocatalytic degradation, they have demonstrated excellent potential for removing contaminants via interfacial chemical transformation and sorption [39, 40].

As co-catalysts, MXenes exhibit excessive conductivity and distinctive layered nanocomposites capable of improving nanocomposites' electrocatalytic and photocatalytic features [41, 42]. Regarding titanium-based MXenes, Ti_3C_2 received incredible attention cause of its potential for photocatalytic degradation of harmful pollutants and its flexible structure, electrical properties, and semimetal nature [43]. Strong interactions between MXenes and surface functionalities like fluoride, hydroxyl, oxygen groups, and other semiconductors could result in effective heterojunctions. MXenes (Ti_3C_2) are efficient adsorbents for the decontamination of dyes from wastewaters because they successfully adsorb harmful contaminants and organic dyes [38, 44], and their porous and layered structures enhance their capacity for adsorption and storage [29, 45]. MXenes have been used to create a number of hybrid structures for photocatalytic applications [46, 47]. For instance, zinc oxide (ZnO) and MXene (Ti_3C_2Tx) were combined to create hybrid structures that resembled rice crusts. These structures had outstanding photocatalytic activity, high surface areas, lattice vacancy barrier formation, and optimal band arrangement. Additionally, the above-mentioned hybrid photocatalytic materials demonstrated outstanding stability as well as recyclability [46].

In this fight against environmental pollution, it has emerged as one of the most popular photocatalytic compounds [48, 49]. MXenes uncovered metallic terminal sites (Titanium of Ti_3C_2 MXenes) have greater capacity for redox reaction than conventional carbonaceous materials [50, 51]. Many active sites are provided by terminal groups having hydroxyl, fluorine, or oxygen groups [52]. Ti_3C_2 MXene's outstanding electrical conductivity accomplishes it a superb electron bound that is simple toward catching light-induced electrons, speeding up that transfer as well as separation of mobile electrons and ions that are light generated [48]. Long-established two-dimensional crystals which are atomic in nature, like layered dihalides (transition metal), graphene, and bimetallic hydroxides, make it challenging to get this advantage. The breakdown of dyes (organic) by photocatalytic material based on graphene also demonstrated that MXene is an environment-relatively stable structure [53–59]. Various organic contaminants are in the environment due to the fast industrialization development [60–66]. Like, methylene blue and rhodamine blue in clothing dyes; antioxidants, benzene, sulphonamides, and tetracycline hydrochlorides in cosmetics; SMZ antibiotics (sulfa) in the medicinal field; *Staphylococcus aureus* and *Pseudomonas aeruginosa* in marine biodeterioration; pesticides in agriculture; carbamazepine; endocrine interferon bisphenol A. The world ecosystem will experience a catastrophe due to the abovementioned contaminants, particularly determined organic contaminants like dioxin, oxole, dichlorodiphenyltrichloroethane, and chlordane. They constitute a remarkable danger for ecosystems and come with unquantifiable dangers for people's health because most of them are poisonous [67–83]. Therefore, finding effective methods for organic pollution degradation has emerged as a major obstacle to the treatment of organic pollutants in wastewater [54, 84, 85].

Organic pollution degradation techniques can be used to purify soil, water of rivers, air, wastewater and another media. Majority of organic contaminants could be removed from process of chemical oxidation technique, biodegradation, combustion and conversion by electrochemical method, etc. Unfortunately, certain minimum level dangerous organic contaminants are tough to remove utilizing the aforementioned techniques along with subordinate contamination would result. Photocatalytic technology, a novel type of advanced oxidation technology, has drawn increasing interest for its effective degradation of dangerous organic pollutants because of its high efficacy, ecological sustainability, relatively inexpensive, and safety [86–93]. Furthermore, a significant source of pollution is heavy metal ions. They are challenging for organisms to break down or metabolize because of their high toxicity [94–100]. As a result, they are continuously compressed to an extent that exceeds the permitted limit, significantly endangering human health [101]. Several technologies, including membrane techniques [102], chemical coagulations [103], adsorption [104] and electrochemical methods [105] became established for removing heavy metallic ions through water. Adsorption therapy is one of these techniques that have received a great attentiveness in the past few years because of their affordability and ease of use [106]. The two-dimensional materials, which possesses become a best adsorbents for environmental cleanup, have two clear benefits over conventional adsorbent materials: wide surface area and plentiful active sites [107–111]. For instance, its compounds with a standard 2D structure probably utilized to eliminate organic dyes, oil from water and heavy metallic ions [112]. However, the absence of distinct surface group types may be a contributing factor to its intrinsic features (like having exclusive one component and a basic terminal function), as well as their limited thermal and chemical stability restrict its usefulness. High levels of hydrophilic nature, chemical stability, expansive terminal groups, adaptable chemical features and a surface negative charge are all displayed by MXene [113, 114].

The industry as a whole is quite concerned about its use in heavy metals. MXenes materials are still in their development and have a lot of room to improve. Researchers are starting to look into both their applications and performance more and more frequently. The majorities of assessments of MXene, according to the literature review is concerned with energy conversion [115], storage and biomedical research [116], and indeed have not achieved numerous investigations on environmental cleanup. This study will therefore examine its use in environment protection in more detail. Since it is challenging to create perfect mechanical interfaces between substantial materials like MXene, the energy boundary for transmission of electron at the interconnection is raised [117]. There is a crucial demand toward discover to tiny materials for solve the above mentioned issues in order to get around MXene's fundamental constraints.

Applying the construction of band gap of MXene for achieve maximum utilization and excellent catalytic achievement [118]. It's interesting to note that MXene's customizable band gap offers a versatile channel that can accomplish the little controlled molecular orbital which are unoccupied with molecular orbital which are highly occupied, as well as facilitate the procedure to further expand its application sectors. The three modifications of heterojunction structure coupled with other semiconductors, element-doped structure with heterojunction, and Schottky junction composition built with other kind of semiconductors are the main emphasis of this review. Lately, a number of techniques, including element doping need to be reported for improve the efficiency of MXene toward photocatalysis, copolymerization with organic molecules or morphology control, precious metal deposition, association with carbon materials, and heterojunction with other semiconductor structures [66, 76, 119–122]. According to earlier studies, Schottky junctions and heterojunctions are among the most straightforward and efficient strategies to improve the internal structure of MXene with electrical features as well as increase its efficacy toward photocatalysis. This article thoroughly describes and dissects the numerous MXene-based Schottky junction and heterojunction techniques. Its design improves light absorption, hastens separation of charge as well as transfer, and prolongs lifespan of carriers. Recent research has shown that linking with other semiconductors can enhance photocatalytic activity [49, 118]. The article provides an overview of the MXene production process, characterization techniques, and impact of termination types on the material's mechanical, magnetic, and optical characteristics. The function of MXene termination in terms of its application is also covered. The function of MXene termination in terms of its application is also covered. Figure 2 summarizes this. The fundamental study for termination at surface of MXene is being examined concurrently.

Here, we provide an outline of the most current study on the topic for connecting additional semiconductors to MXene-based photocatalysts to increase their catalytic activity. Specifically, improved photocatalytic efficacy for environmental contaminants and response of MXene to the overall visible light spectrum is discussed.

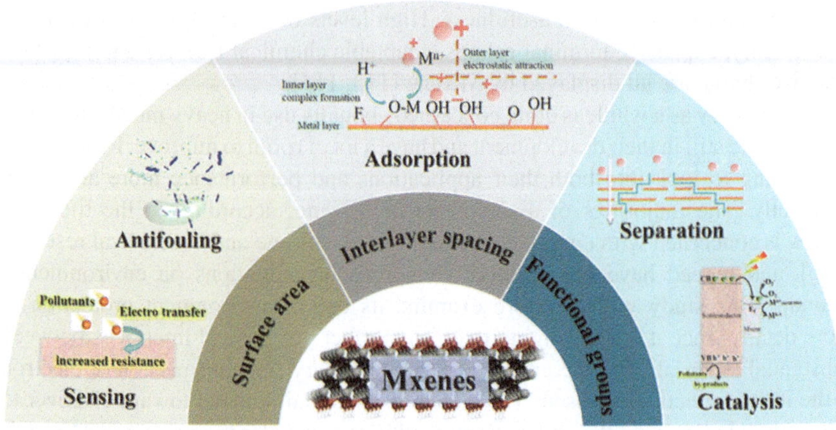

Fig. 2 The research of MXene termination surface. The terminations at surface are showed in yellow color, the applications are showed in blue color and the preparation procedures are highlighted in red (You Wu et al. 2021). Reproduced with permission, © Elsevier

At first, an explanation is given for the pure MXene semiconductors catalytic poor charge usage efficiency. The impacts of different semiconductor coupling surface morphologies, catalytic activity, electronic structures, and particular applications were then extensively examined. At last, useful recommendations for the outline of photocatalytic materials which are based on MXene in the future be made, along with some insightful possibilities toward related study in past years. Also, we encapsulated a number of researches on the heavy metals adsorption by adsorbents which are based on MXene, put out a few workable explanations, and offered some helpful recommendations for the prospective pattern of MXene adsorbents.

2 Methods for Fabrication of MXene by MAX Phase

The evolution of MXene fabrication routes has a big affect on the chemical and electrical features as well as a variety of applications. There are generally three different MXene synthesis methods are top-down, etching, and bottom-up methods. The stacked structures of MXene are produced by the etching process from "A" components of the MAX phases of the ancestor three dimensional layers. The top-down approach might have been acknowledged as enormously popular technique for fabricating MXenes up until this point. When carefully developing the structure using the bottom-up synthesis manufacturing method, as opposed to the top-down manufacturing strategy which often involves such significant variety of precursors. The precursors can be put together using a crystallization process toward a specific two-dimensional sequence to form the structures of MXene. The benefits of bottom-up strategy enable shape, surface terminal functions of MXenes, and precise control

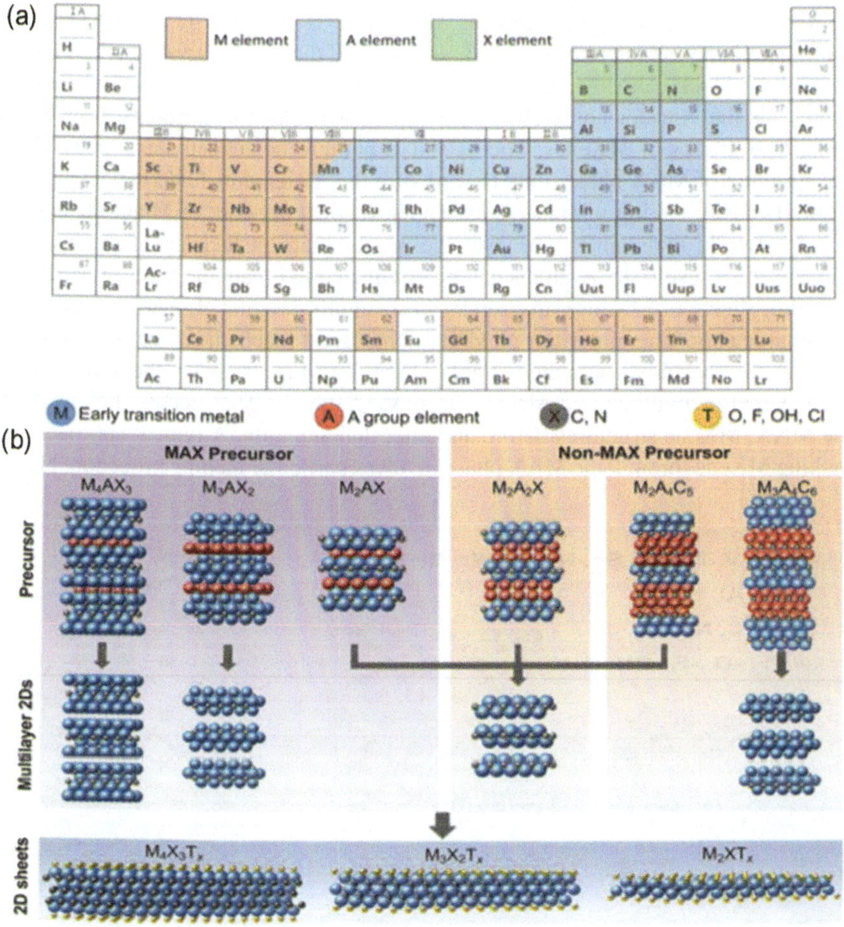

Fig. 3 **a** Periodic table representing the MAX phase elements **b** MXenes have so far reported. M_2XTx, M_3X_2Tx, and M_4X_3Tx are the minimum three possible formulas for MXenes, M is transition metal element and X is Carbide or either Nitride. They could be created by one of three ways: as single M elements; as hard solutions of two different M elements or sequenced double M elements. Carbonitrides are created at the X site by solid solutions (Zhuoheng, B. et al., 2019). Reproduced with permission, © 2019, Elsevier

of the size distribution [123]. Figure 3 shows M_2XTx, M_3X_2Tx, and M_4X_3Tx are the minimum three possible formulas for MXenes, where M is the first row of transition metal and where X is either Carbide or Nitride. They could be created inside one of three routes: as mono elements of M; as hard solutions of slightly two different elements of M or as arranged double elements of M. Compound contains carbon and nitrogen are created at the site of X by hard solutions [124].

MXenes are typically described as Mn-1XnTx layers (n = 1, 2…or 4) created through eliminating interweave "A" elements out of the MAX phase, where M represents the d-block elements, A is the group 3rdA or 4thA elements, and whereas X represents the Nitrides and carbides elements), whereas Tx represents the various functional groups of MXenes. The M and X build to hexagonal shape network in the structure of MAX phase, only toward with the X elements filled the octahedron centrally cages sharing with its edges. The hexagonal network of M and X, not the cubic network, maintained whenever the A atoms are etched out of the layer of MXene. As a result, the A-atoms can be eliminated to create the layer of MnXn-1. The thin sheets of MXenes are usually horizontally orientated, just like its MAX phase precursor. The majorities of MXenes has great mechanical characteristics as well as are probably completely stable [125]. The Fig. 4 shows the synthesis of few-layered or multilayered MXene from MAX phase. The pictures below show the MAX, MXene structures before and after delamination. A typical approach for making MXene through the MAX phase involves particular etching procedure [126].

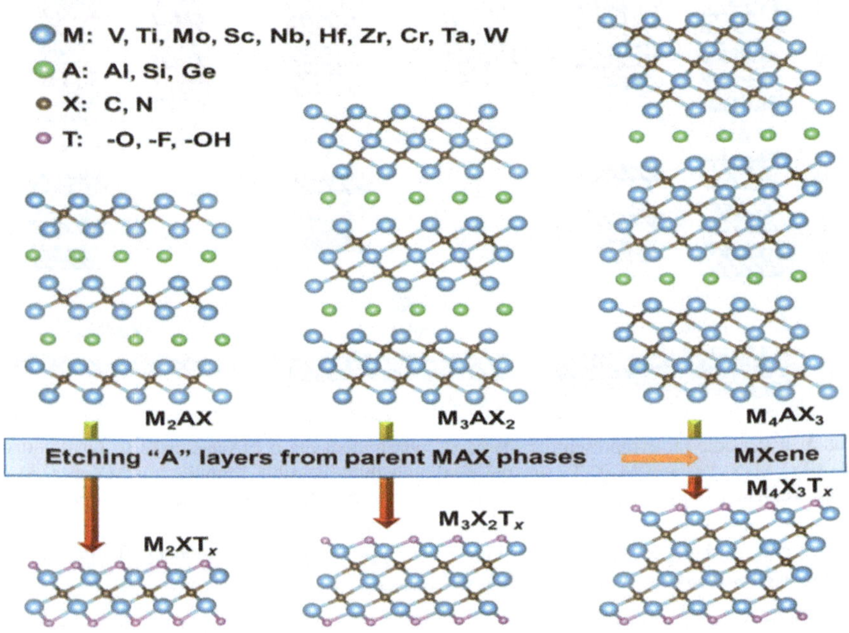

Fig. 4 Shows typical approach for making MXene by MAX phase involves relevant etching The pictures above show the MAX, delaminated-MXene and before delaminated-MXene structures (Kuang, et al., 2020). Reproduced with permission, ©, Elsevier

2.1 *From Etching Method*

MXenes could be synthesized in a numerous different types of routes. Several terminal functions may be combined with the M atoms to minimize their Gibbs free energy at surface and fulfill its coordination spheres by cause of moderations in their etching procedures. As a consequence, the manufacturing of MXenes is significantly influenced by their surface characteristics. In this article, various preparation techniques are covered.

2.1.1 HF Etching Method

Hydrofluoric acid produces Hydrogen gas while removing Al layers from the Ti3AlC2 MAX phase through a straightforward displacement process. Deionized water reacts with the HF strong acid to create Ti_3C_2Tx, where T symbolizes the functionalities of MXenes oxides, fluorides, hydroxyl groups, and H_2. Various MAX phases were successfully deprived into MXenes utilizing the hydrofluoric acid etching method [127, 128]. Since 2011 until now, hydrofluoric acid etching has consistently taken place as an enormously beneficial production method considering the MXenes nanomaterials. The temperature, time duration, and F ions density all matter for fabricating high-quality layers of MXenes in the HF acid etching method. Ti3C2TX creates a superb layered structure with a high concentration of HF acid, which is challenging when working with different acid solutions, according to [129]. The oxygen, fluorine, and hydroxyl functional groups and distinctive surface features were retained in MXenes produced using the HF etching method.

2.1.2 Acid Etching Techniques

Due to the toxic and corrosive characteristics of acid fluoride mixtures, scientists are employing to find methods to avoid using hydrofluoric acid to directly remove the layers of aluminum through MAX phases. The most common technique, termed as earliest hydrofluoric acid etching, substitutes the hydrofluoric acid for the salts of fluoride, like FeF3, potassium fluoride, Lithium fluoride, ammonium fluoride, sodium fluoride and adds hydrochloric acid [130]. AlF3.3H2O is typically created as an unwanted byproduct while synthesizing MXenes through etching the Aluminum or Gallium layers of the MAX phase with the hydrofluoric acid. It is vital to shed light on the elements that lead to this impurity's development in order to synthesize MXenes devoid of it. As a result, customized etching techniques are frequently used. For instance, while carrying out the etching procedure with cobalt fluorides, [131] inferred the conditions that resulted in the creation of the AlF3.3H2O byproduct. The cobalt fluorides/MAX sample's SEM micrograph does not contain any AlF3.3H2O impurities. The intercalation of the cation, which decreases the internal forces between the layers as well as might cause the delamination of layers of the materials under sonication, improves the MXenes interlayer distance created using the modified acid-etching technique. This method allows the creation of few-layer MXene in a sequential manner, cutting down on the time-consuming multiple steps synthesis technique that was previously mentioned [132].

2.1.3 Acid Etching by Fluoride-Based Techniques

In order to avoid the considerable toxicity that hydrofluoric acid etching generates, experts have labored to discover better methods of eliminating the atom layers from MAX. Together with hydrofluoric acid, salts of fluoride (like fluorides of lithium, fluoride of potassium, fluoride of ammonium and fluoride of sodium) and very hard acids could also be utilized for etch out the MAX precursors. Strong acids and fluoride salts have been found to preferentially etch atoms, causing the intercalation of cations including $Sodium^+$, $potassium^+$, $ammonium^+$, and lithium $^+$. Water increases the interlayer gap between MXene layers while decreasing the MXene layer interaction. It is important to keep in mind that the final MXene fragment's size and quality can vary depending on how much fluoride salt and strong acid are present. For instance, the clay method's multilayered Ti_3C_2 (6 M hydrochloric acid/5 M Lithium fluoride) generation calls for a later sonication procedure to delaminate into single nanoflakes which typically yield microscopic faulty MXene flakes [133].

2.1.4 Etching by Molten-Salt

The MAX phases, like Ti_4AlN_3, can also be heated at 550 °C under argon shielding to form Ti_4N_3, which can then be used to generate MXene. It takes 30 min to complete the etching process. $TinCn^{-1}$ is even more stable than $TinNn^{-1}$ but soluble in Hydrofluoric acid or different types of fluoride-based acids that are employed as etchants. Hence, the etching method by molten salts attains the profit of a speedy refining time. Additional washing with DI water and H_2SO_4) and delamination process with TBAOH are required. After delamination process, the XRD patterns of the Ti_4N_3 show that the delaminated Ti_4N_3 has a lesser crystallinity compare with the MXene produced from hydrofluoric acid etching. The final product also exhibits evidence of the TiO_2 phase.

The molten salts etching procedure has the advantage of creating MXenes attains limited stability with the hydrofluoric acid or acids solution which is based on fluorides in contrast with the hydrofluoric acid and acids based on fluoride etching. Contrarily, this method provides the further mentioned drawbacks are the etching procedure uses a lot of energy and heat, the resulting MXenes exhibit less crystallinity and less purity and they attains more number of vacancies and defects [134].

2.1.5 Etching in the Absence of Fluoride

The majority of synthesis methods require hydrofluoric acids or chemicals based on fluoride ions which might also cause the formation of -F with -O terminal groups on interfaces of MXene. Many etching conditions for the synthesis of MXenes have been confirmed. Particularly, -F terminal groups impair the electrochemical properties of supercapacitors based on MXenes [135]. Consequently, to offer satisfactory electrochemical properties, without fluoride manufacturing processes are required. For the manufacture of Ti3C2 MXene, [136] created an alkali-based hydrothermal etching method utilizing NaOH as the etchant. Alkali may be utilized as an etching agent for the MAX phase by cause of their strong interconnection among Al and alkali. Acquiring Multi-layered MXenes with higher purity is until now very difficult. In this case, high concentrations of alkali and high temperatures were used in the Bayer technique to etch out the aluminum layers without harming the Ti3C2 MXene skeleton.

Recently, reported fabricating a Ti_3C_2Tx MXene without fluoride and chloride using etching by electrochemical process [137]. Ti_3C_2Tx was synthesized without the use of any hazardous organic intercalating agents, and delamination was done by using sonication. The resulting MXene nanoparticles possess a density about 3.9 nm and its dispersion inside the water was highly stable. Based on conceptual forecasts with actual discovery, fluoride attached at surface remarkably disrupts the transit of ions that are electrolyte with compromises the sites which are electrochemically active that leads to inefficiency of MXenes for utilize supercapacitors and Lithium ion applications. It is therefore exceedingly desired to produce MXenes utilizing without fluoride techniques [138, 139].

2.2 *Bottom-Up Techniques*

The bottom-up fabrication techniques utilize molecules as antecedents as opposed toward the top-down techniques which use heavy material as their starting point. MXene quantum dots could be created from microscopic antecedents of inorganic and organic substances using bottom-up methods as well. More atomic utilization, morphological and structural controls, and quicker functionalization are only a few of the benefits of bottom-up methods, which result in superior architectures for quantum dots [140–142]. A great framework for bottom-up MXene synthesis is provided by the previous research. Yet, simple, highly effective precursors with less toxicity, mild reactions, outstanding crystalline nature and good yields need to be addressed for huge fabrication. Each bottom-up strategies would probably being utilized to get MXene prepared in later for meet the increasing advantages needs because to their more easy system parameters as opposed to top-down approaches. The work is encouraging because there has been limited research on the bottom-up fabrication techniques for quantum dots of MXene production and emphasis shall been placed heavily on these tactics [143].

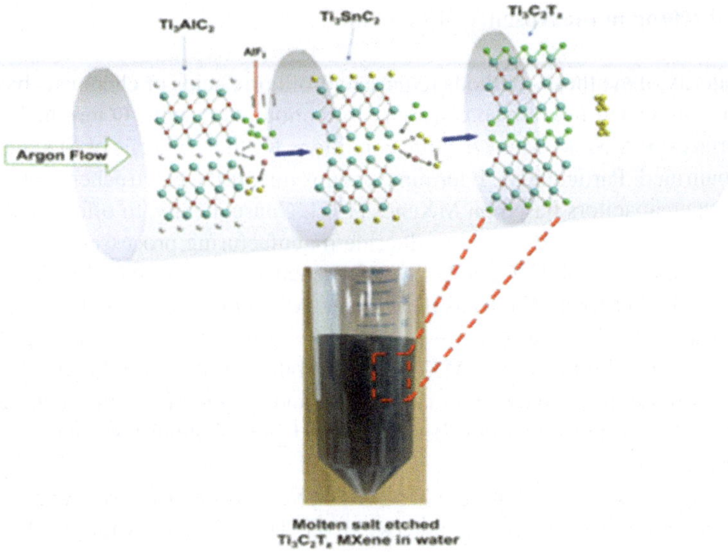

Fig. 5 Shows illustrative for molten salt process for making MXene carbide nanosheets including Molten salt etched Ti_3C_2Tx MXene in water [15]. Reproduced with permission, © Elsevier [296]

2.2.1 Fabrication by Molten-Salt Route

The molten-salt technique had recently become more popular because to its simplicity and speed of reaction. Nanocomposites were created by using molybdenum acetylacetonate as a precursor [144]. They were created a composite made of Mo_2C quantum dots with nanosheets of carbon. Sodium chloride, molybdenum acetyl acetonate and a solution of sucrose was calcined for a period of two hours at 800 degrees Celsius under air atmosphere. The significance of the approach was illustrated by the SEM scan's confirmation of the nanosheet structure. The Mo_2C with carbon nanosheets' homogeneous and ultrathin structural shape was further supported by the images of TEM. The diameter of particle 23 nm deposited upon the layers of carbon with a value of d-spacing was 2.37, which corresponded to Mo_2C with 002 planes, as shown by the HR-TEM picture. The examination by AFM disclosed that the nanosheet was 3.5 nm thick (Fig. 5).

2.2.2 Pyrolysis Technique

As pyrolysis method is a quick, simple and environmentally friendly process, it has become a popular bottom-up approach for MXene. The composites were made by using a straightforward pyrolysis method. In conclusion, the composites were solvothermally synthesized, exposed to high temperatures at 700 °c for 2 h under an Ar atmosphere and after that etched out with acid [137]. This technique is one of the effective, easy and gives high yield with increased crystalline nature and no special conditions needed for the reactions among all top-down techniques. Even so, not much study has been done on the development of MXene. To meet the demands of progressive applications, these strategies will soon be given more consideration. Ti-based MXene was therefore able to function better like an electrocatalytic material owing to its high metallic conductivity [135, 145].

2.3 *Top-Down Synthesis Techniques*

The synthesis technique is top-down approach had a long history of success, particularly development of nanoparticles. These techniques frequently entail the fragmentation of substantial two or three dimensional precursors into vitally small quantum particles [135, 137, 145–148]. Using the top-down method, 3D and 2D precursors became auspiciously transformed in to quantum dots. Hardly any of these mentioned are liquid exfoliating, ball milling, etching, intercalation, hydrothermal, electrochemical, microwave irradiation, and ultrasonication. Many of these top-down techniques create an initial oxygen accommodating functional groups upon catalytic surfaces, which makes it easier for defects to form in the catalysts [149–151]. The large molecules can divide into small quantum particles by reacting at the surface flaws, which act as active sites [152]. Because it may be used at relatively low temperature, this approach is very important. Also, this method allows for the use of a significant quantity of materials and wide-scale fabrication. However, there are downsides, including the lesser amount of yield and essential for determined treatments. The specific strategy of top-down techniques for the fabrication of MXenes is covered in this section.

2.3.1 Ball-Milling Technique

Ball-milling technique is widely utilized top-down technique for lowering nanoparticles size, and it has been used to successfully produce several quantum dots. Many variables, including the kind of wet/dry milling, milling speed, proportion of ball to powdered mass, and milling time, influence the structural and the physical features of the nanomaterials that are formed. The delamination process of MXene and the decrease in particle size are caused due to the development of the functionalities that are bonded together with transition metals precursor during this process, which is controlled by the strength of the functional groups connection. The particle size will be lower the stronger the functionalities interaction. The product contained a solid-state component because of connection of functional groups. This is the perfect technique for creating nanocomposites [153].

2.3.2 Solvothermal Method

MXene is frequently produced via the solvothermal technique because it is easy and more effective than the hydrothermal approach. With the sole exception that the precursor solution is created using non-aqueous organic solvent rather than water, it is very similar to the hydrothermal method. The solvothermal technique is thought to be higher efficient and flexible in comparison with the hydrothermal route because the distribution, size as well as material crystalline structure may be precisely controlled as needed [154]. Also, the desired outcome can be perfect by modifying experimental factors including the temperature, time, and type of solvent during the reaction [155, 156]. The boiling point, the redox reactions, and the polarity of the solvent all had an impact on the width and lateral diameters of the particles. The production of carbon quantum dots from precursor molecules is a drawback of this method. The photonic and catalytic properties are impacted by the unwanted Carbon dots of products [157].

2.3.3 Ultrasonic Method

MXene quantum dots are created using the ultrasonic process, which is free of risk and beneficial to the environment. Stacked and nonstacked nanomaterials are transformed into quantum dots during this process because of the solvent properties. Using the solvent, we used etching exfoliating and mechanical force liquid exfoliation techniques to directly produce MXene by MAX phase. After ultrasonication, the bulk MAX phase particles disintegrate into more manageable fragments with fresh edges and surface locations. These sites were eliminated as aluminum hydroxide ions through solvent treatment of the hydroxyl group. Separate MXene sheets were developing in the meantime [158]. The organic solvents having higher boiling temperatures need significant after-treatment prior to be utilized in the liquid medium by the ultrasonic treatment process. The photocatalytic features of the quantum dots are impacted by extremely high-temperature rotational evaporation as after treatment. Techniques for manufacturing MXene inside the aquatic setting consequently attracted more interest. The prolonged sonication process and limited output result make frequent usage difficult. As a consequence, it is very effective to accelerate the reaction and boost yield by mixing the ultrasonication approach with solvothermal and hydrothermal ones [159, 160].

2.3.4 Hydrothermal Process

A hydrostatic pressure autoclave holding precursors is used in this reaction technique to heat solutions over the water's boiling point. The pH of solution, temperature range, and duration all have a significant impact on how much MXene is produced. Standard reaction temperature ranges of 100–180 °C and a pH range of 6 to 9 are used to create MXene, because temperature and pH have an impact on how long reactions last. The size, characteristics, and depth of a material can also be changed by modifying the hydrothermal reaction's parameters [143, 161–163]. By using various hydrothermal processes, many MXenes have now been created. MXene doped with heteroatoms can also be created by adapting different hydrothermal procedures and using the elements' ensuing precursors. It is critical to notice the etching by hydrothermal process not even only saves the usage of hydrofluoric acid but simply also serves as an efficient technique for the fabrication of nanoflakes [164–167].

2.4 Problems in MXene Synthesis

The synthesis processes have an effect on the purity of MXenes. The MAX phase, which serves as the precursor material in a perfect synthesis method, controls the MXenes product performance. Thus, it is crucial to select a high-quality MAX phase. Also, it's crucial to synthesize the MAX phase with hardly any mechanical milling-related contaminants. However, during the process etching and delamination, the impurities can come off as residue. Also, if they are soluble in acid then they can be easily removed by rinsing. The MAX phase is subjected to harsh temperatures undergoing MXenes etching in order to dissolve and remove the surrounding aluminum layers. The most important task in early procedures, which used harmful HF, was related to the scientist's health. Subsequently, the mild process was created using ideal circumstances as well as less hazardous etching agents. The technique was until reliant on utilizing hydrofluoric acid made from a solution of LiF with Hcl, though. Later innovation makes it possible to create MXenes in a safe and risk-free manner by utilizing ionic liquids or molten salts. However, because the chemicals are expensive, their post processing is quite constrained.

It is critical to control the oxidation of MXenes as well as the processing duration and MAX phase characteristics. The MAX phase's grain sizes are therefore its more significant characteristic. In order to completely eliminate Al, for instance, MAX phase having large size of grains creates MXenes without etched that settle down inside the solution. This 2nd etching pattern is thus necessary. MAX phase having small size of grains, on the other hand, etches very readily but requires a lengthy milling process, producing extra contaminants that occur through the vessel and ball milling. Moreover, the MAX phase having tiny particle proportion generates MXene nanoflakes. Hence, MAX phase having a tiny grain dimension yields MXenes of poor quality.

The delamination technique is also subject to similar consideration. For instance, MXenes with high grain sizes take more time to delaminate and will not delaminate completely. Furthermore, the delaminated flakes would have a big dimension if they weren't sonicated to decrease their size. From this angle, it is difficult to change the material's specifications because different production techniques yield flakes with different properties. Many factors, such as the conductivity, surface-zeta potential, XRD and Raman investigations, must be properly carried in to interpretation in sequence to characterize MXenes as-prepared. Like, different MXene functionalities would produce comparable XRD and Raman spectra because these methods are accurate to the bulk's physical makeup and composition. MXenes nanosheets, however, could have different physical properties. As a result, using complementary and cutting-edge materials characterizations, it is difficult to control and monitor modifications in chemistry as well as size of nanoflakes (surface functionalities).

Another critical stage is the removal of Al etching side products, which disappears when the pH of the solution approaches 6 or 7. A significant amount of the adsorbed organic TMA^+ ions and inorganic Li^+ or Cl^- ions may be challenging to washout either the amount of nanoflake's dispersal is considerable, necessitating multiple washing, filtration, or centrifugation procedures for reaching appropriate pH. Varieties and stoichiometry of MXenes have an impact on how they are synthesized. For instance, it is simple to etch as well as delamination of titanium-based MXene by the solution of LiF with Hcl. The Nb_2CTx MXene, on the other hand, firstly etched out with HF and then delamination is done by TBAOH. This is due to the fact that the LiF/Hcl approach is less efficient for this type of MXene. As a result, Ti3C2Tx manufacturing might be scaled up. Because of their strong oxidation ability and poor stability, certain forms of MXenes, are challenging to etch and delaminate into single flakes. As a result, during these phases, the MXenes produced would have many layers. Due to their oxidative stability, MXenes have a challenge maintaining their structural integrity. Due to the presence of "A" layers, MAX phases have great stability, and removing this layer by etching can significantly reduce the stability of multi-layered structures. The several layers, however, continue to show significant stability to withstand sample preparation for various data analysis and industrial uses. Monolayer or few layered MXene structures are also delaminated, which further destabilizes the structures. But after many hours, these layers oxidize in the presence of oxygen and further degrade with 7 in a few months. The materials need to be stored at a low temperature in a dark, O2-free environment.

All these problems can also effect the material characterization. Like, due to fast oxidation after synthesis the material needs to be directly characterized. Most of the characterization techniques work under surrounding atmosphere. This variation can be going after the Raman and XRD analysis. Therefore, to prevent the MXenes from oxidation, the suitable laser intensity should be selected. About surface phenomenon, the modifications at surface could be analyzed by the XPS technique. By evaluating the sample directly after synthesis, the obtained zeta-potential value is very negative. The XPS analysis after synthesis would examine the without metal oxide fluorine and hydroxide groups with metallic bond. During sample oxidation, the zeta-potential will be near to zero and the metal oxides will be verified by the XPS analysis [168–171].

3 Characterization Methods for MXene-Based Nanomaterials

Various analysis techniques, like X-ray diffraction, scanning electron microscopy, Raman spectroscopy, and transmission electron microscopy are linked to evaluate the geometry and thickness of nanoflakes [172]. Any specimen must be precisely characterized in sequence to verify the status of production of nanomaterials. As many MAX phases can participate in the majority of investigation specimens, data interpretation for the MXene assessment requires establishing the purity of its precursor ingredient MAX phase. It is practically hard to separate and study the possibility of both Ti_3AlC_2 and Ti2AlC phases coexists in the powders of Ti–Al-C precursors. Hence, in order to corroborate the changes in the generated products, characterization of starting materials is required first. As per [173], using XRD after giving the powder a significant amount of texture is the best method for reliably affirming the purity of MAX phases. All other peak locations diminish or disappear when MAX phase completely converts in to MXene, with an exception of the (0001) peak in the XRD study. In addition to the (0001) peak widening, other signs of a wide parameter of c lattice include a decreasing order toward the bottom angles with an expansion of layers in d-spacing. The primary peak for the initial MXene produced by [173] (Ti_3C_2Tx) had drifted by 40–10 upon the XRD arrangement. Yet, visual evaluation is crucial in MXene identification since it is sometimes overlooked that the color of MAX phases changes when they become MXenes. In contrast to MAX phases, all of which are grey in color, MXenes exhibit various color due to their optical features based on their composition and structure. Furthermore, XRD analysis is frequently used to establish the extent to which an etching was completed. Examine at how, after lithium fluoride and hydrochloric acid etching, the (002) crest of Ti_3C_2 converts to superficial angles whereas 39° peak of diffraction (2°) in Ti_3AlC_2 loses its significance. The increased interlayer distance is assigned to replacement of various functional groups like -O, hydroxyl, or -F with Aluminum (Al), which then mixes with water and ions to cause the displacement reaction. The SEM and TEM are also used to measure the transverse size of nanoflakes. Moreover, SEM is regarded as the best way if one needs a visual demonstration of the precursor and a clear proof that MXene is produced. The accordingly structure and morphology indicated by [174], may be seen in ccanning electron microscopy pictures, which were once assumed to really be evidence that MXene manufacturing was effective. Nonetheless, more investigation has shown that the accordion architecture of stacked MXenes varies. The rate of structural evolution is determined by the etchant's strength. It underlined the importance of X-ray Diffractometer and SEM in identifying efficient MXene manufacture. Both tactics are common, but their statistical and graphical data reveal more than what has been before understood. By generating quality images that have been enhanced by up to 5 levels and accurate statistics that are simple to understand, these strategies significantly and probably more enormously contribute to successful MXene analysis. The synthesis and characterization of MXene, Multilayer MXenes are created when stacked ternary carbide of MAX phase powder is M_3AC_2 in this case, dissolved in an aqueous acidic solution that contains Hydrofluoric acid like HF or Hcl with Lithium fluoride). The A layer, like Al is systematically etched and substituted surface terminal groups, i.e., OH, F, or O. MXenes after delamination to create single flake coatings by intercalating water, dimethyl sulphoxide, cations, tetrabutylammonium hydroxide, and other materials into the interlayer gap and then performing sonication. Illustrations of MAX frameworks like M_3AC_2. A representation of the atomic arrangement, digital image of titanium-based MAX phase powdered substance, low and high amplification scanning electron microscopy images for titanium-based MAX phase and an image of HR STEM of Mo_2TiAlC_2. [0]. Figure 7a, represents the usual morphology of a Ti3C_2Tz nanosheet. The poor contrast between the nanosheet and the backdrop in the TEM image indicates that this nanosheet is only one to two layers thick. The supernatant product's UV–Vis spectra are extremely comparable to those of typical MXene products that have been acid-etched. In Fig. 7b, they also carried out X-ray diffraction (XRD) to examine the synthesis of Ti_3C_2Tz; it displays the XRD patterns of Ti_3C_2Tz clay before and after the KOH wash, as well as the XRD of Ti3C2Tz nanosheets from the supernatant. The absence of the Ti_3AlC_2 nonbasal plane peaks in the XRD suggests that Al was successfully etched out of Ti_3AlC_2. The presence of Ti_3C_2Tz's distinctive peaks at 9.4° and 19° validates the element's production. The Ti_3C_2Tz peak seen in the molten salt etching scenario corresponds to the Ti_3C_2Tz typically etched with HF. Additionally, the presence of Sn (in small amounts) may prevent the nanosheets from aligning in close proximity to one another. The reaction mechanism describes an etching procedure that produces Sn as a byproduct, and these big peaks at 30°, 33°, 44°, and 45° are indicative of Sn (Fig. 6).

4 Various MXene Properties

MXenes have a broad and variable surface interaction as a result of quantity in electrons connected to atoms of transition metals. The MAX phase, raw materials and etching as well as delamination and etching techniques, all affect the intriguing features of MXenes, involving all the redox, electrical, mechanical, magnetic, physical, chemical, electrical, and thermal. By regulating its configuration, using various "M" and "X" components, and changing the terminal functionality, it is also possible to modify the MXenes features for the correct approaches. Computer-based theoretical researches have recently been used to interpret the characteristic of MXenes. Below are some characteristics related to energy storage [175].

4.1 Optical Features

The optical properties of MXenes have photo-thermal effect and great transparency. The important property of MXene films has a high surface plasmonic effect. In spite

Fig. 6 Shows the characterization of MXene, **a** SEM of Ti_3AlC_2 MAX phase powder. **b** Before KOH wash $Ti_3C_2T_z$. **c** After KOH wash $Ti_3C_2T_z$. **d** By drop-cast aqueous dispersion $Ti_3C_2T_z$. nanaosheets [15]. Reproduced with permission, © Elsevier

of the verity that clarity diminishes with decrease in the speed of spin coating, it was possible to attain clarity of even higher above 75% in visible and according nearly infrared bands [176]. A V2CTX transmittance research demonstrated that the manufactured film has high transparency with 550 nm and that the transmittance diminishes with increasing film thickness. An 11-nm thick film with a greater transmittance of 89% was achieved. This experiment shows that neither the transmittance nor the absorption coefficient is affected by the heat treatment temperature of the MXenes film. This experiment shows that neither the transmittance nor the absorption coefficients are affected by the temperature range of the MXenes film [177]. Contrary to the conductivity results, the MXene intercalation step of the fabrication resulted in an irreversible chromatic shift. Unfortunately, little research has been done on how manufacturing affects optical properties [178].

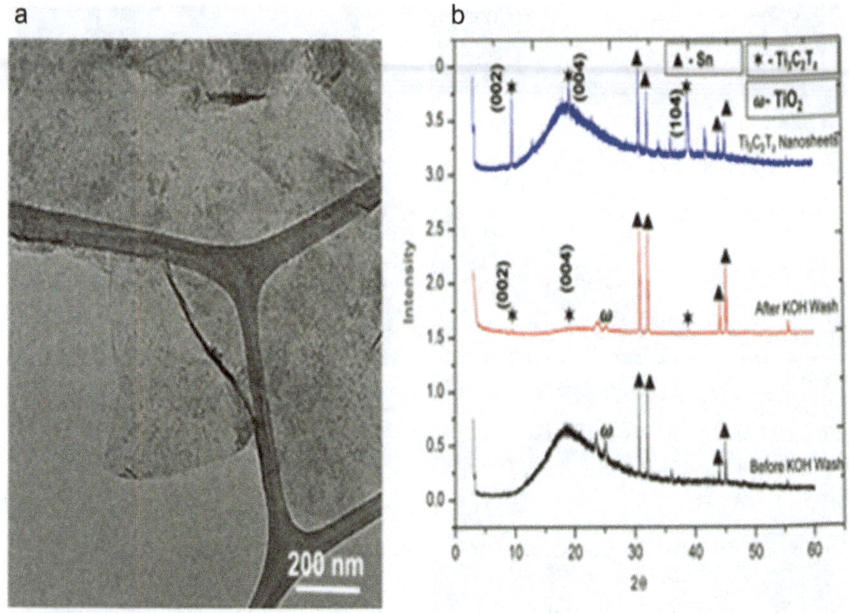

Fig. 7 **a** Shows the TEM images of $Ti_3C_2T_z$ MXene nanosheets. **b** XRD of $Ti_3C_2T_z$ before and after washing with KOH as well as $Ti_3C_2T_z$ nanosheets [15]. Reproduced with permission, © Elsevier

4.2 Structural Features

Because of their uncommon conductivity and superabundant functional groups on their surface, MXenes will work for both of the carriers of intrinsic active or other different types of functional materials for numerous advantages involving energy storage [179]. On the other side, MXenes have limitations because they are very thin 2D-nanomaterials, specifically remarkable capacity to restacking and absence of significant porosity in their structure. Since, the last few years a lot of efforts have been done in the modification of porous structure of MXenes. Still, a number of porous structure of MXene will be sustainable topology and came to be manufactured by various synthetic techniques and utilized in various applications, which disclosed all good performances of modified MXenes. Because of their 2D structures, the porous MXene have also been used for numerous applications, stimulated by cause of their porous structure with varying physico-chemical features. The below table define briefly the functions of pores in their porous structure in the related applications [180, 181] (Table 1).

Table 1 MXenes and their porous structures [181]

Materials	Structures	Synthesis method
Aerogel of rGO with MXenes	Mesoporosity and Macroporosity in aerogel structures	Crosslinking with chemicals and Calcination with freeze dry method
MXene nanoflakes	Mesoporosity in MXene nanoflakes	Oxidative Etching method
3D-films of MXene	Mesoporosity and Macroporosity in 3D-films	Hard template fabrication method
MXene lamellar-liquid–crystal	Vertically aligned mesoporosity and macroporosity	Freeze drying with mechanical shearing assistance fabrication method
MXene-Sponge composite 3D porosity system	Macroporosity in 3D porosity system	Dip-coating method with drying
Aerogels of Super-elastic MXene with PI	Mesoporosity and macro porosity with wide range of size distribution	Freeze drying fabrication method
Aerogel of Ti3C2Tx	Mesoporosity and macro porosity	Freeze drying fabrication method
Lamellar structured Ti3C2Tx and SiCnws foam	Mesoporosity and macro porosity aligned parallel in their structure	Freeze-drying bidirectional fabrication method

4.3 Mechanical Features

The mechanical features of MXenes may change considerably on the basis of surface terminal functionalities. It was explored that the terminal surface groups make weak titanium with carbon bond by withdrawing the charge [182]. Scientists discovered that the Titanium and carbon bond is smaller in Ti3C2Tx as compared to Ti2CTx, which further influence the MXene elastic properties. The researchers also suggest that improve the flexibility by improving its binding strength. For various applications, MXene with Oxygen terminal functionalities should be the first preference for structure nanomaterials due to their great mechanical features [183].

MXene has one unique feature that they have great mechanical strength. In spite of the outstanding modification in the fabrication of MXene nanomaterials, titanium-based MXene has been highly investigated for a number of applications because of their unique features like mechanical capacity, conductivity, and great shielding effect [184]. In various investigations MXene disclosed great mechanical ion capacity of adsorption; covering the way for many investigations in to their feasible utilization in flexible electronics and gas sensors [185]. In the present generation, the electronic appliances growing very fast in our society which causes huge electromagnetic pollution resulting in human health at risk. Because of this hazardous effect, the shielding material with EMI achieves great focus which is lightweight and has unique mechanical strength and a great shielding efficiency. MXene attains great performance in the shielding field with EMI by cause of its unique area at surface, outstanding conductivity with hydrophilicity in nature. In one research, fabrication of hybrid foam of MXene with Carbon from sol–gel method by mixing resorcinol and formaldehyde solution through heat treatment reduction and freeze drying synthesis method. The curing and vacuum accommodated saturation technique was used for the synthesis of MXene with Carbon foam and EMI shielded with epoxy nanocomposites. All types of applications of these formed composites are deeply studied like mechanical strength and conductivity of the nanocomposites. The structural characterization of the fabricated was also studied through characteristic analysis such as SEM, XPS, and Raman [186].

4.4 Electrical and Thermal Features

Ti_3AlC_2 has found with very great thermal oxidation and thermal conductivity. The monolayer of Ti_3C_2Ox has approx 11W/mk thermal conductivity whereas in Ti_3C_2Fx has 108W/mk thermal conductivity [187]. In contrast to previous studies, it is discovered to be roughly 2.84 W/mk for Ti_3C_2Tx films. Ti_3C_2Tx sheets have a 2400 S cm^1 metallic conductivity and are extremely versatile. A particular type of MXene called V2C has been examined, and it was discovered that it possesses conductivity of 3300 S cm^1. Due to its adaptability, it is ideal for uses like electronic gadgets that are worn [188]. The layer thickness and temperature of calcination exhibit a relevant affect upon the conductivities of MXene nanosheets. According to researchers, after heating, the conductivity improved and the resistance of MXene sheets for films with greater thickness was lower compared to thinner films [176].

5 Applications of MXene

MXenes have drawn frequent attention of investigators by cause of its distinctive features, which include outstanding electrical conductivity, very high volumetric-electrochemical capacitance, adjustable plasmonic characteristics as well as their work function, transparency in light with electrochromic inside MXene thin films, excellent thermal resistance, strong mechanical features with the ability to develop extremely stable solutions of colloidal particles in various kinds of solvents that are polar, particularly alcohols or water [126, 189–191]. MXenes have been employed for a number of purposes because of its appealing characteristics, including storage of energy, electronics, components of structure, and cleaning the environment. The storage of energy was the very first and most diverse use of MXenes, but it has now been expanded to encompass supercapacitors and all kinds of batteries [192–194]. MXenes were discovered to be equally efficient as metals for communicating and EMI (electromagnetic interference) screens even in apparent 40–50 nm layers, outperforming graphene or C-based nanocomposite materials. MXenes are used for

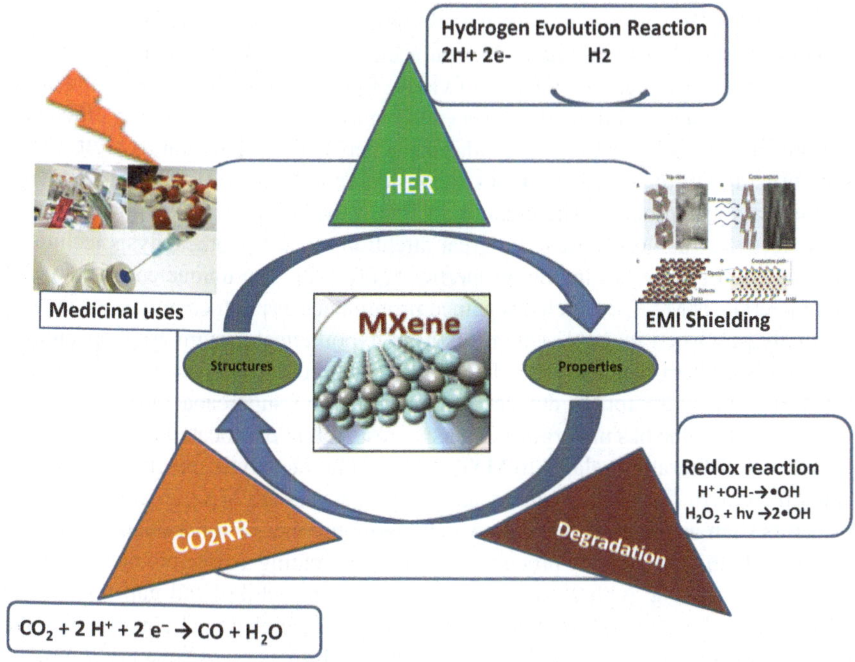

Fig. 8 Shows the different applications of MXene

ecological treatment in a variety of ways, including the elimination of metals that are heavy, desalinization, and the adsorption of pollutants. Electronics, catalysis, and medical use such as photodynamic cancer therapy, dialysate regrowth, implantation of electrodes, and intraocular lens implants are further possible uses for MXenes. MXene sheets replace a Gold or Platinum metallic electrode in healthcare technological advances, which not only improves functionality but also lowers costs. The sector should concentrate on the usage of MXenes in terms of their applications [195–200]. Applications of MXene are summarized in Fig. 8.

5.1 Mechanism for Removal of Pollutants

5.1.1 Photocatalysis

Photocatalysis is a type of catalytic method utilized to eliminate multiple contaminants from the ecosystem and transforms solar energy in to fuel that is chemical in nature. The procedure is divided into several steps and the formation of charge carriers, the separation as well as the transformation of charge particles produced by sunlight toward the surface of the photocatalyst and the activation of redox processes by photoinduced electrons as well as holes are all examples of the generation of charge carriers [201–210]. Figure 8, describes the photocatalytic degradation mechanism of pharmaceutical drug pollutants like paracetamol and ciprofloxacin by advanced oxidation process under the radiation of visible light. Because the fast recombination of charge particles significantly lowers the catalyst's photoactivity and efficiency of transforming solar energy to exothermic energy, the separation as well as the transformation of the charge particles is considered to be the rate-determining stage in photocatalysis process. The creation of innovative 2D materials with distinctive photoelectric structures attracted a great attention toward photocatalysis technologies because of these fascinating properties [211, 212]. The unique category of 2D nanomaterials is MXene which has gained prominence as photocatalyst, since it was originally published in 2011. Due to its high hypothetical gravimetric capacitance as well as excellent conductivity, MXene was firstly researched for the solar energy transformation with storage devices like batteries and supercapacitors [213–215]. From 2014, MXene has undergone extensive research in photocatalytic process, and the amount of literature related to MXene-based photocatalytic material has significantly increased. The efficient photocatalytic process uses of MXenes are mainly due to factors such as the ones that follow: The large number of functionalities created from wet chemically etching process is helpful for creating a close contacted boundaries among MXene with associated semiconductors, while band gap position of MXene could be changed through adjusting the surface chemical composition and the metallic conductivity with electron-accepting abilities of MXene are improved by the extremely conducting metal centre of its multiple-layer framework [216]. As a consequence, MXene has undergone significant research in photocatalysis process for a variety of applications, including conversion of carbon dioxide, water splitting process, oxidation of pollutants, and Nitrogen fixation. It is viewed like feasible alternate toward other 2D nanomaterials [211]. By enhancing the separation of charge carrier and transfer, serving as powerful support, limiting the size of the photocatalyst, and encouraging the adsorption of reactants, MXene could enhance photocatalytic performance in various applications. In order to develop meaningful photocatalytic impacts, MXene, a new class of layered materials, needs to have its many distinct photochemical properties and emerging factors thoroughly researched and examined. These benefits of MXene-based composites may increase photocatalysis for use in practical uses. As a consequence, a detailed analysis of MXene may offer insightful information regarding the creation of hetero-structured MXene-based materials. In this understanding, the fundamental function of MXene in stimulating photocatalytic activity is analyzed, and new developments in the synthesis of heterostructured photocatalysts which are MXene-based are described. Because of the prominent "face-to-face" interacting connection and distinct area of interface being exposed, an efficient mutually beneficial interface might enhance photocatalytic reaction steps and significantly impact the efficiency of heterostructured 2D photocatalysts which are MXene-based. As matter of the fact, a variety of surface groups can be obtained from the MXene structure to form hetero junctions of 2D materials in addition to two-dimensional photocatalytic materials. A heterojunction of 2D MXene-based with near "face to face" interaction has a lot of potential to improve photo-induced separation of charge carrier and transferring through the heterojunction interaction.

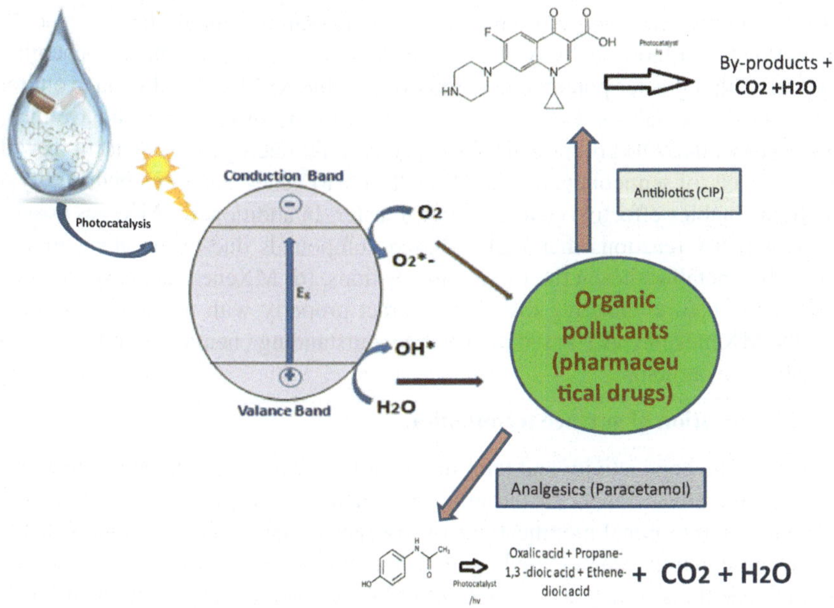

Fig. 9 Photocatalytic degradation mechanism of pharmaceutical drug pollutants like Paracetamol and ciprofloxacin

Due to MXene's large light current and exceptional electronic conductivity, photo-generated electrons produced by the 2D heterogeneous structures photocatalysts could accumulate upon the surface through transfer of electrons. In order to facilitate the quick electrons extraction via 2D materials for improved photocatalytic actions the 2D structured framework of MXene can function as a mediator for electrons [217, 218] (Fig. 9).

5.1.2 Electrocatalysis

(a) MXene electrocatalysts for hydrogen evolution reaction

The advancement in hydrogen energy offers a practical explanation to the present ecological and energy issues. Designing reaction for hydrogen evolution catalysts that have superior stability, outstanding conductivity, and selectivity is essential for the suggested hydrogen system [219]. Highly efficient electrocatalysts can lower the needed excessive potential for hydrogen evolution reaction, hence improving efficiencies [220]. Density functional theory, also known as DFT, can be used to compute the amount of hydrogen adsorbed Gibbs free energy, which is thought of as the modulator of hydrogen evolution reaction function as an initial estimation. When the amount of adsorption Gibbs free energy is quite near temperature neutral, the most effective level of hydrogen evolution reaction is gained. When hydrogen adsorbed Gibbs free energy is quite close to temperature neutral, the best reaction of hydrogen evolution activity is attained. MXenes offer good physico-chemical features with a greater potential in comparison to the NPM and hydrogen evolution reaction electrocatalysts. Examples include (a) MXene surfaces, which have many oxygen as well as OH groups and have capacity of producing solid connections with other surfaces of semiconductor. Additionally, (b) MXenes' superior conductivity in electrical enables effective charge-carrier transfer. In addition, (c) MXenes possess stronger redox reactions than carbon-based compounds due to the abundance of accessible metallic sites at their terminal positions. (d) MXenes also have excellent hydrophilic nature, ensuring that they interact properly with molecules of water. Finally, MXenes in water-based media show outstanding chemical stability as well as stability in structure.

(b) **Modification of surface termination**

Improving the conductivity and termination at the surface of MXene is considered to aid in the advancement of the efficiency of hydrogen evolution reaction. It is prominent that terminal modifications of MXenes optimize the electronic structure which consequently promote hydrogen evolution reaction activities. Thus, substantial theoretical and practical research into MXene terminal modifications has been done recently [221]. The experimental results were supported by surface diagrams of Pourbaix, which showed that this form of MXenes has great surface chemical stability, leading to noticeably increased exchanges in current and improved evolution of hydrogen. Their surface termination with oxygen atoms appears to behave as hydrogen evolution reaction active sites, according to the hydrogen adsorbed Gibbs free energy, which was discovered. These oxygen atoms on the surface helped two-dimensional MXenes and hydrogen make contact. Further investigation indicated that the hydrogen evolution reaction mechanism across these MXenes employed the Heyrovsky process. To investigate their activities of hydrogen evolution reaction, MXenes with various terminal functions like oxygen, hydroxyl, and fluoride groups were utilized. MXene with oxygen terminated surface showed outstanding results compared to MXenes with hydroxyl and fluoride group surface termination [222]. There are a numerous other researches on surface termination modification. On the basis of ultrathin MXene (Ti3C2), having oxygen functionalization enhances the catalyst performance of hydrogen evolution reaction [223]. The MXene with fluoride functionalization on the surface plane was hazardous to reaction of hydrogen evolution which decreases the rate of adsorption of hydrogen. By the reaction of aqueous electrolyte KOH solution with MXene (Ti_3C_2Tx), the Ti_3C_2Ox was formed which shows the decreasing of fluoride termination by the hydroxyl groups. The annealing of resulted Ti_3C_2 (OH)x at temperature of 450 °C under argon atmosphere due to which the hydroxyl groups converts in to oxygen termination through the reaction of dehydration. The resultant MXene with oxygen termination formed the outstanding electrocatalysts for the reaction of hydrogen evolution [224, 225]. This investigation stimulates the framework for changing the surface termination of MXene-based electrocatalytic material to enhance its performance. Briefly, functionalization upon the surface level controls the efficacy of reaction of hydrogen evolution. So, the higher the fluoride termination on surface plane becomes lower the activities of hydrogen evolution reaction for MXenes [226]. For greater efficiency of electrocatalysts for reaction of hydrogen evolution the oxygen termination on surface plane is suitable. The MXenes with other surface functionalities were also fabricated but their electrocatalytic activities for reaction of hydrogen evolution are not yet studied [53, 227, 228].

5.2 *Role of MXenes in Chemisorption of Pollutants*

5.2.1 Chemisorption of Dyes

Dyes are key constituents in textiles, printing, and paper sectors, although their economic usage creates a major risk to the ecology and produces water pollution. As a result, the elimination of the dyes is critical and has essential need for treatment of the environment. Various methods are investigated to date for eliminating the dyes but the adsorption holds the most promise by cause of its low price with preparative application. MXenes need to be recognized as appropriate nanomaterials for the adsorption that remove dyes by cause of its surface having large area, negative charges on the surface, multilayered structure, and high hydrophilic properties. The effective cationic dye adsorption by layered MXene (ML-Ti_3C_2Tx), was originally reported for methylene blue [229]. The report discussed the adsorptive characteristics of MXene regarding anionic and cationic dyes. The results showed that the cationic dyes, i.e., methylene blue showed irreversible bonds with MXene (Ti_3C2Tx), while the anionic dye, i.e., acid blue doesn't have capacity of adsorption with respect to MXene. Electrostatic forces among the ions having positive charges on dye with the negative charge on MXene surfaces was thought to be the cause of the dye which are cations selected adsorption than the dyes which are anions. Later research looked at how dyes adsorb to MXenes which are treated with alkaline solution [230]. The authors used a simple method to treat MXene in a hot solution of alkaline in order to increase the interlayer gap and adjust the surface functionalities. MXene's interlayer gap was increased by 29%, and all of the fluoride functional groups on surface were changed to hydroxyl groups. The use of alkaline solutions improved the methylene blue dye's capacity to absorb and speed up its rate of removal. MXene with LiOH and NaOH among the treated MXenes showed the quickest methylene blue dye absorption, Ti_3C_2Tx with NaOH having the maximum adsorption capacity. With Langmuir as the accepted adsorption model, the appropriate capacity of adsorption was attributed toward the cooperation for functionalization of surface and expanded inter-layer adsorption. In a different research, novel technique for solvothermal treatment of hydrochloric acid with sodium tetrafluoroborate to produce two-dimensional MXenes was developed, including titanium- and niobium-based MXenes [165]. Here, the reaction of hydrochloric acid with NaBF4 produced HF in-situ, allowing for etching of precursors to MAX phase, Nb_2AlC and Ti_3AlC_2. MXenes that were created have a greater specific area on surface than those that were created using the conventional hydrofluoric acid etching technique. While the Nb_2C demonstrated comparatively poorer adsorption for MB and negligible adsorption ability with methylene orange, Ti_3C_2 attained an excellent MB adsorption capacity. It may account for the certainty that Niobium has greater atomic mass compared to titanium. After treating the unaltered MXene with the phytic acid inside hydrothermal environment, the 2D MXene was changed into a rod-like structure [231, 232]. Based on its surface interactions, phytic acid was combined with the MXene in this instance to boost its general amphiphilicity, and the hydrothermal reaction period determined the rod-like morphological structure that characterized the MXene. The composite of MXene with phytic acid demonstrated improved adsorption characteristics for Rhodamine B and methylene blue dyes, with capacities for adsorption of 22 and 42 mg/g, respectively, and with retention capacities of eighty-four and eighty-five percent following constant twelve rounds of adsorption. A growing fascination has been shown in the implementation of MXenes in the removal of colored dyes. There still is more to learn, especially in the area of selective pigment removal using MXenes with designed surface charges.

5.2.2 Chemical Adsorption of Radioactive Ions

After the radioactive substances are used in nuclear-related industries like mines or research in medicine, the radioactive ions are created [233]. Effective nuclear waste removal and management of the environment are required since the leaking of radioactive ions to the nearby groundwater and soil becomes a severe threat. Here, MXenes are shown to be somewhat favorable for absorption of hazardous radioactive metallic ions by cause of its high reduction–oxidation capability on surface with customizable adsorption capabilities. Experiments demonstrated that multilayer V_2CTx MXene was capable to effectively adsorb uranium ions from water-based solutions. The uranium ions absorption capability of 174 mg per g, fast sorption speed, and desired specificity of V_2CTx demonstrated its exceptional adsorption performance. The relevance of functionalization on the surface was highlighted by the DFT computation, which revealed uranium charge particles chose to interact with hydroxyl groups attached toward the vanadium sites inside nanosheets of MXene through bidentate type inner-field compounds. To remove uranium ions from water-based solutions quickly, a variety of aqueous as well as MXene were produced. With a capacity for adsorption of 214 mg/g, the well-hydrated MXene outperformed its dried version in the removal of uranium ions attributed mostly to its greater flexibility and interlayer space. According to this study, Ti_3C_2Tx is a potential choice for the collection and encapsulation of uranium ions. MXenes needs also demonstrated potential adsorption capabilities for the other heavy metallic ions found in the nuclear wastage, including rhenium, thorium, and europium, in addition to uranium [234–238]. The effectiveness of MXene's adsorption on rhenium was assessed. For the improved removal of the perrhenate ions, MXene 3D polyelectrolyte composite was used in the study. MXene with Ti_2CTx was given PDDA, which controlled the charges on the surface and increased stability. The resulted nanocomposites attained the excellent removal efficiency for Rhenium with fast adsorption rate and great selectivity for anions like chloride and sulphate ions, which are present with higher concentration than rhenium. When the MXene was tested for its ability to remove thorium, it was found that its hydrated equivalent had a better capacity for adsorption than the dry version [239]. With the maximal capacity of sorption is 213.2 mg/g, state of sorption equilibria was attained in 720 min. The hydrated MXene demonstrated remarkable selectivity against a variety of competitive charge particles and sorption procedure was strongly dependent on pH having no bearing by the total ionic strength. It was discovered that the sorption mechanism of thorium was inner-sphere interaction resulting from the titanium with hydroxide high efficiency toward thorium. In a different study, the MXene nanocrystals were chemically converted in situ to produce hierarchical nanostructures of titanate [240]. Due to the plentiful convertible the guest cation molecules maintained stratified framework, the produced nanostructures of titanate had a high degree of stability and an adsorption ability of 200 mg/g for europium. The reduction in europium and oxygen gap as well as coordination number is proof that the sequester of europium was attained by the production of inner-sphere complexes at the surface in nano-structured environment. New information on the interactions between radioactive substances with titanates was gained through the discovery of the inner-sphere association caused by the confining as well as coordination effect between titanium with O and titanium with OH.

5.2.3 Chemisorption of Toxic Ions

Because of their persistent and non-degradable character, ions of heavy metals are the most hazardous contaminants in the environment. Since ions of heavy metals have durable toxicity that is active even at minimal level of concentration, they are extremely dangerous pollutants that need to be treated right away. Chemical precipitation method, adsorption method, filtration method, and electro-dialysis process are some of the currently used methods that are thought to be effective at eliminating these metal ions from the environment [241]. Unlike dyes, whose concentration in aquatic environments is quite significant, metal ions have relatively small amounts like parts per million, making their removal very difficult. The techniques derived from chemical or biological processes, still are frequently inefficient at low levels of concentration. The fabricated adsorbent needs sufficiently efficient and economical for removing hazardous metallic ions at the parts per million or parts per billion levels. MXenes and its derivatives are regarded as possible adsorption agents for cations of heavy metals due to their abundance of terminal functionalities with substantial surface areas. According to numerous findings, MXenes strongly bind to different heavy metallic ions. In a research, created a useful MXene for lead ions to be absorbed using the method of exfoliation and the intercalation techniques. In an environment of strong competitive cations that are as calcium and magnesium, the resulting MXene displayed a favorable lead sorption tendency. The absorption equilibrium was reached within 120 s, according to the kinetics results. Having the capacity of adsorption is 4500 kilogrammes water per alkali MXene, MXene was able to efficiently absorb the lead ions. The lead ion exchange processes were made easier by the MXene's activated titanium sites and hydroxyl groups, which were confirmed by experimental data and computer analysis [242]. Based upon active sites on surface, MXenes are also been thought to be efficient toward other harmful ions [117]. One of the factors that could cause eutrophication is the extreme release of phosphate ions into the water ecosystem. In this condition, magnetic nature of ferric oxide was intercalated in to MXene to fabricate a sandwich-like MXene with nanocomposite of iron oxide. For sequestration of phosphate traces, the composite showed outstanding applicability. The composites demonstrated better water treatment capabilities of 2400 and 2100 kg/g in actual and stimulated phosphate contaminated water, when contrasted with commercial adsorption agents. The fabrication of a particular shape of nano ferric oxide where tiny ferric oxide nanoparticles may show intercalation into the MXene internal layers, widening the interlayer gap and activating the overlapping of active layers was credit with the efficient sequestration. MXene was utilized to chemically reduce and remove bromate ions, which are harmful if detected in water used for drinking [243]. Its titanium and carbon active layer might change into titanium dioxide nanoparticles while converting bromate ions in to bromide. The system's pH, temperature, time of contact, and MXene concentration all had an impact on the reduction efficacy of the MXene nanosheets of material. At the pH level of 7 and twenty-five degrees celsius, this method enabled good bromate absorption in less than 50 min. The unique dyes and ions pollutants absorption capabilities of MXenes and their nanocomposites and derivatives, while the materials made from MXene appear to be effective applicants for eliminating harmful ions, there still remain significant obstacles to overcome, such as the need for quick action in the rearranging of the layers of MXene and the targeted absorption of hazardous ions from a complicated structure.

5.2.4 Membrane Separation of Pollutants

Desalination and treatment of wastewater can be accomplished using membrane separation technique. The optimum membranes for desalination processes and treatment of water require great flux, excellent selectivity, good stability, and resistance to fouling and substances with chemicals like chlorine [244]. Additionally, the membrane needs to be stable mechanically and sufficiently thin to optimize permeability to water while maintaining a steady rejection of salt rate. Presently, 2D carbide nanomaterials with ionic and molecular sieve capacities, like graphene and oxide of graphene are intriguing nanomaterials. The decreasing interlayer distance when pressure is applied results in a relatively limited flux for graphene oxide layers, despite their suitability for a broad range of uses. Here, programmable surface chemical composition and controllable interlayer space of MXenes have been shown to be very useful. The membrane for water treatment based on MXene was reported. In order to create the membrane, Titanium-based MXene nanosheets were formed into an independent membrane using a vacuum-based filtration technique [245]. In contrast to graphene or graphene oxide, Ti_3C_2Tx has strong hydrophilic properties that make interlayer water possible and encourage rapid water flux. The resulting membrane displayed strong selectiveness toward the metallic cations with positive charge such as Lithium$^+$, sodium$^+$, potassium$^+$, Mg^{2+}, Al^{3+}, Ca^{2+}, Ni^{2+}, Ca^{2+}, and Methylthioninium$^+$ dyes cations but was impermeable to cations having hydrated radius bigger than the MXene interlayer gap of MXene. Depending on the desired ions hydrating radii and charge, the suggested μm in diameter-thick layer has varying sieve capacities that can achieve a water flux. It is reported the utilization of a porous membranes of MXene that was made utilizing vacuum filtering and a template made of Iron hydroxide nanoparticles. The produced membrane of MXenes demonstrated exceptional water permeation of above 1000 L per square meters per hrs at one bar having a rate of rejection of more than ninety percent for molecules larger than two and a half nanometer. The membrane was stabilized on an anionic oxide of aluminum substrate. Comparable degradation speed membranes accomplished less well than the membrane which is based on MXene [246]. Separations of gas with liquids have demonstrated the potential of the membranes which are graphene oxide based. Coupled graphene oxide with the nanosheets made of MXene could lead to better performance. In an effort to create a ninety nanometer the thick membrane of Ti_3C_2Tx and graphene oxide nanocomposite, researchers attempted to combine graphene oxide and MXene. The lattice period of constructed membrane upon swelling was 14.28, which correlates to an interlayer distance of about 5 Å that permits for the passage of two layers of molecules of water. During the filtration by driven pressure at 5 bars, the membranes of nanocomposite successfully refused the molecules of dyes with hydration radii exceeding 5 Å and the dye molecules with positive charges. The constructed membrane was able to produce an outstanding rate of rejection which was sixty-eight percentage toward methyl red, 99.5% for methylene blue, 93.5% toward Rose Bengal, and one hundred percentage for dazzling blue by cause of its layer structure. This study opened up a modern avenue for the advancement of interlayer nanocomposite based on engineering membrane separation that is effective. In this situation, a 0.2 mega pascal MXene-coated with polyether sulfone ultra filtering membrane for desalination of dye and treatment of wastewater was produced. The constructed membrane composite had a homogeneous distribution of element and layers on the surface that were rough and thick. It's interesting to see that the integrated MXene content affected how often dyes and salts which are inorganic were rejected. Significant rejection rates to the gentian violet and Congo red dyes which are approx 93% and 81% respectively were seen under ideal conditions [247]. Because of the loose lamellar membrane composite framework, the rate of rejection of the inorganic salts was less than 23%. It has been reported that MXene has undergone treatment with silver nanoparticles in an effort to further enhance the fouling-repellent characteristics and boost the aqueous flow of membrane of MXene. Utilizing silver nitrate which is self-reduced on the material nanosheets the MXene with silver nanocomposite membranes with a range of Silver loadings 0 to 35% were created [248, 249]. The Silver and MXene membrane nanocomposite performed best when its average size of pore and thickness were around 2.1 and 470 nm. Additionally, the MXene with silver nanocomposite membrane successfully rejected organic pollutants such as albumin of bovine serum, rhodium blue, and methyl green together with the degradation rate of 100%, 79.9%, and 92.3% demonstrating its suitability for use in these and other applications. Air filtration was also studied with membrane accordance on MXene. A nanosheet of MXene developed by PAN with the membrane fiber about the air filtering was mentioned [250, 251]. Titanium-based nanosheets of MXene were added to fiber of polyaniline, increasing the antibacterial capacity and the PM 2.5 elimination effectiveness toward 99.7% while maintaining a lesser pressure conditions prevail drop with 42 pascals. The vast surface area and easily available terminations on the MXene, which enable quick and successful adsorption for particles of PM 2.5 based on their powerful interaction forces, can be attributed to the significant PM 2.5 removal rate. The membranes based on MXene have capacities for water filtration. Even if the results are encouraging, there are still certain obstacles to be solved before MXenes can fully realize their potential for air and water filtering. The most important issue is MXene's swelling, which is an obstacle for the concurrent filtration of numerous charge particles of different sizes and is present in all two-dimensional materials. Furthermore, this affects the durability and recyclability nature of MXene-based nanofilms. The effective solution to this problem is engineering of interlayer, where the MXenes having d-spacing is adjusted to modify dimensions, allowing the management of swelling and preventing collapse of the structure. This might encourage the creation of membranes which are MXene based with good selectivity, excellent capability, and great cyclability to filtering a large number of various pollutants.

5.2.5 Photocatalytic Degradation of Pollutants

A technology for environmental cleanup that shows promise is photocatalytic destruction of organic contaminants. Several semiconductors which are photoactive have so far demonstrated a promising ability toward photo-degradation of different organic contaminants, including Titanium dioxide, CdS, and g-C3N4. Such individual photocatalysts made of semiconductors are effective, but their practical use is constrained by the photon-excited charged carrier's rapid recombination. Here, MXenes can act as special building blocks for creating cutting-edge co-catalysts that will effectively reduce recombination of charge-carriers while improving the photocatalytic ability to disperse and adsorb [252–254]. It is described that the initial application of MXene for UV-induced removal of AB 80 with methylene blue dyes. For as long as twenty hours without light, MXenes were able to 18% adsorption of methylene blue and none AB 80, but during 5 h of light exposure, 62% of methylene blue and 81% degradation of AB 80 were attained, correspondingly. In this instance, the photocatalytic activity was a direct result of the titanium dioxide that was created following the partial oxidation of the surface of the titanium-based MXene material. While titanium dioxide offered photocatalytic function, Ti3C2 beneath it functioned as an absorbent platform and facilitated quick charge transmission. The composites of TiO_2 with Ti_3C_2 have since received much research for use in photocatalysis. TiO_2/Ti_3C_2 nanocomposites are already created for the photocatalytic degradation of methyl orange dye. 98% of the methyl orange in the nanocomposite was totally mineralized in 30 min, demonstrating the strong photocatalytic degradation capacity of nanocomposite [255]. Additionally, MXene with TiO_2 composite was created, for which the facets of 001 titanium dioxide were produced through the incomplete oxidation of titanium carbide (Ti_3C2) with the aid of Sodium tetrafluoroborate. In this instance, Ti3C2 served as both a counterpart to titanium dioxide with the active ingredient for creation with the 001 facets of a Schottky junction of n-type surface semiconductor [75]. During the light source, the heterojunction achieved a photodegradation speed of 18.8 per min/g for methyl orange dye by realizing an extremely low working function with a spatial dispersion of containing active charge-carrier. Later, the same research reported with 111 facets of TiO_2-x with Ti_3C_2 nanocomposite materials in which two-dimensional Ti_3C_2 nanosheets were converted into Ti^{3+} ions doped with rutile titanium dioxide octahedrons with visible active facets of 111 by oxidation which is hydrothermal, subsequently followed by reducing by the hydrate of hydrazine. The two-dimensional Ti_3C_2 functioned as a channel for transferring holes generated by photons to enhance the efficiency of charge separation, while the visible titanium dioxide with 111 facets enhanced the photocatalytic performance of nanocomposite [256]. In visible light, the hybrid produced improved dye methylene blue degradation is 75% in 150 min. Another use of the titanium dioxide with MXene nanocomposites was for the photodegradation of the drug contaminant carbamazepine [40], which attained degradation efficiency of about 99% in 4 h. Having Kapp value is 0.0304 per minute at pH level 3.0, the rates of photocatalytic procedure followed the model Langmuir–Hinshelwood of kinetic theory. A significant drawback of photocatalysts based on semiconductor that restricts their use over time in the photocatalytic degrade pathway involves recombination of charge carrier. This recombination of charge carriers can be easily reduced by building a semiconductor photocatalyst heterojunction with additional co-catalytic material in a proliferate type arrangement, while also increasing the rate of production of radicals such as free radicals like O_2 and hydroxyl groups, which are liable for the elimination of the targeted contaminant. It was suggested combining the copper oxide and Ti_3C_2 with titanium dioxide systems to create an effective heterojunction of Ti_3C_2/TiO_2 with copper oxide for copper oxide-based photocatalytic material. The composite was made by breaking down of Ti_3C_2 MXene and cupric nitrate mixture in an argon environment, and subsequently attained 99% methyl orange dye efficiency of degradation in 80 min [84]. In a comparable manner, [257] investigated the photocatalytic elimination of salicylic acids utilizing composites made of titanium carbide MXene with oxides of metal including titanium dioxide, silver oxide, and Palladium oxide as well as metals like silver, palladium, and gold nanoparticles. Interestingly, adding nanoparticles of noble metals to MXene did not much improve it beyond acting like metal sites to take electron away by the conduction band of titanium dioxide nanoparticles that had previously been inserted in MXenes to create composites of Ti_2C/TiO_2 with Ag_2O. After 180 min of exposure, the Ti_2C with 3% of TiO_2 and with 1% of Ag_2O arrangement produced the maximum salicylic acid degradation of about 96%. For the photocatalytic process, the full conversion of titanium-based MXenes into titanium dioxide with carbon nanocomposites has also been described. The method of extremely energetic ball milling of two-dimensional MXene produced titanium dioxide with carbon nanosheets. The titanium with carbon materials used in this instance was MXene, and the resulting nanoparticles of titanium dioxide were evenly dispersed on the single or multiple layers of carbide nanosheets. The titanium dioxide with carbon nanosheets demonstrated increased photocatalytic degradation of methylene blue dye by the degrading efficacy is about 86% in 360 min, in contrast to the pristine titanium dioxide [258]. To prevent recombination of charge particles and obtain significant photocatalytic efficacy, sulfides of metals are another important class of accessible and affordable minerals that have been combined with MXenes.

Other photocatalysts have been incorporated into MXenes in addition to titanium dioxide, which is produced from titanium-based MXenes, to create enhanced co-catalyst nanomaterials. A photocatalyst with little usage of solar spectrum, cerium dioxide, has received much research. In this case, a hydrothermal technique was used to combine finally dispersed cerium dioxide nanorods together with Titanium carbide nanosheets to create a cerium dioxide with titanium carbide nanocomposite. The nanocomposite demonstrated improved efficiency for the degradation by photocatalysis process of Rhodium blue dye by UV light radiation in comparison with its pristine different forms, with 75% of Rhodium blue being destroyed after 90 min [259]. Due to its availability, magnetic characteristics, resistance to corrosion, affordable price, and capacity to absorb visible light, hematite makes a desirable photocatalyst. A self-assembly method assisted with ultrasonic was used to create a hematite with titanium carbide MXene nanocomposite. In this instance, the titanium carbide MXene was attached to two-dimensional hematite nanosheets, creating a hybrid structure with strong visible absorbance capacity and excellent degradation efficacy of 98% in 120 min of methylene blue dye [250]. While the hematite-based nanomaterials are a potential class of photocatalytic material, their intrinsic restrict bandgap inhibits the amount of light they can absorb and the amount of photocatalysis they may perform. While the hematite-based nanomaterials are a potential class of photocatalytic material, their intrinsic restrict bandgap inhibits the amount of light they can absorb and the amount of photocatalysis they may perform. In this respect, $BiFeO_3$ was doped with Gd^{3+} and Sn^{4+} to form BGFSO, which was then coupled with two-dimensional MXene to increase surface area and create a begin charging-transfer system. With minimal recombination of charge carrier, the BGFO and 20Sn with MXene nanocomposites displayed complete degradation through photocatalytic activity against CR dyes in a period of 120 min [74]. Similar to this, a sol–gel technique was used to pair up a Lanthanum, Manganese-codoped with $BiFeO_3$ composite and with Ti-based MXene. By visible light radiation treatment, the manufactured composite produced a 92% degradation of Congo red dye in about 10 min. Another kind of photocatalyst recognized for its great plasticity and superior chemical stability is ferrites with spinel structure, including $CuFe_2O_4$. The spinel ferrite nanoparticles having high magnetic properties, however, cause them to aggregate quickly, which lowers the active site proportion [260]. To address this issue, a sol-hydrothermal technique was used to bind $CuFe_2O_4$ nanoparticles upon titanium carbide MXene nanosheets of material. Sulfamethazine, a pharmaceutical contaminant, was degraded photocatalytically by the $CuFe_2O_4$ with MXene photocatalytic material at a rate of 59.4% in visible light, with improved charge-carrier lifespan and photo accesibility [71]. In sequence to develop the light gathering abilities of photocatalytic material based on tungsten, MXenes were additionally employed as intriguing materials for support. The Ag_2WO_4 with Titanium carbide nanocomposite that developed permitted the heterogeneous dispersion of the Ag_2WO_4 catalyst due to the inclusion of the conductivity of Titanium carbide material. In addition to serving like a conductive agent toward lengthening the charge particles lifespan to 34.5 micro seconds, titanium carbide immediately increased the catalytic efficiency and increased the corrosive resistance of Ag_2WO_4. Ag_2WO_4 with titanium carbide nanocomposite showed significant photocatalytic properties [261]. Additionally, a hydrothermal technique was used to create a two-dimensional hybridized photocatalytic material such as Bi_2WO_6 with Nb_2CTx that showed potential photocatalytic properties toward rhodium blue, tetracycline hydrochloride and methylene blue. The 2D interface design in this instance increased photogenerated charge carrier separation which is resulted in degradation efficiencies of 99.8% for rhodium blue, 83.1% for tetracycline hydrochloride and 92.7% for methyl blue [262].

Another instance of self-assemblage was used to create an Ag_3PO_4 with titanium carbide catalyst. In comparison to Ag_3PO_4 graphene nanocomposites have significant photocatalytic rate for the breakdown of results show 2,4-dinitrophenol pollutants were 2.5 folds greater and 10 times larger than the Ag_3PO_4 pristine, correspondingly. In this instance, the close interaction between the hybrids was credited with the higher photocatalytic efficiency, and the integrated Schottky barriers were in position for controlling recombination of charge carriers [263]. The carbon nitride which is graphitic in nature, i.e., g-C_3N_4 is a popular free of metal catalyst with good stability and a suitable bandgap that is one of the most frequently employed photocatalysts. However, the brief charge-carrier lifespan of g-C_3N_4 results in gradually declining photocatalytic efficiency. In this respect, an induced evaporation self-assemblage technique was used to create a visible-light recognizable photocatalytic material made of g-C_3N_4 and titanium carbide. Having completely degraded in about 150 min, the g-C_3N_4 with Ti_3C_2 nanocomposite demonstrated outstanding photocatalytic efficiency against the pharmaceutical contaminant referred to as ciprofloxacin. The degradation curve resembled kinetic of pseudo first order, with the ciprofloxacin degradation occurring 2.2 times more quickly than the original g-C_3N_4 (pristine) when exposed to light that is visible [264]. The use of plasmonic particles, like silver, was recommended to later increase the functionality of g-C_3N_4 with Ti_3C_2 nanocomposite. In this instance, the multilayered MXene composed of Ti_3C_2 served as a support and reducing agent to create nanoparticles of Ag, which later aided in the transit of electrons generated by photons. The silver integrated g-C_3N_4 with titanium carbide nanocomposite realized 81.8% aniline degradation performance attributed to the developed Schottky junction' having increased absorption of light and electron donating property. When compared toward pure g-C_3N_4 (pristine) with titanium carbide, the degradation efficiency by photocatalysis was four to eight times more, accordingly [124].

The conversion of two-dimensional multilayered MXene in to three-dimensional macroscopic volume of the hydrogel has currently expressed excellent potential for photocatalysis [265, 266]. Soft materials based on MXene with variable characteristics can be developed using these intriguing and adaptable platforms when MXenes are combined with hydrogel frameworks [265]. MXenes have also been successfully used in the non-irradiative catalytic elimination and degradation of contaminants from the environment. In this instance, the catalytic framework is set up to enhance the overall efficacy of the oxidizing agents or reducing agents to accomplish complete mineralization and total pollutant degradation. A simple two-step procedure produced palladium with Ti_3C_2Tx MXene hydrogel of graphene that are three-dimensional interlinked porous framework. The palladium with Ti_3C_2Tx MXene hydrogel of graphene exhibited a noticeable porosity design when first generated, allowing it to absorb Ti_3C_2Tx expansion throughout the catalytic procedure. As a result, the palladium with Ti_3C_2Tx MXene hydrogel of graphene, which was mechanically strong, showed significant activity, simple separation, and favorable cyclability [366]. In another instance, a Co_3O_4 with Ti_3C_2 MXene nanocomposite was created utilizing a straightforward solvothermal technique. The Co_3O_4 with Ti_3C_2 nanocomposite showed exceptional methylene blue and rhodium blue dye degradation performance, i.e., 128.91 mg per gram of methylene blue degraded in the 300 min, 47.076 mg per g of rhodium blue degraded in the 100 min and it was able to sustain its catalytic properties for eight successive catalysis cycles [267]. One-pot method was used to create a sandwich type nanocomposite of Co_3O_4 with Ti_3C_2Tx that showed exceptional catalytic removal capacity against bisphenol-A in an environment of peroxymonosulfate oxidant. The Co_3O_4 with 20% of MXene showed the most effective removal performance across the optimal material, degrading 95% of the bisphenol-A under 7 min [76]. A Pd with Ti_3C_2Tx composite was also suggested for the reducing morin and nitro compounds degradation. The procedure whereby Pd nanoparticles were self-reductively formed upon MXene and development of palladium nanoparticles was managed through modifying the duration of reaction. Having rates of reaction 0.089 and 0.180 per second, the Pd with MXene nanocomposite displayed remarkable catalytic breakdown capacity toward nitroaniline and 4-nitrophenol in the presence of sodium borohydride. A period of eight successive cycles, the catalyst still exhibited strong activity, with rates of conversion of above 94 and 91.8% for 4-nitrophenol and 2-nitrophenol, respectively. Morin, a mulberry coloring agent, could be broken down by the Pd with MXene catalyst, but the rate of degradation was not as high as it had been for 4-nitrophenol and 2-nitrophenol [268]. The photocatalysts based upon MXene described for the degradation by photocatalysis of ecosystem contaminants are listed in Table 2 together.

5.2.6 Electrocatalytic Sensors for Pollution Detection Using MXenes

Toxin or pollutant analysis is the initial stage of environmental cleanup. Because of its easy use, sophistication, and specificity, the electrocatalytic technology is a desirable method for direct and indirect investigation of numerous pollutants. The catalytic ingredient of the designed sensor is directly responsible for the electrocatalytic sensor with sensitivity. The sensitivity and specificity of the sensor can be greatly enhanced

Table 2 Efficiency of MXene and composite based on MXene photocatalytic material in catalytic degradation performance for certain pollutants

MXene-based composite	Contaminant	Concentration	Degradation efficiency (%)	Reference
TiO_2 with Ti_3C_2	Methyl orange	20 mg/L	98	348
TiO_2 with Ti_3C_2Tx	Carbamazepine	5 mg/L	98.67	351
Ti_3C_2Tx	Methylene blue	0.012 mg/L	81	320
TiO_2 with Ti_3C_2	Methylene blue	20 mg/L	75	350
Titanium carbide and TiO_2 with Silver	Salicylic acid	100 μM	96	353
Titanium carbide and TiO_2 with copper oxide	Methyl orange	20 mg/L	99	352
CeO_2 with Titanium carbide	Rhodium blue	20 mg/L	75	355
MXene with BGFO-20Sn	CR dye	100 ml	100	357
MXene with $CuFe_2O_4$	Sulphamethoxazole	40 mg/L	59.4	359
Bi_2WO_6 with Nb_2CTx	Rhodium blue, tetracycline hydrochloride and methyl blue	15 mg/L	99.8, 83.1 and 92.7	361
g-C_3N_4 with titanium carbide	Ciprofloxacin	20 mg/L	100	363

by using electrocatalytic substance with strong active surface, enough conductivity, and relevant redox reactivity. MXenes have recently attained a great interest from the sensing world because of their active functionalities, hydrophilic nature, and variable surface chemical composition [269]. In order to gain increased sensitivity and specificity, MXene could also be easily designed or usually combined with other chemical components, like carbon-based compounds and metallic nanoparticles. It was discovered the use of two-dimensional sheets of MXene in the oxidation/reduction-based catalysis method for identifying the bromate charge particles with relation to hazardous ions. The spectroscopic investigation showed that the regular decrease of bromate and an incomplete MXene material oxidation occurred as a result of bromate ions adsorption onto MXene nanosheets. Having a limit of detection of 41 nM for a working aperture of 50 nM with 5 μM the built sensor demonstrated good sensitivity. The research demonstrated the direct applicability of MXene nanosheets in oxidation/reduction-based sensor framework [270]. It was shown how to use PANI with modified titanium carbide layers for the electrochemical analysis of ions containing mercury in a different scenario of dangerous metal detection. Through their co-coupling, the PANI with Ti_3C_2 nanocomposite achieved a uniform identification limit from 0.1 to 20 gL per 1 and a small limit of detection of 0.017

gL per 1. Acetylcholinesterase-based electrocatalytic devices for weak concentration organophosphate pollutant identification which is pesticide suffer from inadequate transfers of electrons. A further investigation suggested using nanosheets made of Ti_3C_2Tx directly to detect carbendazim by electrocatalytic method. The results of the computational simulations demonstrated the significance of Ti_3C_2Tx MXene having fluoride terminal groups for electrochemical detection. The instrument had a low limit of detection of 10.3 nM, highly selective in recognizing fenamiphos presence, metal ions, and ametryn was able to identify the drug with the range from 50 to 100 nM [271]. Furthermore, it suggested an MXene with rGO nanocomposite for carbendazim. The electrolytic transformation of graphene oxide to rGO was done after the MXene and graphene oxide were coupled in solution environment to create the composite. Based on quick charge-transfer routes and more active sites on surface, a conductive framework of rGO packaged with layers of MXene was built as an infrastructure with better catalytic performance and signal generating capability. The MXene with rGO therefore displayed a highly sensitive response of signals against carbendazim 2.0 nM to 10.0 M linear broad range and a lesser limit for the detection of 0.67 nM [272]. The designed electrocatalysts might also function in a matrix of organic material, including samples of cucumber and orange juice. Environmental contaminant detection has also been investigated using structures made of MXene composites. MXenes are a possible substrate for assembling composites that are hybrid due to their excellent conductivity, two-dimensional framework, wide surface area, hydrophilic nature, and mechanical adaptability [273]. Additionally, the multiple layers of MXene and its controlled surface chemical composition enable composites that are based on MXene to exhibit beneficial characteristics, improving both the selectivity and sensitivity of the sensors [274]. In a similar way tyrosinase, which was employed to identify phenol, was immobilized on an MXene framework modified with chitosan. The MXene having advantageous functionalities and two-dimensional interfacial structure made it simple to bind molecules of chitosan, which made it easier for the enzyme tyrosinase to get immobilized. The tyrosinase enzyme having ability to transform ortho-quinone from phenol, which was then reduced electrochemically on surface of electrode to polyhydric phenolic group, provided the basis for the detection process. This increased the electrocatalytic signal of phenol. At a phenol recognition range from about 0.05 to 15.5 mol/L and a detecting range of 12 nmol/L, the manufactured tyrosinase based on MXene biosensor demonstrated excellent analytic capability [275]. The most modern electrocatalytic detectors based on MXene are made to investigate different ecosystem contaminants. However, MXenes have demonstrated remarkable promise in sensing machines; further research should focus on anodic stability toward MXene for the efficient electrocatalytic identification of desired contaminants.

6 Conclusion and Summary

From its initial research, MXenes have demonstrated enormous potential due to variety of applications by cause of their customizable surface chemical features, easiness of functionality, strong hydrophilic nature, and outstanding conductivity. In this above-mentioned overview, we already discussed the current development of MXenes applications for treatment of the environment, beginning with adsorption capacities and separation of membranes, photocatalysis process, and expanding to electrocatalytic detectors specifically made for identifying pollutants. MXene-based nanostructures have interesting adsorption properties due to their huge surface area and -vely charged on surface. The sorption properties of MXene and similar composites need to be investigated for a number of ecological applications including adsorption and photocatalytic elimination of hazardous pigments, ions, and chemicals. MXenes, as opposed to traditional sorbents like carbon and nanomaterials based on graphene, offer a number of benefits in terms of customizable surface chemical composition and high hydrophilic nature, making it simple to surface-absorb contaminants.

The stability of the sheets of MXene continues to be a significant problem for membrane separation and adsorption that has a direct impact on the total life-span of MXene-based adsorption agents. Additionally, the makeup of the surface components has a role in the process of adsorption. As a result, it might be expected that newly developed synthetic pathways and following treatment techniques will improve membrane stability and enable selective adsorption. In order to enhance adsorption of photons and begin charging-carrier transport, it is necessary to construct well-linked interfaces for the usage of MXenes in photocatalytic processes to eliminate contaminants. In this situation, tailored hybrids with enhanced interfacial configurations and better redox activity could provide effective and wide-spectrum photocatalytic material with strong degradation potential. The employment of MXenes in environmental cleanup technology is hampered by their instability. In-depth research is necessary to comprehend oxidation processes in various solvents with ensuing lattice modification of MXenes. Additionally, it is important to concentrate on how the electronic composition of MXenes changes after being doped with many different metals and oxides of metals to create effective photocatalysts, for which the atomic-level analysis approach will be useful. The size of the MXenes flake directly affects the regulation of waterflux in the membrane. Therefore, it is important to utilize both experimental and theoretical techniques in order to uncover the underlying process and achieve an optimal flow rate with a significant salt rate of rejection. Another problem for MXene-based membranes and adsorption agents is the selective accumulation/isolation of hazardous species, which necessitates experimental approaches aimed at building sheets of MXene with a single kind of consistent terminations. Membranes made of MXene have a propensity to expand, which results in poor selectivity and the removal of ions. To prevent expansion or collapse in structure of the material membranes in this situation, interlayer design to adjust d-spacing of the nanosheets of MXene may be a liable option. MXenes produced development in the stream of electrocatalytic sensing networks, one of more current uses. MXenes are recognized as the ideal framework for the electrocatalytic detector because they give better conductivity along with excellent dispersibility, compared to graphene, which provides favorable conductivity at the tradeoff of poor hydrophilic properties. Immediate application of MXene toward electrocatalytic identification of contaminants is currently hampered by its anodic durability. In this circumstance, nanocomposites are made by combining MXenes with electrical-active polymers which are shown to be successful in identifying contaminants like toxic metal ions. It has also been claimed that MXenes can be used with sensors which are enzyme based and are responsive to phenolic contaminants. As the application of MXenes in sensors that are electrochemical remains in its nascent stages, an extensive knowledge of electrochemical performance regarding its surface is necessary. Even though the inclusion of MXenes inside electrocatalytic detectors has increased the sensibility of signal and the restriction of detection for different contaminants, MXenes are typically only used to boost conductivity. Nb_2CTx and V_4C_3Tx, two other MXene segments, should generally also be researched for their adaptability and usefulness in ecological remediation techniques. Additionally, research should focus on composites with hybrid properties derived from MXene that can advance ecological cleanup technique.

References

1. Lapworth, D.J., Baran, N., Stuart, M.E., Ward, R.S.: Emerging organic contaminants in groundwater: a review of sources, fate and occurrence. Environ. Pollut. **163**, 287–303 (2012)
2. Naidu, R., Espana, V.A.A., Liu, Y., Jit, J.: Emerging contaminants in the environment: risk-based analysis for better management. Chemosphere **154**, 350–357 (2016)
3. Sharma, B.M., Bečanová, J., Scheringer, M., Sharma, A., Bharat, G.K., Whitehead, P.G., Nizzetto, L.: Health and ecological risk assessment of emerging contaminants (pharmaceuticals, personal care products, and artificial sweeteners) in surface and groundwater (drinking water) in the Ganges River Basin. India. Sci. Total. Environ. **646**, 1459–1467 (2019)
4. Abdel-Shafy, H.I., Aly, R.O.: Water issue in Egypt: Resources, pollution and protection endeavors. Cent. Eur. J. Med. **8**(1), 3–21 (2002)
5. Anju, A., Ravi S.P., Bechan, S.: Water pollution with special reference to pesticide contamination in India. J. Water Resource Prot. (2010)
6. Wang, Q., Yang, Z.: Industrial water pollution, water environment treatment, and health risks in China. Environ. Pollut. **218**, 358–365 (2016)
7. Karthikeyan, K.G., Meyer, M.T.: Occurrence of antibiotics in wastewater treatment facilities in Wisconsin, USA. Sci. Total. Environ. **361**(1–3), 196–207 (2006)
8. Santos, J.L., Aparicio, I., Alonso, E.: Occurrence and risk assessment of pharmaceutically active compounds in wastewater treatment plants. A case study: seville city (Spain). Environ. Int. **33**(4), 596–601 (2007)
9. Deysher, G., Shuck, C.E., Hantanasirisakul, K., Frey, N.C., Foucher, A.C., Maleski, K., Gogotsi, Y.: Synthesis of Mo4VAlC4 MAX phase and two-dimensional Mo4VC4 MXene with five atomic layers of transition metals. ACS Nano **14**(1), 204–217 (2019)
10. Ihsanullah, I.: MXenes (two-dimensional metal carbides) as emerging nanomaterials for water purification: Progress, challenges and prospects. Chem. Eng. J. **388**, 124340 (2020)
11. Tang, W., Dong, Z., Zhang, R., Yi, X., Yang, K., Jin, M., Cheng, L.: Multifunctional two-dimensional core–shell mxene@ gold nanocomposites for enhanced photo–radio combined therapy in the second biological window. ACS Nano **13**(1), 284–294 (2018)

12. Murugan, C., Sharma, V., Murugan, R.K., Malaimegu, G., Sundaramurthy, A.: Two-dimensional cancer theranostic nanomaterials: Synthesis, surface functionalization and applications in photothermal therapy. J. Control. Release **299**, 1–20 (2019)
13. Shafiee, A., Iravani, S., Varma, R.S.: Graphene and graphene oxide with anticancer applications: challenges and future perspectives. MedCom **3**(1), e118 (2022)
14. Shafiei, N., Nasrollahzadeh, M., Iravani, S.: Green synthesis of silica and silicon nanoparticles and their biomedical and catalytic applications. Comments Inorg. **41**(6), 317–372 (2021)
15. Arole, K., Blivin, J.W., Saha, S., Holta, D.E., Zhao, X., Sarmah, A., Green, M.J.: Water-dispersible Ti3C2Tz MXene nanosheets by molten salt etching. iScience **24**(12), 103403 (2021)
16. Ji, Y., Luo, Y.: New mechanism for photocatalytic reduction of CO_2 on the anatase TiO_2 (101) surface: the essential role of oxygen vacancy. J. Am. Chem. Soc. **138**(49), 15896–15902 (2016)
17. Yang, C., Tan, Q., Li, Q., Zhou, J., Fan, J., Li, B., Lv, K.: 2D/2D Ti3C2 MXene/g-C3N4 nanosheets heterojunction for high efficient CO_2 reduction photocatalyst: dual effects of urea. Appl. Catal. B **268**, 118738 (2020)
18. Rhouati, A., Berkani, M., Vasseghian, Y., Golzadeh, N.: MXene-based electrochemical sensors for detection of environmental pollutants: a comprehensive review. Chemosphere **291**, 132921 (2022)
19. Feng, X., Yu, Z., Sun, Y., Long, R., Shan, M., Li, X., Liu, J.: Review MXenes as a new type of nanomaterial for environmental applications in the photocatalytic degradation of water pollutants. Ceram. Int. **47**(6), 7321–7343 (2021)
20. Iravani, P., Iravani, S., Varma, R.S.: MXene-chitosan composites and their biomedical potentials. Micromachines **13**(9), 1383 (2022)
21. Iravani, S., Varma, R.S.: MXenes and MXene-based materials for tissue engineering and regenerative medicine: recent advances. Adv. Mater. **2**(9), 2906–2917 (2021)
22. Iravani, S., Varma, R.S.: MXenes for cancer therapy and diagnosis: recent advances and current challenges. ACS Biomater. Sci. Eng. **7**(6), 1900–1913 (2021)
23. Sreedhar, A., Noh, J.S.: Recent advances in partially and completely derived 2D Ti3C2 MXene based TiO_2 nanocomposites towards photocatalytic applications: a review Sol Energy **222**, 48–73 (2021)
24. Abbasi, Z., Feizi, S., Taghipour, E., Ghadam, P.: Green synthesis of silver nanoparticles using aqueous extract of dried Juglans regia green husk and examination of its biological properties. Green Process. Synth. **6**(5), 477–485 (2017)
25. Iravani, S., Varma, R.S.: MXenes in photomedicine: advances and prospects. Chem comm **58**(53), 7336–7350 (2022)
26. Rahman, U.U., Humayun, M., Ghani, U., Usman, M., Ullah, H., Khan, A., Khan, A.: MXenes as emerging materials: synthesis, properties, and applications. Molecules **27**(15), 4909 (2022)
27. Shuck, C.E., Ventura-Martinez, K., Goad, A., Uzun, S., Shekhirev, M., Gogotsi, Y.: Safe synthesis of MAX and MXene: guidelines to reduce risk during synthesis. J. Chem. Health Saf. **28**(5), 326–338 (2021)
28. Sun, Y., Meng, X., Dall'Agnese, Y., Dall'Agnese, C., Duan, S., Gao, Y., Wang, X. F.: 2D MXenes as co-catalysts in photocatalysis: synthetic methods. Nanomicro Lett **11**, 1–22 (2019)
29. Wei, Y., Zhang, P., Soomro, R.A., Zhu, Q., Xu, B.: Advances in the synthesis of 2D MXenes. Adv. Mater. Lett. **33**(39), 2103148 (2021)
30. Wu, Q., Li, N., Wang, Y., Xu, Y., Wu, J., Jia, G., Cui, X.: Ultrasensitive and selective determination of carcinoembryonic antigen using multifunctional ultrathin amino-functionalized Ti3C2-MXene nanosheets. Anal. Chem. **92**(4), 3354–3360 (2020)
31. Wyatt, B.C., Rosenkranz, A., Anasori, B.: 2D MXenes: tunable mechanical and tribological properties. Adv. Mater. **33**(17), 2007973 (2021)
32. Iravani, S., Varma, R.S.: MXenes in Cancer Nanotheranostics. J. Nanomater. **12**, 3360 (2022)
33. You, Z., Liao, Y., Li, X., Fan, J., Xiang, Q.: State-of-the-art recent progress in MXene-based photocatalysts: a comprehensive review. Nanoscale **13**(21), 9463–9504 (2021)
34. Kuang, P., Low, J., Cheng, B., Yu, J., Fan, J.: MXene-based photocatalysts. J. Mater. Sci. Technol. **56**, 18–44 (2020)

35. Im, J.K., Sohn, E.J., Kim, S., Jang, M., Son, A., Zoh, K.D., Yoon, Y.: Review of MXene-based nanocomposites for photocatalysis. Chemosphere **270**, 129478 (2021)
36. Khatami, M., Iravani, P., Jamalipour Soufi, G., Iravani, S.: MXenes for antimicrobial and antiviral applications: Recent advances. Mater. Technol. **37**(11), 1890–1905 (2022)
37. Khatami, M., Iravani, S.: MXenes and MXene-based materials for the removal of water pollutants: challenges and opportunities. Comments Inorg. Chem. **41**(4), 213–248 (2021)
38. Zhang, S., Bilal, M., Adeel, M., Barceló, D., Iqbal, H.M.: MXene-based designer nanomaterials and their exploitation to mitigate hazardous pollutants from environmental matrices. Chemosphere **283**, 131293 (2021)
39. Bai, X., Hou, S., Wang, X., Hao, D., Sun, B., Jia, T., Ni, B.J.: Mechanism of surface and interface engineering under diverse dimensional combinations: the construction of efficient nanostructured MXene-based photocatalysts. Catal. Sci. Technol. Catal. Sci. Technol. **11**(15), 5028–5049 (2021)
40. Wojciechowski, T., Rozmysłowska-Wojciechowska, A., Matyszczak, G., Wrzecionek, M., Olszyna, A., Peter, A., Jastrzebska, A.: Ti2C MXene modified with ceramic oxide and noble metal nanoparticles: synthesis, morphostructural properties, and high photocatalytic activity. Inorg. Chem. **58**(11), 7602–7614 (2019)
41. Makola, L.C., Moeno, S., Ouma, C.N., Sharma, A., Vo, D.V.N., Dlamini, L.N.: Facile fabrication of a metal-free 2D–2D Nb2CTx@ g-C3N4 MXene-based Schottky-heterojunction with the potential application in photocatalytic processes. J. Alloys Compd. **916**, 165459 (2022)
42. Sun, Y., Zhou, Y., Liu, Y., Wu, Q., Zhu, M., Huang, H., Kang, Z.: A photoactive process cascaded electrocatalysis for enhanced methanol oxidation over Pt-MXene-TiO 2 composite. Nano Res. **13**, 2683–2690 (2020)
43. Hasija, V., Raizada, P., Sudhaik, A., Sharma, K., Kumar, A., Singh, P., Thakur, V.K.: Recent advances in noble metal free doped graphitic carbon nitride based nanohybrids for photocatalysis of organic contaminants in water: a review. Appl. Mater. Today **15**, 494–524 (2019)
44. Soni, V., Singh, P., Quang, H.H.P., Khan, A.A.P., Bajpai, A., Van Le, Q., Raizada, P.: Emerging architecture titanium carbide (Ti3C2Tx) MXene based photocatalyst toward degradation of hazardous pollutants: Recent progress and perspectives. Chemosphere, 133541 (2022)
45. Ta, Q.T.H., Tran, N.M., Noh, J.S.: Rice crust-like ZnO/Ti3C2Tx MXene hybrid structures for improved photocatalytic activity. Catalysts **10**(10), 1140 (2020)
46. Ihsanullah, I.: MXenes as next-generation materials for the photocatalytic degradation of pharmaceuticals in water. J. Environ. Chem. Eng. 107381 (2022)
47. Cai, T., Wang, L., Liu, Y., Zhang, S., Dong, W., Chen, H., Luo, S.: Ag3PO4/Ti3C2 MXene interface materials as a Schottky catalyst with enhanced photocatalytic activities and anti-photocorrosion performance. Appl. Catal. B **239**, 545–554 (2018)
48. Prasad, C., Yang, X., Liu, Q., Tang, H., Rammohan, A., Zulfiqar, S., Shah, S.: Recent advances in MXenes supported semiconductors based photocatalysts: Properties, synthesis and photocatalytic applications. J. Ind. Eng. Chem. **85**, 1–33 (2020)
49. Ghidiu, M., Lukatskaya, M.R., Zhao, M.Q., Gogotsi, Y., Barsoum, M.W.: Conductive two-dimensional titanium carbide 'clay'with high volumetric capacitance. Nature **516**(7529), 78–81 (2014)
50. Wang, Z., Zhang, N., Yu, M., Liu, J., Wang, S., Qiu, J.: Boosting redox activity on MXene-induced multifunctional collaborative interface in high Li2S loading cathode for high-energy Li-S and metallic Li-free rechargeable batteries. J. Energy Chem. **37**, 183–191 (2019)
51. Ai, Z., Shao, Y., Chang, B., Huang, B., Wu, Y., Hao, X.: Effective orientation control of photogenerated carrier separation via rational design of a Ti3C2 (TiO2)@ CdS/MoS2 photocatalytic system. Appl. Catal. B **242**, 202–208 (2019)
52. Zhou, J., Zha, X., Chen, F.Y., Ye, Q., Eklund, P., Du, S., Huang, Q.: A two-dimensional zirconium carbide by selective etching of Al3C3 from nanolaminated Zr3Al3C5. Angew. Chem., Int. Ed. Engl. **55**(16), 5008–5013 (2016)
53. Mashtalir, O., Cook, K.M., Mochalin, V.N., Crowe, M., Barsoum, M.W., Gogotsi, Y.: Dye adsorption and decomposition on two-dimensional titanium carbide in aqueous media. J. Mater. Chem. A **2**(35), 14334–14338 (2014)

54. Wang, K., Zhou, Y., Xu, W., Huang, D., Wang, Z., Hong, M.: Fabrication and thermal stability of two dimensional carbide Ti3C2 nanosheets. Ceram. Int. **42**(7), 8419–8424 (2016)
55. Zhang, C.J., Pinilla, S., McEvoy, N., Cullen, C.P., Anasori, B., Long, E., Nicolosi, V.: Oxidation stability of colloidal two-dimensional titanium carbides (MXenes). Chem. Mater. **29**(11), 4848–4856 (2017)
56. Wu, X., Wang, Z., Yu, M., Xiu, L., Qiu, J.: Stabilizing the MXenes by carbon nanoplating for developing hierarchical nanohybrids with efficient lithium storage and hydrogen evolution capability. Adv. Mater. **29**(24), 1607017 (2017)
57. Ng, V.M.H., Huang, H., Zhou, K., Lee, P.S., Que, W., Xu, J.Z., Kong, L.B.: Recent progress in layered transition metal carbides and/or nitrides (MXenes) and their composites: synthesis and applications. J. Mater. Chem. A **5**(7), 3039–3068 (2017)
58. Ni, W., Shi, L.: Layer-structured carbonaceous materials for advanced Li-ion and Na-ion batteries: beyond graphene. J. Vac. Sci. **37**(4), 040803 (2019)
59. Shao, L., Chen, G.Q.: Water footprint assessment for wastewater treatment: method, indicator, and application. Environ. Sci. Technol. **47**(14), 7787–7794 (2013)
60. Jasper, J.T., Yang, Y., Hoffmann, M.R.: Toxic byproduct formation during electrochemical treatment of latrine wastewater. Environ. Sci. Technol. **51**(12), 7111–7119 (2017)
61. Luo, Y., Guo, W., Ngo, H.H., Nghiem, L.D., Hai, F.I., Zhang, J., Wang, X.C.: A review on the occurrence of micropollutants in the aquatic environment and their fate and removal during wastewater treatment. Sci. Total. Environ. **473**, 619–641 (2014)
62. Gavrilescu, M., Demnerová, K., Aamand, J., Agathos, S., Fava, F.: Emerging pollutants in the environment: present and future challenges in biomonitoring, ecological risks and bioremediation. N. Biotechnol. **32**(1), 147–156 (2015)
63. Carolin, C.F., Kumar, P.S., Saravanan, A., Joshiba, G.J., Naushad, M.: Efficient techniques for the removal of toxic heavy metals from aquatic environment: a review. J. Environ. Chem. Eng. **5**(3), 2782–2799 (2017)
64. Lv, H., Ji, G., Yang, Z., Liu, Y., Zhang, X., Liu, W., Zhang, H.: Enhancement photocatalytic activity of the graphite-like C3N4 coated hollow pencil-like ZnO. J. Colloid Sci. **450**, 381–387 (2015)
65. Quan, B., Liu, W., Liu, Y., Zheng, Y., Yang, G., Ji, G.: Quasi-noble-metal graphene quantum dots deposited stannic oxide with oxygen vacancies: synthesis and enhanced photocatalytic properties. J. Colloid Sci. **481**, 13–19 (2016)
66. Wu, Y., Li, X., Yang, Q., Wang, D., Yao, F., Cao, J., Li, X.: Mxene-modulated dual-heterojunction generation on a metal-organic framework (MOF) via surface constitution reconstruction for enhanced photocatalytic activity. J. Chem. Eng. **390**, 124519 (2020)
67. Kim, S., Yu, M., Yoon, Y.: Fouling and retention mechanisms of selected cationic and anionic dyes in a Ti3C2T x MXene-ultrafiltration hybrid system. ACS Appl. Mater. Interfaces **12**(14), 16557 16565 (2020)
68. Cai, C., Wang, R., Liu, S., Yan, X., Zhang, L., Wang, M., Jiao, T.: Synthesis of self-assembled phytic acid-MXene nanocomposites via a facile hydrothermal approach with elevated dye adsorption capacities. Colloids Surf A Physicochem Eng. Asp. **589**, 124468 (2020)
69. Liu, T., Liu, X., Graham, N., Yu, W., Sun, K.: Two-dimensional MXene incorporated graphene oxide composite membrane with enhanced water purification performance. J. Membr. Sci. **593**, 117431 (2020)
70. Zhu, Z., Xiang, M., Shan, L., He, T., Zhang, P.: Effect of temperature on methylene blue removal with novel 2D-Magnetism titanium carbide. J. Solid State Chem. **280**, 120989 (2019)
71. Cui, C., Guo, R., Xiao, H., Ren, E., Song, Q., Xiang, C., Jiang, S.: Bi2WO6/Nb2CTx MXene hybrid nanosheets with enhanced visible-light-driven photocatalytic activity for organic pollutants degradation. Appl. Surf. Sci. **505**, 144595 (2020)
72. Liu, X., Chen, C.: Mxene enhanced the photocatalytic activity of ZnO nanorods under visible light. Mater. Lett. **261**, 127127 (2020)
73. Feng, H., Wang, W., Zhang, M., Zhu, S., Wang, Q., Liu, J., Chen, S.: 2D titanium carbide-based nanocomposites for photocatalytic bacteriostatic applications. Appl. Catal. B **266**, 118609 (2020)

74. Cao, Y., Fang, Y., Lei, X., Tan, B., Hu, X., Liu, B., Chen, Q.: Fabrication of novel CuFe2O4/MXene hierarchical heterostructures for enhanced photocatalytic degradation of sulfonamides under visible light. J. Hazard. Mater. **387**, 122021 (2020)
75. Shahzad, A., Rasool, K., Nawaz, M., Miran, W., Jang, J., Moztahida, M., Lee, D.S.: Heterostructural TiO2/Ti3C2Tx (MXene) for photocatalytic degradation of antiepileptic drug carbamazepine. J. Chem. Eng. **349**, 748–755 (2018)
76. Liu, Y., Luo, R., Li, Y., Qi, J., Wang, C., Li, J., Wang, L.: Sandwich-like Co3O4/MXene composite with enhanced catalytic performance for Bisphenol A degradation. J. Chem. Eng. **347**, 731–740 (2018)
77. Zhou, L., Zhang, X., Ma, L., Gao, J., Jiang, Y.: Acetylcholinesterase/chitosan-transition metal carbides nanocomposites-based biosensor for the organophosphate pesticides detection. Biochem. Eng. J. **128**, 243–249 (2017)
78. Huang, G., Li, S., Liu, L., Zhu, L., Wang, Q.: Ti3C2 MXene-modified Bi2WO6 nanoplates for efficient photodegradation of volatile organic compounds. Appl. Surf. Sci. **503**, 144183 (2020)
79. Zhang, S., Liao, S., Qi, F., Liu, R., Xiao, T., Hu, J., Min, Y.: Direct deposition of two-dimensional MXene nanosheets on commercially available filter for fast and efficient dye removal. J. Hazard. Mater. **384**, 121367 (2020)
80. Ding, M., Chen, W., Xu, H., Shen, Z., Lin, T., Hu, K., Xie, Z.: Novel α-Fe2O3/MXene nanocomposite as heterogeneous activator of peroxymonosulfate for the degradation of salicylic acid. J. Hazard. Mater. **382**, 121064 (2020)
81. Du, Y., Yu, B., Wei, L., Wang, Y., Zhang, X., Ye, S.: Efficient removal of Pb (II) by Ti3C2T x powder modified with a silane coupling agent. J. Mater. Sci. **54**(20), 13283–13297 (2019)
82. Ding, X., Li, Y., Li, C., Wang, W., Wang, L., Feng, L., Han, D.: 2D visible-light-driven TiO2@ Ti3C2/g-C3N4 ternary heterostructure for high photocatalytic activity. J. Mater. Sci. **54**(13), 9385–9396 (2019)
83. Houas, A., Lachheb, H., Ksibi, M., Elaloui, E., Guillard, C., Herrmann, J.M.: Photocatalytic degradation pathway of methylene blue in water. Appl. Catal. B **31**(2), 145–157 (2001)
84. Li, J., Wang, S., Du, Y., Liao, W.: Enhanced photocatalytic performance of TiO2@ C nanosheets derived from two-dimensional Ti2CTx. Ceram. Int. **44**(6), 7042–7046 (2018)
85. Lu, Y., Yao, M., Zhou, A., Hu, Q., Wang, L.: Preparation and photocatalytic performance of Ti3C2/TiO2/CuO ternary nanocomposites. J. Nanomater. (2017)
86. Samimi, M., Shahriari Moghadam, M.: Phenol biodegradation by bacterial strain O-CH1 isolated from seashore. Glob. J. Environ. Sci. Manag. **6**(1), 109–118 (2020)
87. Ma, S., Jing, J., Liu, P., Li, Z., Jin, W., Xie, B., Zhao, Y.: High selectivity and effectiveness for removal of tetracycline and its related drug resistance in food wastewater through schwertmannite/graphene oxide catalyzed photo-Fenton-like oxidation. J. Hazard. Mater. **392**, 122437 (2020)
88. de Matos, D.B., Barbosa, M.P.R., Leite, O.M., Steter, J.R., Lima, N.S., Torres, N.H., Cavalcanti, E.B.: Characterization of a tubular electrochemical reactor for the degradation of the commercial diuron herbicide. Environ. Technol. **41**(10), 1307–1321 (2020)
89. Zhang, S., Li, B., Wang, X., Zhao, G., Hu, B., Lu, Z., Wang, X.: Recent developments of two-dimensional graphene-based composites in visible-light photocatalysis for eliminating persistent organic pollutants from wastewater. J. Chem. Eng. **390**, 124642 (2020)
90. Yu, M., Wang, J., Tang, L., Feng, C., Liu, H., Zhang, H., Xie, Q.: Intimate coupling of photocatalysis and biodegradation for wastewater treatment: mechanisms, recent advances and environmental applications. Water Res. **175**, 115673 (2020)
91. Guan, G., Ye, E., You, M., Li, Z.: Hybridized 2D nanomaterials toward highly efficient photocatalysis for degrading pollutants: current status and future perspectives. Small **16**(19), 1907087 (2020)
92. Long, Z., Li, Q., Wei, T., Zhang, G., Ren, Z.: Historical development and prospects of photocatalysts for pollutant removal in water. J. Hazard. Mater. **395**, 122599 (2020)
93. Ling, L., Fan, M., Wang, B., Zhang, R.: Application of computational chemistry in understanding the mechanisms of mercury removal technologies: a review. Energy Environ. Sci. **8**(11), 3109–3133 (2015)

94. Mergola, L., Scorrano, S., Bloise, E., Di Bello, M.P., Catalano, M., Vasapollo, G., Del Sole, R.: Novel polymeric sorbents based on imprinted Hg (II)-diphenylcarbazone complexes for mercury removal from drinking water. Polym. J. **48**(1), 73–79 (2016)
95. Yardim, M.F., Budinova, T., Ekinci, E., Petrov, N., Razvigorova, M., Minkova, V.: Removal of mercury (II) from aqueous solution by activated carbon obtained from furfural. Chemosphere **52**(5), 835–841 (2003)
96. Li, J., Huang, J.T., Gong, Y.: Factorized adaptation for deep neural network. Proc. IEEE Int. Conf. Acoust. Speech Signal Process, 5537–5541(2014)
97. Namasivayam, C., Kadirvelu, K.: Uptake of mercury (II) from wastewater by activated carbon from an unwanted agricultural solid by-product: coirpith. C. **37**(1), 79–84 (1999)
98. McNutt, M.: Mercury and health. Science **341**(6153), 1430–1430 (2013)
99. Feng, X., Yu, Z., Long, R., Li, X., Shao, L., Zeng, H., Zuo, Y.: Self-assembling 2D/2D (MXene/LDH) materials achieve ultra-high adsorption of heavy metals Ni2+ through terminal group modification. Sep. Purif. Technol. **253**, 117525 (2020)
100. Vunain, E., Mishra, A.K., Mamba, B.B.: Dendrimers, mesoporous silicas and chitosan-based nanosorbents for the removal of heavy-metal ions: a review. Int. J. Biol. Macromol. **86**, 570–586 (2016)
101. Moradi, G., Zinadini, S., Rajabi, L., Derakhshan, A.A.: Removal of heavy metal ions using a new high performance nanofiltration membrane modified with curcumin boehmite nanoparticles. J. Chem. Eng. **390**, 124546 (2020)
102. Shao, B., Wang, J., Liu, Z., Zeng, G., Tang, L., Liang, Q., Yuan, X.: Ti 3 C 2 T x MXene decorated black phosphorus nanosheets with improved visible-light photocatalytic activity: experimental and theoretical studies. J. Mater. Chem. A **8**(10), 5171–5185 (2020)
103. Ahmad, S.Z.N., Salleh, W.N.W., Ismail, A.F., Yusof, N., Yusop, M.Z.M., Aziz, F.: Adsorptive removal of heavy metal ions using graphene-based nanomaterials: toxicity, roles of functional groups and mechanisms. Chemosphere **248**, 126008 (2020)
104. Xu, T., Zhou, Y., Lei, X., Hu, B., Chen, H., Yu, G.: Study on highly efficient Cr (VI) removal from wastewater by sinusoidal alternating current coagulation. J. Environ. Manage. **249**, 109322 (2019)
105. Gusain, R., Kumar, N., Ray, S.S.: Recent advances in carbon nanomaterial-based adsorbents for water purification. Coord. Chem. Rev. **405**, 213111 (2020)
106. Zhu, W., Chen, X., Sun, P., Li, J., Zhai, P., Zheng, Y., Zhang, Z.: Facile hydrothermal synthesis of hierarchical porous priceite (Ca 4 B 10 O 19· 7H 2 O) microspheres as high-efficiency adsorbents for heavy metal ions removal. CrystEngComm **21**(46), 7141–7154 (2019)
107. Luo, J., Fu, K., Sun, M., Yin, K., Wang, D., Liu, X., Crittenden, J.C.: Phase-mediated heavy metal adsorption from aqueous solutions using two-dimensional layered MoS2. ACS Appl. Mater. Interfaces **11**(42), 38789–38797 (2019)
108. Feng, X., Yu, Z., Long, R., Sun, Y., Wang, M., Li, X., Zeng, G.: Polydopamine intimate contacted two-dimensional/two-dimensional ultrathin nylon basement membrane supported RGO/PDA/MXene composite material for oil-water separation and dye removal. Sep. Purif. Technol. SEP **247**, 116945 (2020)
109. Gan, D., Huang, Q., Dou, J., Huang, H., Chen, J., Liu, M., Wei, Y.: Bioinspired functionalization of MXenes (Ti3C2TX) with amino acids for efficient removal of heavy metal ions. Appl. Surf. Sci. **504**, 144603 (2020)
110. Fard, A.K., Mckay, G., Chamoun, R., Rhadfi, T., Preud'Homme, H., Atieh, M.A.: Barium removal from synthetic natural and produced water using MXene as two dimensional (2-D) nanosheet adsorbent. J. Chem. Eng. **317**, 331–342 (2017)
111. Wang, C., He, C., Luo, Y.H., Su, S., Wang, J.Y., Hong, D.L., Sun, B.W.: Efficient mercury chloride capture by ultrathin 2D metal-organic framework nanosheets. J. Chem. Eng. **379**, 122337 (2020)
112. Zhang, A., Liu, R., Tian, J., Huang, W., Liu, J.: MXene-based nanocomposites for energy conversion and storage applications. Chem. Eur. J. **26**(29), 6342–6359 (2020)
113. Naguib, M., Mochalin, V.N., Barsoum, M.W., Gogotsi, Y.: 25th anniversary article: MXenes: a new family of two-dimensional materials. J. Adv. Mater. **26**(7), 992–1005 (2014)

114. Huang, H., Jiang, R., Feng, Y., Ouyang, H., Zhou, N., Zhang, X., Wei, Y.: Recent development and prospects of surface modification and biomedical applications of MXenes. Nanoscale **12**(3), 1325–1338 (2020)
115. Yao, Z., Sun, H., Sui, H., Liu, X.: Construction of BPQDs/Ti3C2@ TiO2 composites with favorable charge transfer channels for enhanced photocatalytic activity under visible light irradiation. J. Nanomater. **10**(3), 452 (2020)
116. Sinopoli, A., Othman, Z., Rasool, K., Mahmoud, K.A.: Electrocatalytic/photocatalytic properties and aqueous media applications of 2D transition metal carbides (MXenes). Curr. Opin. Solid State Mater. Sci. **23**(5), 100760 (2019)
117. Huang, H., Song, Y., Li, N., Chen, D., Xu, Q., Li, H., Lu, J.: One-step in-situ preparation of N-doped TiO2@ C derived from Ti3C2 MXene for enhanced visible-light driven photodegradation. Appl. Catal. B **251**, 154–161 (2019)
118. Li, Y., Chen, X., Sun, Y., Meng, X., Dall'Agnese, Y., Chen, G., Wang, X.F.: Chlorosome-like molecular aggregation of chlorophyll derivative on Ti3C2Tx MXene nanosheets for efficient noble metal-free photocatalytic hydrogen evolution. Adv. Mater. Interfaces **7**(8), 1902080 (2020)
119. Gong, L.G., Qi, X.X., Yu, K., Gao, J.Q., Zhou, B.B., Yang, G.Y.: Covalent conductive polymer chain and organic ligand ethylenediamine modified MXene-like-{AlW 12 O 40} compounds for fully symmetric supercapacitors, electrochemical sensors and photocatalysis mechanisms. J. Mater. Chem. A **8**(11), 5709–5720 (2020)
120. Li, N., Peng, J., Ong, W.J., Ma, T., Zhang, P., Jiang, J., Zhang, C.J.: MXenes: an emerging platform for wearable electronics and looking beyond. Matters **4**(2), 377–407 (2021)
121. Wang, Z.: A review on MXene: synthesis, properties and applications on alkali metal ion batteries. IOP Conf. Ser.: Earth Environ. Sci. (042030) (2021)
122. Naguib, M., Kurtoglu, M., Presser, V., Lu, J., Niu, J., Heon, M., Barsoum, M.W.: Two-dimensional nanocrystals produced by exfoliation of Ti3AlC2. Adv. Mater. **23**(37), 4248–4253 (2011)
123. Song, F., Li, G., Zhu, Y., Wu, Z., Xie, X., Zhang, N.: Rising from the horizon: three-dimensional functional architectures assembled with MXene nanosheets. J. Mater. Chem. A **8**(36), 18538–18559 (2020)
124. Alhabeb, M., Maleski, K., Mathis, T.S., Sarycheva, A., Hatter, C.B., Uzun, S., Levitt, A., Gogotsi, Y.: Selective etching of silicon from Ti3SiC2 (MAX) to obtain 2D titanium carbide (MXene). Angew. Chem. **130**, 5542–5546 (2018)
125. Anasori, B., Lukatskaya, M.R., Gogotsi, Y.: 2D metal carbides and nitrides (MXenes) for energy storage. Nat. Rev. Mater. **2**(2), 1–17 (2017)
126. Alhabeb, M., Maleski, K., Anasori, B., Lelyukh, P., Clark, L., Sin, S., Gogotsi, Y.: Guidelines for synthesis and processing of two-dimensional titanium carbide (Ti3C2T x MXene). Chem. Mater. **29**(18), 7633–7644 (2017)
127. Kim, Y.J., Kim, S.J., Seo, D., Chae, Y., Anayee, M., Lee, Y., Jung, H.T.: Etching mechanism of monoatomic aluminum layers during MXene synthesis. Chem. Mater. **33**(16), 6346–6355 (2021)
128. Wang, X., Garnero, C., Rochard, G., Magne, D., Morisset, S., Hurand, S., Celerier, S.: A new etching environment (FeF 3/HCl) for the synthesis of two-dimensional titanium carbide MXenes: a route towards selective reactivity vs. water. J. Mater, Chem. A **5**(41), 22012–22023 (2017)
129. Cockreham, C.B., Zhang, X., Li, H., Hammond-Pereira, E., Sun, J., Saunders, S.R., Wu, D.: Inhibition of AlF3· 3H2O Impurity Formation in Ti3C2T x MXene Synthesis under a Unique CoF x/HCl Etching Environment. ACS Appl. Energy Mater. **2**(11), 8145–8152 (2019)
130. Gogotsi, Y., Anasori, B.: The rise of MXenes. ACS Nano **13**(8), 8491–8494 (2019)
131. Luo, J., Matios, E., Wang, H., Tao, X., Li, W.: Interfacial structure design of MXene-based nanomaterials for electrochemical energy storage and conversion. INFOMAT. **2**(6), 1057–1076 (2020)
132. Li, T., Yao, L., Liu, Q., Gu, J., Luo, R., Li, J., Zhang, D.: Fluorine-free synthesis of high-purity Ti3C2Tx (T= OH, O) via alkali treatment. Angew. Chem., Int. Ed. Engl. **57**(21), 6115–6119 (2018)

133. Chen, J., Chen, M., Zhou, W., Xu, X., Liu, B., Zhang, W., Wong, C.: Simplified synthesis of fluoride-free Ti3C2T x via electrochemical etching toward high-performance electrochemical capacitors. ACS Nano **16**(2), 2461–2470 (2022)
134. Lukatskaya, M.R., Bak, S.M., Yu, X., Yang, X.Q., Barsoum, M.W., Gogotsi, Y.: Probing the mechanism of high capacitance in two-dimensional titanium carbide using in-situ X-Ray absorption spectroscopy. Adv. Energy Mater. **10**(BNL-108018–2015-JA) (2015)
135. Tang, Q., Zhou, Z., Shen, P.: Are MXenes promising anode materials for Li ion batteries? Computational studies on electronic properties and Li storage capability of Ti3C2 and Ti3C2X2 (X= F, OH) monolayer. J. Am. Chem. Soc. **134**(40), 16909–16916 (2012)
136. Yin, W., Liu, X., Zhang, X., Gao, X., Colvin, V.L., Zhang, Y., Yu, W.W.: Synthesis of tungsten disulfide and molybdenum disulfide quantum dots and their applications. Chem. Mater. **32**(11), 4409–4424 (2020)
137. Xu, Q., Cai, W., Li, W., Sreeprasad, T.S., He, Z., Ong, W.J., Li, N.: Two-dimensional quantum dots: fundamentals, photoluminescence mechanism and their energy and environmental applications. Mater. Today Energy **10**, 222–240 (2018)
138. Xu, F., Zhang, L., Ding, X., Cong, M., Jin, Y., Chen, L., Gao, Y.: Selective electroreduction of dinitrogen to ammonia on a molecular iron phthalocyanine/O-MWCNT catalyst under ambient conditions. Chem comm **55**(94), 14111–14114 (2019)
139. Chen, J., Cheng, H., Ding, L.X., Wang, H.: Competing hydrogen evolution reaction: a challenge in electrocatalytic nitrogen fixation. Mater. Chem. Front. **5**(16), 5954–5969 (2021)
140. Shao, B., Liu, Z., Zeng, G., Wang, H., Liang, Q., He, Q., Song, B.: Two-dimensional transition metal carbide and nitride (MXene) derived quantum dots (QDs): synthesis, properties, applications and prospects. J. Mater. Chem. A **8**(16), 7508–7535 (2020)
141. Cheng, H., Ding, L.X., Chen, G.F., Zhang, L., Xue, J., Wang, H.: Molybdenum carbide nanodots enable efficient electrocatalytic nitrogen fixation under ambient conditions. Adv. Mater. **30**(46), 1803694 (2018)
142. Zhang, J., Kong, N., Uzun, S., Levitt, A., Seyedin, S., Lynch, P.A., Razal, J.M.: Scalable manufacturing of free-standing, strong Ti3C2Tx MXene films with outstanding conductivity. Adv. Mater. **32**(23), 2001093 (2020)
143. Yu, H., Wang, Y., Jing, Y., Ma, J., Du, C.F., Yan, Q.: Surface modified MXene-based nanocomposites for electrochemical energy conversion and storage. Small **15**(25), 1901503 (2019)
144. Yaseen, M., Humayun, M., Khan, A., Usman, M., Ullah, H., Tahir, A.A., Ullah, H.: Preparation, functionalization, modification, and applications of nanostructured gold: a critical review. Energies **14**(5), 1278 (2021)
145. Din, I.U., Nasir, Q., Garba, M.D., Alharthi, A.I., Alotaibi, M.A., Usman, M.: A review of preparation methods for heterogeneous catalysts. Mini Rev Org Chem **19**(1), 92–110 (2022)
146. Han, Y., Chen, Y., Wang, N., He, Z.: Magnesium doped carbon quantum dots synthesized by mechanical ball milling and displayed Fe3+ sensing. Mater. Technol. **34**(6), 336–342 (2019)
147. Zdrazil, L., Zahradnicek, R., Mohan, R., Sedlacek, P., Nejdl, L., Schmiedova, V., Hubalek, J.: Preparation of graphene quantum dots through liquid phase exfoliation method. J. Lumin. **204**, 203–208 (2018)
148. Mohanty, B., Giri, L., Jena, B.K.: MXene-derived quantum dots for energy conversion and storage applications. Energy Fuels **35**(18), 14304–14324 (2021)
149. Xu, Y., Wang, X., Zhang, W.L., Lv, F., Guo, S.: Recent progress in two-dimensional inorganic quantum dots. Chem. Soc. Rev. **47**(2), 586–625 (2018)
150. Zhang, T., Jiang, X., Li, G., Yao, Q., Lee, J.Y.: A red-phosphorous-assisted ball-milling synthesis of few-layered Ti3C2Tx (MXene) Nanodot Composite. ChemNanoMat **4**(1), 56–60 (2018)
151. Chen, X., Sun, X., Xu, W., Pan, G., Zhou, D., Zhu, J., Song, H.: Ratiometric photoluminescence sensing based on Ti 3 C 2 MXene quantum dots as an intracellular pH sensor. Nanoscale **10**(3), 1111–1118 (2018)
152. Feng, Y., Zhou, F., Deng, Q., Peng, C.: Solvothermal synthesis of in situ nitrogen-doped Ti3C2 MXene fluorescent quantum dots for selective Cu2+ detection. Ceram. Int. **46**(6), 8320–8327 (2020)

153. Xu, G., Niu, Y., Yang, X., Jin, Z., Wang, Y., Xu, Y., Niu, H.: Preparation of Ti3C2Tx MXene-derived quantum dots with white/blue-emitting photoluminescence and electrochemiluminescence. Adv. Opt. Mater. **6**(24), 1800951 (2018)
154. Mohanty, B., Mitra, A., Jena, B., Jena, B.K.: MoS2 quantum dots as efficient electrocatalyst for hydrogen evolution reaction over a wide pH range. Energy Fuels **34**(8), 10268–10275 (2020)
155. Yu, X., Cai, X., Cui, H., Lee, S.W., Yu, X.F., Liu, B.: Fluorine-free preparation of titanium carbide MXene quantum dots with high near-infrared photothermal performances for cancer therapy. Nanoscale **9**(45), 17859–17864 (2017)
156. Huang, D., Xie, Y., Lu, D., Wang, Z., Wang, J., Yu, H., Zhang, H.: Demonstration of a white laser with V2C MXene-based quantum dots. Adv. Mater. **31**(24), 1901117 (2019)
157. Wang, Z., Xuan, J., Zhao, Z., Li, Q., Geng, F.: Versatile cutting method for producing fluorescent ultrasmall MXene sheets. ACS Nano **11**(11), 11559–11565 (2017)
158. Yang, X., Jia, Q., Duan, F., Hu, B., Wang, M., He, L., Zhang, Z.: Multiwall carbon nanotubes loaded with MoS2 quantum dots and MXene quantum dots: Non–Pt bifunctional catalyst for the methanol oxidation and oxygen reduction reactions in alkaline solution. Appl. Surf. Sci. **464**, 78–87 (2019)
159. Xue, Q., Zhang, H., Zhu, M., Pei, Z., Li, H., Wang, Z., Zhi, C.: Photoluminescent Ti3C2 MXene quantum dots for multicolor cellular imaging. Adv. Mater. **29**(15), 1604847 (2017)
160. Wang, J., Zhang, Z., Yan, X., Zhang, S., Wu, Z., Zhuang, Z., Han, W.Q.: Rational design of porous N-Ti 3 C 2 MXene@ CNT microspheres for high cycling stability in Li–S battery. Nano-Micro Lett. **12**, 1–14 (2020)
161. Qin, Y., Wang, Z., Liu, N., Sun, Y., Han, D., Liu, Y., Kang, Z.: High-yield fabrication of Ti 3 C 2 T x MXene quantum dots and their electrochemiluminescence behavior. Nanoscale **10**(29), 14000–14004 (2018)
162. Guan, Q., Ma, J., Yang, W., Zhang, R., Zhang, X., Dong, X., Xu, Q.: Highly fluorescent Ti 3 C 2 MXene quantum dots for macrophage labeling and Cu 2+ ion sensing. Nanoscale **11**(30), 14123–14133 (2019)
163. Yan, X., Ma, J., Yu, K., Li, J., Yang, L., Liu, J., Cai, L.: Highly green fluorescent Nb2C MXene quantum dots for Cu2+ ion sensing and cell imaging. Chin. Chem. Lett. **31**(12), 3173–3177 (2020)
164. Peng, C., Wei, P., Chen, X., Zhang, Y., Zhu, F., Cao, Y., Peng, F.: A hydrothermal etching route to synthesis of 2D MXene (Ti3C2, Nb2C): Enhanced exfoliation and improved adsorption performance. Ceram. Int. **44**(15), 18886–18893 (2018)
165. Shekhirev, M., Shuck, C.E., Sarycheva, A., Gogotsi, Y.: Characterization of MXenes at every step, from their precursors to single flakes and assembled films. Prog. Mater. Sci. **120**, 100757 (2021)
166. Naguib, M., Gogotsi, Y.: Synthesis of two-dimensional materials by selective extraction. Acc. Chem. Res. **48**(1), 128–135 (2015)
167. Ronchi, R.M., Arantes, J.T., Santos, S.F.: Synthesis, structure, properties and applications of MXenes: current status and perspectives. Ceram. Int. **45**(15), 18167–18188 (2019)
168. Wozniak, J., Jastrzębska, A., Olszyna, A.: Challenges and opportunities in tailoring MAX phases as a starting materials for MXenes development. Mater. Technol. **37**(11), 1639–1650 (2022)
169. Baig, M.M., Gul, I.H., Baig, S.M., Shahzad, F.: 2D MXenes: Synthesis, properties, and electrochemical energy storage for supercapacitors–A review. J. Electroanal. Chem. **904**, 115920 (2022)
170. Ying, G., Kota, S., Dillon, A.D., Fafarman, A.T., Barsoum, M.W.: Conductive transparent V2CTx (MXene) films. FlatChem **8**, 25–30 (2018)
171. Naguib, M., Mochalin, V.N., Barsoum, M.W., Gogotsi, Y.: 25th anniversary article: MXenes: a new family of two-dimensional materials. Adv. Mater. **26**(7), 992–1005 (2014)
172. Bhat, A., Anwer, S., Bhat, K.S., Mohideen, M.I.H., Liao, K., Qurashi, A.: Prospects challenges and stability of 2D MXenes for clean energy conversion and storage applications. NPJ 2D Mater. Appl. **5**(1), 61 (2021)

173. Fu, B., Sun, J., Wang, C., Shang, C., Xu, L., Li, J., Zhang, H.: MXenes: Synthesis, optical properties, and applications in ultrafast photonics. Small **17**(11), 2006054 (2021)
174. Zhang, C., Ma, Y., Zhang, X., Abdolhosseinzadeh, S., Sheng, H., Lan, W., Nüesch, F.: Two-dimensional transition metal carbides and nitrides (MXenes): synthesis, properties, and electrochemical energy storage applications. Energy Environ. **3**(1), 29–55 (2020)
175. Li, X., Ran, F., Yang, F., Long, J., Shao, L.: Advances in MXene films: synthesis, assembly, and applications. Trans. Tianjin Univ. **27**, 217–247 (2021)
176. Zhang, J., Zhao, Y., Guo, X., Chen, C., Dong, C.L., Liu, R.S., Wang, G.: Single platinum atoms immobilized on an MXene as an efficient catalyst for the hydrogen evolution reaction. Nat. Catal. **1**(12), 985–992 (2018)
177. Yun, Q., Lu, Q., Zhang, X., Tan, C., Zhang, H.: Three-dimensional architectures constructed from transition-metal dichalcogenide nanomaterials for electrochemical energy storage and conversion. Angew. Chem., Int. Ed. Engl. **57**(3), 626–646 (2018).
178. Bu, F., Zagho, M.M., Ibrahim, Y., Ma, B., Elzatahry, A., Zhao, D.: Porous MXenes: Synthesis, structures, and applications. Nano Today **30**, 100803 (2020)
179. Magnuson, M., Halim, J., Näslund, L.Å.: Chemical bonding in carbide MXene nanosheets. J. Electron Spectrosc. Relat. Phenom. **224**, 27–32 (2018)
180. Zha, X.H., Luo, K., Li, Q., Huang, Q., He, J., Wen, X., Du, S.: Role of the surface effect on the structural, electronic and mechanical properties of the carbide MXenes. EPL **111**(2), 26007 (2015)
181. Lipatov, A., Lu, H., Alhabeb, M., Anasori, B., Gruverman, A., Gogotsi, Y., Sinitskii, A.: Elastic properties of 2D Ti3C2T x MXene monolayers and bilayers. Sci. Adv. **4**(6), eaat0491 (2018)
182. Fu, Z., Wang, N., Legut, D., Si, C., Zhang, Q., Du, S., Zhang, R.: Rational design of flexible two-dimensional MXenes with multiple functionalities. Chem. Rev. **119**(23), 11980–12031 (2019)
183. Wang, L., Qiu, H., Song, P., Zhang, Y., Lu, Y., Liang, C., Gu, J.: 3D Ti3C2Tx MXene/C hybrid foam/epoxy nanocomposites with superior electromagnetic interference shielding performances and robust mechanical properties. Compos. Part A Appl. **123**, 293–300 (2019)
184. Meng, F., Seredych, M., Chen, C., Gura, V., Mikhalovsky, S., Sandeman, S., Gogotsi, Y.: MXene sorbents for removal of urea from dialysate: A step toward the wearable artificial kidney. ACS Nano **12**(10), 10518–10528 (2018)
185. Shan, Q., Mu, X., Alhabeb, M., Shuck, C.E., Pang, D., Zhao, X., Dall'Agnese, Y.: Two-dimensional vanadium carbide (V2C) MXene as electrode for supercapacitors with aqueous electrolytes. Electrochem. commun. **96**, 103–107 (2018)
186. Shuck, C.E., Gogotsi, Y.: Taking MXenes from the lab to commercial products. J. Chem. Eng. **401**, 125786 (2020)
187. Salles, P., Pinto, D., Hantanasirisakul, K., Maleski, K., Shuck, C.E., Gogotsi, Y.: Electrochromic effect in titanium carbide MXene thin films produced by dip-coating. Adv. Funct. Mater. **29**(17), 1809223 (2019)
188. Seredych, M., Shuck, C.E., Pinto, D., Alhabeb, M., Precetti, E., Deysher, G., Gogotsi, Y.: High-temperature behavior and surface chemistry of carbide MXenes studied by thermal analysis. Chem. Mater. **31**(9), 3324–3332 (2019)
189. Lipatov, A., Alhabeb, M., Lu, H., Zhao, S., Loes, M.J., Vorobeva, N.S., Sinitskii, A.: Electrical and elastic properties of individual single-layer Nb4C3Tx MXene flakes. Adv. Electron. Mater. **6**(4), 1901382 (2020)
190. Anasori, B., Gogotsi, Y.: Introduction to 2D transition metal carbides and nitrides (MXenes). 2D Metal Carbides and Nitrides (MXenes) Structure, Properties and Applications, 3–12 (2019)
191. Ashraf, M., Shah, S.S., Khan, I., Aziz, M.A., Ullah, N., Khan, M., Tahir, M.N.: A high-performance asymmetric supercapacitor based on tungsten oxide nanoplates and highly reduced graphene oxide electrodes. Chem. Eur. J. **27**(23), 6973–6984 (2021)
192. Sarycheva, A., Polemi, A., Liu, Y., Dandekar, K., Anasori, B., Gogotsi, Y.: 2D titanium carbide (MXene) for wireless communication. Sci. Adv. **4**(9), eaau0920 (2018)

193. Yun, T., Kim, H., Iqbal, A., Cho, Y.S., Lee, G.S., Kim, M.K., Koo, C.M.: Electromagnetic shielding of monolayer MXene assemblies. Adv. Mater. **32**(9), 1906769 (2020)
194. Zhang, Y., Wang, L., Zhang, N., Zhou, Z.: Adsorptive environmental applications of MXene nanomaterials: a review. RSC Adv. **8**(36), 19895–19905 (2018)
195. Huang, K., Li, Z., Lin, J., Han, G., Huang, P.: Two-dimensional transition metal carbides and nitrides (MXenes) for biomedical applications. Chem. Soc. Rev. **47**(14), 5109–5124 (2018)
196. Lee, K.H., Zhang, Y.Z., Jiang, Q., Kim, H., Alkenawi, A.A., Alshareef, H.N.: Ultrasound-driven two-dimensional Ti3C2T x MXene hydrogel generator. ACS Nano **14**(3), 3199–3207 (2020)
197. Ward, E.J., Lacey, J., Crua, C., Dymond, M.K., Maleski, K., Hantanasirisakul, K., Sandeman, S.: 2D titanium carbide (Ti3C2Tx) in accommodating intraocular lens design. Adv. Funct. Mater. **30**(47), 2000841 (2020)
198. Raziq, F., Hayat, A., Humayun, M., Mane, S.K.B., Faheem, M.B., Ali, A., Qiao, L.: Photocatalytic solar fuel production and environmental remediation through experimental and DFT based research on CdSe-QDs-coupled P-doped-g-C3N4 composites. Appl. Catal. B **270**, 118867 (2020)
199. Raziq, F., He, J., Gan, J., Humayun, M., Faheem, M.B., Iqbal, A., Qiao, L.: Promoting visible-light photocatalytic activities for carbon nitride based 0D/2D/2D hybrid system: beyond the conventional 4-electron mechanism. Appl. Catal. B **270**, 118870 (2020)
200. Raziq, F., Humayun, M., Ali, A., Wang, T., Khan, A., Fu, Q., Qiao, L.: Synthesis of S-Doped porous g-C3N4 by using ionic liquids and subsequently coupled with Au-TiO2 for exceptional cocatalyst-free visible-light catalytic activities. Appl. Catal. B **237**, 1082–1090 (2018)
201. Li, Z., Qu, Y., Hu, K., Humayun, M., Chen, S., Jing, L.: Improved photoelectrocatalytic activities of BiOCl with high stability for water oxidation and MO degradation by coupling RGO and modifying phosphate groups to prolong carrier lifetime. Appl. Catal. B **203**, 355–362 (2017)
202. Shan, J., Raziq, F., Humayun, M., Zhou, W., Qu, Y., Wang, G., Li, Y.: Improved charge separation and surface activation via boron-doped layered polyhedron SrTiO3 for co-catalyst free photocatalytic CO2 conversion. Appl. Catal. B **219**, 10–17 (2017)
203. Chen, S., Yan, R., Zhang, X., Hu, K., Li, Z., Humayun, M., Jing, L.: Photogenerated electron modulation to dominantly induce efficient 2, 4-dichlorophenol degradation on BiOBr nanoplates with different phosphate modification. Appl. Catal. B **209**, 320–328 (2017)
204. Raziq, F., Qu, Y., Humayun, M., Zada, A., Yu, H., Jing, L.: Synthesis of SnO2/BP codoped g-C3N4 nanocomposites as efficient cocatalyst-free visible-light photocatalysts for CO_2 conversion and pollutant degradation. Appl. Catal. B **201**, 486–494 (2017)
205. Humayun, M., Zada, A., Li, Z., Xie, M., Zhang, X., Qu, Y., Jing, L.: Enhanced visible-light activities of porous BiFeO3 by coupling with nanocrystalline TiO2 and mechanism. Appl. Catal. B **180**, 219–226 (2016)
206. Humayun, M., Sun, N., Raziq, F., Zhang, X., Yan, R., Li, Z., Jing, L.: Synthesis of ZnO/Bi-doped porous LaFeO3 nanocomposites as highly efficient nano-photocatalysts dependent on the enhanced utilization of visible-light-excited electrons. Appl. Catal. B **231**, 23–33 (2018)
207. Usman, M., Zhu, J., Chuiyang, K., Arslan, M.T., Khan, A., Galadima, A., Yamani, Z.H.: Propene adsorption-chemisorption behaviors on H-SAPO-34 zeolite catalysts at different temperatures. Catalysts **9**(11), 919 (2019)
208. Kuang, P., Low, J., Cheng, B., Yu, J., Fan, J.: MXene-based photocatalysts. J. Mater. Sci. Technol **56**, 18–44 (2020)
209. Li, Y., Jin, Z., Zhang, L., Fan, K.: Controllable design of Zn-Ni-P on g-C3N4 for efficient photocatalytic hydrogen production. Chinese J. Catal. **40**(3), 390–402 (2019)
210. Xu, F., Zhang, D., Liao, Y., Wang, G., Shi, X., Zhang, H., Xiang, Q.: Synthesis and photocatalytic H2-production activity of plasma-treated Ti3C2Tx MXene modified graphitic carbon nitride. J. Am. Ceram. **103**(2), 849–858 (2020)
211. Zhan, X., Si, C., Zhou, J., Sun, Z.: MXene and MXene-based composites: synthesis, properties and environment-related applications. Nanoscale Horiz **5**(2), 235–258 (2020)

212. Khan, I., Baig, N., Ali, S., Usman, M., Khan, S.A., Saeed, K.: Progress in layered cathode and anode nanoarchitectures for charge storage devices: challenges and future perspective. Energy Stor. **35**, 443–469 (2021)
213. Li, Z., Wu, Y.: 2D early transition metal carbides (MXenes) for catalysis. Small **15**(29), 1804736 (2019)
214. Cheng, L., Li, X., Zhang, H., Xiang, Q.: Two-dimensional transition metal MXene-based photocatalysts for solar fuel generation. J. Phys. Chem. Lett. **10**(12), 3488–3494 (2019)
215. Liu, Q., Ai, L., Jiang, J.: MXene-derived TiO 2@ C/gC 3 N 4 heterojunctions for highly efficient nitrogen photofixation. J. Mater. Chem. A **6**(9), 4102–4110 (2018)
216. Sun, H., Li, L., Humayun, M., Zhang, H., Bo, Y., Ao, X., Xiong, Y.: Achieving highly efficient pH-universal hydrogen evolution by superhydrophilic amorphous/crystalline Rh (OH) 3/NiTe coaxial nanorod array electrode. Appl. Catal. B **305**, 121088 (2022)
217. Yang, S., Zhang, P., Wang, F., Ricciardulli, A.G., Lohe, M.R., Blom, P.W., Feng, X.: Fluoride-free synthesis of two-dimensional titanium carbide (MXene) using a binary aqueous system. Angew. Chem. **130**(47), 15717–15721 (2018)
218. Gao, G., O'Mullane, A.P., Du, A.: 2D MXenes: a new family of promising catalysts for the hydrogen evolution reaction. ACS Catal. **7**(1), 494–500 (2017)
219. Bai, S., Yang, M., Jiang, J., He, X., Zou, J., Xiong, Z., Liu, S.: Recent advances of MXenes as electrocatalysts for hydrogen evolution reaction. NPJ 2D Mater. Appl. **5**(1), 78 (2021)
220. Cao, S., Shen, B., Tong, T., Fu, J., Yu, J.: 2D/2D heterojunction of ultrathin MXene/Bi2WO6 nanosheets for improved photocatalytic CO2 reduction. Adv. Funct. Mater. **28**(21), 1800136 (2018)
221. Pandey, M., Thygesen, K.S.: Two-dimensional MXenes as catalysts for electrochemical hydrogen evolution: a computational screening study. J. Phys. Chem. C **121**(25), 13593–13598 (2017)
222. Jiang, Y., Sun, T., Xie, X., Jiang, W., Li, J., Tian, B., Su, C.: Oxygen-functionalized ultrathin Ti3C2Tx MXene for enhanced electrocatalytic hydrogen evolution. Chemsuschem **12**(7), 1368–1373 (2019)
223. Handoko, A.D., Fredrickson, K.D., Anasori, B., Convey, K.W., Johnson, L.R., Gogotsi, Y., Seh, Z.W.: Tuning the basal plane functionalization of two-dimensional metal carbides (MXenes) to control hydrogen evolution activity. ACS Appl. Energy Mater. **1**(1), 173–180 (2017)
224. Mohammed, T.M., Soomro, R.A., Han, X., Zhu, Q., Wei, Y., Xu, B.: Application of MXenes in environmental remediation technologies. Nano Converg. **8**(1) (2021)
225. Mara, D., Horan, N.J.: Handbook of Water and Wastewater Microbiology. Elsevier (Eds.) (2003)
226. Handoko, A.D., Steinmann, S.N., Seh, Z.W.: Theory-guided materials design: two-dimensional MXenes in electro-and photocatalysis. Nanoscale Horiz **4**(4), 809–827 (2019)
227. Wei, Z., Peigen, Z., Wubian, T., Xia, Q., Yamei, Z., ZhengMing, S.: Alkali treated Ti3C2Tx MXenes and their dye adsorption performance. Mater. Chem. Phys. **206**, 270–276 (2018)
228. Li, K., Zou, G., Jiao, T., Xing, R., Zhang, L., Zhou, J., Peng, Q.: Self-assembled MXene-based nanocomposites via layer-by-layer strategy for elevated adsorption capacities. Colloids Surf. **553**, 105–113 (2018)
229. Cai, C., Wang, R., Liu, S., Yan, X., Zhang, L., Wang, M., Jiao, T.: Synthesis of self-assembled phytic acid-MXene nanocomposites via a facile hydrothermal approach with elevated dye adsorption capacities. Colloids Surf. **589**, 124468 (2020)
230. Rahman, R.A., Ibrahium, H.A., Hung, Y.T.: Liquid radioactive wastes treatment: a review. Water **3**(2), 551–565 (2011)
231. Hwang, S.K., Kang, S.M., Rethinasabapathy, M., Roh, C., Huh, Y.S.: MXene: An emerging two-dimensional layered material for removal of radioactive pollutants. J. Chem. Eng. **397**, 125428 (2020)
232. Wang, L., Yuan, L., Chen, K., Zhang, Y., Deng, Q., Du, S., Shi, W.: Loading actinides in multilayered structures for nuclear waste treatment: the first case study of uranium capture with vanadium carbide MXene. ACS Appl. Mater. Interfaces **8**(25), 16396–16403 (2016)

233. Wang, L., Tao, W., Yuan, L., Liu, Z., Huang, Q., Chai, Z., Shi, W.: Rational control of the interlayer space inside two-dimensional titanium carbides for highly efficient uranium removal and imprisonment. Chem comm **53**(89), 12084–12087 (2017)
234. Wang, L., Song, H., Yuan, L., Li, Z., Zhang, Y., Gibson, J.K., Shi, W.: Efficient U (VI) reduction and sequestration by Ti2CT x MXene. Environ. Sci. Technol. **52**(18), 10748–10756 (2018)
235. Wang, L., Song, H., Yuan, L., Li, Z., Zhang, P., Gibson, J.K., Shi, W.: Effective removal of anionic Re (VII) by surface-modified Ti2CT x MXene nanocomposites: implications for Tc (VII) sequestration. Environ. Sci. Technol. **53**(7), 3739–3747 (2019)
236. Li, S., Wang, L., Peng, J., Zhai, M., Shi, W.: Efficient thorium (IV) removal by two-dimensional Ti2CTx MXene from aqueous solution. J. Chem. Eng. **366**, 192–199 (2019)
237. Zhang, P., Wang, L., Yuan, L.Y., Lan, J.H., Chai, Z.F., Shi, W.Q.: Sorption of Eu (III) on MXene-derived titanate structures: the effect of nano-confined space. J. Chem. Eng. **370**, 1200–1209 (2019)
238. Carolin, C.F., Kumar, P.S., Saravanan, A., Joshiba, G.J., Naushad, M.: J. Environ. Chem. Eng. (2017)
239. Peng, Q., Guo, J., Zhang, Q., Xiang, J., Liu, B., Zhou, A., Tian, Y.: Unique lead adsorption behavior of activated hydroxyl group in two-dimensional titanium carbide. J. Am. Chem. Soc. **136**(11), 4113–4116 (2014)
240. Zhang, Q., Teng, J., Zou, G., Peng, Q., Du, Q., Jiao, T., Xiang, J.: Efficient phosphate sequestration for water purification by unique sandwich-like MXene/magnetic iron oxide nanocomposites. Nanoscale **8**(13), 7085–7093 (2016)
241. Goh, P.S., Ismail, A.F.: A review on inorganic membranes for desalination and wastewater treatment. Desalination **434**, 60–80 (2018)
242. Ren, C.E., Hatzell, K.B., Alhabeb, M., Ling, Z., Mahmoud, K.A., Gogotsi, Y.: Charge-and size-selective ion sieving through Ti3C2T x MXene membranes. J. Phys. Chem. Lett. **6**(20), 4026–4031 (2015)
243. Ding, L., Wei, Y., Wang, Y., Chen, H., Caro, J., Wang, H.: A two-dimensional lamellar membrane: MXene nanosheet stacks. Angew. Chem. Int. Ed. **56**(7), 1825–1829 (2017)
244. Kang, K.M., Kim, D.W., Ren, C.E., Cho, K.M., Kim, S.J., Choi, J.H., Jung, H.T.: Selective molecular separation on Ti3C2T x–graphene oxide membranes during pressure-driven filtration: comparison with graphene oxide and MXenes. ACS Appl. Mater. Interfaces **9**(51), 44687–44694 (2017)
245. Han, R., Ma, X., Xie, Y., Teng, D., Zhang, S.: Preparation of a new 2D MXene/PES composite membrane with excellent hydrophilicity and high flux. RSC Adv. **7**(89), 56204–56210 (2017)
246. Pandey, R.P., Rasool, K., Madhavan, V.E., Aïssa, B., Gogotsi, Y., Mahmoud, K.A.: Ultrahigh-flux and fouling-resistant membranes based on layered silver/MXene (Ti 3 C 2 T x) nanosheets. J. Mater. Chem. A **6**(8), 3522–3533 (2018)
247. Gao, X., Li, Z.K., Xue, J., Qian, Y., Zhang, L.Z., Caro, J., Wang, H.: Titanium carbide Ti3C2Tx (MXene) enhanced PAN nanofiber membrane for air purification. J. Membr. Sci. **586**, 162–169 (2019)
248. Tunesi, M.M., Soomro, R.A., Han, X., Zhu, Q., Wei, Y., Xu, B.: Application of MXenes in environmental remediation technologies. Nano Converg. **8**(1), 1–19 (2021)
249. Babu, B., Akkinepally, B., Shim, J., Yoo, K.: Facile one-step synthesis of pellet-press-assisted saddle-curl-edge-like g-C3N4 nanosheets for improved visible-light photocatalytic activity. Ceram. Int. **45**(12), 15178–15187 (2019)
250. Jamble, S.N., Ghoderao, K.P., Kale, R.B.: Effect of temperature on structural and optical properties of solvothermal assisted CdS nanowires with enhanced photocatalytic degradation under natural sunlight irradiation. Res. Chem. Intermed. **45**, 1381–1399 (2019)
251. Gao, Y., Wang, L., Zhou, A., Li, Z., Chen, J., Bala, H., Cao, X.: Hydrothermal synthesis of TiO2/Ti3C2 nanocomposites with enhanced photocatalytic activity. Mater. Lett. **150**, 62–64 (2015)
252. Peng, C., Yang, X., Li, Y., Yu, H., Wang, H., Peng, F.: Hybrids of two-dimensional Ti3C2 and TiO2 exposing 001 facets toward enhanced photocatalytic activity. ACS Appl. Mater. Interfaces **8**(9), 6051–6060 (2016)

253. Peng, C., Wang, H., Yu, H., Peng, F.: (111) TiO2-x/Ti3C2: Synergy of active facets, interfacial charge transfer and Ti3+ doping for enhance photocatalytic activity. Mater. Res. Bull. **89**, 16–25 (2017)
254. Liu, T., Zhou, T., Yao, Y., Zhang, F., Liu, L., Liu, Y., Leng, J.: Stimulus methods of multi-functional shape memory polymer nanocomposites: A review. Compos. Part A Appl. **100**, 20–30 (2017)
255. Zhou, W., Zhu, J., Wang, F., Cao, M., Zhao, T.: One-step synthesis of Ceria/Ti3C2 nanocomposites with enhanced photocatalytic activity. Mater. Lett. **206**, 237–240 (2017)
256. Zhang, H., Li, M., Zhu, C., Tang, Q., Kang, P., Cao, J.: Preparation of magnetic α-Fe2O3/ZnFe2O4@ Ti3C2 MXene with excellent photocatalytic performance. Ceram. Int. **46**(1), 81–88 (2020)
257. Tariq, A., Ali, S.I., Akinwande, D., Rizwan, S.: Efficient visible-light photocatalysis of 2D-MXene nanohybrids with Gd3+-and Sn4+-codoped bismuth ferrite. ACS Omega **3**(10), 13828–13836 (2018)
258. Fang, Y., Cao, Y., Chen, Q.: Synthesis of an Ag2WO4/Ti3C2 Schottky composite by electrostatic traction and its photocatalytic activity. Ceram. Int. **45**(17), 22298–22307 (2019)
259. Cai, T., Wang, L., Liu, Y., Zhang, S., Dong, W., Chen, H., Luo, S.: Ag3PO4/Ti3C2 MXene interface materials as a Schottky catalyst with enhanced photocatalytic activities and anti-photocorrosion performance. Appl. Catal. B., 545–554 (2018)
260. Liu, N., Lu, N., Su, Y., Wang, P., Quan, X.: Fabrication of g-C3N4/Ti3C2 composite and its visible-light photocatalytic capability for ciprofloxacin degradation. Sep. Purif. Technol. **211**, 782–789 (2019)
261. Ding, X., Li, C., Wang, L., Feng, L., Han, D., Wang, W.: Fabrication of hierarchical g-C3N4/MXene-AgNPs nanocomposites with enhanced photocatalytic performances. Mater. Lett. **247**, 174–177 (2019)
262. Bao, Z., Lu, C., Cao, X., Zhang, P., Yang, L., Zhang, H., Sun, Z.: Role of MXene surface terminations in electrochemical energy storage: A review. Chin. Chem. Lett. **32**(9), 2648–2658 (2021)
263. Zhu, Y., Wu, Z., Xie, X., Zhang, N.: A retrospective on MXene-based composites for solar fuel production. Pure Appl. Chem. **92**(12), 1953–1969 (2020)
264. Xie, X., Wu, Z., Zhang, N.: Robust and easily retrievable Pd/Ti3C2Tx⊂ graphene hydrogels for efficient catalytic hydrogenation of nitroaromatic compounds. Chin. Chem. Lett. **31**(4), 1014–1017 (2020)
265. Luo, S., Wang, R., Yin, J., Jiao, T., Chen, K., Zou, G., Peng, Q.: Preparation and dye degradation performances of self-assembled MXene-Co3O4 nanocomposites synthesized via solvothermal approach. ACS Omega **4**(2), 3946–3953 (2019)
266. Yin, J., Zhang, L., Jiao, T., Zou, G., Bai, Z., Chen, Y., Peng, Q.: Highly efficient catalytic performances of nitro compounds and morin via self-assembled MXene-Pd nanocomposites synthesized through self-reduction strategy. Nanomater. **9**(7), 1009 (2019)
267. Soomro, R.A., Jawaid, S., Zhu, Q., Abbas, Z., Xu, B.: A mini-review on MXenes as versatile substrate for advanced sensors. Chin. Chem. Lett. **31**(4), 922–930 (2020)
268. Rasheed, P.A., Pandey, R.P., Rasool, K., Mahmoud, K.A.: Ultra-sensitive electrocatalytic detection of bromate in drinking water based on Nafion/Ti3C2Tx (MXene) modified glassy carbon electrode. Sens. Actuators B Chem. **265**, 652–659 (2018)
269. Cheng, H., Yang, J.: Preparation of Ti3C2-PANI composite as sensor for electrochemical determination of mercury ions in water. Int. J. Electrochem. **15**, 2295–2306 (2020)
270. Jiang, Y., Zhang, X., Pei, L., Yue, S., Ma, L., Zhou, L., Gao, J.: Silver nanoparticles modified two-dimensional transition metal carbides as nanocarriers to fabricate acetycholinesterase-based electrochemical biosensor. J. Chem. Eng.. **339**, 547–556 (2018)
271. Wu, D., Wu, M., Yang, J., Zhang, H., Xie, K., Lin, C.T., Fu, L.: Delaminated Ti3C2Tx (MXene) for electrochemical carbendazim sensing. Mater. Lett. **236**, 412–415 (2019)
272. Xie, Y., Gao, F., Tu, X., Ma, X., Xu, Q., Dai, R., Lu, L.: Facile synthesis of MXene/electrochemically reduced graphene oxide composites and their application for electrochemical sensing of carbendazim. J. Electrochem. Soc. **166**(16), B1673 (2019)

273. Soomro, R.A., Jawaid, S., Kalawar, N.H., Tunesi, M., Karakuş, S., Kilislioğlu, A., Willander, M.: In-situ engineered MXene-TiO2/BiVO4 hybrid as an efficient photoelectrochemical platform for sensitive detection of soluble CD44 proteins. Biosens. Bioelectron. **166**, 112439 (2020)
274. Kalambate, P.K., Gadhari, N.S., Li, X., Rao, Z., Navale, S.T., Shen, Y., Huang, Y.: Recent advances in MXene–based electrochemical sensors and biosensors. TrAC Trends Anal. Chem. **120**, 115643 (2019)
275. Wu, L., Lu, X., Wu, Z.S., Dong, Y., Wang, X., Zheng, S., Chen, J.: 2D transition metal carbide MXene as a robust biosensing platform for enzyme immobilization and ultrasensitive detection of phenol. Biosens. Bioelectron. **107**, 69–75 (2018)

2D Metal Carbides and Nitrides (MXenes) in Water Treatment

Ankita Yadav, Swati Agarwal, and Suphiya Khan

Abstract One of the biggest health concerns in the world is the rise in pollutants like hazardous refractory impurities, organic compound dyes, pharmaceuticals, and pesticides that are released into water supplies as a result of population growth and global industrialization, due to its distinctive and versatile qualities, such as surface functionality, high photothermal efficiency, and tunable interlayer spacing. The scientific community is very interested in the potential benefits of MXenes for water purification because of their high conductivity, hydrophilicity, and catalytic activity. MXenes showed significant sorption selectivity and effective reduction capabilities for several water contaminants. Adsorbents, electrodes, desalination membranes, and antibacterial agents for purifying water are just some of the ways MXenes have advanced in recent years. To shed insight on the development of MXene-based membranes, idyllic nominees for water purification, and their practical separation applications, future views and concerns are also discussed.

Keywords MXenes · Desalination · Adsorbent · Two-dimensional materials · Water purification

1 Introduction

Water, a valuable natural resource, directly affects the circumstances of human survival. However, the issue of water contamination follows and worsens due to the nation's economy and society developing so quickly [1]. The public is concerned about a variety of organic pollutants released into our rivers, lakes, and seas without any efficient treatment since they negatively impact the water quality [2–4]. According to estimates, World Water Day highlights that about 2.5 billion people

A. Yadav · S. Khan (✉)
Department of Bioscience and Biotechnology, Banasthali Vidyapith, Tonk 304022, Rajasthan, India
e-mail: suphiyakhan@gmail.com

S. Agarwal
Drumlins Water Technologies Pvt. Ltd, Jaipur 302005, Rajasthan, India

N. Talreja et al. (eds.), *Two-dimensional Hybrid Composites*, Engineering Materials,
https://doi.org/10.1007/978-981-99-8010-9_5

lack the most basic sanitation, putting them at an increased risk of contracting infectious diseases and dying prematurely [5, 6]. The World Health Organisation (WHO) estimates that diseases and decreasing water quality could cause one newborn baby death worldwide every minute. Surprisingly, this water issue is quite serious, and the growing water quality issue has been seriously disrupting the hydrological cycle and posing a serious threat to humanity's ability to obtain clean and safe water [7]. Therefore, maintaining livelihoods and protecting people from organic contaminants in the water system and water-borne diseases requires efficient and affordable water treatment technology.

Chemical precipitation, flocculation, electrolysis, coagulation, reverse osmosis, membrane separation, ion exchange, evaporation, and adsorption are examples of traditional water treatment methods that have been tried and failed to remove complex and recalcitrant organic pollutants from water streams [8]. The methods have several problems, including low removal efficiency, high sewage sludge production, inefficiency with regard to cost and energy, the development of toxic by-products, and the addition of potentially dangerous chemicals into the ecosystem [9]. The most efficient and cost-effective technique has been discovered to combine adsorption, desalination, and photocatalysis. Furthermore, they can be used as adsorbents to adsorb compounds from contaminated solutions [10]. To name a few of these cutting-edge functional nanomaterials, there are some metal–organic frameworks (MOFs), zeolites, carbon nanostructures, and ceramics. Developing useful nanomaterial-based devices can help us enhance applications for environmental remediation [11]. However, their use is restricted by poor selectivity, a low clearance rate, exorbitant costs, and other drawbacks [12, 13].

Researchers also employ two-dimensional (2D) nanomaterials because of their distinct structures and functionalities to solve environmental issues. A wide range of substances, including oxides, halides, hydroxides, transition metal chalcogenides, and graphene and its derivatives, are included in the family of 2D nanomaterials [14]. 2D nanomaterials are very thin in structure and have lateral dimensions (tens of nanometers to a few micrometers) [15]. The enormous lateral dimensions and nanoscale thickness of 2D nanomaterials make them appropriate for various environmental remediation applications, including adsorption, sensing, and catalysis [16]. Graphene-based materials are commonly used for the environmental remediation process. It can be used as sorbents, membrane separators, and sometimes as photocatalytic materials for waste purification [17, 18]. Historically, GO and GO derivatives are the most exploited 2D materials for removing various contaminants from aqueous solutions [19, 20]. However, the possible applications of graphene are limited by how easily it can be stacked again to create graphite [21].

MXenes are the family of transition metal carbides, nitrides, and carbonitride compounds. The general formula used for MXenes is $Mn^{+1}X_nT_x$ (n = 1–3), where M represents the transition metal group, X is used for carbon or nitrogen, and T_x is used for surface terminal groups such as –OH, –O– or F. A new kind of 2D nanomaterial, MXenes, has recently emerged from this group [22]. So far, different MXenes have been synthesized successfully, which are $Ti_3C_2T_x$, Ti_3CNT_x, $Ti_4C_3T_x$, Ti_2CT_x, $Nb_4C_3T_x$ [23], V_2CT_x, Nb_2CT_x [24], and $Nb_4C_3T_x$ [25].

Other MXene compositions have been created, such as $Mo_2Ti_2C_3$, $(V_{0.5}, Cr_{0.5})_3C_2$, and $(Ti_{0.5}, Nb_{0.5})_2C$, all of which feature more than one transition metal in the M layers [26]. However, applying selective acid etching techniques to make nitride MXenes is still fascinatingly difficult. The only material that stands out is titanium carbonitride, Ti_3CN, which has been researched for various uses [25]. In-depth research into nitride and carbonitride MXene for environmental remediation applications is thought to be hindered by this.

MXenes are ideal for membrane separation and waste removal because of their beneficial properties. MXenes contain abundant, non-toxic components, including Ti, N, and C [25]. MXenes are, therefore, prospective and effective photothermal materials and adsorbents that can be employed for environmental remediation [26, 27]. The titanium-based MXene $Ti_3C_2T_x$ has the most promising usage possibilities in water treatment [28]. The specialized fields include heavy metal removal [29, 30], organic dye removal [31], removal of radionuclides [32], and bacteria removal [33], among others. This chapter describes what are MXenes and their derivatives, and their application in the treatment of water. The first section of the chapter initially describes the methods for MXene synthesis, including etching, delamination, and surface modification. In the second section, the important properties of MXene in terms of conductivity, selectivity, and stability were discussed. Finally, we spoke about how MXene may be used to purify water by removing radionuclides, dyes, heavy metals, and other organic and inorganic impurities by adsorption and membrane separation.

2 MAX Phases and Synthesis of MXenes

A-layers, a layered material, join 2D $Mn^{+1}X_n$ in MAX phases. Elements that can be used are Al or Si due to the M-A bond's higher degree of fragility and comparative failure compared to the M-X bond. As a result, the MAX phases' A-group atoms are stripped out to generate MXenes (Fig. 1a) [34, 35]. Their increased strength and stiffness enhance the performance over time of MXenes for industrial applications like water treatment.

2.1 *Conventional Synthesis Procedure*

After wet etching to form MXenes from a precursor MAX phase [36], bulk crystals are exfoliated into single- or multilayer MXene sheets using an etchant, typically hydrofluoric acid (HF). To explain the synthesis of MXenes, we can use the following equations:

$$Mn^{+1}AX_n(s) + 3HF\,(aq) => Mn^{+1}Xn(s) + AF_3(s) + 3/2\,H_2(g) \quad (1)$$

$$Mn^{+1}X_n(s) + 2H_2O(l) => Mn^{+1}Xn(OH)_2(s) + H_2(g) \quad (2)$$

$$Mn^{+1}X_n(s) + 2HF(aq) => Mn^{+1}XnF_2(s) + H_2(g) \quad (3)$$

(1) As a result of the interaction between HF (or fluoride ions) and the A layers of MAX phases, MXene layers are formed.
(2) All A atoms in MAX phases have been swapped out for hydrophilic -OH, -O, and/or -F groups.
(3) When the connections among the $Mn^{+1}X_n$ layers break down, MXenes, which resemble graphite in their loose packing, develop.

MXene having negatively charged surfaces resulting from forming different functional groups (−OH, −O, and/or −F) results in solid dispersions (synthesized flakes with diameters between 1 and 10 m). For the creation of Ti_3C_2Tx MXene, which was originally synthesized in 2011, Ti_3AlC_2 is the most popular MAX precursor [37–39]. For the synthesis of Mo_2C, Ti_2C, Ti_3SiC_2, and some other MXenes, the HF etching approach has been commonly utilized [40]. However, there are risks associated with consumption and handling when synthesizing MXenes using HF etchants. Additionally, this method fails to produce stable nitride-based MXenes ($Tin^{+1}N_n$) [41]. Therefore, creating more eco-friendly techniques for the synthesis of MXene is required.

2.2 *Novel Approaches for Synthesizing MXene*

Softer and less damaging etchants have recently been developed in contrast to HF to exfoliate the classic and evolving MAX phases [41]. These methods may be broken down into three classes: (a) hydrothermal procedures, (b) MXene synthesis without the use of fluoride, and (c) MXene synthesis utilizing gentler fluorine salts. Multi-layered, early transition metal carbides may be made using these techniques, having a wide range of compositions and tunable physical and chemical characteristics (Fig. 1b) [42]. For instance, $Ti_3C_2T_x$ MXene monolayers made employing a mixture of lithium fluoride (LiF) and HCl are of greater quality and have defect-free surfaces than MXenes formed with normal HF etching (Table 1). $Ti_3C_2T_x$, utilized to remove electrically negative compounds from waterways, can be created using milder etchant solutions such as ammonium bifluoride (NH_4HF_2) [43]. Hydrothermal processes remove the risk of exposure to toxic HF vapors. Hydrothermal etching of Ti_3AlC_2 sheets with etchants such as NH_4F results in $Ti_3C_2T_x$ MXene [44]. Researchers have shown that hydrothermal etching using $NaBF_4$ and HCl may produce Ti_3C_2 and Nb_2C MXenes without needing HF [45, 46]. When compared to HF-etched MXenes, those generated by hydrothermal methods are markedly superior in terms of their ability to remove A groups, the purity of their bilayers, the simplicity with which they may be exfoliated, and the value of their c-lattice parameter [45]. MXene

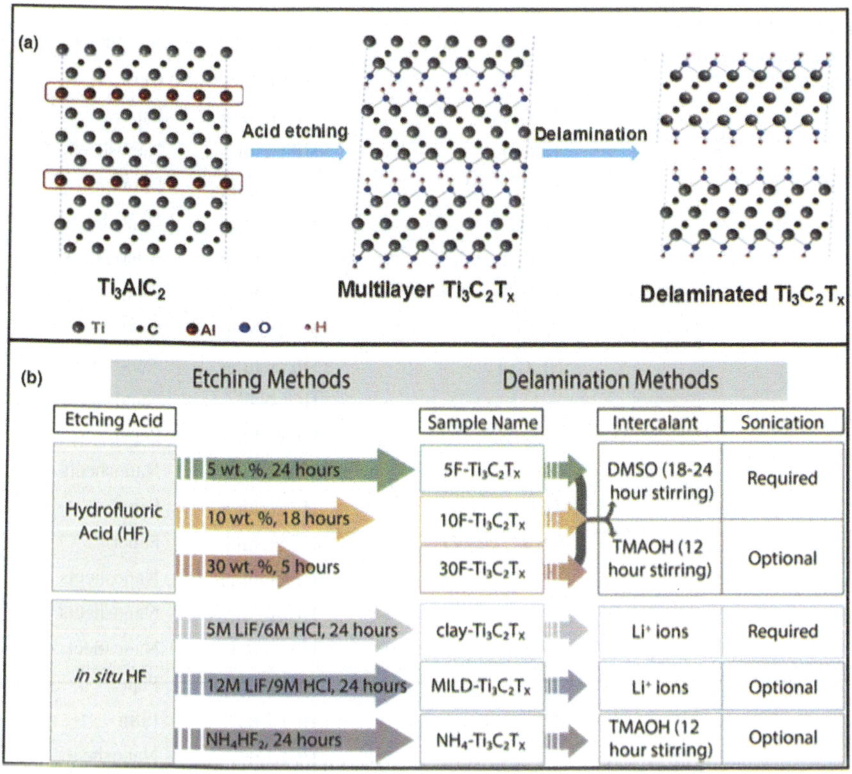

Fig. 1 **a** Diagram showing the changes in Ti_3AlC_2's atomic structure as it is exfoliated from the MAX phase precursor (Ti_3AlC) to ML-Ti_3C_2Tx and then, after cation intercalation, to DL-Ti_3C_2Tx. (Adapted from Ref. [50] and reproduced with permission of Copyright! 2019 Materials Today, **b** Synthesis approaches for preparing DL-Ti_3C_2Tx utilizing direct HF and in situ HF techniques (Ref. [35] and reproduced with permission of Copyright! 2019 from Materials Today

may now be synthesized without the use of HF thanks to advances in the field [47] that have enabled different methods, such as melting salt, salty-templated, and chemically deposited vapor. After soaking Ti_3AlC_2 in 1 M NaOH for 100 h at 80 degrees [48], Al-layers from MAX phases may be produced by hydrothermally treating the material in 1 M H_2SO_4 for two hours at 80 degrees. Due to its length, however, scalability must take precedence over optimization. Ti_3AlC_2 anodic corrosion is a possible alternative approach; however, it has not been widely used yet [49]. More research is needed to find a safer way to synthesize MXenes while still achieving the desired controlled surface functionality, high surface area, and effective physiochemical strength.

Table 1 MXene hybrid composition, synthesis, and structure [56, 57]

MXene hybrid	Synthesis	Morphology
$Ti_3C_2T_x$	HCl-LiF	Clay
$Ti_3C_2T_x$	HF etching	Paper
$Ti_3C_2T_x$	HF etching	Nanosheets
Ti_3C_2-C_3N_4	HF etching	Film
Mo_2CT_x	HF etching	Nanosheets
$Ti_3C_2T_x$-EG	HF etching	Nanosheets
$Ti_3C_2T_x$-CNT	HF etching	Film
Ti_2CO_x	HF etching	Particles
$Hf_3C_2T_z$	HF etching	Nanosheets
V_2C	HCl-NaF	Nanosheets
Mo_2TiC_2	HF etching	Paper
$Nb_2C_3T_x$	HF etching	Nanosheets
Ti_3C_2	HF etching	Paper
Nb_2CT_x-CNT	HF etching	Paper
V_2CT_x	HF etching	Nanosheets
NaV_2CT_x-HC	HF etching	Nanosheets
Sulphur-T_2C	HF etching	Nanosheets
$Ti_3C_2T_x$-polypyrrole	HCl-LiF	Paper
d-$Ti_3C_2T_x$-glycine	HCl-LiF	Film
$Ti_3C_2T_x$-rGO	HCl-LiF	Nanosheets
$Na_{0.23}TiO_2$-Ti_3C_2	HF etching	Sandwich
$Ti_3C_2T_x$-CMK_5	HF etching	Nanosheets
$Li_4Ti_5O_{12}$-$Ti_3C_2T_x$	HF etching	Nanosheets
$Ti_3C_2T_x$-SnS_2	HF etching	Sandwich
BiOCl-$Ti_3C_2T_x$	HF etching	Nanosheets
SnS-$Ti_3C_2T_x$	HF etching	Nanosheets

2.3 Delamination

In many cases, MXene nanosheets reassemble into neat columns after being etched, a phenomenon attributed to van der Waals forces. These ordered arrangements reduce an electrode's salt adsorption capacity (SAC) by limiting the intercalating activity in the interlayer region. Delamination may be accomplished by increasing the distance between layers and using dimethyl sulfoxide, tetramethylammonium hydroxide, or another organic solvent [51]. Solvents of this kind can affect interlayer modification by osmotic swelling, ion exchange, and, ultimately, exfoliation. The attraction forces among the layers are considerably reduced, leading to delamination and an excessive amount of liquid being sucked into the original structure [52]. If further delamination is needed, the MXene flake size will be lowered, but this process might introduce new structural defects [45]. Recent publications provide a more thorough overview of MXene synthesis and delamination techniques [41, 53, 54]. SEM pictures of the MAX phase, MXene, and Cu-MXene materials are shown in (Fig. 2.) SEM pictures of the MAX phase at lower and higher magnifications are shown in (Fig. 2a, b). It was discovered that the MAX phase sample has a multilayer structure. The SEM pictures of the MXene sample are shown in (Fig. 2c, d). After the MAX phase was etched, as shown in the figure, separated layers or increased interlayer spacing between the two layers was seen. A photograph of the MXene sample at greater magnification revealed the inter-layered space between the two layers (Fig. 2d). The Cu-MXene material is shown in (Fig. 2e, f). As can be seen from the figure, sparkling particles and separated layers or interlayered gaps between the two levels are present [55].

3 Properties of MXenes

3.1 *Theoretical Capacity*

A significant specific capacitance is found in $Ti_3C_2T_x$ MXene, particularly due to the high number of pseudocapacitive sites; as a result of the oxygen functional groups being protonated, Ti's valence phase, which is bonded to the oxygen surface categories, shifts continuously. Because of this, MXenes in an acidic electrolyte exhibit the following pseudocapacitive charge storage properties: According to a recent study [55].

$$Ti_3C_2Ox(OH)yFz + dH + +de \ll Ti_3C_2Ox_{-d}(OH)_{U+d}F_z$$

The highest theoretical capacity for Ti_3C_2Tx in the potential range of 0.6–0 V is around 615 C gl [55]. For a voltage range of 0.55 V, the experimentally obtained values are close to 135 C g^1, much lower than the predicted capacity. The limited potential range may be to blame for stifled redox processes or underutilized active sites. Electrochemical studies of MXenes typically use platinum or gold current detectors [55]. Both possible consequences are water splitting in the relevant potential range and a drop in Coulombic efficiency owing to the repeated charge–discharge process. A large potential window of 1 V was achieved by employing glassy carbon to serve as the MXene electrode's current collectors to get around this. This is due to the excellent overpotential that glassy carbon exhibits for the hydrogen evolution reaction, which permits the investigation of the inherent properties of various materials within the targeted potential range without causing water to be split. The 90-nm-thick electrodes could attain a specific capacitance of up to 450 F gl, equivalent to a very high volumetric capacitance of about 1500 F cm^3 [58]. The optical capacitance of MXene can be increased by altering the surface chemistry and/or doping heteroatoms. For example, adaptable and unsupported N-doped $Ti_3C_2T_x$ films were successfully made via solvothermal processing. The resulting nitrogen-doped

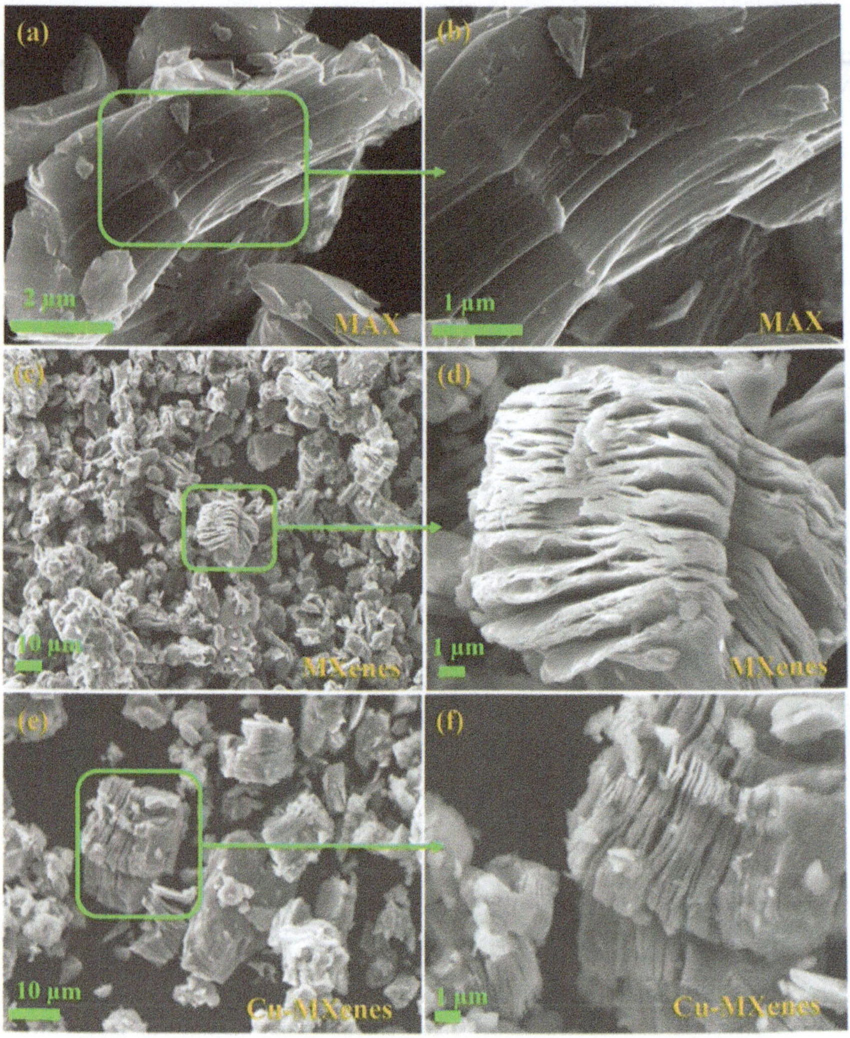

Fig. 2 SEM images of the prepared **a**, **b** MAX phase, **c**, **d**) MXene, and **e**, **f** Cu-MXene materials. (Adapted from Ref. [55] and reproduced with permission of Copyright! 2023 Materials Chemistry and Physics

$Ti_3C_2T_x$ films broke the previous record for all known MXene-based materials with an extraordinarily high capacitance of 2836 F cm^3 (927 F g^1) at 5 mV s^1. [59].

3.2 Electronic Band Structure

MXene's most significant characteristic is its metallic behavior, which is similar to MAX phases and has an electron density that is well-fixed and close to the Fermi level [60]. By creating more Ti-X bonds, the metallic behavior can be tailored. MXenes can be produced by changing the functional terminal groups in a narrow band-gap semiconductor. The degree to which surface functionalization modifies the electrical composition of MXene is a major factor in determining its electronic characteristics. Because their oxidation states are equivalent and both allow for the donation and acceptance of a single electron, the electrical structures of MXene are similarly affected by the F and OH groups. On the other hand, O groups act differently because they receive two electrons in the equilibrium state [61]. The band structure topology of MXene can be trivial or non-trivial depending on their spin–orbit interaction.

MXene can also be divided into semiconducting, metallic, and metallic-semi-metallic forms based on electrical conductivity. $Ti_3C_2T_x$ filtrated film possesses a lot of chemical diversity and strong intrinsic electronic and ionic conductivities compared to graphene [60]. $Ti_3C_2T_x$ filtrated film possesses a lot of chemical diversity and strong intrinsic electronic and ionic conductivities compared to graphene. $Ti_3C_2T_x$'s electrical conductivity is strongly influenced by its morphological and surface properties since high conductivity can frequently result from excellent contact between tiny individual flakes and big flake sizes. The metallic conductivity of the $Ti_3C_2T_x$ film has been shown experimentally to be higher than that of any other type of 2D metal sulfide or hydroxide [60]. For comparison, graphene has a conductivity of 6000 S cm^1, and restacked MoS_2 has a conductivity of 1.37 9 102 S cm^1, while the conductivity of severely flawed HF-etched $Ti_3C_2T_x$ powders is less than 1000 S cm^1 [62].

3.3 Morphology and Surface Chemistry of MXenes

MXene morphologies are significantly influenced by the etching technique and etchant concentration. Beginning with the MAX phase, the creation of accordion-like m-MXene utilizing the HF etching approach; a greater HF concentration corresponds to more significant openings of MXene lamellae. These moderate etching methods (mild LiF-HCl and fluoride-based methods) remove the "A" element despite having a comparable morphology to the MAX phase. As a result, XRD and EDX studies should be made instead of relying solely on the accordion-like morphology to determine whether Al was successfully removed from the MAX [45]. Several studies have demonstrated that MXene surface terminations are randomly distributed rather than concentrated in one area with a particular set of functions [63]. Moreover, there are no nearby OH terminations.

3.4 Mechanical Properties

Excellent mechanical flexibility can be seen in delaminated MXene nanosheets, especially in the MXene monolayer. Young's modulus of a monolayer or bilayer nanosheet has proven difficult to measure; however, it can be easily determined by filtering a delaminated MXene solution via a vacuum. To study the elastic properties of MXenes (monolayer and bilayer $Ti_3C_2T_x$), a recent technique, nanoindentation, with the help of an atomic force microscope tip, has been utilized. Testing yielded force–displacement curves that were used to determine the effective Young's modulus of a single layer of $Ti_3C_2T_x$, which was found to be 0.33 TPa [64]. The experimentally determined Young's modulus of monolayer Ti_3C_2 in free space is 502 GPa, which is not too far off from the expected value [65]. Strains of 9.5% under biaxial stress and 18% and 17% under uniaxial tension were within the strain tolerances of bare 2D Ti_3C_2 because of the stretching and contracting of the Ti-C bonds [65]. However, polymers such as chitosan and polyethylene, when added to $Ti_3C_2T_x$ films, may improve their mechanical characteristics. For instance, when PVA was intercalated into $Ti_3C_2T_x$, the resulting composite films showed excellent mechanical and electronic conductivity and could withstand 5000 times their own weight. In light of these findings, developing conductive polymer/MXene composites for reinforcing composites has great promise [66].

3.5 Thermal Stability Properties

Applications involving solution or thin-film storage need more investigation into MXene's thermal stability. It was discovered through thermogravimetric and mass spectrometry analyses that the environment and chemical makeup of MXenes significantly impacted their thermal stability [65]. A thermogravimetric investigation shows that a substantial amount of mass is lost beyond 800 degrees Celsius in an argon (Ar) environment when $Ti_3C_2T_x$ is converted to TiC. When heated in an oxygen atmosphere during annealing, however, $Ti_3C_2T_x$.

Additionally, MXene's exceptional thermal conductivity is advantageous for electronic devices. Lattice size affects MXene's thermal conductivity. For instance, it is projected that the thermal conductivity of HF_2CO_2 in a 5-lm flake will be 86.25 W m^1 K^1, and it will climb to 131.2 W m^1 K^1 in a 100-lm flake. HF_2CO_2 has a thermal expansion coefficient of 6.094 9 106 K^1 at room temperature [67]. By doping MXene with n-atoms, single-layer Mo_2C's thermal conductivity is raised from 48.4 to 64.7 W m^1 K^1 in the armchair direction at room temperature [68].

4 Application of MXene in Water Purification

MXene has emerged as a popular option for water treatment, particularly heavy metals, radionuclides, excess salt, and the elimination of bacterial contaminants due to its potent adsorption properties (Table 2). $Ti_3C_2T_x$ MXene's enhanced ion reduction and adsorption capabilities following surface modification present yet another potential avenue for water treatment. As a result, $Ti_3C_2T_x$ MXene and composite nanomaterials based on $Ti_3C_2T_x$ MXene are perfect adsorbents for water contaminants [69].

4.1 *$Ti_3C_2T_x$ MXene in Removing Heavy Metal Ions*

Heavy metal industrial discharges into waterways have now raised concerns on a global scale. Toxic metals found in industrial effluent include lead, nickel, copper, chromium, mercury, and others that tend to build up in living things and cannot be broken down [70]. These have a substantial effect on organisms over time as well. Therefore, it is essential to purge water sources of heavy metal ions. Recently, several researchers added polymers to the outer layer of MXenes to eliminate ions of heavy metals using several processes, such as physiological acceptance, hydrogen bonding, and magnetic attraction. MXene-based polymer composites can be made using $Ti_3C_2T_x$ and levodopa (DOPA) to remove Cu^{2+} under moderate reaction conditions [71]. Using an intercalation peeling technique, we created a new form of 2D Ti3C2Tx nanosheet that effectively removes hazardous Cr(VI). More importantly, the amount of Cr (VI) that was successfully transformed into a less dangerous Cr(III) form was substantially below the WHO-recommended level for drinking water [72].

In brief, HF acid was used to produce the $Ti_3C_2T_x$ MXene, which was then mixed with MXene in a Tris buffer solution to provide a final pH of 8.5 for the suspension. After combining the MXene suspension and PODA solution, a centrifuge was used to separate the components before being washed with deionized water. It was demonstrated that $Ti_3C_2T_x$-PDOPA had a significantly higher adsorption capacity for Cu^{2+} than $Ti_3C_2T_x$. The functional groups of DOPA contributed to the electrostatic contact process that allowed for the absorption of additional Cu^{2+} from water [73]. The new surface modification strategy's considerable increase in adsorption capacity. These results demonstrated that when Cu^{2+} concentrations increased, additional sites for adsorption filled up, and the number of optimum adsorption capacity sites increased as the adsorption process gradually neared saturation [69]. To decrease Cr(VI), intercalated $Ti_3C_2T_x$ with DMSO after preparing $Ti_3C_2T_x$ using HF acid at various doses. $Ti_3C_2T_x$ 10% nanosheets, which have the highest specific surface area, showed the greatest removal performance of 250 mg g^{-1} [74]. The reaction speed slowed when the active site and Cr(VI) concentrations dropped. In addition, the rate at which Cr(VI) was removed increased when the concentration of Ti_3C_2Tx 10% nanosheets rose from 0.2 gL^{-1} to 1.0 gL^{-1}. The Cr(VI) clearance rate was likewise

Table 2 Different MXenes used in water treatment via several methods

MXenes	Mode of Purification	Contaminant	Experiment setup	Outcome	References
alk-MXene$Ti_3C_2(OH/ONa)_xF_{2-x}$	Adsorption	Pb(II)	Batch adsorption pH 5–7	The alk-MXene exhibits large sorption capacities, fast kinetic, extremely trace Pb(II) effluent, and reversible adsorption properties	[75]
(2-D) $Ti_3C_2T_x$	Adsorption	Methylene blue (MB) dye	Aqueous 0.05 mg/mL MB solution was kept in contact with suspended $Ti_3C_2T_x$ in the dark	The adsorption capacity of Ti3C2Tx for MB is ~ 39 mg/g	[81]
(2-D) $Ti_3C_2T_x$ nanosheets	Reductive removal	Cr(VI)	Reduction pH 5	$Ti_3C_2T_x$ nanosheets showed Cr removal capacity of 250 mg/g	[31]
(2-D) $Ti_3C_2T_x$ nanosheets	Filtration	Alkaline earth (Li^+, Na^+, K^+, Mg^{2+}, Ca^{2+}), other metal (Ni^{2+} and Al^{3+}), and methylthioninium$^+$ (MB^+) dye cations	Selective sieving pH 6.05	Ti_3C_2Tx membranes exhibit strong selectivity towards metal cations that are single-, double-, or triple-charged, as well as dye ions of various sizes.	[99]
$Ti_3C_2T_x$	Delamination	*Escherichia coli* (*E. coli*) and *Bacillus subtilis* (*B. subtilis*)	Batch assay for antibacterial activity	More than 98% of both types of bacteria's viability was lost after 4 h of exposure to 200 g/mL Ti_3C_2Tx, which was discovered to have concentration-dependent antibacterial action	[98]
Urchined rutile TiO_2-C (u-RTC) nanocomposite	Adsorption	Cr(VI)	MXene was introduced into 50 ml Cr(VI) solution at 200°C for 12 h in an incubator shaker	u-RTC displays a high Cr(VI) adsorption capacity of ~225 mg/g	[100]

(continued)

Table 2 (continued)

MXenes	Mode of Purification	Contaminant	Experiment setup	Outcome	References
(2D) multilayered V_2CTx nanosheets	Adsorption	U(VI)	Batch sorption experiment	V_2CTx nanosheets displays a high U(VI) adsorption capacity of 174 mg/g	[101]
(2-D) $Ti_3C_2T_x$ nanosheets	Adsorption	Barium	Batch experiment 1–50 mg/L barium concentration pH 3–4 for 120 min at 25°C	The maximum adsorption capacity occurs at pH 6 with a value of 1.7 mg/g	[102]
2D a-Fe_2O_3 doped Ti3C2 MXene	Photocatalysis	Rhodamine B	20 mg of a-Fe_2O_3/Ti_3C_2 composite was dispersed into 100 mL of 10 mg/l RhB aqueous solution	a-Fe_2O_3/Ti_3C_2MXene composite had excellent photocatalytic activity and recycling stability	[103]
PANI/$Ti_3C_2T_x$ nanosheets	Adsorption	U(VI)	Batch experiment	The maximum adsorption capacity occurs at pH 5 with a value of 102.8 mg/g	[45]
$Ti_3C_2T_x$ nanosheets	Delamination	*Escherichia coli* (*E. coli*) and *Bacillus subtilis* (*B. subtilis*)	Terrific broth (TB) culture media treated with 100 μg/ml of 0.09, 0.35, 0.57, and 4.40 μm wide Ti3C2Tx MXene nanosheets. The strain was then grown aerobically at 37 °C at 150 rpm for ~18 h	Ti_3C_2Tx MXene nanosheets damage the bacterial cell wall significantly in less than 3 h, causing the DNA of the bacterium to be released from the cytosol, followed by the bacteria cells being dispersed	[104]

(continued)

Table 2 (continued)

MXenes	Mode of Purification	Contaminant	Experiment setup	Outcome	References
$Ti_3C_2T_x$	Adsorption	Cs^+	Batch adsorption method 10 mg of Ti_3C_2Tx was added into 10 mL of 5 mg L − 1 Cs + solution at pH 6 and placed on a rocking mixer for 30 min	The adsorbent reached the steady state within 1 min with the maximum Cs^+ adsorption capacity of 25.4 mg g^{-1} at room temperature	[105]
$Ti_3C_2T_x$	Filtration	Oil/water separation	1% v/v oil-in-water emulsion was prepared	The membrane shows high separation efficiency for oil/water emulsions, of over 99%, and a high water permeation flux of over 450 L per m^2 per h per bar	[83]
α-Fe_2O_3/ $ZnFe_2O_4$@Ti_3C_z	Photocatalysis	Rhodamine B	20 mg of the obtained photocatalyst was added into 100 mL of 10 mg L − 1 RhB 300 W Xe lamp (HSX-F300, NBeT, China) with the UV cutoff filter (λ > 400 nm) was used as light source	α-Fe_2O_3/$ZnFe_2O_4$@Ti_3C_2 exhibits the terrific photocatalytic activity and reusability in degradation RhB	[90]

(continued)

Table 2 (continued)

MXenes	Mode of Purification	Contaminant	Experiment setup	Outcome	References
NiFe2O4/MXene	Photocatalysis	*Escherichia coli* (*E. coli*)	To test the antibacterial activity, DI-water dispersed solutions (150 lg/ml) of NiFe/MXene were spread on agar plates	An excellent antibacterial activity against *E. coli* bacteria was observed	[106]
$Ti_3C_2T_x$	Adsorption	Pharmaceutical waste (amitriptyline (AMT), verapamil, carbamazepine, 17 α-ethinyl estradiol, ibuprofen, and diclofenac)	40 mg/l $Ti_3C_2T_x$ was added into 10 μM of selected pharmaceutical compounds, each at three different pH (3.5, 7, and 10.5) conditions for 3 days	The adsorption capacity for $Ti_3C_2T_x$ showed the highest value (58.7 mg/g) at pH 7 MXene indicated insignificant reduced adsorption capacity with the number of recycles	[107]
$Bi_2O_2CO_3$@MXene	Filtration	Oil/water separation	The volume ratio of 1:100 was used to mix oil and water as well as the surfactant (SDS) with a concentration of 0.15 mg/ml	N-$Bi_2O_2CO_3$@MXene/PES composite membrane realized a nearly 100% rejection ratio for oil/water emulsions from wastewater	[108]
$Ti_3C_2T_x$	Adsorption	Radionuclides U(VI) and Eu(III)	–	According to the findings, Ti_3C_2Tx displayed great effectiveness in removal (279.57 mg U/g and 73.99 mg Eu/g), remarkable selection, and good durability. The reaction reached equilibrium in approximately 20 min	[109]

(continued)

Table 2 (continued)

MXenes	Mode of Purification	Contaminant	Experiment setup	Outcome	References
$Ti_3C_2T_x$	Size-selective sieving coupled with adsorption	Antibiotics (azithromycin)	Molecular sieving and electrostatic repulsion	Satisfactorily high selectivity of antibiotics (~99.0% for azithromycin) with unprecedentedly high pure water permeance (~26.0 L m^{-2} h^{-1} bar^{-1})	[110]
Ti_3C_2Tx MXene@metal–organic frameworks nanosheets	Desalination and adsorption	Salts (Na_2SO_4, $MgSO_4$, $MgCl_2$, and NaCl) Heavy metals (Ni^{2+}, Cd^{2+}, Mn^{2+}, Cu^{2+}, and Zn^{2+})	–	Nanocomposite membrane showed high salt rejection for Na_2SO_4, $MgSO_4$, $MgCl_2$, and NaCl. Concurrently, the rejection rate of five different types of heavy metal ions (Ni^{2+}, Cd^{2+}, Mn^{2+}, Cu^{2+}, and Zn^{2+}) was tested and denoted more than a 95.2 ± 0.5% rejection rate for all of them, notably high for Mn^{2+} (97.6 ± 0.4%)	[111]

correlated with pH level. The rate of elimination increases as the pH decreases. The rate of Cr(VI) elimination decreased significantly as pH increased [30]. The effective removal of various heavy metal ions from an aqueous environment using MXene-based adsorbents. Improved adsorption capability against the removal of harmful heavy metals in water treatment methods was revealed by MXene-based adsorbents. These investigations suggested using MXene-based adsorbents to remove dangerous pollutants and contaminants from an aqueous environment effectively.

4.2 *Ti_3C_2Tx MXene in Removing Organic Dyes*

Dye use has been commonplace in several manufacturing sectors, most notably the printing, textile, and paper industries. The environment has been severely harmed by dye wastewater [74]. Literature indicates that dyes are highly toxic to both plants and animals, being both mutagenic and carcinogenic. Numerous nanocomposite materials have been discovered in recent years for dye adsorption. Due to their excellent adsorption and catalytic capabilities, several dyes can be removed using MXene and MXene-based materials such as methylene blue (MB), methyl orange (MO), rhodamine B (Rh B), etc. (Fig. 3) [75]. Developed a self-forming Ti_3C_2Tx MXene nanocomposite that was altered by cubed Co_3O_4 ions using the solvothermal method. The Ti_3C_2Tx sheets were well-organized composite structures with evenly scattered Co3O4 particles on their surfaces and internal layers, which facilitated the breakdown by catalysis of the MB and RhB dyes [76]. Wide surfaces of the MXene-Co_3O_4 composite, which were favorable for the adsorption of dye molecules, were principally responsible for its remarkable performance in the catalytic breakdown of MB and RhB [77].

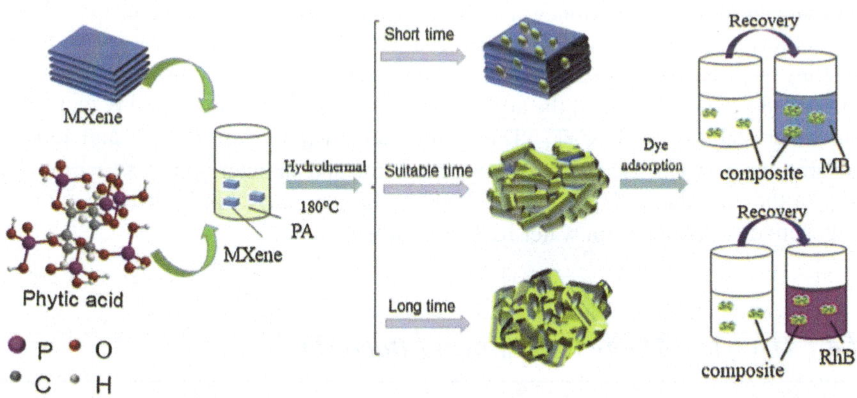

Fig. 3 MB adsorption characteristics that can be recycled in a schematic representation of the synthetic process for PA-MXene composites using a hydrothermal technique [75] Copyright! 2020 Colloids and Surfaces A: Physicochemical and Engineering Aspects

Dye adsorption findings are shown for both conventional MXene (t-Ti_3C_2) made using HF acid etching and hydrothermally synthesized MXene (h-Ti_3C_2). Neither h-Ti_3C_2 nor t- Ti_3C_2 could easily adsorb the anionic dye MO, as the negatively charged Ti_3C_2 generated by wet chemical methods repelled MO. On cationic dyes (like MB), the produced Ti_3C_2 shows a noticeable adsorption effect [78, 79]. The outcomes demonstrated that Ti_3C_2 created using a hydrothermal approach had greater adsorption performance than Ti_3C_2 prepared using a conventional HF etching process. This might be because h- Ti_3C_2 16 h had a greater specific surface area than t- Ti_3C_2 12 h. Nb_2C had lesser adsorption than Ti_3C_2, in addition. According to the results above, $Ti_3C_2T_x$ MXene has a high dye removal efficiency [80, 81]. Consequently, MXenes have recently made strides in dye removal from wastewater through their use as adsorbents, membranes, photocatalytic agents, and catalysts.

4.3 Ti_3C_2Tx MXene in Removing Radionuclides

As the nuclear energy sector has expanded, so has the amount of attention paid to the dangers posed by radioactive waste. $Ti_3C_2T_x$ MXene has the potential to be employed as a novel adsorbent for the elimination of radionuclides such as palladium (Pd(II)), uranium (U(VI)), cesium (Cs^+), thorium (Th(IV), and europium (Eu(III)) due to its unusual structures and characteristics [82]. Ion exchange and electrostatic contact between radionuclides and the surface of MXene are the primary methods by which MXene eliminates radionuclides. According to research [65]. Removal of Cs^+, Ba^{2+}, and Sr^{2+} using Ti_3C_2Tx is effective. The adsorption of radionuclides depends on various factors, such as the pH of the solution, temperature, contact time, and its ion concentration [83]. Cations are strongly adsorbed to the negatively charged surface of $Ti_3C_2T_x$ MXene. Researchers have identified ion exchange as the dominant mechanism for ion adsorption. Carboxyl-functionalized $Ti_3C_2T_x$ MXene (TCCH) is very efficient in removing U(VI) and Eu(III). This illustration shows the spontaneous grafting of a diazonium salt with a carboxyl-terminal group onto the surface of MXene [84]. The addition of the aryl diazonium salt has now considerably increased the water stability of $Ti_3C_2T_x$. The results demonstrated that U(VI) and Eu(III) adsorption on TCCH used electrostatic interaction and an inner-sphere structure [32, 85]. In the end, the $Ti_3C_2T_x$ MXene and its intricate parts can be used to remove radioactive pollutants from water in the real world.

4.4 Ti_3C_2Tx MXene in Removing Bacteria

With their potential antibacterial properties, MXenes pave the way for creating disinfection-crediting water treatment and desalination process units. More than two logs of bacteria were shown to be inactivated by MXenes, making them competitive with conventional water purification techniques such as slow sand filtration.

However, only microorganisms have been used to date in testing antibacterial capabilities. MXenes' potential to inactivate pathogens, including viruses and protozoan parasites, should be further investigated. The MXene nanosheet is versatile and can be employed in a suspension of delaminated nanosheets, functionalized onto a membrane surface, or even to construct a membrane [69]. $Ti_3C_2T_x$ MXene was first characterized as an antibacterial agent for water treatment, and it is obvious that bacterial activity is substantially reduced with increasing Ti_3C_2Tx concentration. The two bacteria's actions were also considerably different from graphene oxide (GO) at various concentrations, showing that Ti_3C_2Tx exhibited superior antibacterial characteristics [86]. Additionally, the size and atomic-level organization of nanosheets affect their efficacy. Future research is expected to clarify how different synthesis stages, like etchants and intercalants, may produce materials with higher antimicrobial activity. MXene's antimicrobial properties are a welcome bonus to an already superior technology for purifying and desalinating water. [87]. MXenes functioning on a membrane surface have anti-biofouling qualities that show promise for enhancing the performance of current treatment systems and decreasing the requirement for pre-treatment in RO systems.

4.5 Ti_3C_2Tx MXene in Oil/Water Separation

Oily effluent from factory outflows and oil spills can be dangerous for the environment and people. The odd layered structure and adjustable interfacial reactivity of MXenes made them attractive candidates for use as nanoscale fillers in polymers to enhance their properties [88]. In one study, light, strong, and hydrophilic polyimide/MXene 3D structures were produced by thermal imidization and freezing and drying polyamide acid/MXene suspensions (Fig. 4). Through the etching of $Ti_3C_2T_x$ and ultrasonic exfoliation, the MXene (Ti_3C_2Tx) dispersion was created [89]. It was found that the powerful interactions between polyimide sequences and MXene nanosheets produced extremely porous hydrophilic aerogels with a small density that had improved compressible action, raised absorption capacity, and effectively separated oil and water. The designed aerogel displayed exceptional fatigue resistance and very high elasticity after 50 compression-release cycles [81]. Significant absorption capabilities of 18 to 58 times their own weight were also discovered for a range of organic liquids. These modern designs are thermally stable, can swiftly separate chloroform, liquid soybean, oil, and liquid paraffin from the water–oil system, and may be used repeatedly in difficult settings [50, 90]. The design of low-cost, elastic barriers utilized normal print paper as the substrate and Ti_3C_2Tx MXene as the functionality component. A membrane with a contact angle for oil underwater of 137 degrees was created using a simple coating process using MXene ink. Researchers used these membranes to achieve oil/water emulsion separation efficiencies of over 99% and permeable water flows of over 450 L per m^2 per hour at 1 bar [89]. $Ti_3C_2T_x$ MXene 2D nanosheet carbides were deposited on porous polyvinylidene fluoride membranes using vacuum filtering to make MXene membranes [90]. The multilayer

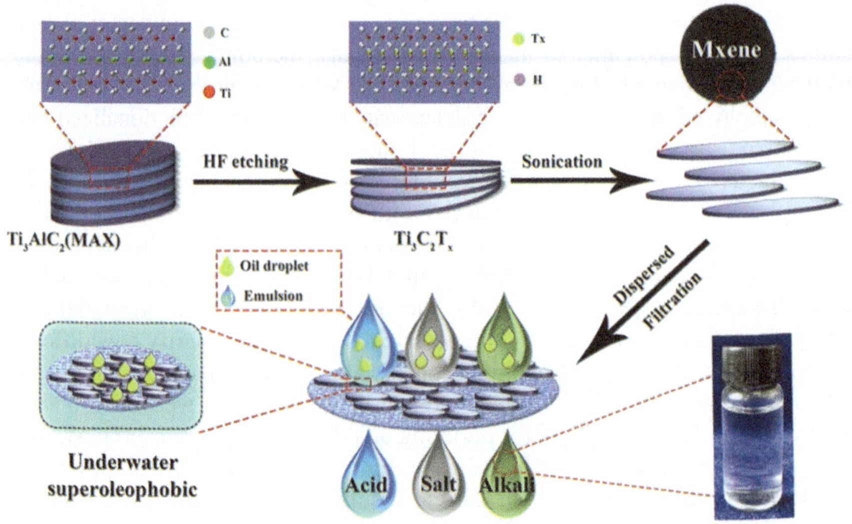

Fig. 4 The fabrication technique of the Ti_3C_2Tx MXene membrane for separating oil-in-water emulsions. Reproduced with permission from Ref 80 Copyright! 2020 Journal of Colloid and Interface Science

MXene membranes also successfully separated a variety of solid fluids, including mixtures of raw oil and water, and they displayed a remarkable penetration flux of 887 L m^2 h^1 bar1 and a significant separating rate of 99.4%. These membranes could extract oil droplets from water in caustic settings [91].

4.6 *Ti_3C_2Tx MXene in the Removing Pharmaceutical Compounds*

A hybrid photocatalyst based on $Ti_3C_2T_x$ MXene nanosheets was created to remove carbamazepine (Fig. 5) [74]. The calculated Kapp value of carbamazepine under UV light, 0.0304 min^1, was higher than measured in daylight. The breakdown capacity was drastically decreased under low pH (3.0–5.0). During photocatalytic breakdown, the carbamazepine molecule was damaged by hydroxyl and oxygen radicals. $Ti_3C_2T_x$ MXene's potential for the adsorption of several pharmaceuticals was evaluated [92]. Adsorption of amitriptyline was greatest (58.7 mg/g) at a pH of 7 due to electrostatic repulsion among the positively charged amitriptyline and the negatively charged MXene. To increase adsorption activity, MXene was also sonicated at different frequencies (0, 28, and 580 kHz); the highest adsorption capacities were found in the following order: 580 kHz (172 mg/g) > 0 kHz (138 mg/g) > 28 kHz (214 mg/g) [92]. Lower frequencies exhibited the highest activity levels, particularly by generating larger cavitation waves. Adsorption ability was inhibited by the presence of ambient

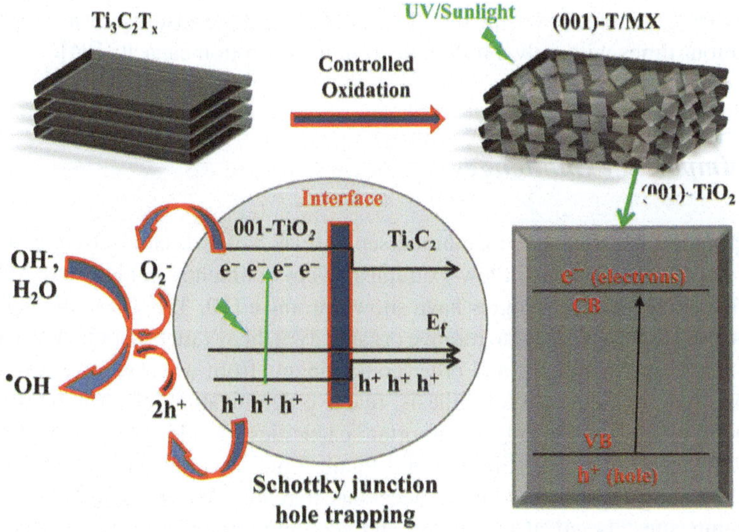

Fig. 5 Photocatalytic degradation of carbamazepine via Hetero-structural MXene. Reproduced with permission from Ref. [74] Copyright! 2018 Chemical Engineering Journal

inorganics but bolstered by the presence of background organics, such as spontaneous organic matter. In addition, the cationic surfactant cetylpyridinium chloride showed that it competed with medicinal molecules, reducing adsorption efficiency [92, 93].

5 Regeneration and Reuse Capability of MXenes

Numerous investigations have shown that MXenes and their derivatives can be recycled or regenerated after the adsorption of contaminants. After Hg(II) ions were adsorbed, the magnetic Ti_3C_2Tx nanocomposite was once again employed. Notably, the removal efficiencies exceeded 90% for four straight cycles [29]. This reusability is attributable to the composite's increased stability brought on by the encapsulation of Fe_2O_3 nanoparticles. MXene Fe_2O_3 was washed with ethanol and dried at 70 °C after MB adsorption. The MXene Fe_2O_3 was reused for MB adsorption five times [29]. MXene's removal effectiveness toward MB remained at around 77% after 5 cycles but with some decrease. MXene-COOH@(PEI/PAA)n can be reused 8 times in a row after MB adsorption, with only a little decrease in capacity. Following the adsorption and removal of Cu(II) ions, delaminated MXene (Ti_3C_2Tx-DL) was restored using a solution of calcium nitrate and nitric acid. The third cycle, however, saw a 30% reduction in adsorption capacity. This decrease in adsorption capacity may be brought on by the partial oxidation of $Ti_3C_2T_x$ into TiO_2 nanoparticles and the inadequate desorption of Cu(II) ions during adsorption. $Ti_3C_2T_x$-KH_570's adsorbed

Pb(II) ions were removed using 0.5 M Na2EDTA, and the adsorbent was effectively reused four times with only a little reduction in adsorption capacity [94].

5.1 Important Challenges

Pervaporation desalination, membrane separations, solar desalinization, and capacious deionization are just a few examples of environmental fields where MXenes and MXene-based technologies have shown promise [10, 35]. These 2D materials may be used as an adsorbent to remove organic dyes, heavy metals, radioactive pollutants, medicinal chemicals, and other contaminants from water and wastewater, as well as in membranes, photosynthesis, reactive deionization, electrolytic division, and membranes. They also have important capacities for adsorption, selection, and reuse. However, there are still several difficult topics that demand serious thought. Further, scaling-up efforts would be necessary to realize MXenes and MXene-based nanocomposites' practical usefulness and economic viability in large-scale water treatment applications [95, 96]. These nanostructures can be stacked into lamellar membranes to increase selectivity and permeance in water desalination. However, as 2D membranes tend to expand in water, improving their stability in aqueous solutions remains a significant challenge. Significant difficulties also come from the MXene manufacturing process's poor yield and expensive cost. Traditional synthetic methods can use potentially hazardous and toxic chemical materials, raising safety and environmental issues [97]. Thus, researchers should focus on developing simple, affordable, productive, and environmentally friendly methods for producing MXenes and substances made from them on a large scale. The surface chemistry of MXenes and the methods by which these 2D materials remove contaminants have to be evaluated analytically and systematically [98]. Aggregation of MXenes, which might lower these materials' adsorption efficiency and surface area, is another significant problem. Some examined MXenes displayed antimicrobial properties against harmful bacteria, underscoring their enticing possibilities for water purification and anti-biofouling films. To identify the antibacterial mechanisms of MXenes, however, additional analyses and evaluations need to be done [85].

6 Conclusion and Future Outlooks

Due to their higher surface area, notable thermal and electrical conductivity, antimicrobial abilities, high stability in chemicals, substantial absorption selection, hydrophilic nature, and remarkable conductivity, MXenes have demonstrated a variety of important applications in water purification, magnetic shielding, photoelectrochemical catalysts, and energy storage areas. These materials have attracted a lot of research attention because of how easily they may be surface functionalized for various uses, such as ecological adhesion, medical devices, chemical deterioration, and energy storage. Additionally, they provide the benefits of flexible, changeable portions and a programmable minimum nanolayer depth. MXenes with surface functionality are the next generation of water filtration and purification materials because of their wide range of possible applications, increased stability, and high recycling rate. Developing new MXene structures and identifying several major MXene applications for environmental remediation is essential, even though Ti3C2Tx has been widely investigated in wastewater and water treatment.

References

1. Veettil, A.V., Mishra, A.K.: Water security assessment using blue and green water footprint concepts. J. Hydrol. **542**, 589–602 (2016)
2. Khazaei, M., Arai, M., Sasaki, T., Estili, M., Sakka, Y.: Two- dimensional molybdenum carbides: potential thermoelectric materials of the MXene family. Phys. Chem. Chem. Phys. **16**, 7841–7849 (2014)
3. Liang, X., Rangom, Y., Kwok, C.Y., Pang, Q., Nazar, L.F.: Inter- woven MXene nanosheet/carbon-nanotube composites as Li–S cathode hosts. Adv. Mater. **29**, 1603040 (2017)
4. Domènech, B., Muñoz, M., Muraviev, D.N., Macanás, J.: Polymer-stabilized palladium nanoparticles for catalytic membranes: ad hoc polymer fabrication. Nanoscale Res. Lett. **6**, 1–5 (2011)
5. Mourdikoudis, S., Montes-García, V., Rodal-Cedeira, S., Winckelmans, N., Pérez-Juste, I., Wu, H., Pastoriza-Santos, I.: Highly porous palladium nanodendrites: wet-chemical synthesis, electron tomography and catalytic activity. Dalton Trans. **48**(11), 3758–3767 (2019)
6. Baghbamidi, S.E., Hassankhani, A., Sanchooli, E., Sadeghzadeh, S.M.: The reduction of 4-nitrophenol and 2-nitroaniline by palladium catalyst based on a KCC-1/IL in aqueous solution. Appl. Organomet. Chem. **32**(4), e4251 (2018)
7. Carolin, C.F., Kumar, P.S., Saravanan, A., Joshiba, G.J., Naushad, M.: Efficient techniques for the removal of toxic heavy metals from aquatic environment: a review. J. Environ. Chem. Eng. **5**(3), 2782–2799 (2017)
8. Praveen, P.K., Ganguly, S., Wakchaure, R., Para, P.A., Mahajan, T., Qadri, K., Dalai, N.: Water-borne diseases and its effect on domestic animals and human health: a Review. Int. J. Emerg. Technol. Adv. Eng. **6**(1), 242–245 (2016)
9. Shariful, M.I., Sharif, S.B., Lee, J.J.L., Habiba, U., Ang, B.C., Amalina, M.A.: Adsorption of divalent heavy metal ion by mesoporous-high surface area chitosan/poly (ethylene oxide) nanofibrous membrane. Carbohydr. Polym. **157**, 57–64 (2017)
10. Ye, S., Zeng, G., Wu, H., Zhang, C., Liang, J., Dai, J., Cheng, M.: Co-occurrence and interactions of pollutants, and their impacts on soil remediation—a review. Crit. Rev. Environ. Sci. Technol. **47**(16), 1528–1553 (2017)
11. Gao, L., Li, C., Huang, W., Mei, S., Lin, H., Ou, Q., Zhang, H.: MXene/polymer membranes: synthesis, properties, and emerging applications. Chem. Mater. **32**(5), 1703–1747 (2020)
12. Dieckmann, M.S., Gray, K.A.: A comparison of the degradation of 4-nitrophenol via direct and sensitized photocatalysis in TiO2 slurries. Water Res. **30**(5), 1169–1183 (1996)
13. Thebo, K.H., Qian, X., Wei, Q., Zhang, Q., Cheng, H.-M., Ren, W.: Reduced graphene oxide/metal oxide nanoparticles composite membranes for highly efficient molecular separation. J. Mater. Sci. Technol. **34**, 1481–1486 (2018)
14. Aziz, N., Hanafiah, M.M.: Application of life cycle assessment for desalination: Progress, challenges and future directions. Environ. Pollut. **268**, 115948 (2021)
15. Gude, V.G.: Desalination and sustainability—an appraisal and current perspective. Water Res. **89**, 87–106 (2016)

16. Tan, C., Cao, X., Wu, X.J., He, Q., Yang, J., Zhang, X., Zhang, H.: Recent advances in ultrathin two-dimensional nanomaterials. Chem. Rev. **117**(9), 6225–6331 (2017)
17. Anand, A., Unnikrishnan, B., Mao, J.Y., Lin, H.J., Huang, C.C.: Graphene-based nanofiltration membranes for improving salt rejection, water flux and antifouling–a review. Desalination **429**, 119–133 (2018)
18. Nag, A., Mitra, A., Mukhopadhyay, S.C.: Graphene and its sensor-based applications: a review. Sens. Actuators A., 177–194 (2018)
19. Wang, J., Chen, Z., Chen, B.: Adsorption of polycyclic aromatic hydrocarbons by graphene and graphene oxide nanosheets. Environ. Sci. Technol. **48**(9), 4817–4825 (2014)
20. Ersan, G., Apul, O.G., Perreault, F., Karanfil, T.: Adsorption of organic contaminants by graphene nanosheets: a review. Water Res. **126**, 385–398 (2017)
21. Khazaei, M., Arai, M., Sasaki, T., Chung, C.-Y., Venkatara-manan, N.S., Estili, M., Sakka, Y., Kawazoe, Y.: Novel electronic and magnetic properties of two-dimensional transition metal car- bides and nitrides. Adv. Funct. Mater. **23**, 2185–2192 (2013)
22. Lei, J.C., Zhang, X., Zhou, Z.: Recent advances in MXene: preparation, properties, and applications. Front. Phys. **10**, 276–286 (2015)
23. Naguib, M., Mashtalir, O., Carle, J., Presser, V., Lu, J., Hultman, L., Barsoum, M.W.: Two-dimensional transition metal carbides. ACS Nano **6**(2), 1322–1331 (2012)
24. Lukatskaya, M.R., Mashtalir, O., Ren, C.E., Dall'Agnese, Y., Rozier, P., Taberna, P.L., Gogotsi, Y.: Cation intercalation and high volumetric capacitance of two-dimensional titanium carbide. Science. **341**(6153), 1502–1505 (2013)
25. Rasool, K., Pandey, R.P., Rasheed, P.A., Buczek, S., Gogotsi, Y., Mahmoud, K.A.: Water treatment and environmental remediation applications of two-dimensional metal carbides (MXenes). Mater. Today **30**, 80–102 (2019)
26. Anasori, B., Lukatskaya, M.R., Gogotsi, Y.: 2D metal carbides and nitrides (MXenes) for energy storage. Nat. Rev. Mater. **2**(2), 1–17 (2017)
27. Jun, B.M., Kim, S., Heo, J., Park, C.M., Her, N., Jang, M., Yoon, Y.: Review of MXenes as new nanomaterials for energy storage/delivery and selected environmental applications. Nano Res. **12**, 471–487 (2019)
28. Liu, T., Liu, X., Graham, N., Yu, W., Sun, K.: Two-dimensional MXene incorporated graphene oxide composite membrane with enhanced water purification performance. J. Membr. Sci. **593**, 117431 (2020)
29. Jiang, H., Wang, Z., Yang, Q., Tan, L., Dong, L., Dong, M.: Ultrathin Ti 3 C 2 T x (MXene) nanosheet-wrapped NiSe 2 octahedral crystal for enhanced supercapacitor performance and synergetic electrocatalytic water splitting. Nano-Micro Lett. **11**, 1–14 (2019)
30. Shahzad, A., Rasool, K., Miran, W., Nawaz, M., Jang, J., Mahmoud, K.A., Lee, D.S.: Two-dimensional Ti3C2T x MXene nanosheets for efficient copper removal from water. ACS Sustain. Chem. Eng. **5**(12), 11481–11488 (2017)
31. Ying, Y., Liu, Y., Wang, X., Mao, Y., Cao, W., Hu, P., Peng, X.: Two-dimensional titanium carbide for efficiently reductive removal of highly toxic chromium (VI) from water. ACS Appl. Mater. Interfaces **7**(3), 1795–1803 (2015)
32. Jun, B.M., Jang, M., Park, C.M., Han, J., Yoon, Y.: Selective adsorption of Cs+ by MXene (Ti3C2Tx) from model low-level radioactive wastewater. Nucl. Eng. Technol. **52**(6), 1201–1207 (2020)
33. Wang, L., Song, H., Yuan, L., Li, Z., Zhang, P., Gibson, J.K., Shi, W.: Effective removal of anionic Re (VII) by surface-modified Ti2CT x MXene nanocomposites: implications for Tc (VII) sequestration. Environ. Sci. Technol. **53**(7), 3739–3747 (2019)
34. Jastrzębska, A.M., Karwowska, E., Wojciechowski, T., Ziemkowska, W., Rozmysłowska, A., Chlubny, L., Olszyna, A.: The atomic structure of Ti 2 C and Ti 3 C 2 MXenes is responsible for their antibacterial activity toward E. coli bacteria. J. Mater. Eng. Perform. **28**, 1272–1277 (2019)
35. Meshkian, R., Näslund, L.Å., Halim, J., Lu, J., Barsoum, M.W., Rosen, J.: Synthesis of two-dimensional molybdenum carbide, Mo2C, from the gallium based atomic laminate Mo2Ga2C. Scr. Mater. **108**, 147–150 (2015)

36. Hwang, S.K., Kang, S.M., Rethinasabapathy, M., Roh, C., Huh, Y.S.: MXene: an emerging two-dimensional layered material for removal of radioactive pollutants. Chem. Eng. J. **397**, 125428 (2020)
37. Ronchi, R.M., Arantes, J.T., Santos, S.F.: Synthesis, structure, properties and applications of MXenes: current status and perspectives. Ceram. Int. **45**(15), 18167–18188 (2019)
38. Kayali, E., VahidMohammadi, A., Orangi, J., Beidaghi, M.: Controlling the dimensions of 2D MXenes for ultrahigh-rate pseudocapacitive energy storage. ACS Appl. Mater. Interfaces **10**(31), 25949–25954 (2018)
39. Xiong, D., Li, X., Bai, Z., Lu, S.: Recent advances in layered Ti3C2Tx MXene for electrochemical energy storage. Small **14**(17), 1703419 (2018)
40. Levitt, A.S., Alhabeb, M., Hatter, C.B., Sarycheva, A., Dion, G., Gogotsi, Y.: Electrospun MXene/carbon nanofibers as supercapacitor electrodes. J. Mater. Chem. A. **7**(1), 269–277 (2019)
41. Tesfaye, A.T., Mashtalir, O., Naguib, M., Barsoum, M.W., Gogotsi, Y., Djenizian, T.: Anodized Ti3SiC2 as an anode material for Li-ion microbatteries. ACS Appl. Mater. Interfaces **8**(26), 16670–16676 (2016)
42. Malik, R.: Maxing out water desalination with MXenes. Joule **2**(4), 591–593 (2018)
43. Jeon, M., Jun, B.M., Kim, S., Jang, M., Park, C.M., Snyder, S.A., Yoon, Y.: A review on MXene-based nanomaterials as adsorbents in aqueous solution. Chemosphere **261**, 127781 (2020)
44. Halim, J., Lukatskaya, M.R., Cook, K.M., Lu, J., Smith, C.R., Näslund, L.Å., Barsoum, M.W.: Transparent conductive two-dimensional titanium carbide epitaxial thin films. Chem. Mater. **26**(7), 2374–2381 (2014)
45. Gu, P., Xing, J., Wen, T., Zhang, R., Wang, J., Zhao, G., Wang, X.: Experimental and theoretical calculation investigation on efficient Pb (II) adsorption on etched Ti_3AlC_2 nanofibers and nanosheets. Environ. Sci. Nano **5**(4), 946–955 (2018)
46. Alhabeb, M., Maleski, K., Anasori, B., Lelyukh, P., Clark, L., Sin, S., Gogotsi, Y.: Guidelines for synthesis and processing of two-dimensional titanium carbide (Ti3C2T x MXene). Chem. Mater. **29**(18), 7633–7644 (2017)
47. Guo, Y., Jin, S., Wang, L., He, P., Hu, Q., Fan, L.Z., Zhou, A.: Synthesis of two-dimensional carbide Mo2CTx MXene by hydrothermal etching with fluorides and its thermal stability. Ceram. Int. **46**(11), 19550–19556 (2020)
48. Gogotsi, Y.: Transition metal carbides go 2D. Nat. Mater. **14**(11), 1079–1080 (2015)
49. Verger, L., Xu, C., Natu, V., Cheng, H.M., Ren, W., Barsoum, M.W.: Overview of the synthesis of MXenes and other ultrathin 2D transition metal carbides and nitrides. Curr. Opin. Solid State Mater. Sci. **23**(3), 149–163 (2019)
50. Nasrallah, G.K., Al-Asmakh, M., Rasool, K., Mahmoud, K.A.: Ecotoxicological assessment of Ti3C2Tx (Mxene) using a zebrafish embryo model. Environ. Sci. Nano **5**, 1002–1011 (2018)
51. Natu, V., Pai, R., Sokol, M., Carey, M., Kalra, V., Barsoum, M.W.: 2D Ti_3C_2Tz MXene synthesized by water-free etching of Ti_3AlC_2 in polar organic solvents. Chem. **6**(3), 616–630 (2020)
52. Hantanasirisakul, K., Zhao, M.Q., Urbankowski, P., Halim, J., Anasori, B., Kota, S., Gogotsi, Y.: Fabrication of Ti3C2Tx MXene transparent thin films with tunable optoelectronic properties. Adv. Electron. Mater. **2**(6), 1600050 (2016)
53. Ma, R., Sasaki, T.: Acc. Chem. Res. Two-dimensional oxide and hydroxide nanosheets: controllable high-quality exfoliation, molecular assembly, and exploration of functionality. Acc. Chem. Res. **48**(1), 136–143 (2015)
54. Ihsanullah, I.: MXenes (two-dimensional metal carbides) as emerging nanomaterials for water purification: progress, challenges and prospects. Chem. Eng. J. **388**, 124340 (2020)
55. Talreja, N., Ashfaq, M., Chauhan, D., Mangalaraja, R.V.: Cu-MXene: A potential biocide for the next-generation biomedical application. Mater. Chem. Phys. **294**, 127029 (2023)
56. Ding, L., Wei, Y., Wang, Y., Chen, H., Caro, J., Wang, H.: A two- dimensional lamellar membrane: MXene nanosheet stacks. Angew. Chem. Int. Ed. **56**, 1825–1829 (2017)

57. Sun, Y., Li, S., Zhuang, Y., Liu, G., Xing, W., Jing, W..: Adjustable interlayer spacing of ultrathin MXene-derived membranes for ion rejection. J. Memb. Sci. **591**, 117350 (2019)
58. Elumalai, S., Yoshimura, M., Ogawa, M.: Simultaneous delamination and rutile formation on the surface of Ti_3C_2Tx MXene for copper adsorption. Chem. Asian J. **15**(7), 1044–1051 (2020)
59. Lukatskaya, M.R., Kota, S., Lin, Z., Zhao, M.Q., Shpigel, N., Levi, M.D., Gogotsi, Y.: Ultra-high-rate pseudocapacitive energy storage in two-dimensional transition metal carbides. Nat. Energy **2**(8), 1–6 (2017)
60. Benck, J.D., Pinaud, B.A., Gorlin, Y., Jaramillo, T.F.: Substrate selection for fundamental studies of electrocatalysts and photoelectrodes: inert potential windows in acidic, neutral, and basic electrolyte. PLoS ONE **9**(10), e107942 (2014)
61. Yang, C., Tang, Y., Tian, Y., Luo, Y., Faraz Ud Din, M., Yin, X., Que, W.: Flexible nitrogen-doped 2D titanium carbides (MXene) films constructed by an ex situ solvothermal method with extraordinary volumetric capacitance. Adv. Energy Mater. **8**(31), 1802087 (2018)
62. Huang, K., Li, Z., Lin, J., Han, G., Huang, P.: Two-dimensional transition metal carbides and nitrides (MXenes) for biomedical applications. Chem. Soc. Rev. **47**(14), 5109–5124 (2018)
63. Pang, J., Mendes, R.G., Bachmatiuk, A., Zhao, L., Ta, H.Q., Gemming, T., Rummeli, M.H.: Applications of 2D MXenes in energy conversion and storage systems. Chem. Soc. Rev. **48**(1), 72–133 (2019)
64. Dillon, A.D., Ghidiu, M.J., Krick, A.L., Griggs, J., May, S.J., Gogotsi, Y., Fafarman, A.T.: Highly conductive optical quality solution-processed films of 2D titanium carbide. Adv. Funct. Mater. **26**(23), 4162–4168 (2016)
65. Hope, M.A., Forse, A.C., Griffith, K.J., Lukatskaya, M.R., Ghidiu, M., Gogotsi, Y., Grey, C.P.: NMR reveals the surface functionalisation of Ti 3 C 2 MXene. Phys. Chem. Chem. Phys. **18**(7), 5099–5102 (2016)
66. Lipatov, A., Lu, H., Alhabeb, M., Anasori, B., Gruverman, A., Gogotsi, Y., Sinitskii, A.: Elastic properties of 2D Ti3C2Tx MXene monolayers and bilayers. Sci. Adv. **4**, eaat0491 (2018)
67. Wang, H., Wu, Y., Yuan, X., Zeng, G., Zhou, J., Wang, X., Chew, J.W.: Clay-inspired MXene-based electrochemical devices and photo-electrocatalyst: state-of-the-art progresses and challenges. Adv. Mater. **30**(12), 1704561 (2018)
68. Ling, Z., Ren, C.E., Zhao, M.Q., Yang, J., Giammarco, J.M., Qiu, J., Gogotsi, Y.: Flexible and conductive MXene films and nanocomposites with high capacitance. Proc. Natl. Acad. Sci. **111**(47), 16676–16681 (2014)
69. Mashtalir, O., Lukatskaya, M.R., Kolesnikov, A.I., Raymundo-Pinero, E., Naguib, M., Barsoum, M.W., Gogotsi, Y.: The effect of hydrazine intercalation on the structure and capacitance of 2D titanium carbide (MXene). Nanoscale **8**(17), 9128–9133 (2016)
70. Zha, X.H., Yin, J., Zhou, Y., Huang, Q., Luo, K., Lang, J., Du, S.: Intrinsic structural, electrical, thermal, and mechanical properties of the promising conductor Mo2C MXene. J. Phys. Chem. C **120**(28), 15082–15088 (2016)
71. Wang, Y., Niu, B., Zhang, X., Lei, Y., Zhong, P., Ma, X.: Ti3C2Tx MXene: an emerging two-dimensional layered material in water treatment. ECS J Solid State Sci Technol. **10**(4), 047002 (2021)
72. Abbas, A., Al-Amer, A.M., Laoui, T., Al-Marri, M.J., Nasser, M.S., Khraisheh, M., Atieh, M.A.: Heavy metal removal from aqueous solution by advanced carbon nanotubes: critical review of adsorption applications. Sep. Purif. Technol. **157**, 141–161 (2016)
73. Gan, L.Y., Huang, D., Schwingenschlögl, U.: Oxygen adsorption and dissociation during the oxidation of monolayer Ti 2 C. J. Mater. Chem. A. **1**(43), 13672–13678 (2013)
74. Shahzad, A., Rasool, K., Nawaz, M., Miran, W., Jang, J., Moztahida, M., Mahmoud, K.A., Lee, D.S.: Heterostructural $TiO_2/Ti_3C_2T_x$ (Mxene) for photocatalytic degradation of antiepileptic drug carbamazepine. Chem. Eng. J. **349**, 748–755 (2018)
75. Peng, Q., Guo, J., Zhang, Q., Xiang, J., Liu, B., Zhou, A., ... Tian, Y.: Unique lead adsorption behavior of activated hydroxyl group in two-dimensional titanium carbide. J. Am. Chem. Soc. **136**(11), 4113–4116 (2014)

76. Tang, J., Zhang, Y.F., Liu, Y., Li, Y., Hu, H.: Efficient ion-enhanced adsorption of congo red on polyacrolein from aqueous solution: experiments, characterization and mechanism studies. Sep. Purif. Technol. **252**, 117445 (2020)
77. Kang, Z., Khan, M.A., Gong, Y., Javed, R., Xu, Y., Ye, D., Zhang, J.: Recent progress of MXenes and MXene-based nanomaterials for the electrocatalytic hydrogen evolution reaction. J. Mater. Chem. A **9**(10), 6089–6108 (2021)
78. Cai, C., Wang, R., Liu, S., Yan, X., Zhang, L., Wang, M., Jiao, T.: (Synthesis of self-assembled phytic acid-MXene nanocomposites via a facile hydrothermal approach with elevated dye adsorption capacities. Colloids Surf A Physicochem Eng. Asp **589**, 124468 (2020)
79. Pandey, R.P., Rasool, K., Madhavan, V.E., Aïssa, B., Gogotsi, Y., Mahmoud, K.A.: Ultrahigh-flux and fouling-resistant membranes based on layered silver/MXene (Ti 3 C 2 T x) nanosheets. J. Mater. Chem. A. **6**(8), 3522–3533 (2018)
80. Wei, S., Xie, Y., Xing, Y., Wang, L., Ye, H., Xiong, X., Wang, S., Han, K.: Two- dimensional Graphene Oxide/MXene composite lamellar membranes for efficient solvent permeation and molecular separation. J. Membr. Sci. **582**, 414–422 (2019)
81. Gao, H., Chen, W., Xu, C., Liu, S., Tong, X., Chen, Y.: Two-Dimensional Ti3C2Tx MXene/GO hybrid membranes for highly efficient osmotic power generation. Environ. Sci. Technol. **54**, 2931–2940 (2020)
82. Xie, X., Chen, C., Zhang, N., Tang, Z.-R., Jiang, J., Xu, Y.-J.: Microstructure and surface control of mxene films for water purification. Nat. Sustain. **2**, 856–862 (2019)
83. Zhang, H., Wang, Z., Shen, Y., Mu, P., Wang, Q., Li, J.: Ultrathin 2D $Ti_3C_2T_x$ MXene membrane for effective separation of oil-in-water emulsions in acidic, alkaline, and salty environment. J. Colloid Interface Sci. **561**, 861–869 (2020)
84. Jun, B.M., Park, C.M., Heo, J., Yoon, Y.: Adsorption of Ba2+ and Sr2+ on Ti3C2Tx MXene in model fracking wastewater. J. Environ. Manag. **256**, 109940 (2020)
85. Hassan, M.U., Lee, S., Mehran, M.T., Shahzad, F., Husnain, S.M., Ryu, H.J.: Post- decontamination treatment of MXene after Adsorbing Cs from contaminated water with the enhanced thermal stability to form a stable radioactive waste matrix. J. Nucl. Mater. **543**, 152566 (2021)
86. Zirehpour, A., Rahimpour, A., Arabi Shamsabadi, A., Sharifian, M.G., Soroush, M.: Mitigation of thin-film composite membrane biofouling via immobilizing nano-sized biocidal reservoirs in the membrane active layer. Environ. Sci. Technol. **51**, 5511–5522 (2017)
87. Rasool, K., Helal, M., Ali, A., Ren, C.E., Gogotsi, Y., Mahmoud, K.A.: Antibacterial activity of Ti3C2T x MXene. ACS Nano **10**(3), 3674–3684 (2016)
88. Dixit, F., Zimmermann, K., Dutta, R., Prakash, N.J., Barbeau, B., Mohseni, M., Kandasubramanian, B.: Application of MXenes for water treatment and energy-efficient desalination: a review. Hazard. Mater. **423**, 127050 (2022)
89. Li, Z.-K., Liu, Y., Li, L., Wei, Y., Caro, J., Wang, H.: Ultra-thin titanium carbide (Mxene) sheet membranes for high-efficient oil/water emulsions separation. J. Membr. Sci. **592**, 117361 (2019)
90. Saththasivam, J., Wanga, K., Yiming, W., Liu, Z., Mahmoud, K.A.: A Flexible Ti3C2Tx (Mxene)/paper membrane for efficient oil/water separation. RSC Adv. **9**, 16296–16304 (2019)
91. He, S., Zhan, Y., Hu, J., Zhang, G., Zhao, S., Feng, Q., Yang, W.: Chemically Stable Two-dimensional MXene@UIO-66-$(COOH)_2$ composite lamellar membrane for multi-component pollutant-oil-water emulsion separation. Compos. B Eng. **197**, 108188 (2020)
92. Moghaddasi, A.A., Sobolčiak, P., Popelka, A., Krupa, I.: Separation of Water/Oil emulsions by an electrospun copolyamide mat covered with a 2D $Ti_3C_2T_x$ MXene. Materials. **13**, 3171 (2020)
93. Kim, S., Gholamirad, F., Yu, M., Park, C.M., Jang, A., Jang, M., Taheri-Qazvini, N., Yoon, Y.: Enhanced adsorption performance for selected pharmaceutical compounds by sonicated Ti3C2TX MXene. Chem. Eng. J. **406**, 126789 (2021)
94. Jang, J., Shahzad, A., Woo, S.H., Lee, D.S.: Magnetic Ti3C2Tx (Mxene) for diclofenac degradation via the ultraviolet/chlorine advanced oxidation process. Environ. Res. **182**, 108990 (2020)

95. Du, Y., Yu, B., Wei, L., Wang, Y., Zhang, X., Ye, S.: Efficient removal of Pb (II) by Ti3C2T x powder modified with a silane coupling agent. J. Mater. Sci. **54**(20), 13283–13297 (2019)
96. Wang, L., Tao, W., Yuan, L., Liu, Z., Huang, Q., Chai, Z., Shi, W.: Rational control of the interlayer space inside two-dimensional titanium carbides for highly efficient uranium removal and imprisonment. ChemComm. **53**(89), 12084–12087 (2017)
97. Zavabeti, A., Jannat, A., Zhong, L., Haidry, A.A., Yao, Z., Ou, J.Z.: Two-dimensional materials in large-areas: synthesis, properties and applications. Nanomicro Lett. **12**, 1–34 (2020)
98. Champagne, A., Charlier, J.C.: Physical properties of 2D MXenes: from a theoretical perspective. Mater. Today Phys. **3**(3), 032006 (2020)
99. Ren, C.E., Hatzell, K.B., Alhabeb, M., Ling, Z., Mahmoud, K.A., Gogotsi, Y.: Charge-and size-selective ion sieving through Ti3C2T x MXene membranes. J. Phys. Chem. Lett. **6**(20), 4026–4031 (2015)
100. Zou, G., Guo, J., Peng, Q., Zhou, A., Zhang, Q., Liu, B.: Synthesis of urchin-like rutile titania carbon nanocomposites by iron-facilitated phase transformation of MXene for environmental remediation. J. Mater. Chem. A. **4**(2), 489–499 (2016)
101. Naguib, M., Mashtalir, O., Lukatskaya, M.R., Dyatkin, B., Zhang, C., Presser, V., Barsoum, M.W.: One-step synthesis of nanocrystalline transition metal oxides on thin sheets of disordered graphitic carbon by oxidation of MXenes. ChemComm. **50**(56), 7420–7423 (2014)
102. Fard, A.K., Mckay, G., Chamoun, R., Rhadfi, T., Preud, H., Atieh, M.A.: Barium removal from synthetic natural and produced water using MXene as two dimensional (2-D) nanosheet adsorbent. Chem. Eng. Sci. **317**, 331–342 (2017)
103. Zhang, H., Li, M., Cao, J., Tang, Q., Kang, P., Zhu, C., Ma, M.: 2D a-Fe2O3 doped Ti3C2 MXene composite with enhanced visible light photocatalytic activity for degradation of Rhodamine B. Ceram. Int. **44**(16), 19958–19962 (2018)
104. Arabi Shamsabadi, A., Anasori, B., Soroush, M.: Antimicrobial mode-of-action of colloidal $Ti_3C_2T_x$ MXene nanosheets. ACS Sustain. Chem. Eng. **6**(12), 16586–16596 (2018)
105. Khan, A.R., Husnain, S.M., Shahzad, F., Mujtaba-ul-Hassan, S., Mehmood, M., Ahmad, J., Rahman, S.: Two-dimensional transition metal carbide (Ti 3 C 2 T x) as an efficient adsorbent to remove cesium (Cs^+). Dalton Trans. **48**(31), 11803–11812 (2019)
106. Rasheed, T., Rasheed, A., Munir, S., Ajmal, S., Shahzad, Z.M., Alsafari, I.A., Shakir, I.: A cost-effective approach to synthesize $NiFe_2O_4$/MXene heterostructures for enhanced photodegradation performance and antibacterial activity. Adv. Powder Technol. **32**(7), 2248–2257 (2021)
107. Ghanbari, R., Zare, E.N., Paiva-Santos, A.C., Rabiee, N.: Ti_3C_2Tx MXene@ MOF decorated polyvinylidene fluoride membrane for the remediation of heavy metals ions and desalination. Chemosphere **311**, 137191 (2023)
108. Lin, Q., Zeng, G., Yan, G., Luo, J., Cheng, X., Zhao, Z., Li, H.: Self-cleaning photocatalytic MXene composite membrane for synergistically enhanced water treatment: oil/water separation and dyes removal. Chem. Eng. J. **427**, 131668 (2022)
109. Zhang, D., Zhao, B., Liu, L., Tang, H., Wang, X., Yu, S.: Insights into enhanced elimination of U (VI) and Eu (III) by amidoxime-functionalized Ti3C2Tx MXenes. Sep. Purif. Technol. **294**, 121179 (2022)
110. Zhang, H., Zheng, Y., Zhou, H., Zhu, S., Yang, J.: Nanocellulose-intercalated MXene NF membrane with enhanced swelling resistance for highly efficient antibiotics separation. Sep. Purif. Technol. **305**, 122425 (2023)
111. Wang, L., Yuan, L., Chen, K., Zhang, Y., Deng, Q., Du, S., Shi, W.: Loading actinides in multi layered structures for nuclear waste treatment: the first case study of uranium capture with vanadium carbide MXene. ACS Appl. Mater. Interfaces **8**(25), 16396–16403 (2016)

Two-Dimensional (2D) Hybrid Nanocomposites for Environmental Sensing Applications

Nigel Twi-Yeboah, Dacosta Osei, William H. Dontoh, and Michael K. Danquah

Abstract This chapter focuses on the development and application of 2D hybrid nanocomposites for environmental sensing. The unique properties of 2D materials, which include graphene, have made them promising candidates for environmental sensing applications. In particular, the incorporation of 2D materials into hybrid nanocomposites has made the manipulation of their optical, electrical, and mechanical properties for specific sensing applications, possible. An advantage of 2D hybrid nanocomposites for environmental sensing is their high sensitivity to various analytes, including gases, humidity, and chemicals. For instance, graphene-based hybrid nanocomposites have been proven to show high sensitivity to trace gases such as ammonia, nitrogen dioxide, and sulfur dioxide. In addition, 2D hybrid nanocomposites offer several advantages, including ease of fabrication and low cost. These materials can be synthesized using various methods such as chemical vapor deposition, solution processing, and physical vapor deposition. The chapter also discusses the challenges associated with the development of 2D hybrid nanocomposites, including the need for improved selectivity, stability, and reproducibility. Some future perspectives on the potential applications of 2D hybrid nanocomposites in environmental sensing are discussed, including their use in air and water quality monitoring and food safety.

Keywords 2D nanocomposites · Hybrid nanomaterials · Material synthesis and characterization · Environmental sensing

---

N. Twi-Yeboah · D. Osei · W. H. Dontoh
Chemical and Petroleum Engineering Department, University of Kansas, 1450 Jayhawk Boulevard, 230, Lawrence, KS 66045, USA

M. K. Danquah (✉)
Department of Chemical and Biomolecular Engineering, The University of Tennessee, Knoxville, TN 37996, USA
e-mail: michael-danquah@utc.edu

N. Talreja et al. (eds.), *Two-dimensional Hybrid Composites*, Engineering Materials,
https://doi.org/10.1007/978-981-99-8010-9_6

1 Introduction

Environmental sensing is the process whereby chemical, physical, and biological parameters are measured and/or detected. This process is very significant, as it gives a quantitative and qualitative indication of the quality of the environment. Some of the factors that are measured include temperature, water quality, air quality, humidity, pollutant and pathogen concentration [1]. Being aware of the quality of the environment protects lives, health and prevents pollution of the environment [2].

Nanotechnology application in environmental sensing has brought about great improvements to the conventional methods and materials. This is mainly due to unique nanomaterial characteristics, both physical and chemical. Nanomaterials applied in this process are able to make detections at higher resolution scales, therefore improving accuracy. They also have fast response times and higher sensitivity than the conventional materials [3–5].

One of the most recent discoveries in the study of environmental sensing is the use of 2D hybrid nanocomposites/2D hybrid nanocomposites which are produced when 2D nanomaterials (materials on the nanoscale, existing in a two-dimensional structure) are combined with other nanomaterials. These secondary materials can be in the form of metals, metal oxides and sometimes enzymes, to enhance detection abilities of 2D nanomaterials [6]. The two-dimensional nature gives the advantage of a high surface area to volume ratio (which allows for a greater efficiency in the interaction of the sensing material and the target analytes), tunable electronic properties, and improved mechanical strength (which ensures stability and durability in diverse environmental conditions) [7]. The following are some of the significance of 2D hybrid nanocomposites in environmental sensing applications, based on their properties:

- Improved sensitivity: The high surface area to volume ratio, improves the exposure of the active sites of the sensors to the parameters, being measured.
- Handling and cost-effective: Since sensors in this form are miniaturized, they are easy to handle since they do not need to take up so much space in the area where the detection is being done. Costs involved in manufacturing and transporting products are also reduced, since there is very little material used, and a lot can be transported at once.
- Low detection limits: Sensors in this form are able to detect pollutants or contaminants that may exist in very low concentrations due to high sensitivity.
- Tunability: Using a 2D hybrid nanocomposite introduces the flexibility in tuning electronic and optical properties simply by changing the composition, size, or type, based on the function needed to be carried out.
- Response time: The sensitivity of 2D hybrid nanocomposite enables fast response time, making it possible for accurate real time detection and monitoring.
- Wide range: 2D hybrid nanocomposite sensors allow for detection of wide range of analytes.

Generally, the use of 2D hybrid nanocomposites has brought a major advancement in the field of environmental sensing, due to their unique properties as mentioned. Harnessing the potential of these materials and their applications will improve sustainability in this field for the benefit of the ecosystem. In this chapter, methods of synthesis of 2D hybrid nanocomposite sensors have been discussed. Applications of these types of sensors in environmental sensing, such as gas sensing, biosensing, humidity sensing, temperature sensing, and pressure sensing are outlined. Moreover, the advantages of 2D hybrid nanocomposite sensors in terms of their performances and properties are highlighted. Furthermore, challenges involved in the manufacturing and application of these sensors are thoroughly addressed, with possible strategies to overcome the challenges and future outlook of their applications.

2 Synthesis Methods of 2D Hybrid Nanocomposites for Environmental Sensing

Synthesis methods of 2D hybrid nanocomposites for environmental sensing come in different forms which include chemical vapor deposition (CVD), liquid-phase exfoliation (LPE), Electrochemical deposition or Electroplating and Hybridization approaches for 2D materials.

2.1 Chemical Vapor Deposition (CVD)

CVD is vacuum deposition technique which is advanced and utilized to create solid materials with great performance and quality. In CVD, the introduction of gases that are reactive are introduced into a chamber for coating or filming. The substrate is interacted with volatile precursors (gases) that react or degrade on the surface of the substrate to generate thin deposits of film [8]. In CVD applications, when a precursor gas or gasses are injected into a reactor, chemical reactions cause the precursor gases' desired constituents to form a thin coating on the surface. The substrates are initially cleaned and placed in the chamber which is followed by evacuation of the chamber, allowing a low-pressure field and eventually followed by the introduction of the precursor or reactive gases for deposition to occur. According to [9, 10], high entropy oxides can be made by physical vapor deposition, which makes it easier to do so at low temperatures. However, chemical vapor deposition synthesis makes it easier to make conformal layers on a 3D substrate. The CVD method uses transition metals like nickel, copper, platinum, gallium, and palladium that create a catalyst layer for large-scale graphene synthesis. A carbon-containing slurry is added to the reaction after the material has been placed on the substrate and etched in order to create high-grade graphene. In CVD applications, production of high purity materials is obtained due to its occurrence in a controlled vacuum and the method allows for extremely fine-grained manipulation of material properties through excellent control over substrate thickness and homogeneity. Excellent adhesion between the deposited material and the substrate is also possible with CVD, resulting in coatings that are long-lasting and robust.

2.2 *Liquid-Phase Exfoliation (LPE)*

LPE describes a series of methods which directly exfoliate bulk graphite into thin graphene [11] in a liquid medium. LPE, like CVD, is a promising approach for producing bulk graphene where the thickness and/or exfoliation can be controlled. Choosing a suitable solvent or dispersion that may interact with graphite and diminish the interlayer forces holding the graphite layers together is the first step in the LPE process.

Layered crystals are changed into two-dimensional nanocomposites in a solution processing approach during LPE applications. One of the most promising methods to accomplish mass manufacturing at a very cheap cost is the conversion of graphite into graphene via LPE. Because of its outstanding qualities, including mechanical toughness and high electrical and thermal conductivity, graphene has attracted a lot of attention lately. But in order to realize graphene's full potential for a variety of uses, it must be produced in huge quantities and at a reasonable cost. Liquid-phase exfoliation is a fundamental method for producing 2D materials like graphene in bulk with equilibrium between price and performance, and it is currently commonly used by both the academic and industrial sectors [12]. Liquid-phase exfoliation has been employed for high performance and successful graphene production. In the LPE method, a variety of solvents are used to decrease the interlayer forces holding the graphene layers together, allowing them to split and suspend in the liquid phase. Recent and/or cutting-edge LPE methods that have been developed perform very well in the creation of two-dimensional carbon-based nanostructures [11]. The creation of a graphite precursor, ideally graphite flakes, is the first step in the LPE process. The chosen solvent is then combined with these flakes to create a stable suspension. Ultrasonic treatment or shear pressures are applied to the suspension, which helps to exfoliate the graphite layers into graphene sheets.

Various exfoliation methods such as ultrasonic, shear, electrochemical, and functionalization-assisted exfoliation have been proposed to be the four basic categories for liquid-phase exfoliation. For ultrasonic exfoliation, the presence of different stabilizers and their influence on the resurfacing process are considered. The effects of the electrolyte type, concentration, and voltage are explored in relation to electrochemical exfoliation.

Different approaches that generate graphene by directly ultrasonically abrading bulk graphite in liquid media are referred to as direct ultrasonic exfoliation in liquid. Two methods that influence the performance and yield of exfoliation are the liquid media that is used for the exfoliation and energy involved. In order to counteract the vdW interactions between layers, an adequate solvent aids reduction of energy around neighboring layers of the graphite [13–15].

2.3 *Electrochemical Deposition*

Modified electrodes have been essential for the creation of various electroanalytical devices with improved performance (e.g. sensitivity and selectivity) [16]. Research on such electrodes has been conducted to understand their charge transfer/transport properties. Also, these electrodes undergo surface alterations to add certain features that enhance their electrochemical performance. They have been thoroughly investigated and manufactured. Electricity is used to facilitate the chemical reaction, which occurs at the contact between solid and liquid. It is well known that electrochemical deposition is an easy way to create complicated structures and catalytic surfaces, and it has a great deal of potential to be a breakthrough technique in nanotechnology. By adjusting the experimental synthesis conditions, the electrochemical method of synthesis enables the creation of nanostructures with clearly defined morphologies and a range of sizes without the need for a template [17]. The ability of the electrodes to improve the sensitivity and selectivity of electroanalytical devices is one of the main benefits in electrochemical deposition applications. Target analytes can be detected and quantified with extreme precision by adding various types of modifiers to the electrode surface, such as nanoparticles and polymers.

Metal nanoparticle preparation by electrochemical deposition is effective, but it is typically less popular than wet-chemical techniques. This method has many benefits, especially in terms of the speed of synthesis, and the absence of undesirable products, yet it can occasionally reveal some limitations in terms of the nanomaterial dimensions and permitted morphologies [18]. Additionally, direct deposition of the modifier layer onto the electrode enables the achievement of a greater adhesion [19]. Numerous electrochemical techniques are used in electrodeposition [17] (Fig. 6.1).

For electrodeposition application in metal–organic framework synthesis, some electrochemical deposition approaches for MOF synthesis include cathodic deposition, anodic deposition, and electrophoretic deposition. Using anodic and cathodic

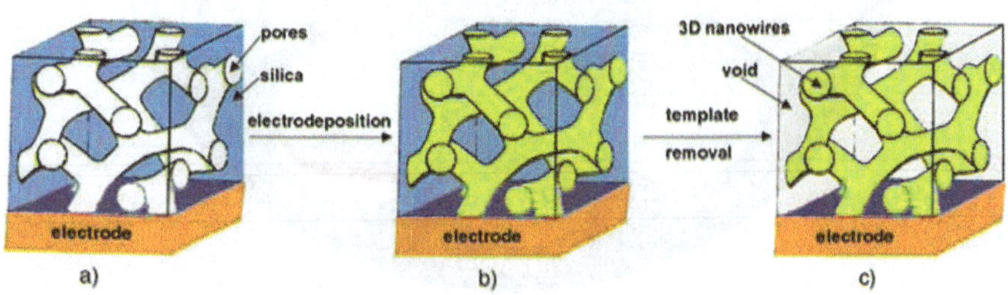

Fig. 6.1 Diagram illustrating how a templated electrodeposition approach can create 3D continuous macroscopic metal or semiconductor nanowire networks [20]

deposition methods, the metal–organic framework is directly created during an in-place process on the substrate surface. The electrochemical synthesis of MOFs is used in the anodic and cathodic deposition processes, allowing for the controlled development of MOF films or coatings on a conductive substrate. A metal substrate is used as the anode in an electrochemical cell during anodic deposition, and the substrate is submerged in an electrolyte solution that contains the desired MOF's organic linkers and metal ions. A metal oxide layer is created on the substrate surface as a result of the metal ions oxidizing when an electric current is applied. However, in cathodic deposition, metal ions and organic linkers are reduced at the cathode, the electrochemical cell's negatively charged electrode. Similar to anodic deposition, the substrate in this technique is submerged in a solution containing metal ions and organic linkers. But at the cathode, the metal ions and organic linkers are concurrently reduced rather than producing a layer of metal oxide.

These methods are preferable to electrophoretic deposition, ex-situ synthesis, which seeks to deposit a MOF that has already been created onto a targeted substrate. The simplest and most common method is anodic oxidation, where the anode functions as a metal source and supports the production of metal–organic framework films. Additionally, the electrodeposition method can create multi-metallic MOF films, broadening its range of applications. Moreover, multiple metals are incorporated into the metal–organic framework to produce films with improved characteristics and new opportunities for diverse technological breakthroughs (Fig. 6.2).

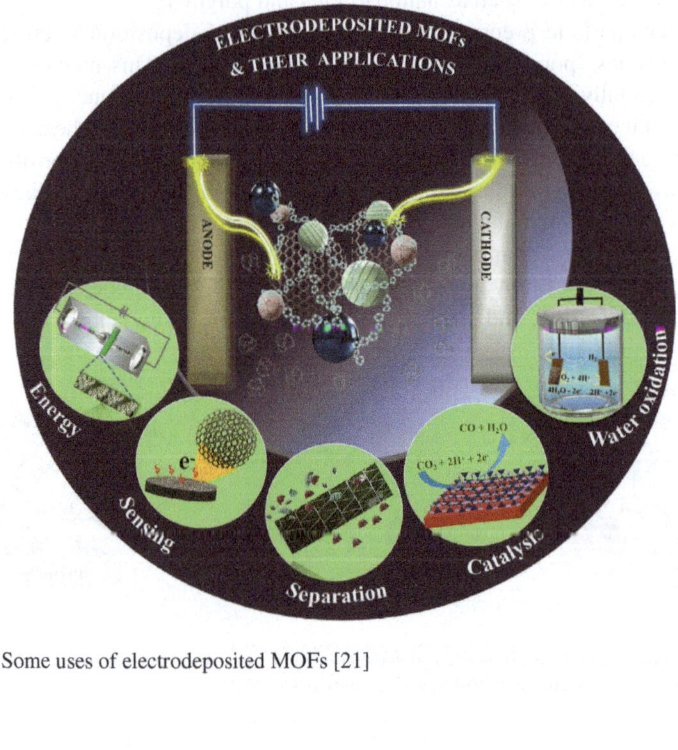

Fig. 6.2 Some uses of electrodeposited MOFs [21]

2.4 Hybridization Approaches for 2D Materials

Through hybridization of functional species onto and/or into other 2D materials, the properties of the nanocomposite materials are greatly enhanced with improved performances. In hybridization approaches for 2D nanomaterials, in order to significantly increase the properties of nanocomposite materials and improve overall performance, the concept of hybridization entails incorporating functional species onto or into other 2D materials. By utilizing diverse 2D nanomaterials and functionalized hybridization [22] reported numerous hybridization. There has been an increase in interest in adding particular species to 2D nanosheets, either directly or indirectly, in order to give hybridized nanostructures new or better characteristics and synergistic capabilities that would enable them to perform well in cutting-edge applications.

Through their methodical production, diverse 2D nanosheets with various atomic structures and compositions have been thoroughly studied and reports have indicated enhanced properties and performance through sp^2 and sp^3 hybridization [23, 24].

3 Environmental Sensing Applications of 2D Hybrid Nanocomposites

3.1 Biosensing

2-D hybrid nanocomposites can be applied in various fields, including biosensing (nano biosensors), especially in the detection of pathogens, which pose danger to human health. Nano biosensors are able to detect pathogens by using nucleic acids, functional proteins, antibodies, and cell-based biosensors as elements of recognition, to attach themselves to pathogen. This initiates a physicochemical process which is then detected by a transducing element. Recent advancements in nanotechnology have led to the production of nanobiosensors with high sensitivity and selectivity. Signals from the physicochemical interaction between the recognition elements and the pathogen are converted by the transducing element into electrical signals which can be measured and quantified. This makes it possible for pathogens to be detected and monitored with high sensitivity and selectivity [25, 26]. In a study by Muniandy et al. [27] the effectiveness of reduced graphene oxide-titanium dioxide was demonstrated. From their findings, reduced graphene oxide-titanium dioxide as a hybrid nanocomposite with recognition elements, showed good selectivity, sensitivity, and a wide range for detecting Salmonella, a food pathogen. This paper proved the capabilities of hybrid nanocomposites in improving food safety and public health by enhancing pathogen detection techniques.

3.2 Pressure Sensing

2D-hybrid nanocomposite sensors can be used to monitor pressure in the environment. There is great potential for a variety of applications when 2D-hybrid nanocomposite sensors are incorporated into pressure monitoring systems. These sensors can be used for accurate pressure monitoring in reaction vessels, pipelines, and process units in the chemical sector, ensuring the best process control and safety. They can be implemented into touch-sensitive electronics, making pressure-sensitive interactions possible and enhancing user experience. Additionally, by making it possible to accurately monitor things like intracranial pressure, breathing patterns, and blood pressure, among other things, in the medical field, these sensors have the potential to transform healthcare. These sensors' small size and adaptability enable their easy integration into wearable technology, enabling real-time health monitoring and the early identification of life-threatening illnesses. The working principle of pressure sensors can be categorized under four types, based on how pressure is detected, transmitted, and quantified. The four ways are through piezoresistive, capacitive, triboelectric, and piezoelectric means. The piezoresistive type of pressure sensors, detect pressure as a change in contact area and distribution of internal sensing materials, after being subjected to an external force. This change corresponds to an alteration in electrical signal, which is then transmitted and measured as pressure change. On the other hand, the capacitive type of pressure sensors operates by detecting changes in the distance between two electrodes on either side of the sensors, when subjected to and external force. These changes alter the capacitance of the device, which is transmitted and quantified as pressure change. Triboelectric types of pressure sensors detect external forces or pressure using the periodic mechanical energy of contacting and separating electrodes, to generate electrostatic charges. The charges can then be used to detect external forces, which are also as a result of changes in pressure. Lastly, the piezoelectric types of sensors detect pressure through polarization of charges, when external forces create uneven charges within the sensor. The polarization effect enables the detection and measurement of the pressure changes [28]. In an experimental study by Pham et al. [29] the sensitivity and detection limit of ZnO/chlorine-trapped doped bilayer graphene were investigated. The researchers reported that doping graphene with chlorine radicals using a heavy p-type chlorine trap, was responsible for the significant improvement in the sensitivity. This hybridized 2D nanocomposite pressure sensor shows promising applications in chemical, electronic, and medical fields.

3.3 Humidity Sensing

Humidity sensors work by detecting the amount of moisture of water vapor in air. It may be expressed as absolute or relative humidity where absolute humidity measure the actual amount of moisture in the air and relative humidity compares the moisture

Table 6.1 Sensing techniques for humidity sensors

Sensor technique	Sensing mechanism/working principle	References
Capacitive	Uses changes in dielectric constant of hygroscopic dielectric materials in the sensor (which leads to changes in capacitance) to sense changes in humidity	Zhang et al. [30] Zhang et al. [31]
Resistive	Uses changes in electrical impedance as hygroscopic media present in the sensor, absorb water molecules, to detect humidity changes	Zhang et al. [32] Cavallo et al. [33]
Thermal conductivity	Uses changes in temperature gradients (induced by moisture content) between two temperature-sensitive elements in a sensing channel (exposed to the measured air) and a reference channel (exposed to a dry reference gas)	Okcan and Akin [34]
Optical	Uses changes in refractive index in sensing materials, induced by humidity, and causing a variation in transmitted optical intensity	Lee and Lee [35] Rao et al. [36]
Gravimetric	Uses changes in mass of moisture adsorbed unto the sensing material's surface, as a result of increase in relative humidity	Lee and Lee [35] Zheng et al. [37]

present in the air and the maximum moisture holding capacity of the air at a given temperature. Based on sensing techniques, humidity sensor types include capacitive, resistive, thermal conductivity, optical and gravimetric techniques. Table 6.1 shows the working principles of these types of techniques used in humidity sensing.

2D hybrid nanocomposites have shown very promising applications in humidity sensing. For example, Zhang and co-workers, conducted a study on the use of tin oxide/graphene hybrid nanocomposite. The sensitivity of this hybrid nanocomposite sensor was tested by exposing it to a broad range of relative humidity (11% to 97%), at room temperature. They found that this sensor had rapid response and recovery attributes over the full range of humidity measurement [30]. Another study conducted by Zhang and co-workers, investigated the humidity sensing capabilities of molybdenum-disulfide-modified tin oxide nanocomposite. It was observed that this hybrid sensor had better sensing characteristics than pure molybdenum-disulfide and pure tin oxide sensors [38].

3.4 Gas Sensing

In order to monitor and identify the presence of different gas contaminants in the environment, gas sensors are essential. To ensure precise and trustworthy measurements, these sensors make use of cutting-edge technologies and materials. 2D-hybrid nanocomposite sensors are one such technique, and they have demonstrated encouraging results in gas detection. 2D-hybrid nanocomposite sensors can be used to detect the presence of gas pollutants in the environment as a result of natural and anthropogenic activities. Some of the gases of great concern are H_2S, CO, CO_2, H_2, NH_3, and SO_2.

According to Dhall and co-workers, there are four main types of gas sensors, based on their sensing mechanisms. They are, infrared sensors, metal semiconductor sensors, fluorescence sensors, and electrochemical sensors [39]. Infrared gas sensors, which are also known as non-dispersive infrared sensors, monitor gases through absorption of radiations in the infrared region. In these types of gas sensors, the gas is collected in a chamber where infrared light with known intensity through it. There is a detector on the opposite side of the chamber which measures the intensity of the infrared light after interacting with the gas molecules. The changes in wavelength and intensity, as a result of the interactions, are characteristic of the gas molecules, and this information is compared to a reference to determine the identity and concentration of the molecules in the sample gas [40]. Metal semiconductor gas sensors work based on the principle of changes in electrical resistances before and after gas adsorption. The sensing materials used are metal oxide semiconductors. A redox reaction occurs when the gas comes into contact with the metal oxide semiconductor, causing a change in its resistance or conductivity. This change is detected and translated into detecting the identity and concentration of the gas [41]. Fluorescence gas sensors detect gases by measuring changes in the length of fluorescent waves after they interact with gas molecules that have been adsorbed by the sensor [42]. Lastly, electrochemical gas sensors have electrodes submerged in electrolytic solutions, which undergo an electrochemical reaction with the sample gas after adsorption. The reaction produces electrons and consequently, current, which is proportional to the gas concentration [43]. Yu and co-workers synthesized an α-Fe_2O_3-graphene hybrid nanocomposite gas sensor. From their study, they observed that, for 50 ppm of gaseous acetone, α-Fe_2O_3-graphene, as a hybrid gas sensor had better response than pure α-Fe_2O_3 as a sensor (about 2.2 times higher). This was attributed to the porous nature of the hybrid sensor, increased specific surface area, and the p-n heterostructure of the hybrid [44].

The SnO_2/rGO hybrid, which was researched by Yin et al. in 2014, is another hybrid gas sensor that is noteworthy. With an exceptionally quick response time of under 7 s, this hybrid gas sensor displayed excellent sensitivity. In this study, it was observed that the increase in the specific area of SnO_2 nanoparticles was also as a result of the reduced graphene oxide (rGO) inclusion. This greatly improved the sensor's ability to detect gases [45].

3.5 Temperature Sensing

In many different fields and applications, temperature sensing is essential. It is essential for observing and regulating procedures, ensuring safety of personnel, and increasing productivity. Temperature sensors, for instance, are widely used in automotive applications to track engine temperature in order to avoid overheating. Temperature sensors are used in the healthcare sector to give precise temperature measurements for patient care using medical devices such as thermometers and incubators. Temperature sensors are also used in environmental monitoring to monitor and analyze temperature variations in ecosystems, weather prediction, and climate change research. Temperature sensors have been included into a variety of smart home technology products in recent years for effective energy management and comfort control. Based on current temperature readings, these sensors allow automated HVAC (heating, ventilation, and air conditioning) system modifications. Furthermore, industrial automation primarily relies on temperature sensing for accurate management and monitoring of manufacturing processes, ensuring the quality of the finished product and the dependability of the equipment. Miniaturized temperature sensors have grown in popularity as a result of technological breakthroughs. These sensors have a compact form and deliver precise temperature readings in restricted locations. They have uses in industries where size and weight are important considerations, such as wearable technology, electronics, and aerospace.

Temperature sensors generally work on the principle of detecting changes in physical properties in the sensors, that are sensitive to temperature. Some of these changes include resistance changes, capacitance changes, and thermal expansion/contraction. Based on the sensing mechanisms, temperature sensing types may include thermistors, resistance temperature detectors, thermocouples, and infrared temperature sensors.

Thermistors usually have semiconductor materials, such as metal oxides as their sensing materials. In this type of sensor, temperature changes cause a non-linear change in resistance, which is used as the feed to sense or monitor the temperature of the environment. Resistance temperature detectors work on similar principles as thermistors. However, their sensing materials are mainly metals. They are also more suitable for industrial application than environmental application [46]. Thermocouples on the other hand, have two different metal wires that are joined together at one end. A temperature difference between the measuring point and the joint end causes the generation of voltage, in a proportional manner (Seebeck effect). The voltage serves as the feed signal, to measure the change in temperature [47].

Unlike the other types mentioned before, infrared temperature sensors detect temperature without direct contact. Infrared radiation temperature sensors detect temperature by using photodetectors to measure the thermal radiation emitted by objects in the environment. This radiation is proportional to temperature changes [48].

Tung and co-workers, conducted a study on hybrid films of graphene and carbon nano-tubes for temperature sensing. In this study, carbon nanotube and graphene were coupled through polyionic liquid-mediated hybridization. It was observed that the resulting hybrid nanocomposite temperature sensor (CNT/PIL/graphene) had very good output voltage response to temperature changes ranging from 25 to 40 °C. The output voltage and the temperature change had very strong correlation with an R-square value of 0.9949 [49].

4 Performance and Properties of 2D Hybrid Nanocomposites for Environmental Sensing

Graphene has generated recent attention in nanoscience and bio-nanotechnology. Graphene's high specific surface area allows for a high molecular loading for enhanced detection sensitivity. Due to this, Gr and GrO are excellent building blocks in novel nanocomposites and sensors. Graphene has been extensively exploited as a chemical building block with a variety of materials in developing highly sensitive and focused electrochemical sensing microdevices.

There are many benefits to using 2D materials in detection and sensing techniques. First, single-layer materials work well for 2D materials because they provide an astonishingly high surface-area-to-volume ratio [50]. Due to their capacity to display an astoundingly high surface-area-to-volume ratio, single-layer materials are recognized for their outstanding applicability in the world of two-dimensional (2D) materials. Their outstanding performance and wide range of applications are largely due to this essential quality.

The relationship between a material's interior volume and external surface area is referred to as the surface-area-to-volume ratio. This ratio takes on special significance when applied to single-layer materials, which are made up of just one atomic layer. This is due to the fact that these materials have an incredibly thin structure, which gives them a surface area that is much larger than their volume. Numerous benefits and new opportunities in numerous research and application areas are made possible by the high surface-area-to-volume ratio. In the case of catalysis, when chemical reactions take place at a material's surface, a larger surface area enables increased catalytic activity. The large surface area of single-layer materials like graphene makes catalytic processes possible and encourages desired reaction rates.

Similar to this, single-layer materials offer a significant benefit in the area of energy storage. Improved energy storage capacity, quicker charge/discharge rates, and improved overall performance are all results of the increased surface area in energy storage devices like batteries and supercapacitors. The search for cutting-edge energy storage solutions for a variety of uses, from portable gadgets to electric cars and grid-scale energy storage systems, is greatly aided by this feature. Additionally, sensors and electrical devices greatly benefit from the high surface-area-to-volume ratio of single-layer materials. Increased sensitivity and selectivity occur from the increased number of active sites available on the bigger surface area for sensing or interacting with other materials. Precision detection and effective signal transduction are crucial in a variety of industries, including environmental monitoring, medical diagnostics, and electronic gadgets.

In conclusion, the remarkable high surface-area-to-volume ratio that single-layer materials exhibit is the key to their extraordinary qualities and adaptability. Due to their special property, these materials have higher catalytic activity, better energy storage capacities, increased sensor sensitivity, and a variety of other uses. As scientists continue to investigate and utilize the possibilities of these materials, they open the door for exciting developments and discoveries across a variety of scientific

Table 6.2 A comparison of platinum/palladium hydrogen sensor and other works [55]

Properties	Samples			
	NPs with a pure Pt/palladium core	Pt–graphene composite	Palladium cube–graphene hybrid	Pt/palladium core–shell–graphene hybrid
Range of detection (ppm)	1000–40,000	100–10,000	6–10,000	1–40,000
Response to 1% hydrogen at RT (20 °C)	16%	6%	18%	36%
Response/recovery time (min)	5/6	8/20	7/6	3/1.2
Repeatability	Poor	Acceptable	Good	Very good

and technical fields. Consequently a number of reactive sites are offered between the analytes and substance [51, 52]. Additionally, managing structural flaws can be used to alter the conductivity of 2D materials [53]. Also, due to their flexibility and mechanical strength, 2D nanomaterials can be used with cutting-edge technologies including wearable electronics, metal electrodes, and ultrathin silicon channels [54]. The performance of sensors is greatly influenced by the kind being considered, the transduction mechanism, and the 2D material characteristics. In general, a high surface area is offered by two-dimensional materials to the analytes which enables high sensitivity and small detection threshold.

The surface of the material is additionally equipped with active regions that promote interaction with the target species. Conductivity improves the sensitivity in graphene and other 2D materials, especially in sensors that can be electrically transduced. Table 6.2 compares some of the characteristics of platinum or palladium core–shell-graphene hybrid hydrogen sensors with those of other works [55].

5 Challenges and Future of Direction of 2D Hybrid Nanocomposites or Environmental Sensing

2D hybrid nanocomposites have proven to be a reliable technological source for environmental sensing, however, there are some limitations and applications that need improvement. Wafer-scale 2D material growth and deposition may be appropriate. Even so, contamination and flaws still fall short of full compliance with industry norms for manufacture. The European 2D materials pilot line's overarching objective is to overcome these significant manufacturing challenges. Taking note of these difficulties, the European 2D Materials Pilot Line has established a vital goal to get over these obstacles and open the door for the widespread production of high-quality 2D materials. The manufacture of broad-area films on substrates, usually silicon wafers, is required for the wafer-scale growth and deposition of 2D materials. This strategy benefits from scalability and is compatible with current semiconductor manufacturing procedures. Using well-known methods like chemical vapor deposition (CVD), molecular beam epitaxy (MBE), or physical vapor deposition (PVD), scientists may deposit atomically thin layers over sizable surface areas, opening the door to the possibility of mass production of 2D materials. However, contamination and faults might appear during this production process, posing serious difficulties. Impurities in the growing environment or leftover components from the substrate can be the source of contaminants. These imperfections could lower the quality of the material and have an impact on its electrical, optical, and mechanical characteristics. The performance of the material can also be hampered by flaws like dislocations, vacancies, or grain boundaries, which therefore restrict the applications that it can be used for. The European 2D Materials Pilot Line seeks to address these problems and guarantee the manufacture of top-notch 2D materials by overcoming manufacturing difficulties. Researchers, engineers, and business partners come together as part of this endeavor to develop cutting-edge technologies and establish reliable manufacturing processes.

The goal of the pilot line is to establish industry-compliant standards for large-scale manufacturing by using cutting-edge characterization techniques, process optimization, and quality control measures. The main goal of the European 2D Materials Pilot Line is to create a dependable and expandable manufacturing platform for the growth and deposition of 2D materials on wafers. This platform would make it possible to produce materials of the highest quality with the fewest contaminants and faults, satisfying the exacting standards of sectors including electronics, optoelectronics, energy, and sensing. The pilot line strives to increase both the fundamental understanding of 2D materials and their smooth incorporation into current production processes by overcoming the obstacles caused by contamination and defects. Since the inception of 2D materials studies, about 18 years ago, literature has proven various applications of sensors using two-dimensional materials. There have not been many commercially successful 2D product developments. In view of this, a reliable design, affordability, and scalability are some fundamental qualities of competing in the market. Some promising two-dimensional materials for manufacturing include graphene and transition metal dichalcogenide.

Additionally, although research on graphene-based skin sensors has made outstanding progress, there are still several problems that nanomaterials cannot yet resolve. With a few notable exceptions, the majority of 2D sensing derivatives require biomolecule modifications because they are typically unable to detect the target analytes on their own. Due to their poor stability and higher production costs, the characteristics limit their applicability. Although various biomedical devices from 2D material have shown small-scale biocompatibility, fundamental obstacle to their practical usage is in vivo biocompatibility. It has been demonstrated that mechanical elements with biomimetic qualities or coatings made of biocompatible materials can lower immune reactions and inflammation. For potential commercialization, wearable and implantable devices must be stable and functioning over time. There have been few publications over the years about the cytotoxicity of Gr and/or TMDs based sensors' cutaneous toxicity. Skin irritation and allergic reaction are generally the most common scenarios, but the exact cause is yet unknown. The current difficulty facing 2D materials in bio-sensing is the need to increase device repeatability.

New sensing platforms are developed at a faster pace, and the field is rapidly growing. The future lies in multiplatform sensing, which combines several materials. In addition to 2D-2D blocks, additional low-dimensional materials can also be produced into cross-linked heterostructures, which can offer previously unimaginable properties that individual 2D equivalents cannot attain [56]. It is observed that the desorption of gases occurs slowly due to strong adsorption of gas by two-dimensional substances. Typically, this is because the desorption of gases necessitates external assistance. As a result, it is important to resolve this limitation of the practical use of two-dimensional materials which hinders sensing performance.

The majority of the gas sensing tests using 2D materials must also be conducted in a controlled environment because it has been reported that, the sensitivity values are affected by the presence of other gases in real-world environmental conditions. In light of this, future research must focus on developing 2D material-based ultra-high selectivity gas sensing devices. This problem can be solved by altering the material's surface with the appropriate modifiers. Future development in this area will require a thorough apprehension of fundamental characteristics at nanoscale of 2D materials, as well as the impact of gas adsorption on electrical properties for assessing these materials' potential for gas detection.

6 Conclusion

In conclusion, due to their nanomaterial characteristics, nanocomposite sensors in general, have the ability to detect and monitor at high resolutions. The development of 2D hybrid nanocomposites, has introduced greater opportunities for developing advanced materials with tailored properties and multiple functionalities. 2D hybrid nanocomposite sensors have the advantage of merging sensing characteristics of two different materials and adding improvements such as high surface area to volume ratio, tunable electronic properties, and improved mechanical strength. These characteristics enhance the performance of the sensors particularly in areas of sensitivity, selectivity, stability, repeatability, and limit of detection, better than conventional sensing materials. In this paper, the sensing mechanisms involved in biosensing, pressure sensing, humidity sensing, gas sensing, and temperature sensing have been outlined, as well as the significance of 2D hybrid nanocomposite sensors in these applications. Works from literature have been cited to compare these performances to those of conventional sensing materials. In future, synthesis and scalability techniques should be keenly investigated to produce sensors with more uniformity and stability for effective sensing applications. Also, cheaper but more efficient materials and synthesis processes should be investigated and applied in order to reduce the cost of manufacturing and production of 2D hybrid nanocomposite sensors, making its application economically viable and commercially possible. Further research and developmental works in this area have great potential for the fabrication of more innovative composites that can cause massive improvement in industries such as biomedical, automotive, aerospace, electrical, and electronics.

References

1. Lein, J.K.: Environmental Sensing. In: Lein, J.K. (ed.) Environmental Sensing: Analytical Techniques for Earth Observation, pp. 23–49. Springer, New York, NY (2012)
2. Bogue, R.: Environmental sensing: strategies, technologies and applications. Sens. Rev. **28**, 275–282 (2008). https://doi.org/10.1108/02602280810902550
3. Wang, L., Ma, W., Xu, L., Chen, W., Zhu, Y., Xu, C., Kotov, N.A.: Nanoparticle-based environmental sensors. Mater. Sci. Eng. R. Rep. **70**, 265–274 (2010). https://doi.org/10.1016/j.mser.2010.06.012
4. Su, S., Wu, W., Gao, J., Lu, J., Fan, C.: Nanomaterials-based sensors for applications in environmental monitoring. J. Mater. Chem. **22**, 18101–18110 (2012). https://doi.org/10.1039/C2JM33284A
5. Munonde, T.S., Nomngongo, P.N.: Nanocomposites for electrochemical sensors and their applications on the detection of trace metals in environmental water samples. Sensors **21**, 131 (2021). https://doi.org/10.3390/s21010131
6. Khan, M.A., Ramzan, F., Ali, M., Zubair, M., Mehmood, M.Q., Massoud, Y.: Emerging two-dimensional materials-based electrochemical sensors for human health and environment applications. Nanomaterials **13**, 780 (2023). https://doi.org/10.3390/nano13040780
7. Baig, N.: Two-dimensional nanomaterials: a critical review of recent progress, properties, applications, and future directions. Compos. Part A: Appl. Sci. Manuf. **165**, 107362 (2023). https://doi.org/10.1016/j.compositesa.2022.107362
8. Martin, P.M. ed: Chapter 1—Deposition Technologies: An Overview. In: Handbook of Deposition Technologies for Films and Coatings (Third Edition), pp. 1–31. William Andrew Publishing, Boston (2010)
9. Braun, J.L., Rost, C.M., Lim, M., Giri, A., Olson, D.H., Kotsonis, G.N., Stan, G., Brenner, D.W., Maria, J.-P., Hopkins, P.E.: Charge-induced disorder controls the thermal conductivity of entropy-stabilized oxides. Adv. Mater. **30**, e1805004 (2018). https://doi.org/10.1002/adma.201805004
10. Sharma, Y., Musico, B.L., Gao, X., Hua, C., May, A.F., Herklotz, A., Rastogi, A., Mandrus, D., Yan, J., Lee, H.N., Chisholm, M.F., Keppens, V., Ward, T.Z.: Single-crystal high entropy perovskite oxide epitaxial films. Phys. Rev. Mater. **2**, 060404 (2018). https://doi.org/10.1103/PhysRevMaterials.2.060404
11. Amiri, A., Naraghi, M., Ahmadi, G., Soleymaniha, M., Shanbedi, M.: A review on liquid-phase exfoliation for scalable production of pure graphene, wrinkled, crumpled and functionalized graphene and challenges. FlatChem. **8**, 40–71 (2018). https://doi.org/10.1016/j.flatc.2018.03.004
12. Li, Z., Young, R.J., Backes, C., Zhao, W., Zhang, X., Zhukov, A.A., Tillotson, E., Conlan, A.P., Ding, F., Haigh, S.J., Novoselov, K.S., Coleman, J.N.: Mechanisms of liquid-phase exfoliation for the production of graphene. ACS Nano **14**, 10976–10985 (2020). https://doi.org/10.1021/acsnano.0c03916
13. Blake, P., Brimicombe, P.D., Nair, R.R., Booth, T.J., Jiang, D., Schedin, F., Ponomarenko, L.A., Morozov, S.V., Gleeson, H.F., Hill, E.W., Geim, A.K., Novoselov, K.S.: Graphene-based liquid crystal device. Nano Lett. **8**, 1704–1708 (2008). https://doi.org/10.1021/nl080649i
14. Hernandez, Y., Nicolosi, V., Lotya, M., Blighe, F.M., Sun, Z., De, S., McGovern, I.T., Holland, B., Byrne, M., Gun'Ko, Y.K., Boland, J.J., Niraj, P., Duesberg, G., Krishnamurthy, S., Goodhue, R., Hutchison, J., Scardaci, V., Ferrari, A.C., Coleman, J.N.: High-yield production of graphene

by liquid-phase exfoliation of graphite. Nature Nanotech. **3**, 563–568 (2008). https://doi.org/10.1038/nnano.2008.215
15. Intermolecular and Surface Forces, 3rd Edition. https://www.researchgate.net/publication/292720794_Intermolecular_and_Surface_Forces_3rd_Edition
16. Zen, J.-M., Senthil Kumar, A., Tsai, D.-M.: Recent updates of chemically modified electrodes in analytical chemistry. Electroanalysis **15**, 1073–1087 (2003). https://doi.org/10.1002/elan.200390130
17. Tonelli, D., Scavetta, E., Gualandi, I.: Electrochemical deposition of nanomaterials for electrochemical sensing. Sensors **19**, 1186 (2019). https://doi.org/10.3390/s19051186
18. Cioffi, N., Colaianni, L., Ieva, E., Pilolli, R., Ditaranto, N., Angione, M.D., Cotrone, S., Buchholt, K., Spetz, A.L., Sabbatini, L., Torsi, L.: Electrosynthesis and characterization of gold nanoparticles for electronic capacitance sensing of pollutants. Electrochim. Acta **56**, 3713–3720 (2011). https://doi.org/10.1016/j.electacta.2010.12.105
19. Scavetta, E., Casagrande, A., Gualandi, I., Tonelli, D.: Analytical performances of Ni LDH films electrochemically deposited on Pt surfaces: Phenol and glucose detection. J. Electroanal. Chem. **722–723**, 15–22 (2014). https://doi.org/10.1016/j.jelechem.2014.03.018
20. Wang, D., Luo, H., Kou, R., Gil, M.P., Xiao, S., Golub, V.O., Yang, Z., Brinker, C.J., Lu, Y.: A general route to macroscopic hierarchical 3D nanowire networks. Angew. Chem. Int. Ed. Engl. **43**, 6169–6173 (2004). https://doi.org/10.1002/anie.200460535
21. Mariella Babu, A., Varghese, A.: Electrochemical deposition for metal organic frameworks: advanced energy, catalysis, sensing and separation applications. J. Electroanal. Chem. **937**, 117417 (2023). https://doi.org/10.1016/j.jelechem.2023.117417
22. Guan, G., Han, M.-Y.: Functionalized hybridization of 2D nanomaterials. advanced science. **6** (2019). https://doi.org/10.1002/advs.201901837
23. Zhang, X., Xie, H., Liu, Z., Tan, C., Luo, Z., Li, H., Lin, J., Sun, L., Chen, W., Xu, Z., Xie, L., Huang, W., Zhang, H.: Black phosphorus quantum dots. Angew. Chem. Int. Ed. Engl. **54**, 3653–3657 (2015). https://doi.org/10.1002/anie.201409400
24. Zheng, Y., Liu, J., Liang, J., Jaroniec, M., Qiao, S.: Graphitic carbon nitride materials: controllable synthesis and applications in fuel cells and photocatalysis. Energy Environ. Sci. **5**, 6717–6731 (2012). https://doi.org/10.1039/c2ee03479d
25. Choi, C.: Integrated nanobiosensor technology for biomedical application. NDD. 1 (2012). https://doi.org/10.2147/NDD.S26422
26. Poltronieri, P., Mezzolla, V., Primiceri, E., Maruccio, G.: Biosensors for the detection of food pathogens. Foods. **3**, 511–526 (2014). https://doi.org/10.3390/foods3030511
27. Muniandy, S., Teh, S.J., Appaturi, J.N., Thong, K.L., Lai, C.W., Ibrahim, F., Leo, B.F.: A reduced graphene oxide-titanium dioxide nanocomposite based electrochemical aptasensor for rapid and sensitive detection of Salmonella enterica. Bioelectrochemistry **127**, 136–144 (2019). https://doi.org/10.1016/j.bioelechem.2019.02.005
28. Lai, Q.-T., Sun, Q.-J., Tang, Z., Tang, X.-G., Zhao, X.-H.: Conjugated polymer-based nanocomposites for pressure sensors. Molecules **28**, 1627 (2023). https://doi.org/10.3390/molecules28041627
29. Pham, V.P., Nguyen, M.T., Park, J.W., Kwak, S.S., Nguyen, D.H.T., Mun, M.K., Phan, H.D., Kim, D.S., Kim, K.H., Lee, N.-E., Yeom, G.Y.: Chlorine-trapped CVD bilayer graphene for resistive pressure sensor with high detection limit and high sensitivity. 2D Mater. **4**, 025049 (2017). https://doi.org/10.1088/2053-1583/aa6390
30. Zhang, D., Chang, H., Li, P., Liu, R., Xue, Q.: Fabrication and characterization of an ultrasensitive humidity sensor based on metal oxide/graphene hybrid nanocomposite. Sens. Actuators, B Chem. **225**, 233–240 (2016). https://doi.org/10.1016/j.snb.2015.11.024
31. Zhang, D., Tong, J., Xia, B., Xue, Q.: Ultrahigh performance humidity sensor based on layer-by-layer self-assembly of graphene oxide/polyelectrolyte nanocomposite film. Sens. Actuators, B Chem. **203**, 263–270 (2014). https://doi.org/10.1016/j.snb.2014.06.116
32. Zhang, D., Tong, J., Xia, B.: Humidity-sensing properties of chemically reduced graphene oxide/polymer nanocomposite film sensor based on layer-by-layer nano self-assembly. Sens. Actuators, B Chem. **197**, 66–72 (2014). https://doi.org/10.1016/j.snb.2014.02.078

33. Cavallo, P., Acevedo, D.F., Fuertes, M.C., Soler-Illia, G.J.A.A., Barbero, C.A.: Understanding the sensing mechanism of polyaniline resistive sensors. Effect of humidity on sensing of organic volatiles. Sens. Actuators B: Chem. **210**, 574–580 (2015). https://doi.org/10.1016/j.snb.2015.01.029
34. Okcan, B., Akin, T.: A thermal conductivity based humidity sensor in a standard CMOS process. In: 17th IEEE International Conference on Micro Electro Mechanical Systems. Maastricht MEMS 2004 Technical Digest, pp. 552–555 (2004)
35. Lee, C.-Y., Lee, G.-B.: Humidity sensors: a review. Sens. Lett. **3**, 1–15 (2005). https://doi.org/10.1166/sl.2005.001
36. Rao, X., Zhao, L., Xu, L., Wang, Y., Liu, K., Wang, Y., Chen, G.Y., Liu, T., Wang, Y.: Review of optical humidity sensors. Sensors **21**, 8049 (2021). https://doi.org/10.3390/s21238049
37. Zheng, Z., Zhang, G., Wang, X., Kong, X.: Comparative study of gravimetric humidity sensor platforms based on CMUT and QCM. Micromachines **13**, 1651 (2022). https://doi.org/10.3390/mi13101651
38. Zhang, D., Sun, Y., Li, P., Zhang, Y.: Facile fabrication of MoS2-modified SnO2 hybrid nanocomposite for ultrasensitive humidity sensing. ACS Appl. Mater. Interfaces **8**, 14142–14149 (2016). https://doi.org/10.1021/acsami.6b02206
39. Dhall, S., Mehta, B.R., Tyagi, A.K., Sood, K.: A review on environmental gas sensors: materials and technologies. Sens. Int. **2**, 100116 (2021). https://doi.org/10.1016/j.sintl.2021.100116
40. Sagberg, H., Fismen, B., Sandven, K., Nordbryhn, P., Aakvaag, N., Borgen, L., Tschudi, J., Bakke, K.A., Johansen, I.-R.: Wireless Infrared Gas Sensor. (2012). https://doi.org/10.1364/AIO.2012.AW1B.3
41. Nikolic, M.V., Milovanovic, V., Vasiljevic, Z.Z., Stamenkovic, Z.: Semiconductor gas sensors: materials, technology, design, and application. Sensors (Basel). **20**, 6694 (2020). https://doi.org/10.3390/s20226694
42. Liu, Y., Tang, Y., Barashkov, N.N., Irgibaeva, I.S., Lam, J.W.Y., Hu, R., Birimzhanova, D., Yu, Y., Tang, B.Z.: Fluorescent chemosensor for detection and quantitation of carbon dioxide gas. J. Am. Chem. Soc. **132**, 13951–13953 (2010). https://doi.org/10.1021/ja103947j
43. Korotcenkov, G., Stetter, J., Zeng, X., Liu, Y., Tang, Y.: Electrochemical gas sensors: fundamentals, fabrication and parameters. Presented at the June 1 (2011)
44. Yu, X., Cheng, C., Feng, S., Jia, X., Song, H.: Porous α-Fe2O3 nanorods@graphite nanocomposites with improved high temperature gas sensitive properties. J. Alloy. Compd. **784**, 1261–1269 (2019). https://doi.org/10.1016/j.jallcom.2019.01.099
45. Yin, L., Chen, D., Cui, X., Ge, L., Yang, J., Yu, L., Zhang, B., Zhang, R., Shao, G.: Normal-pressure microwave rapid synthesis of hierarchical SnO2@rGO nanostructures with superhigh surface areas as high-quality gas-sensing and electrochemical active materials. Nanoscale **6**, 13690–13700 (2014). https://doi.org/10.1039/C4NR04374J
46. Paul, S., Saikia, A., Majhi, V., Pandey, V.K.: Chapter 3 - Transducers and amplifiers. In: Paul, S., Saikia, A., Majhi, V., and Pandey, V.K. (eds.) Introduction to Biomedical Instrumentation and Its Applications, pp. 87–167. Academic Press (2022)
47. Yadav, B., Srivastava, R., Singh, S., Kumar, A., Yadav, A.: Temperature sensors based on semiconducting oxides: an overview (2012)
48. Diwanji, M.M., Hisvankar, S.M., Khandelwal, C.S.: Temperature Measurement using infrared contactless thermal gun. In: 2020 International Conference on Smart Innovations in Design, Environment, Management, Planning and Computing (ICSIDEMPC), pp. 134–137 (2020)
49. Tung, T.T., Pham-Huu, C., Janowska, I., Kim, T., Castro, M., Feller, J.-F.: Hybrid films of graphene and carbon nanotubes for high performance chemical and temperature sensing applications. Small **11**, 3485 (2015)
50. Novoselov, K.S., Geim, A.K., Morozov, S.V., Jiang, D., Zhang, Y., Dubonos, S.V., Grigorieva, I.V., Firsov, A.A.: Electric field effect in atomically thin carbon films. Science **306**, 666–669 (2004). https://doi.org/10.1126/science.1102896
51. Rao, C.N.R., Sood, A.K., Subrahmanyam, K.S., Govindaraj, A.: Graphene: the new two-dimensional nanomaterial. Angew. Chem. Int. Ed. **48**, 7752–7777 (2009). https://doi.org/10.1002/anie.200901678

52. Varghese, S.S., Varghese, S.H., Swaminathan, S., Singh, K.K., Mittal, V.: Two-dimensional materials for sensing: graphene and beyond. Electronics **4**, 651–687 (2015). https://doi.org/10.3390/electronics4030651
53. Late, D.J., Huang, Y.-K., Liu, B., Acharya, J., Shirodkar, S.N., Luo, J., Yan, A., Charles, D., Waghmare, U.V., Dravid, V.P., Rao, C.N.R.: Sensing behavior of atomically thin-layered MoS2 transistors. ACS Nano **7**, 4879–4891 (2013). https://doi.org/10.1021/nn400026u
54. Sulleiro, M.V., Dominguez-Alfaro, A., Alegret, N., Silvestri, A., Gómez, I.J.: 2D materials towards sensing technology: from fundamentals to applications. Sensing Bio-Sensing Res. **38**, 100540 (2022). https://doi.org/10.1016/j.sbsr.2022.100540
55. Phan, D.-T., Uddin, A.S.M.I., Chung, G.-S.: A large detectable-range, high-response and fast-response resistivity hydrogen sensor based on Pt/Pd core–shell hybrid with graphene. Sens. Actuators, B Chem. **220**, 962–967 (2015). https://doi.org/10.1016/j.snb.2015.06.029
56. Xie, H., Li, Z., Cheng, L., Haidry, A.A., Tao, J., Xu, Y., Xu, K., Ou, J.Z.: Recent advances in the fabrication of 2D metal oxides. iScience. **25**, 103598 (2022). https://doi.org/10.1016/j.isci.2021.103598

Graphene-Based Nanocomposites in Electrochemical Sensing

Untika Pengsomjit, Fatima Alabdo, Waleed Alahmad, Pakorn Varanusupakul, and Charoenkwan Kraiya

Abstract The use of graphene nanocomposites in the fabrication and creation of electrochemical sensors is discussed in this chapter. Graphene derivatives have garnered significant interest in the last decades due to their exceptional electrical, mechanical, and thermal characteristics, which have made them a preferred choice in the development of sensor electrodes. Moreover, their electrochemical properties make them ideal candidates for electrochemical applications, especially in sensing due to their electrocatalytic activity, highly effective surface area, excellent electrical conductivity, adsorption capability, and high porosity. Over the past few years, there has been significant progress in the development of graphene-based nanocomposites. This chapter aims to provide an overview of these works and their findings for those with prior knowledge of the subject matter. Also, it elucidates the characteristics of graphene nanocomposites and their potential use as electrochemical sensors.

Keywords Graphene · Exfoliated graphene · Nanocomposition · Electrochemistry · Sensors

1 Introduction

Two-dimensional substance called graphene (Gr) is made of carbon atoms organized in a honeycomb lattice, which provides it with extraordinary properties, such as large surface area, high conductivity, and exceptional mechanical strength [1]. These

U. Pengsomjit · W. Alahmad · P. Varanusupakul · C. Kraiya (✉)
Department of Chemistry, Faculty of Science, Chulalongkorn University, Phayathai Road, Pathumwan, Bangkok 10330, Thailand
e-mail: charoenkwan.k@chula.ac.th

U. Pengsomjit · C. Kraiya
Electrochemistry and Optical Spectroscopy Center of Excellence, Department of Chemistry, Faculty of Science, Chulalongkorn University, Phayathai Road, Pathumwan, Bangkok 10330, Thailand

F. Alabdo
Department of Chemistry and Physics, Faculty of Science, Idlib University, Idlib, Syria

N. Talreja et al. (eds.), *Two-dimensional Hybrid Composites*, Engineering Materials,
https://doi.org/10.1007/978-981-99-8010-9_7

unique features of graphene make it an ideal material for use in electrochemical sensing applications [2]. The characteristics of graphene and graphene oxide are identical, except that graphene oxide (GO) contains –CO, –COOH, –OH, and epoxy groups. These groups increase interlayer distance, resulting in a more hydrophilic layer. As a result, GO is an oxygenated graphene sheet. As GO is partially reduced, it forms a sheet similar to graphene called reduced graphene oxide (rGO) [3].

Several approaches for producing graphene materials have been investigated during the last two decades. Chemical vapor deposition, mechanical exfoliation, liquid-phase exfoliation helped by solvents or surfactants, ultrasonically exfoliated graphite, epitaxial growth, and synthesis of compounds intercalated with graphite are the common methods for graphene production [4]. Graphene-based nanocomposites, formed by incorporating graphene with other materials, have emerged as promising candidates for electrochemical sensing applications, as they can enhance the selectivity, sensitivity, and stability of sensors [5].

In recent years, extensive research has been conducted to develop graphene-based nanocomposites for various electrochemical sensing applications, including glucose sensing [6], DNA detection [7], environmental monitoring [8], and biomedical diagnostics [9]. These nanocomposites exhibit exceptionally large surface area, electrical conductivity, excellent chemical stability, and high mechanical strength, making them ideal for constructing high-performance electrochemical sensors. A variety of methods can be used to synthesize graphene-based nanocomposites. The most common process involves the initial synthesis of graphene, followed by nanocomposites based on graphene. Such methods include the and the self-assembly method one-step hydrothermal method. Other methods are presented in Tables 7.1 and 7.2. In this chapter, the most common methods for synthesizing graphene and graphene-based nanocomposites have been summarized. The application of these nanocomposites to construct electrochemical sensors has been discussed. This chapter serves as a single point of reference for researchers in graphene science, materials science, biosensors, chemistry, nanotechnology, and electrochemistry. A wide range of audiences from a variety of research fields will find this chapter of particular interest and stimulate further interest in graphene-based biosensors.

2 Methods to Synthesize Graphene and Its Derivatives

Ever since graphene was discovered in 2004, scientists have been able to develop various methods for producing graphene. These can be split into approaches that start from the top and those that start from the bottom. The top-down approach divides a set of bulk graphite into its derivatives of graphene, and the bottom-up approach assembles the components (carbon) into a whole system (graphene or graphite) [10]. A variety of top-down techniques are available, such as mechanical, and chemical. The carbon nanotubes can be broken down by electrochemically oxidizing or reducing

Table 7.1 Synthesizing graphene@nanocomposites methods

Method	Summary
Solution mixing	Disperse the graphene in a suitable solvent by ultrasonic or magnetic stirring. Add the nanofiller suspension to the polymer and mix them together using stir or shear mixing and wash away any remaining solvent with water and dry the product off [24]
Hydrothermal/ solvothermal method	Involve heating materials under high pressure and temperatures to make nanocomposites. Use aqueous and non-aqueous solution in hydrothermal and solvothermal methods, respectively [25]
Self-assembly	Organized formation of a system's constituents into a distinct pattern or structure due to precise local interactions among materials. Occurs spontaneously without the need for external intervention or manipulation [26]
Microwave irradiation method	A rapid method for heating reactants to high temperatures with minimal energy usage, while also promoting close interaction among the different components, reducing reaction times, and creating a cleaner reaction environment [27, 28]
Electrochemical deposition	An approach in which metal ions from a solution are deposited onto an electrically conductive surface, resulting in the formation of a solid metal film [29]
In-situ polymerization	Monomer liquid mixed with filler and polymerization can be conducted through heating, initiating with a catalyst, or utilizing radiation [30]

them, as well as by chemically and thermally splitting them apart. Bottom-up techniques include catalytic processes, heat treatments, and chemical processes [11]. Herein, we briefly summarized the most common methods as follows:

2.1 *Micromechanical Exfoliation*

In this technique, physically separating graphene sheets from highly ordered pyrolytic graphite (HOPG) surface is achieved through the application of micromechanical cleavage, which uses adhesive tape [12]. This adhesive substrate separates graphene layers directly by applying force to them and is perfect for the creation of exceptionally high-quality graphene layers. To remove the graphene layer, tape is applied to the acetone substrate and peeled off with new tape. The graphene layers that remain are likely to be twisted into multiple flakes by repeated peeling. This synthesis technology of graphene is desirable because of its affordability and effectiveness, and the produced graphene varies in size and thickness. However, due to its high manual labor requirement, long production time, and lack of control, this technique is not appropriate for mass production. In order to make this technique feasible and scalable for industrial applications, these undesirable factors must be addressed.

Table. 7.2 Different approaches used for synthesizing graphene@nanocomposites

Graphene@nanocomposites	Fabrication method	Application	Refs.
Gr–MnO_2	Microwave irradiation method	Supercapacitors	Yan et al. [34]
Co–Al LDH-NS/GO	Self-assembly	Supercapacitors	Wang et al. [35]
Gr-ZrO_2	Electrochemical deposition	Electrochemical sensor	Du et al. [36]
TiO_2–rGO	Hydrothermal method	Photocatalytic activity	Gu et al. [37]
rGO/Ala/PANI	In-situ polymerization	Electrochemical sensor	Akhtar et al. [38]
rGO-TiO_2/PPy	Hydrothermal method, In-situ polymerization	Electrochemical sensor	Deivanayaki et al. [39]
PDDA-Gr-(Co–Ni)	Solvothermal method	Electrochemical sensor	Li et al. [40]
SnO_2-rGO	Hydrothermal method	Electrochemical sensor	Zhang et al. [41]
ZnO/PPy/PANI/GO	Solution mixing	Electrochemical sensor	Kumar et al. [42]
GO–AgNPs	Self-assembly	Anti-bacterial activity	Liu et al. [43]
NiNPs/erGO	Electrochemical deposition	Electrochemical sensor	Kurt Urhan et al. [44]
GO$ZnFe_2O_4$/ PANi	In-situ polymerization	Supercapacitors	Alsulami et al. [45]
Pd-PPy/NGE	In-situ polymerization	Electrocatalysts	Xie et al. [46]
QOTM/Gr	Solvothermal method	Electrochemical sensor	Li et al. [47]
rGO-Co_3O_4@Au	Hydrothermal method	Electrochemical sensor	Shahid et al. [48]
PB-PEDOT-LSG	Solution mixing	Electrochemical sensor	Machhindra and Yen [49]
rGO-WS_2@Fe_3O_4	Microwave irradiation method	Electrochemical sensor	Rana et al. [50]
TiO_2–Gr	Hydrothermal method	Photocatalytic catalyst	Diang et al. [51]
TiO_2–rGO	Microwave irradiation method	Photocatalytic activity	Liu et al. [52]
CeO_2-Gr	Solvothermal method	Electrochemical sensor	Yang et al. [53]
CoFe/NGr	Hydrothermal method	Electrochemical sensor	Hassan et al. [54]
MgO/Gr/Ta	Electrochemical deposition	Biosensor	Zhao et al. [55]

(continued)

Table. 7.2 (continued)

Graphene@nanocomposites	Fabrication method	Application	Refs.
PANI/rGO	In-situ polymerization	Gas sensor	Jee et al. [56]
Piperazine/Phytic acid/GO	Self-assembly	Flame retardants	Fang et al. [57]
NiS@GO	Hydrothermal method	Electrochemical sensor	Awan et al. [58]
Pt/NGr	Microwave irradiation method	Electrochemical sensor	Anuar et al. [59]
Ag-AuNPs/RGO	Solvothermal method	Electrochemical sensor	Zhao et al. [60]
PCA-RGO/Au NPs	Solution mixing	Electrochemical sensor	Pifferi et al. [61]
AuNP@Gr/MOF_5	Hydrothermal method	Electrochemical sensor	Ganash and Alotaibi [62]
CG: CuNPs	Aerosol-assisted capillary compression method	Electrochemical sensor	Alencar et al. [63]
rGO-Nb_2O_5	Solvothermal method	Electrochemical sensor	Durai and Badhulika [64]
Ag/rGO	Solution mixing	Electrochemical sensor	Shen et al. [65]

2.2 *Electrochemical Exfoliation*

In this method, a two-electrode electrochemical cell is used to synthesize the graphene. One electrode is made of graphite rods, plates, or powder. Another one is made from platinum, copper, or even graphite material. These electrodes are placed in either non-aqueous or aqueous electrolytes, and then a voltage is applied to cause electrode expansion, resulting in a floated layer of graphene on top of the electrolyte solution. As a result of the applied voltage, intercalating ions can penetrate between the graphite sheets and exfoliate the graphite into anodic graphene. Based on the nature of the used electrolytes (e.g., H_2SO_4, $(NH_4)_2SO_4$, Na_2SO_4, …etc.) the degree of intercalation, and the size of the intercalating ion; the electrochemical exfoliation method produces either graphene oxides or few-layered graphene [4]. Anodic exfoliation is not the only method available for obtaining graphene; cathodic exfoliation can also be used alone or in combination with ultrasonic treatment and thermal expansion in organic solvents [5]. As a result of this approach, high-quality few-layer graphene sheets with a wide range of thicknesses and sizes could be produced.

2.3 Liquid-Phase Exfoliation

Graphene is most commonly produced with liquid-phase exfoliation, which has a favorable quality: cost ratio for large quantities. For example, graphene nanosheets with good quality were successfully synthesized through the exfoliation of graphite in a mixture of water and DMF [13]. Despite the use of different polar and nonpolar organic solvents, the highest possible yield has not been achieved. It remains a challenge, however, to produce multilayer graphene with high resolution using this method.

2.4 Chemical Synthesis

The chemical synthesis method is one of the best proper methods for the graphene production and its derivatives. By producing a colloidal suspension, graphene can be modified from graphite and graphite intercalation compounds [14]. It is worth noting that graphene is a polycyclic aromatic hydrocarbon containing nanocomposites that can be controlled to deliver different graphene structures. Chemical exfoliation comprises two steps. At first, the interlayer forces (van der Waals) are reduced to increase the interlayer spacing, forming graphene-intercalated compounds. By rapid heating or sonication, graphene is produced with a single to few layers in the second step. On the contrary, graphene oxide can be easily prepared by using Hummers' method. In this method, graphite is oxidized with strong oxidizing agents such as $NaNO_3$ and $KMnO_4$ in H_3PO_4/H_2SO_4 [15].

2.5 Chemical Vapor Deposition (CVD)

CVD is a chemical reaction in which molecules are heated and converted to gaseous forms, which is known as the precursor. The CVD approach for large-scale graphene production employs transition metals such as copper and nickel to generate a metallic catalyst layer on the base sheet. After the material has been chemically etched and deposited on the substrate, a carbon-containing slurry is poured into the reaction chamber to make high-quality graphene [16]. There are several different types of CVD procedures depending on the precursors, material quality, structure required, thermal, plasma-assisted CVD, cold wall, and hot wall. In CVD process reactors, such as hot wall reactors, the temperature is typically constant throughout, but in cold wall systems, the walls are never heated. Graphene is mainly produced on Cu thin films using a cold wall method.

2.6 Arc Discharge Method

Graphene may be synthesized using an arc discharge process in the presence of He, H_2, or N_2 in 2–3 layers with flake sizes ranging from 100 to 200 nm, which is eminent to that produced using chemical techniques [17]. Favorable conditions for growing graphene on the inner wall are high current (>100 A), high gas pressure (>200 Torr), and high voltage (>50 V). The combination of CO_2 and He during the arc discharge process to synthesize graphene has also been used as well [18]. The arc discharge process offers a potential path toward scalable and clean graphene manufacturing, but it is fraught with difficulties, including graphene size restrictions, high cost, and limited yield.

2.7 Laser-Scribed Graphene

Graphene can be produced through the exposure of a polymeric precursor to laser radiation, commonly known as laser-induced graphene (LIG) or laser-scribed graphene (LSG). This is accomplished by photothermal conversion the hybridization of carbon atoms from sp^3 to sp^2 [19]. The laser-scribing process begins with depositing a layer of carbon-containing material, such as a polymer, onto a substrate. A laser beam is then focused onto the material, which heats up and decomposes, leaving behind a pattern of graphene. The laser energy removes the carbon atoms from the polymer, leaving a pattern of graphene behind. The graphene pattern is highly conductive and can be used as a circuit or electrode [20]. One advantage of the laser-scribing technique is that it is a simple and scalable process, suitable for large-scale production. It also allows for precise control over the graphene pattern, enabling the creation of complex and intricate designs. Overall, the laser-scribing technique is an auspicious method for creating high-quality graphene patterns on a variety of surfaces, opening up new possibilities for graphene-based apparatus and technologies.

3 Methods to Synthesize Graphene@nanocomposites

Even though graphene has many great qualities, pure graphene sheets still have a limited number of useful applications due to some of their characteristics such as the very poor stability of pure graphene layers or the low electronic double-layer capacitance [21]. Therefore, to achieve the desired characteristics of graphene for extensive practical applications, pure graphene sheets combined with appropriate nanocomposites were developed. Nanocomposites are made up of matrices such as organic polymers and metal nanoparticles. The addition of these inclusions not only results in an enhancement of the material's properties but also makes the material's surface functionalized with a variety of enzymes and proteins [22]. To synthesize graphene@nanocomposites, many ways are currently used such as solution mixing, electrochemical deposition, self-assembly, hydrothermal/solvothermal method, microwave irradiation method, and in-situ polymerization [23, 24]. Table 7.1 provides a brief summary of the various methods used for synthesizing graphene@nanocomposites.

Jasuja et al. studied the solvent-dispersible uncapped gold nanoparticles on rGO sheets synthesis using microwave irradiation [31]. The material was successfully synthesized and the gold nanoparticles formation diagram in the absence and presence of GO is shown in Fig. 7.1a. Another method used by Wang et al. [32] is in-situ polymerization. Figure 7.1b displays the step of preparing PA6T-Graphene composite. Then, one-step electrochemical deposition was utilized by Sangili et al. for another graphene@nanocomposite synthesis [33]. Figure 7.1c demonstrates the fabrication of AuNP/rGO/Ab/BSA/Ag/GCE via a one-step electrochemical deposition of layer-by-layer assembly of AuNP/rGO-based electrochemical immunoassay for endometriosis detection. For other methods, Table 7.2 shows some examples of different synthesizing graphene@nanocomposites methods.

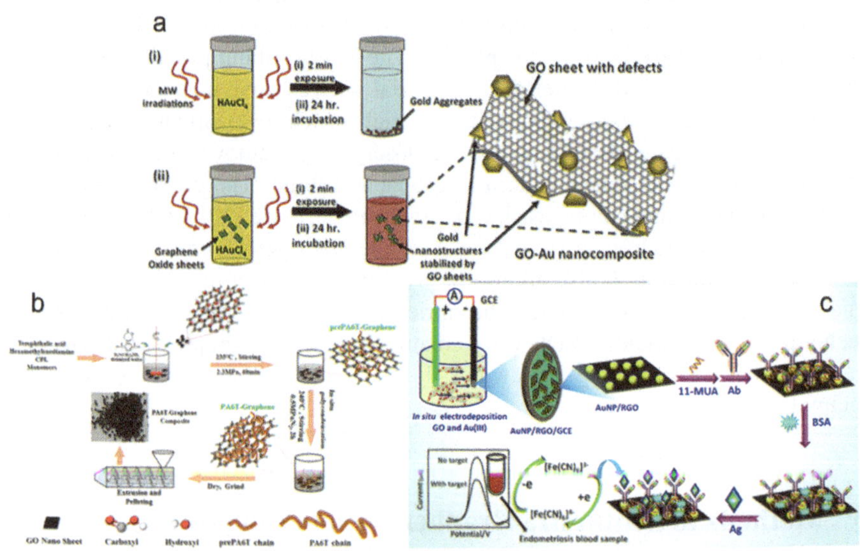

Fig. 7.1 **a** in-situ synthesis scheme of multiple shaped bare-surfaced Au NPs on GO using microwave irradiation method (reprinted from ref [31], Copyright © 2010, American Chemical Society) **b** PA6T-graphene nanocomposite fabrication step via in-situ polymerization (reprinted from ref [32], Copyright © 2019 Elsevier) **c** fabrication method of AuNP/rGO/Ab/BSA/Ag/GCE via electrochemical deposition (reprinted from ref [33], Copyright © 2020, American Chemical Society)

4 Application in Electrochemical Sensing

Currently, numerous electrochemical sensors are extensively utilized in a wide range of applications so improving the electrochemical sensor's functionality is necessary for good selectivity and sensitivity of the devices. And modifying sensors with graphene-based nanocomposites is one way to achieve that. By incorporating graphene, electron transfer speeds up, allowing for direct electrochemistry and biological sensing. Additionally, diverse compositions and structures of graphene derivatives can alter their electrochemical performance for better outcomes. Therefore, in this section, we provide a summary and example of works using different graphene-based nanocomposites for various electrochemical sensors.

4.1 Graphene-Based Nanocomposites Electrochemical Sensors for Heavy Metal Ion

Heavy metal ions such as cadmium, lead, and mercury, along with semimetals like arsenic, exhibit high toxicity and can pose severe health risks and environmental problems. Even though heavy metals are naturally occurring from biogeochemical processes, their presence in the environment is mainly due to human activities such as emissions released into the air from various sources like automobiles or coal-burning plants and process wastes from mining or industry [66, 67]. These extremely harmful contaminants are non-biodegradable and can cause negative impacts on the immune, central nervous, and reproductive systems, even at very minimal concentrations [68]. Therefore, developing precise and selective methods to measure heavy metal contamination in food, water, and the environment is crucial.

Electrochemical detectors show great potential for evaluating the amount of heavy metals in many media due to their selectivity, sensitivity, low cost, durability, and ability to be deployed in the field. As for graphene and its nanocomposites, they have displayed promising potential in the removal of toxic heavy metals from different aqueous solutions [69, 70] and now received more attention for use as modifying materials in electrochemical sensors [71]. Table 7.3 provides a summary of electrochemical sensors fabricated from graphene and its nanocomposites for various metal ions detection.

Xu et al. developed amino-functionalized $MgFe_2O_4$/rGO composite-modified GCE for the electrochemical detection of Cu^{2+} [84]. The composite was produced through a one-pot method, and it was characterized by FT-IR, XRD, SEM, and electrochemical methods. Using the optimized conditions, the modified electrodes detected Cu^{2+} with SWASV technique and obtained a detection limit of 0.2 nM with a linear range of 2–1000 nM. Another work was presented by Rahman et al. [77] about the detection of Hg^{2+} using platinum electrode modified with GO-Silver Nanowire (AgNW) Nanocomposites. The joint impact of conductive AgNW and GO has been shown to considerably enhance electron transportation and sensing capability for

Table 7.3 Overview of electrochemical sensors modified with graphene-based nanocomposites for heavy metal detection

Electrode	Analyte	Detection technique	Linear range	LOD	Refs.
Gr/GO/GCE	Cd^{2+}	DPV	0–10 μM	0.087 μM	[72]
Gr/MOF/GCE	As^{3+}	DPASV	2.67–334 nM	0.801 nM	[73]
LSG/PB-PEDOT/GCE	Cd^{2+}	DPV	1–10,000 nM	0.85 nM	[49]
AgNPs/rGO/GCE	Cd^{2+} Pb^{2+} Cu^{2+} Hg^{2+}	SWASV	0.05–3.5 μM 0.05–2.5 μM 0.05–3.5 μM 0.5–3 μM	0.254 μM 0.141 μM 0.178 μM 0.285 μM	[74]
Au NPs/rGO/CPE	As^{3+}	ASV	13.35–266.95 nM	1.74 nM	[75]
Au NPs/rGO/GCE	Fe^{3+}	DPV	0.03–3 μM	3.5 nM	[76]
AuNP@Gr/MOF_5/CPE	Cd^{2+}	SWV	0.01–30 μM	0.005 μM	[62]
GO/AgNW/Pt	Hg^{2+}	SWASV	1–70 nM	0.10 nM	[77]
dendritic Au/GO/GCE	Fe^{3+}	DPV	0.007–1 μM	1.5 nM	[78]
rGO/AuNP/Nafion/Au electrode	Cu^{2+}	DPASV	0.02–1 μM	4 nM	[79]
Fe_3O_4/rGO/SPE	As^{3+}	DPASV	0.027–4 μM	1.33 nM	[80]
Fe_3O_4/rGO/GCE	As^{3+}	SWASV	0.00013–67.4 nM	0.0016 nM	[81]
ZnO–rGO/SPE	Cd^{2+} Pb^{2+}	DPSV	0.089–0.71 μM 0.048–0.39 μM	1.42 nM 0.821 nM	[82]
$MnFe_2O_4$/GO/GCE	Pb^{2+}	SWASV	0.2–1.1 μM	0.0883 μM	[83]
1,6-hexanediamine functionalized $MgFe_2O_4$/rGO/GCE	Cu^{2+}	SWASV	2–1000 nM	0.2 nM	[84]
Co-doped ZnO/rGO/GCE	Cd^{2+} Pb^{2+}	DPV	0.089–0.8 μM 0.048–0.43 μM	8.36 nM 4 nM	[85]
Curcumin/MnO_2/Gr/GCE	Hg^{2+}	DPV	0.249–5.982 μM	0.096 μM	[86]
Co_3O_4/rGO/chitosan/GCE	Pb^{2+}	DPASV	1–200 nM	0.35 nM	[87]
benzothiazole-2-carboxaldehyde/Fe_3O_4/GO/GCE	Cd^{2+} Pb^{2+}	SWASV	0.7–800 nM 0.3 nM–430 nM	0.3 nM 0.1 nM	[88]
Au@MoS_2/rGO-AuPd@Fe-MOFs	Pb^{2+}	Amperometry	5.0 pM–2.0 μM	0.07 pM	[89]
PrGO/AuNPs/Sal-Cys/GCE	Cd^{2+} Pb^{2+}	SWASV	1–10 nM 1–10 nM	0.06 nM 0.04 nM	[90]
PCA-rGO@Au NPs/SPCE	As^{3+}	ASV	0.1–100 μM	3.6 μM	[61]
GO/UiO-67@PtNPs/GCE	As^{3+}	SWASV	2.7–33.4 nM	0.48 nM	[91]
PANI/GO/GCE	Pb^{2+}	SWASV	0.2–3500 nM	0.04 nM	[92]
rGO/Ala/PANI/GCE	Cd^{2+} Pb^{2+} Cu^{2+}	SWASV	0.08–100 nM 0.08–100 nM 0.08–100 nM	0.03 nM 0.063 nM 0.045 nM	[38]
T-GO-C/GCE	Hg^{2+} Cd^{6+}	SWV	0.025–2.99 μM 0.096–11.5 μM	0.005 μM 0.385 μM	[93]

(continued)

Table 7.3 (continued)

Electrode	Analyte	Detection technique	Linear range	LOD	Refs.
CS/GO-IIP/GCE	Cu^{2+}	DPASV	0.5–100 μM	0.15 μM	[94]
ChOx/DWCNTs-Gr/SPE	As^{5+}	SWV	0.013–0.134 μM	0.004 μM	[95]
Apt/CS/rGO/TiO_2/GCE	Pb^{2+}	DPV	0.005–4.83 nM	0.002 nM	[96]
Apt/CS/rGO/MWCNT/AuNP/GCE	Pb^{2+}	DPV	0.05–200 nM	0.007 nM	[97]
GO – poly(dimethylsiloxane)/SPCE	Pb^{2+}	SWASV	1.21–377.05 nM	0.34 pM	[98]
poly-L-lysine/chitosan/rGO	Cd^{2+} Pb^{2+} Cu^{2+}	DPASV	0.44–88.96 nM 0.24–48.26 nM 0.79–157.37 nM	0.089 nM 0.097 nM 0.31 nM	[99]
PVA/chitosan-TRG/GC	Pb^{2+}	SWASV	4.83–241 nM	0.242 nM	[100]
Gr-CD/PPy/SPCE	Hg^{2+}	DPV	1–300 nM	0.47 nM	[101]
CNW:rGO/PA6	Hg^{2+}	DPV	2.5–200 μM	5.2 nM	[102]
Ru(II)-tris(bipy)-GO/AChE/Pt	Cd^{2+} As^{3+}	Amperometry	0.02–0.7 μM 0.05–0.8 μM	0.07 μM 0.03 μM	[103]
ZIF-8@DMG/β-CD/rGO/GCE	Ni^{2+}	DPASV	0.01–1.0 μM	0.005 μM	[104]
Sb film/GO/SPE	Cd^{2+} Pb^{2+} Cu^{2+} Hg^{2+}	SWASV	0.3–1.5 μM 0.1–1.3 μM 0.3–1.5 μM 0.1–1.3 μM	0.054 μM 0.026 μM 0.06 μM 0.066 μM	[105]
Bi film/Gr/GCE	Cd^{2+} Pb^{2+}	SWASV	0.062–1.068 μM 0.024–0.579 μM	4.18 nM 1.98 nM	[106]

Hg^{2+}. Under the optimized conditions, the sensor gave a linearity in the range of 1–70 nM and a detection limit of 0.1 nM for Hg^{2+} detection using SWASV. Next, Priya et al. developed the penetrable nature of reduced graphene/AuNPs/modified L-cysteine nanocomposite coated on GCE for simultaneous electrochemical detection of Pb^{2+} and Cd^{2+} using SWASV [90]. A cooperative impact of properties of the electrode materials used could explain why the PrGO/AuNPs/Sal-Cys/GC electrode performed better. GO pores could increase electrode surface area, AuNPs act as electrocatalysts [107] and electronegative groups in Sal-Cys facilitate effective chelation of target metal ions. With an optimized medium, the linearity was in the range of 1–10 nM for both Cd^{2+} and Pb^{2+}, while the limits of detection were 0.06 nM and 0.04 nM for Cd^{2+} and Pb^{2+}, respectively.

An ion-imprinted sensor based on chitosan-GO composite polymer-modified GCE was developed for Cu^{2+} detection by Wei et al. [94]. Figure 7.2a displays the CS/GO-IIP sensor fabrication and chemical processes. The sensor was characterized by Raman spectroscopy, FT-IR, SEM, EDS, AFM, and electrochemical measurements. The modified electrodes were used to detect Cu^{2+} with differential pulse anodic stripping voltammetry and achieved a detection limit of 0.15 μM with a linear range of 0.5–100 μM. Ru et al. then demonstrated the determination of the As^{3+} method using

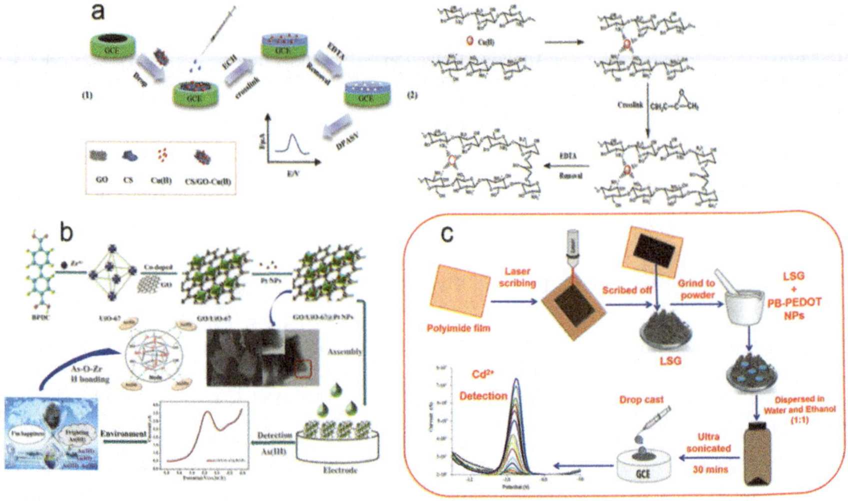

Fig. 7.2 Electrochemical sensors modified with graphene-based nanocomposites for heavy metal detection **a** Diagram of the fabrication procedure for CS/GO-IIP sensor (1) and corresponding chemical reactions (2). (reprinted from ref [94], Copyright © 2019 Elsevier) **b** Schematic illustrates the fabrication process for as^{3+} electrochemical detection using GO/UiO-67@PtNPs.(reprinted from ref [91], Copyright © 2022 Elsevier) **c** Schematic representation of LSG fabrication and PB-PEDOT nanoparticle synthesis for cadmium ion sensing (reprinted from ref [49])

SWASV with graphene oxide/octahedral UiO-67metal-organic framework/ platinum nanoparticles modified GCE [91]. With the sensor (Fig. 7.2b), As^{3+} showed a linear response in the range of 2.7–33.4 nM, and the detection limit is 0.48 nM. Another electrochemical sensor for Cd^{2+} detection was introduced by Machhindra and Yen [49]. Prussian blue (PB), poly(3,4-ethylenedioxythiophene) polystyrene sulfonate (PEDOT)-loaded laser-scribed graphene (LSG) nanocomposite was successfully synthesized and coated on GCE. The combined materials gave the advantages of large surface area from LSG and high selectivity and sensitivity and rapid detection ability from PB nanoparticles. Under the optimum conditions, the modified electrode was used with differential pulse voltammetry to detect Cd^{2+} and obtained an LOD equals to 0.85 nM with a linear range of 1–10,000 nM.

4.2 *Graphene-Based Nanocomposites Electrochemical Sensors for Drug Detection*

Currently, drugs have a significant presence in our routine due to their intentional use for healing as well as non-medical purposes. These drugs fall under three main categories: therapeutic, legal, and illicit. Medications that are primarily recommended by medical practitioners to manage various ailments are referred to as therapeutic drugs. Legal drugs are commonly used in commercial goods. Upon ingestion, these substances induce psychoactive effects within the body. And final group, illicit drugs, are primarily used for recreational purposes. These drugs adversely affect the central nervous system and consequently contribute to a variety of health problems that persist over extended periods [108, 109]. In spite of the fact that the last category of drugs was initially developed for pharmaceutical purposes, their potential for harmful misuse ultimately outweighed their therapeutic use. Furthermore, the problem of drug overdose in numerous countries has become more severe due to insufficient regulations. Therefore, developing precise and selective techniques for the quantification of drugs from various sources is necessary.

Electrochemical sensors are now transforming the way drug-detection techniques are being conducted in various fields, especially because of their cost-effectiveness, ease of usage, and portability when compared to traditional chromatography techniques. As for graphene, when compared to other nanomaterials, they offer a much higher surface-to-weight ratio. This property has prompted the development of more sensitive and dependable sensors. Moreover, due to the remarkable qualities of graphene and its ability to be customized for specific molecule targeting, graphene-based sensors have been a highly favored sensing material in the scientific community. Table 7.4 offers a review of electrochemical graphene-based and its nanocomposite sensors for various drug detection.

Wang et al. proposed a new sensor for the simultaneous determination of 4-aminophenol (4-AP) and acetaminophen (AC) based on the nanohybrid of palladium-rGO modified with AuNPs nanocomposites [123]. The experiment results indicated that the prepared nanohybrid of Au/Pd/rGO (Fig. 7.3a) gave a remarkable level of electrocatalytic activity for the simultaneous oxidation and reduction of AC and 4-AP. A linear range and detection limit obtained for both drugs is 1.00–250.00 μM, 0.30 μM for AC and 1.00–300.00 μM, 0.12 μM for 4-AP. Another method for AC detection was introduced by Rokhsefid and Shishehbore [131]. Their work is about developing a method for the simultaneous determination of acetaminophen and tramadol. AuNPs/GN) nanocomposite was successfully synthesized and mixed with graphite powder and 4-hydroxyl-2-(triphenylphosphonio)phenolate (HTP) to make a carbon paste electrode (CPE). The electrode was used to measure TRA and AC simultaneously with DPV and obtained a 0.82 μM detection limit in a linear range of 1.0–100.0 μM for TRA and a linear range of 40.0–120.0 mM for AC. Aflatoonian et al. developed a new type of electrochemical sensor using a composite of Gr and Co_3O_4 as a modified material on the screen-printed electrode (SPE) for detecting tramadol with DPV [117]. The electrochemical oxidation of TRA was found to be more durable using Gr/ Co_3O_4/SPE with a more negative potential. Under the optimized condition, the linear range of tramadol is 0.1–500.0 μM with the limit of detection at 0.03 μM.

Baghayeri et al. developed a highly efficient electrochemical sensor which was made of two layers of silver nanoparticle-decorated graphene for the methadone measurement [113]. Two layers of Gr/AgNPs nanocomposite on GCE were fabricated by two-steps: Gr drop-casting and electrochemical deposition of AgNPs. The synergetic effects between Gr and AgNPs help increase the electrocatalytic activity

Table 7.4 Overview of electrochemical sensors modified with graphene-based nanocomposites for drug detection

Electrode	Analyte	Detection technique	Linear range	LOD	Refs.
GO-MWCNTs/CPE	TRA	DPV	2.00–1,100,000 nM	0.15 nM	Mohamed et al. [110]
erGO/GCE	AC	DPV	0.0219–2.30 μM	0.007 μM	Ahmed et al. [111]
GO/G/ITO	SMX	DPV	0.1–50 μM	0.06 μM	Yeh et al. [112]
$(Gr/AgNPs)_2$/GCE	MTD	DPV	1.0–200.0 μM	0.12 μM	Baghayeri et al. [113]
Pd–Ag/rGO/GCE	AC	DPV	1.2–30 nM	3.26 nM	Veera Manohara Reddy et al. [114]
GO-Fe/ZnO/SPCE	CPZ	DPV	0.02–172.74 μM 222.48–1047.74 μM	0.02 μM	Sebastian et al. [115]
GO-Co_3O_4/CPE	MO	DPV	5 nM - 600 nM	0.54 nM	Atta et al. [116]
Gr/ Co_3O_4/SPE	TRA	DPV	0.1–500.0 μM	0.03 μM	Aflatoonian et al. [117]
Fe_3O_4 NPs-ErGO/GCE	Rutin	CV	6–80,000 nM	4.0 nM	He et al. [118]
MWCNTs/GO/PG/CPE	OME	DPV	0.2–600 nM 600–100,000 nM	0.01 nM	Mohamed et al. [119]
GO–Fe_3O_4@SiO_2/SPE	MO	DPV	1.0–100 μM	0.75 μM	Hadi and Fariba Garkani [120]
Fe_3O_4@PPy/ErGO/GCE	MET	SWV	0.005–200 μM	0.001 μM	Riahifar et al. [121]
CeO_2NP/rGO/GCE	MET	SWV	25.0–166.6 μM	8.7 μM	Anvari et al. [122]
CS/Au/Pd/rGO/GCE	AC 4-AP	CV	1.00–250.00 μM 1.00–300.00 μM	0.30 μM 0.12 μM	Wang et al. [123]
CoOOH-rGO/SPCE	CNZ	DPV	0.1–350 μM	0.038 μM	Sachdev and Matai [124]
rGO/β-CD/GCE	IVM	DPV	0.571–45.7 nM	0.286 nM	Silva et al. [125]
TiO_2@CuO-N-rGO/poly (L-Cys) / GCE	FNZ	DPV	1–50,000 nM	0.3 nM	Sohouli et al. [126]
rGO-CB-CTS/GCE	PAR DA	SWV	2.8–19 μM 3.2–32 μM	0.053 μM 0.2 μM	Baccarin et al. [127]

(continued)

Table 7.4 (continued)

Electrode	Analyte	Detection technique	Linear range	LOD	Refs.
prGO-ANSA/Au/MWCNT-CPE	NFX	DPV	0.03–1.0 μM 1.0–50.0 μM	0.016 μM	Liu et al. [128]
rGO/Cu-poly(Ala)/GCE	MO	DPV	0.050 μm–80 μM	0.047 μM	Kumary et al. [129]
poly(CTAB)/GO/GCE	MO	DPV	0.50–60 μM	0.36 μM	Abraham et al. [130]
HTP–AuNPs/GN/CPE	TRA AC	DPV	1.0–100.0 μM 40.0–120.0 mM	0.82 μM –	Rokhsefid and Shishehbore [131]
TGA@CdSe/GO/PGE	MO MTD	DPV	0.05–350 μM 0.1–323 μM	0.01 μM 0.26 μM	Nazari and Es' haghi [132]
Gr/Pd NPs/MIP/SPE	Cocaine	SWV	100–500 μM	50 μM	Florea et al. [133]
Gr–Ag NPs-MIP/CPE	TRA	CV	0.004–10,000 μM	0.002 μM	Bagheri et al. [134]
MIP/GNW/GO/GCE	CEF	DPV	20.0–950.0 nM	7.1 nM	Dehghani et al. [135]
KT-MIM/MOFs@Gr/SPE	Ketamine	DPV	0.1–40,000 nM	0.04 nM	Fu et al. [136]
NSG/GCE	MO	DPV	2.00–1030 μM	0.91 μM	Prakash et al. [137]
FSG/SPE	COD	SWV	0.02–200 μM	0.006 μM	Mohamed et al. [138]
Gr-NC/ISM/SPE	CHX	Potentiometry	1–1000 μM	0.476 μM	Magdy et al. [139]
3D printed Gr–PLA	Cocaine	SWV	20–100 μM	6 μM	Rocha et al. [140]
LI-PGr	MET	DPV	6.70–201 μM	2.08 μM	Saisahas et al. [141]
NPG/Pt	Oxytocin	SWSV	0.1–10 nM 15–95 nM	0.04 nM	Thomas and Balachandran [142]
Gr–$(PHL)^{2+}$-$(NB)^{-}$ ISE	PHL	Potentiometry	0.2–10,000 μM	0.1 μM	Abd-Rabboh et al. [143]
Zeo-GO	Ketamine	CV	0.001–5 nM	0.001 nM	Narang et al. [144]

AC: acetaminophen, 4-AP: 4-aminophenol, CPZ: chlorpromazine, FNZ: flunitrazepam, OME: omeprazole, TRA: tramadol, PAR: paracetamol, DA: dopamine, NFX: norfloxacin, DCF: diclofenac, CEF: cefixime, MO: morphine, COD: codeine, PHL: pholcodine, MTD: methadone, MET: methamphetamine, CHX: chlorhexidine digluconate, SMX: sulfamethoxazole, CNZ: clonazepam, IVM: ivermectin

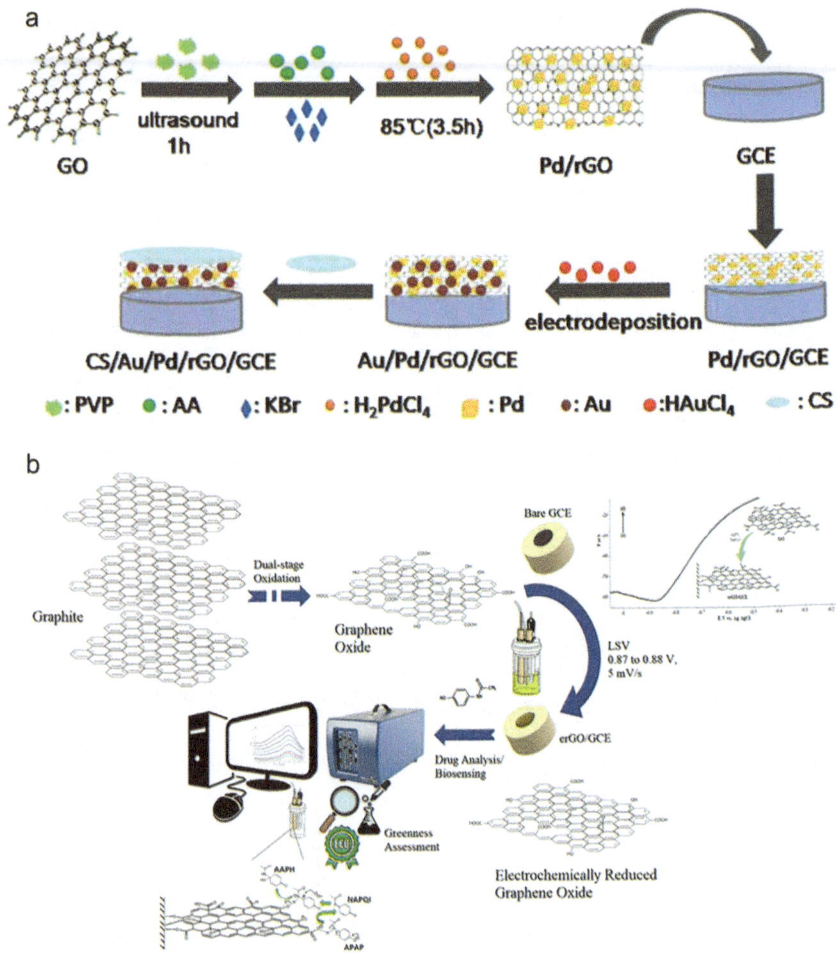

Fig. 7.3 Electrochemical sensors modified with graphene-based nanocomposites for drug detection **a** Schematic illustration of the fabrication of the CS/Au/Pd/rGO sensor for simultaneous determination of 4-aminophenol and acetaminophen. (reprinted from ref [123], copyright © 2017 Elsevier) **b** Schematic illustration of the fabrication process of erGO/GCE sensor for analysis of acetaminophen drug. (reprinted from ref [111])

toward methadone oxidation. Under the optimized conditions, the sensor gave a linear response in the range of 1.0–200.0 μM, and the limit of detection is 0.12 μM using DPV. Another acetaminophen analysis was proposed by Ahmed et al. [111]. The GO was synthesized by a modified Tours' method and used in the electrochemical fabrication of electrochemically reduced graphene oxide (erGO)/GCE sensor via LSV technique to improve the deposition (Fig. 7.3b). The sensor was used to detect AC with DPV and achieved a LOD of 0.007 μM with a working range of 0.0219–2.30 μM.

4.3 Graphene-Based Nanocomposites Electrochemical Sensors for Pesticides

Pesticides are extensively utilized in agricultural practices worldwide for the purpose of managing weeds and pests and maintaining high food production. However, these chemicals have the potential to be toxic to humans. Overusing and incorrectly disposing of pesticides can lead to substantial contamination of both food and the environment due to pesticide's ability to dissolve easily in water but difficult to decompose which can lead to harmful health effects [145]. Therefore, developing sensitive, selective, and convenient sensors for pesticide determination in food or environment is required.

Compared to traditional analytical techniques for detection of pesticides like chromatography or mass spectrometry, electrochemical sensors have more advantages in cost-effectiveness, easy operation, and portability. On the other hand, graphene is an exceptional absorber of pesticides and widely used as a sorbent material because of its substantial surface area, [22]. Moreover, its exceptional conductivity and narrow band gap facilitate the flow of electrons [146] which makes them more interesting as materials for electrochemical sensor modification. A summary of electrochemical sensors synthesized from graphene and its nanocomposites for numerous pesticide detection is provided in Table 7.5.

A new sensor for the detection of fenitrothion based on the modification of GCE with cerium oxide@rGO nanocomposite was introduced by Ensafi et al. [154]. From the study, this nanocomposite has a synergetic effect on the fenitrothion oxidation by reducing the working potential closer to 0.00 V versus Ag/AgCl and increases in the current responses. The sensor showed a linearity in the range of 0.025–2.00 μM with an LOD of 0.003 μM for fenitrothion detection using DPV. A similar pesticide detection method was developed by Shahid et al. [48]. Figure 7.4a shows rGO-Co_3O_4@Au nanocomposite was synthesized by a one-pot hydrothermal method and then was used as a modified material on GCE for hydrazine determination with amperometry technique. The nanocomposite of rGO- Co_3O_4@Au demonstrated commendable electrocatalytic properties for the process of hydrazine oxidation in phosphate buffer with a pH value of 7.2. The sensor achieved a detection limit of 0.443 μM with a linear range of 10–620 μM.

Yan et al. took the advantages of the reduced graphene oxide nanosheets and gold nanoparticles to link them covalently to ferrocene-terminated dendrimer coated on GCE (AuNPs/FcDr/rGO/GCE) sensor for dichlorvos detection [159]. The combination of AuNPs and rGO nanosheets has a cooperative effect on the acceleration of electron transfer from FcDr to GCE. This leads to an increased electrochemical signal in the AuNPs/FcDr/rGO/GCE sensor and an overall improvement in the sensor's ability to detect an analyte. From the results, the developed sensor has a wide linear range from 0.43 to 218.4 μM and a detection limit of 0.21 μM using the DPV technique. Another sensor for dichlorvos detection was prepared by Zhang et al. [161]. A layer-by-layer assembly method was employed to drip-coat graphene nanofragments that had been modified with chitosan and acetylcholinesterase (AChE) onto a GCE's

Table 7.5 Overview of electrochemical sensors modified with graphene-based nanocomposites for pesticides detection

Electrode	Analyte	Detection technique	Linear range	LOD	Refs.
GS/GCE	Carbendazim	DPV	5–1570 nM	0.78 nM	Wei et al. [147]
SGr/GCE	Hydrazine	Amperometry	0.5–6 μM	0.25 μM	Saritha Rani et al. [148]
rGO-Au–Pd/CPE	Parathion	SWASV	0.01–11.2 μM	0.008 μM	Jahromi et al. [149]
AuNPs/erGO/AuE	Hydrazine	DPV	4–1000 μM 2000–8000 μM	0.074 μM	Amin et al. [150]
BNQDs/GO/GCE	Methyl parathion Diazinon Chlorpyrifos	DPV	1–10,000 pM 1–10,000 pM 1–10,000 pM	0.31 pM 0.06 pM 0.03 pM	Yola [151]
rGO-$CuFeS_2$/SPCE	Methyl paraoxon	DPV	0.073–801.5 μM	0.005 μM	Rajaji et al. [152]
rGO-CMF/SPCE	Fenitrothion	DPV	0.03–1133.8 μM	0.008 μM	Velusamy et al. [153]
CeO_2/rGO/GCE	Fenitrothion	DPV	0.025–2.00 μM	0.003 μM	Ensafi et al. [154]
Gr@NiFeSP/GCE	Paraoxon Ethy	SWV	0.012–10 μM	0.004 μM	Aghaie et al. [155]
rGO/Co_3O_4@Au/GCE	Hydrazine	Amperometry	10–620 μM	0.443 μM	Shahid et al. [48]
FS@rGO/GCE	Fenitrothion	DPV	5–1000 nM	0.19 nM	Özcan et al. [156]
ZrP/GO/GCE	Fenitrothion	Amperometry	0.008–26 μM	0.001 μM	Kokulnathan et al. [157]
TiO_2/GO@UiO-66/GCE	Paraoxon Chlorpyrifos	SWV	1.0–100.0 nM 5.0–300.0 nM	0.2 nM 1.0 nM	Karimian et al. [158]
AuNPs/FcDr/rGO/GCE	Dichlorvos	DPV	0.45–281.4 μM	0.21 μM	Yan et al. [159]
MIP/GO/GCE	Profenofos	SWV	0.05–3500 μM	0.005 μM	Khalifa and Abdallah [160]
AChE/Gr&CS/GCE	Dichlorvos	DPV	0.1–100,000 nM	0.054 nM	Zhang et al. [161]
CNFs/GO/CS-GO/AChE/SPCE	Chlorpyrifos	SWV	2.5–1000 nM	2.2 nM	Thet Tun et al. [162]
AChE/GA/CIS/rGO/SPCE	Chlorpyrifos	LSV	1.43–1341 nM	0.066 nM	Itsoponpan et al. [163]

(continued)

Table 7.5 (continued)

Electrode	Analyte	Detection technique	Linear range	LOD	Refs.
AChE–CS/3DG–CuO NFs/GCE	Malathion	SWV	3–46,665 pM	0.93 pM	Bao et al. [164]
AChE/$Ti_3C_2T_x$-CS/Gr/GCE	Dichlorvos	DPV	0.023–11.31 μM	0.014 μM	Wang et al. [165]
poly(FBThF)/Ag-rGO-NH_2/AChE/GCE	Malathion Trichlorfon	CV	0.3–30 nM 0.08–8 nM	0.097 nM 0.004 nM	Zhang et al. [166]
$(PDDA/ERGO)_5$/$(MNP/PSS)_5$ /ITO	Carbofuran	DPV	0.83–11.4 μM	0.407 μM	Miyazaki et al. [167]
AuNPs-rGO@NF	Hydrazine	Amperometry	0.2–200 μM	0.056 μM	Wng et al. [168]
ZrO_2/rGO/MoS_2-Au/AuE	Fenitrothion	SWV	0.018–21.6 μM	0.008 μM	Qi et al. [169]
$MoTe_2$ NPs/rGO/ITO	Profenofos	PEC	3–26,764,000 pM	0.883 pM	Ding et al. [170]

surface (Fig. 7.4b). The sensor gave a linearity in the range of 0.1–100,000 nM while the LOD was 0.054 nM for dichlorvos detection using DPV.

4.4 Graphene-Based Nanocomposites Electrochemical Sensors for Hydrogen Peroxide (H_2O_2)

H_2O_2 has significant applications in clinical, food, pharmaceutical, and environmental monitoring fields. It serves as a mediator due to its exceptional oxidizing and reducing properties [171]. H_2O_2 is produced as a byproduct when different oxidases are involved in biochemical reactions taking place inside living organisms. H_2O_2 functions as a crucial signaling molecule that regulates various biological signal transduction processes including vascular remodeling, immune cell activation, stomatal closure, apoptosis, and root growth [172]. Additionally, it also holds a significant physiological function as an indicator of oxidative stress in aging and illness, and as a protective agent in the event of pathogenic invasion so H_2O_2 eruption can activate various types of vital signaling proteins that impact cell growth, potentially causing various types of disorders in the body [173, 174]. Therefore, it is critical to observe the level of H_2O_2 and the development of simple, low-cost, fast, selective, and sensitive sensors are crucial.

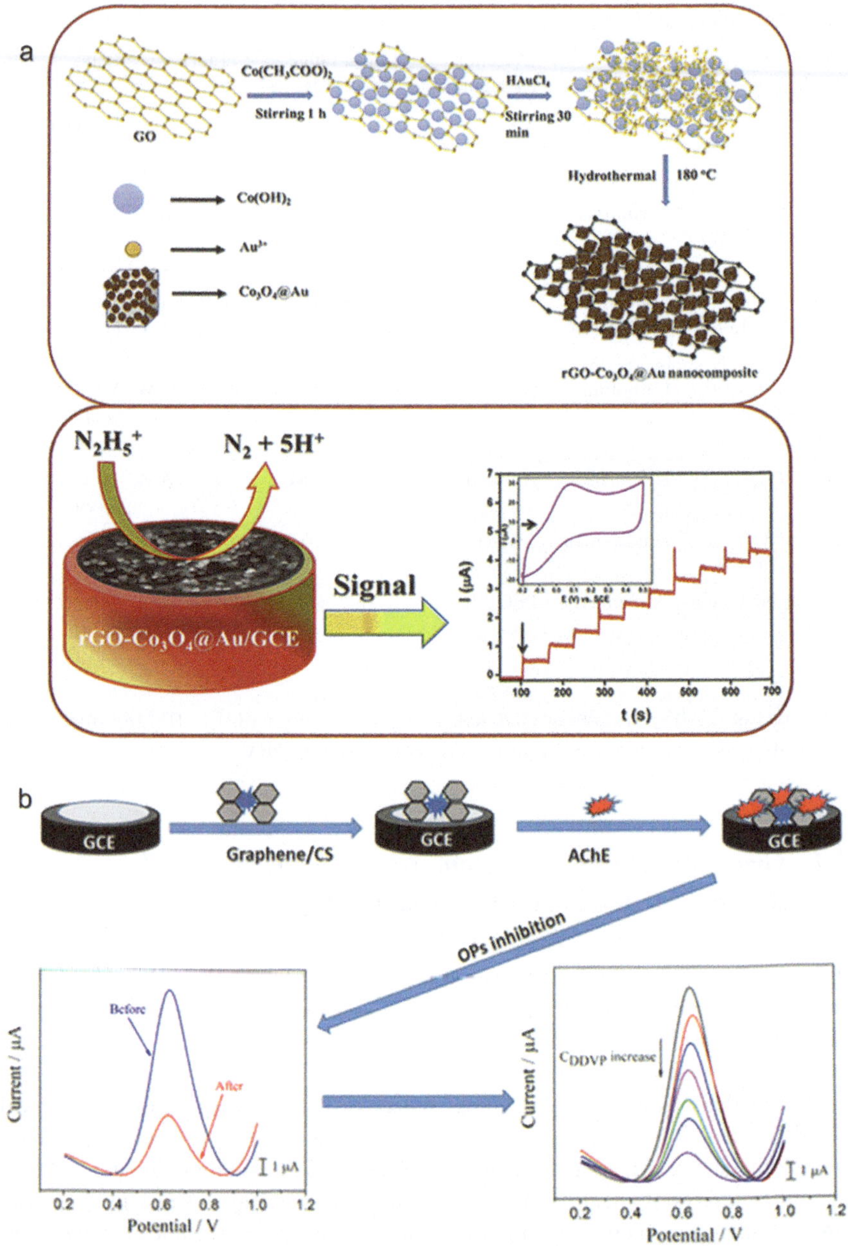

Fig. 7.4 Electrochemical sensors modified with graphene-based nanocomposites for pesticides detection **a** A scheme depicting how cobalt oxide nanocubes with gold nanoparticles interleaved into reduced graphene oxide are fabricated via hydrothermal methods for hydrazine determination. (reprinted from ref [48], copyright © 2017 Elsevier) **b** Schematic diagram of sensor modification with acetylcholinesterase/chitosan/graphene nanocomposites and Dichlorvos detection principle (reprinted from ref [161] with permission from Wiley)

Graphene-based materials have become an attractive option for detecting H_2O_2 due to their exceptional strength, high surface area, and remarkable electrical conductivity [171]. Moreover, incorporating nanocomposites onto the graphene surface can substantially enhance the active surface area, leading to improved adsorption and accelerated electron transfer mechanisms between the electrode and analytes. Table 7.6 shows electrochemical sensors for hydrogen peroxide fabricated from graphene-based nanocomposites.

Liu et al. introduced a simple electrochemical detection of the hydrogen peroxide method using a rGO/AuNPs composite-based paper sensor [193]. GO and AuNPs were reduced by using L-ascorbic acid. Then the nanocomposites were drop-coat on the paper substrate and heated to dry. The coated paper was made into an electrode as shown in Fig. 7.5a and used to determine H_2O_2 with the amperometry technique. The performance of the paper sensor in detecting H_2O_2 was good and can detect in the linear range of 8.53–17.35 mM with a LOD of 0.015 mM. Another method for H_2O_2 detection was proposed by Guan et al. [196]. Manganese dioxide-graphene nanosheets (MnO_2-GNSs) composite synthesis was achieved by a one-step hydrothermal method (Fig. 7.5b). The MnO_2-GNSs modified GCE displayed a remarkable level of electrocatalytic activity for H_2O_2 oxidation due to the conductivity of GNSs and the synergistic effect of the catalytic capacity of MnO_2 particles. Under the optimized conditions, the sensor gave a wide linear range of 0.5–350 μM with a low LOD at 0.19 μM using the amperometry technique.

Qjan et al. developed a simple and green non-enzymatic sensor for H_2O_2 detection using Cu_2O/Ag-rGO nanocomposites as electrode-modifying material [201]. The results displayed that the sensor is selective and sensitive and can detect in a wide linear range from 1 to 310 μM with a LOD of 0.34 μM. Another work for H_2O_2 detection was developed by Shen et al. [65]. Graphene doped with AgNP composites was synthesized via an in-situ reduction and laser etching method as shown in Fig. 7.5c. The sensor showed high sensitivity and exhibited a linear response from 1 to 110 μM and a LOD of 0.24 μM.

5 Conclusion

Graphene-based nanocomposites have shown great potential in electrochemical sensing applications. The unique properties of graphene, such as excellent mechanical, large surface area, and strength high and conductivity, make it an ideal material for developing high-performance sensors. Several methods have been developed to synthesize graphene such as electrochemical exfoliation, chemical vapor deposition, and laser-scribing approach. The incorporation of graphene with other materials to form the nanocomposites has further improved the selectivity and sensitivity of the sensors. These nanocomposites have demonstrated excellent performance in detecting various analytes such as heavy metals, drugs, and hydrogen peroxide, among others. Overall, graphene-based nanocomposites have revolutionized the field of electrochemical sensing, and their use is expected to increase in the future. With

Table 7.6 Overview of electrochemical sensors modified with graphene-based nanocomposites for the detection of hydrogen peroxide

Electrode	Detection technique	Linear range	LOD	Refs.
erGO–AgNCs/GCE	Amperometry	0.02–10 mM	0.003 mM	Zhong et al. [175]
rGO–AuNP/GCE	Amperometry	0.25–22.5 mM	0.006 mM	Zhang et al. [176]
MnO_2 NTs/rGO NCs/GCE	Amperometry	0.1–30 mM	0.001 mM	Mahmoudian et al. [177]
AgNPs–CNT–rGO/GCE	Amperometry	0.01–10 mM	0.001 mM	Zhang et al. [178]
Pt/Gr/GCE	Amperometry	1–1477 μM	0.50 μM	Liu et al. [179]
AgNWs/Gr/GCE	Amperometry	10.0–34.3 μM	1.0 μM	Zhang and Wang [180]
HRP/CeO_2-rGO/GCE	CV	0.1–500 μM	0.021 μM	Radhakrishnan and Kim [181]
Gr-Co_3O_4NPs/GCE	Amperometry	0.2–211.5 μM	0.06 μM	Karuppiah et al. [182]
AgNPs/TWEEN/GO/GCE	Amperometry	0.02–23.1 mM	0.009 mM	Yang et al. [183]
rGO–PtNPs/GCE	Amperometry	0.05–750.6 μM	0.016 μM	Palanisamy et al. [184]
Ag@Pt-Gr/GCE	Amperometry	5.0–12,400 μM	0.9 μM	Liu et al. [185]
Pd-NPs/BGrs/GCE	Amperometry	4–13,500 μM	1.5 μM	Wang et al. [186]
Pd/$ZnFe_2O_4$/rGO/GCE	Amperometry	25–10,200 μM	2.12 μM	Ning et al. [187]
rGO-Nf@Ag6/GCE	Amperometry	1–10 μM	0.535 μM	Yusoff et al. [188]
rGO/Cu Fe_2O_4/GCE	Amperometry	1–11,000 μM	0.35 μM	Karthikeyan et al. [189]
Cu_2O/rGO paper electrode	Amperometry	1–1470 μM	0. 37 μM	Cheng et al. [190]
CoS/rGO/GCE	Amperometry	0.1–2542.4 μM	0.042 μM	Kubendhiran et al. [191]
rGO/nAPAMSs/GCE	Amperometry	5–4000 μM	0.008 μM	Bai et al. [192]
GO/AuNP-paper electrode	Amperometry	8.53–17.35 mM	0.015 mM	Liu et al. [193]
$NiCo_2S_4$/rGO/GCE	Amperometry	25–11,250 μM	0.19 μM	Wang et al. [194]
rGO/AgNPs/GCe	Amperometry	2–20,000 μM	0.73 μM	Salazar et al. [195]
MnO_2-GNS/GCE	Amperometry	0.5–350 μM	0.19 μM	Guan et al. [196]
GNS@FeOOH/GCE	Amperometry	0.25–1200 μM	0.08 μM	Chen et al. [197]
rGO-AuNPs/ITO	Amperometry	25–3000 μM	6.5 μM	Patella et al. [198]

(continued)

Table 7.6 (continued)

Electrode	Detection technique	Linear range	LOD	Refs.
$CuCo_2O_4$/rGO/GCE	Amperometry	30–5010 μM	0.08 μM	Jiang et al. [199]
PANI/Gr/ITO	CV	100–500 μM	–	Verma et al. [200]
CG:Cu/FTO	Amperometry	32–803 μM	0.64 μM	Alencar et al. [63]
Cu_2O/Ag-rGO/GCE	Amperometry	1–310 μM	0.34 μM	Qian et al. [201]
NiS@GO/PGE	CV	0.1–1000 μM	0.059 μM	Awan et al. [58]
Ag/rGO/SPE	Amperometry	1–110 μM	0.24 μM	[65]
Ag/ Co_3O_4/rGO/GCE	Amperometry	0.5–6480 μM	0.17 μM	Zhang et al. [202]

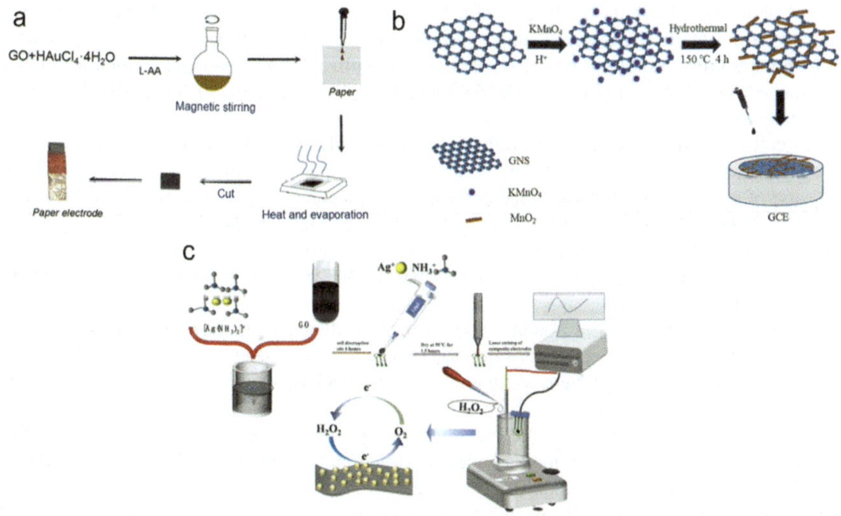

Fig. 7.5 Electrochemical sensors modified with graphene-based nanocomposites for hydrogen peroxide detection **a** Illustration for the fabrication of rGO/AuNP-based paper sensor. (reprinted from ref [193], copyright ©2018 Taylor & Francis) **b** Diagram for the preparation of MnO_2-GNSs/GCE. (reprinted from ref [196], copyright © 2019 Elsevier) **c** Ag/rGO electrode preparation and electrochemical portable workstation detection (reprinted from ref [65], Copyright © 2023, Elsevier)

further advancements in nanotechnology, it is expected that these sensors will become even more sensitive, selective, and reliable, leading to their widespread adoption in different fields, including food safety, medical diagnosis, and environmental analysis.

Acknowledgements This research project is supported by the Second Century Fund (C2F), Chulalongkorn University, and the Rachadapisek Sompot Fund for Postdoctoral Fellowship, Chulalongkorn University.

Declaration of Competing Interest The authors declare that they have no known competing financial interests or personal relationships that could have appeared to influence the work reported in this paper.

References

1. Das, S., Ghosh, C.K., Sarkar, C.K., Roy, S.: Facile synthesis of multi-layer graphene by electrochemical exfoliation using organic solvent **7**, 497–508 (2018).https://doi.org/10.1515/ntrev-2018-0094
2. Chen, D., Tang, L., Li, J.: Graphene-based materials in electrochemistry. Chem. Soc. Rev. **39**, 3157–3180 (2010). https://doi.org/10.1039/B923596E
3. Yao, J., Wang, H., Chen, M., Yang, M.: Recent advances in graphene-based nanomaterials: properties, toxicity and applications in chemistry, biology and medicine. Microchim. Acta **186**, 395 (2019). https://doi.org/10.1007/s00604-019-3458-x
4. Alabdo, F., Alahmad, W., Pengsomjit, U., Halabi, M., Varanusupakul, P., Kraiya, C.: An environmentally friendly and simple method for producing multi-layer exfoliated graphene in mass production from pencil graphite and its utilization for removing cadmium from an aqueous medium. Carbon Lett. **33**, 455–465 (2023). https://doi.org/10.1007/s42823-022-00435-6
5. Reddy, Y.V.M., Shin, J.H., Palakollu, V.N., Sravani, B., Choi, C.-H., Park, K., Kim, S.-K., Madhavi, G., Park, J.P., Shetti, P.: Strategies, advances, and challenges associated with the use of graphene-based nanocomposites for electrochemical biosensors. Adv. Coll. Interface. Sci. **304**, 102664 (2022). https://doi.org/10.1016/j.cis.2022.102664
6. Zhang, C., Zhang, Z., Yang, Q., Chen, W.: Graphene-based electrochemical glucose sensors: fabrication and sensing properties. Electroanalysis **30**, 2504–2524 (2018). https://doi.org/10.1002/elan.201800522
7. Singh, A., Sinsinbar, G., Choudhary, M., Kumar, V., Pasricha, R., Verma, H.N., Singh, S.P., Arora, K.: Graphene oxide-chitosan nanocomposite based electrochemical DNA biosensor for detection of typhoid. Sens. Actuators, B Chem. **185**, 675–684 (2013). https://doi.org/10.1016/j.snb.2013.05.014
8. Tanwar, S., Mathur, D.: Graphene-based nanocomposites as sensing elements for the electrochemical detection of pesticides: a review. J. Solid State Electrochem. **25**, 2145–2159 (2021). https://doi.org/10.1007/s10008-021-04990-2
9. Irkham, I., Ibrahim, A.U., Pwavodi, P.C., Al-Turjman, F., Hartati, Y.W.: Smart graphene-based electrochemical nanobiosensor for clinical diagnosis: review. Sensors (2023)
10. Tiwari, S.K., Sahoo, S., Wang, N., Huczko, A.: Graphene research and their outputs: status and prospect. J. Sci.: Adv. Mater. Dev. **5**, 10–29 (2020). https://doi.org/10.1016/j.jsamd.2020.01.006
11. Avouris, P., Dimitrakopoulos, C.: Graphene: synthesis and applications. Mater. Today **15**, 86–97 (2012). https://doi.org/10.1016/s1369-7021(12)70044-5
12. Sumdani, M.G., Islam, M.R., Yahaya, A.N.A., Safie, S.I.: Recent advances of the graphite exfoliation processes and structural modification of graphene: a review. J. Nanopart. Res. **23**, 253 (2021). https://doi.org/10.1007/s11051-021-05371-6
13. Zhu, L., Zhao, X., Li, Y., Yu, X., Li, C., Zhang, Q.: High-quality production of graphene by liquid-phase exfoliation of expanded graphite. Mater. Chem. Phys. **137**, 984–990 (2013). https://doi.org/10.1016/j.matchemphys.2012.11.012
14. Bhuyan, M.S.A., Uddin, M.N., Islam, M.M., Bipasha, F.A., Hossain, S.S.: Synthesis of graphene. Int. Nano Lett. **6**, 65–83 (2016). https://doi.org/10.1007/s40089-015-0176-1
15. Guerrero-Contreras, J., Caballero-Briones, F.: Graphene oxide powders with different oxidation degree, prepared by synthesis variations of the Hummers method. Mater. Chem. Phys. **153**, 209–220 (2015). https://doi.org/10.1016/j.matchemphys.2015.01.005

16. Mattevi, C., Kim, H., Chhowalla, M.: A review of chemical vapour deposition of graphene on copper. J. Mater. Chem. **21**, 3324–3334 (2011). https://doi.org/10.1039/c0jm02126a
17. Subrahmanyam, K.S., Panchakarla, L.S., Govindaraj, A., Rao, C.N.R.: Simple method of preparing graphene flakes by an arc-discharge method. The J. Phys. Chem. C **113**, 4257–4259 (2009). https://doi.org/10.1021/jp900791y
18. Wu, Z.S., Ren, W., Gao, L., Zhao, J., Chen, Z., Liu, B., Tang, D., Yu, B., Jiang, C., Cheng, H.M.: Synthesis of graphene sheets with high electrical conductivity and good thermal stability by hydrogen arc discharge exfoliation. ACS Nano **3**, 411–417 (2009). https://doi.org/10.1021/nn900020u
19. Vivaldi, F.M., Dallinger, A., Bonini, A., Poma, N., Sembranti, L., Biagini, D., Salvo, P., Greco, F., Di Francesco, F.: Three-dimensional (3D) laser-induced graphene: structure, properties, and application to chemical sensing. ACS Appl. Mater. Interfaces **13**, 30245–30260 (2021). https://doi.org/10.1021/acsami.1c05614
20. Lin, J., Peng, Z., Liu, Y., Ruiz-Zepeda, F., Ye, R., Samuel, E.L., Yacaman, M.J., Yakobson, B.I., Tour, M.: Laser-induced porous graphene films from commercial polymers. Nat. Commun. **5**, 5714 (2014). https://doi.org/10.1038/ncomms6714
21. Viswanathan, P., Ramaraj, R.: Chapter 2—functionalized graphene nanocomposites for electrochemical sensors. In: Pandikumar, A., Rameshkumar, P. (eds.) Graphene-Based Electrochemical Sensors for Biomolecules, Elsevier, pp. 43–65 (2019)
22. Krishnan, S.K., Singh, E., Singh, P., Meyyappan, M., Nalwa, H.S.: A review on graphene-based nanocomposites for electrochemical and fluorescent biosensors. RSC Adv. **9**, 8778–8881 (2019). https://doi.org/10.1039/C8RA09577A
23. Khan, M., Tahir, M.N., Adil, S.F., Khan, H.U., Siddiqui, M.R.H., Al-warthan, A.A., Tremel, W.: Graphene based metal and metal oxide nanocomposites: synthesis, properties and their applications. J. Mater. Chem. A **3**, 18753–18808 (2015). https://doi.org/10.1039/C5TA02240A
24. Bilisik, K., Akter, M.: Graphene nanocomposites: a review on processes, properties, and applications. J. Ind. Text. **51**, 3718S-3766S (2022). https://doi.org/10.1177/15280837211024252
25. Feng, S.H., Li, G.H.: Chapter 4 - Hydrothermal and Solvothermal Syntheses. In: Xu, R., Xu, Y. (eds.) Modern Inorganic Synthetic Chemistry (Second Edition), pp. 73–104. Elsevier, Amsterdam (2017)
26. Genix, A.-C., Oberdisse, J.: Nanoparticle self-assembly: from interactions in suspension to polymer nanocomposites. Soft Matter **14**, 5161–5179 (2018). https://doi.org/10.1039/C8SM00430G
27. Qin, W., Chen, T., Pan, L., Niu, L., Hu, B., Li, D., Li, J., Sun, Z.: MoS2-reduced graphene oxide composites via microwave assisted synthesis for sodium ion battery anode with improved capacity and cycling performance. Electrochim. Acta **153**, 55–61 (2015). https://doi.org/10.1016/j.electacta.2014.11.034
28. Wang, Y., Sun, W., Li, H.: Microwave-assisted synthesis of graphene nanocomposites: recent developments on lithium-ion batteries. Rep. Electrochem. (2015). https://doi.org/10.2147/rie.S65118
29. Randa, A.-K.: Electrochemical synthesis of nanocomposites. In: Adel, M.A.M., Teresa, D.G. (eds.) Electrodeposition of Composite Materials, IntechOpen, Rijeka p. Ch. 1 (2016)
30. Tariq, W., Ali, F., Arslan, C., Nasir, A., Gillani, S.H., Rehman, A.: Synthesis and applications of graphene and graphene-based nanocomposites: conventional to artificial intelligence approaches. Front. Environ. Chem. **3** (2022). https://doi.org/10.3389/fenvc.2022.890408
31. Jasuja, K., Linn, J., Melton, S., Berry, V.: Microwave-reduced uncapped metal nanoparticles on graphene: tuning catalytic electrical, and Raman properties. The J. Phys. Chem. Lett. **1**, 1853–1860 (2010). https://doi.org/10.1021/jz100580x
32. Wang, Z., Tong, X., Yang, J.-C., Wang, X.-J., Zhang, M.-l., Zhang, G., Long, S.-R., Yang, J.: Improved strength and toughness of semi-aromatic polyamide 6T-co-6(PA6T/6)/GO composites via in situ polymerization. Compos. Sci. Technol. **175**, 6–17 (2019). https://doi.org/10.1016/j.compscitech.2019.02.026

33. Sangili, A., Kalyani, T., Chen, S.-M., Nanda, A., Jana, K.: Label-free electrochemical immunosensor based on one-step electrochemical deposition of AuNP-RGO nanocomposites for detection of endometriosis marker CA 125. ACS Appl. Bio Mater. **3**, 7620–7630 (2020). https://doi.org/10.1021/acsabm.0c00821
34. Yan, J., Fan, Z., Wei, T., Qian, W., Zhang, M., Wei, F.: Fast and reversible surface redox reaction of graphene–MnO2 composites as supercapacitor electrodes. Carbon **48**, 3825–3833 (2010). https://doi.org/10.1016/j.carbon.2010.06.047
35. Wang, L., Wang, D., Dong, X.Y., Zhang, Z.J., Pei, X.F., Chen, X.J., Chen, B., Jin, J.: Layered assembly of graphene oxide and Co–Al layered double hydroxide nanosheets as electrode materials for supercapacitors. Chem. Commun. **47**, 3556–3558 (2011). https://doi.org/10.1039/C0CC05420H
36. Du, D., Liu, J., Zhang, X., Cui, X., Lin, Y.: One-step electrochemical deposition of a graphene-ZrO2 nanocomposite: preparation, characterization and application for detection of organophosphorus agents. J. Mater. Chem. **21**, 8032–8037 (2011). https://doi.org/10.1039/C1JM10696A
37. Gu, Y., Xing, M., Zhang, J.: Synthesis and photocatalytic activity of graphene based doped TiO2 nanocomposites. Appl. Surf. Sci. **319**, 8–15 (2014). https://doi.org/10.1016/j.apsusc.2014.04.182
38. Akhtar, M., Tahir, A., Zulfiqar, S., Hanif, F., Warsi, M.F., Agboola, P.O., Shakir, I.: Ternary hybrid of polyaniline-alanine-reduced graphene oxide for electrochemical sensing of heavy metal ions. Synthetic Metals **265** (2020). https://doi.org/10.1016/j.synthmet.2020.116410
39. Deivanayaki, S., Jayamurugan, P., Ashokan, S., Gopala Krishnan, V., Yogeswari, B., Ubaidullah, M., Pandit, B., Sarma, G.V.S.S., Narsetti, H.K.: Growth of non-enzymatic cholesterol biosensor using TiO2 decorated graphene oxide with bare GCE and PPy-GCE. J. Ind. Chem. Soc. **100** (2023). https://doi.org/10.1016/j.jics.2023.100906
40. Li, L., Zheng, H., Guo, L., Qu, L., Yu, L.: Construction of novel electrochemical sensors based on bimetallic nanoparticle functionalized graphene for determination of sunset yellow in soft drink. J. Electroanal. Chem. **833**, 393–400 (2019). https://doi.org/10.1016/j.jelechem.2018.11.059
41. Zhang, C., Li, L., Ju, J., Chen, W.: Electrochemical sensor based on graphene-supported tin oxide nanoclusters for nonenzymatic detection of hydrogen peroxide. Electrochim. Acta **210**, 181–189 (2016). https://doi.org/10.1016/j.electacta.2016.05.151
42. Kumar, A., Gupta, G.H., Singh, G., More, N., Keerthana, M., Sharma, A., Jawade, D., Balu, A., Kapusetti, G.: Ultrahigh sensitive graphene oxide/conducting polymer composite based biosensor for cholesterol and bilirubin detection. Biosens. Bioelectron.: X **13** (2023). https://doi.org/10.1016/j.biosx.2022.100290
43. Liu, L., Liu, J., Wang, Y., Yan, X., Sun, D.D.: Facile synthesis of monodispersed silver nanoparticles on graphene oxide sheets with enhanced antibacterial activity. New J. Chem. **35**, 1418–1423 (2011). https://doi.org/10.1039/C1NJ20076C
44. Kurt Urhan, B., Demir, Ü., Öznülüer Özer, T., Öztürk Doğan, H.: Electrochemical fabrication of Ni nanoparticles-decorated electrochemically reduced graphene oxide composite electrode for non-enzymatic glucose detection. Thin Solid Films **693** (2020). https://doi.org/10.1016/j.tsf.2019.137695
45. Alsulami, Q.A., Alharbi, L.M., Keshk, S.M.A.S.: Synthesis of a graphene oxide/ZnFe2O4/polyaniline nanocomposite and its structural and electrochemical characterization for supercapacitor application. Int. J. Energy Res. **46**, 2438–2445 (2022). https://doi.org/10.1002/er.7318
46. Xie, S., Deng, L., Huang, H., Yuan, J., Xu, J., Yue, R.: One-pot synthesis of porous Pd-polypyrrole/nitrogen-doped graphene nanocomposite as highly efficient catalyst for electrooxidation of alcohols. J. Colloid Interface Sci. **608**, 3130–3140 (2022). https://doi.org/10.1016/j.jcis.2021.11.039
47. Li, Y., Zhang, S., Dai, H., Hong, Z., Lin, Y.: An enzyme-free photoelectrochemical sensing of concanavalin A based on graphene-supported TiO2 mesocrystal. Sens. Actuators, B Chem. **232**, 226–233 (2016). https://doi.org/10.1016/j.snb.2016.03.143

48. Shahid, M.M., Rameshkumar, P., Basirunc, W.J., Wijayantha, U., Chiu, W.S., Khiew, P.S., Huang, N.M.: An electrochemical sensing platform of cobalt oxide@gold nanocubes interleaved reduced graphene oxide for the selective determination of hydrazine. Electrochim. Acta **259**, 606–616 (2018). https://doi.org/10.1016/j.electacta.2017.10.157
49. Machhindra, L.A., Yen, Y.-K.: A highly sensitive electrochemical sensor for cd2+ detection based on prussian blue-PEDOT-loaded laser-scribed graphene-modified glassy carbon electrode. Chemosensors **10**, (2022). https://doi.org/10.3390/chemosensors10060209
50. Rana, D.S., Kalia, S., Kumar, R., Thakur, N., Singh, D., Singh, K.: Microwave-assisted facile synthesis of layered reduced graphene oxide-tungsten disulfide sandwiched Fe3O4 nanocomposite as effective and sensitive sensor for detection of dopamine. Mater. Chem. Phys. **287**, 126283 (2022). https://doi.org/10.1016/j.matchemphys.2022.126283
51. Jiang, M., Zhang, M., Wang, L., Fei, Y., Wang, S., Núñez-Delgado, A., Bokhari, A., Race, M., Khataee, A., Jaromír Klemeš, J., Xing, L., Han, N.: Photocatalytic degradation of xanthate in flotation plant tailings by TiO2/graphene nanocomposites. Chem. Eng. J. **431**, 134104 (2022). https://doi.org/10.1016/j.cej.2021.134104
52. Liu, X., Pan, L., Lv, T., Zhu, G., Lu, T., Sun, Z., Sun, C.: Microwave-assisted synthesis of TiO2-reduced graphene oxide composites for the photocatalytic reduction of Cr(vi). RSC Adv. **1**, 1245–1249 (2011). https://doi.org/10.1039/C1RA00298H
53. Yang, M., Chen, Y., Wang, H., Zou, Y., Wu, P., Zou, J., Jiang, J.: Solvothermal preparation of CeO2 nanoparticles–graphene nanocomposites as an electrochemical sensor for sensitive detecting pentachlorophenol. Carbon Lett. **32**, 1277–1285 (2022). https://doi.org/10.1007/s42823-022-00353-7
54. Hassan, M., Jiang, Y., Bo, X., Zhou, M.: Sensitive nonenzymatic detection of hydrogen peroxide at nitrogen-doped graphene supported-CoFe nanoparticles. Talanta **188**, 339–348 (2018). https://doi.org/10.1016/j.talanta.2018.06.003
55. Zhao, L., Li, H., Gao, S., Li, M., Xu, S., Li, C., Guo, W., Qu, C., Yang, B.: MgO nanobelt-modified graphene-tantalum wire electrode for the simultaneous determination of ascorbic acid, dopamine and uric acid. Electrochim. Acta **168**, 191–198 (2015). https://doi.org/10.1016/j.electacta.2015.03.215
56. Jee, C.E., Chow, M.K., Yeap, S.P.: Fabrication of flexible carbon dioxide gas sensor with conductive polymer/reduced graphene oxide hybrids: Effects of substrate type and mass ratio. J. Appl. Polym. Sci. **140**, e53415 (2023). https://doi.org/10.1002/app.53415
57. Fang, F., Ran, S., Fang, Z., Song, P., Wang, H.: Improved flame resistance and thermomechanical properties of epoxy resin nanocomposites from functionalized graphene oxide via self-assembly in water. Compos. B Eng. **165**, 406–416 (2019). https://doi.org/10.1016/j.compositesb.2019.01.086
58. Awan, M., Nasir, M., Iqbal, S., Manshad, K., Hayat, A., Ajab, H.: Fabrication of novel electroactive nickel sulfide@graphene oxide nanocomposite integrated transduction platform for non-enzymatic electrochemical sensing of hydrogen Peroxide in environmental and biological samples. Synthetic Metals **292** (2023). https://doi.org/10.1016/j.synthmet.2022.117240
59. Anuar, N.S., Basirun, W.J., Ladan, M., Shalauddin, M., Mehmood, M.S.: Fabrication of platinum nitrogen-doped graphene nanocomposite modified electrode for the electrochemical detection of acetaminophen. Sens. Actuators, B Chem. **266**, 375–383 (2018). https://doi.org/10.1016/j.snb.2018.03.138
60. Zhao, L., Wang, Y., Zhao, X., Deng, Y., Li, Q., Xia, Y.: Green preparation of Ag-Au bimetallic nanoparticles supported on graphene with alginate for non-enzymatic hydrogen peroxide detection. Nanomaterials **8**, 507 (2018)
61. Pifferi, V., Testolin, A., Ingrosso, C., Curri, M.L., Palchetti, I., Falciola, L.: Au nanoparticles decorated graphene-based hybrid nanocomposite for As(III) electroanalytical detection. Chemosensors **10** (2022). https://doi.org/10.3390/chemosensors10020067
62. Ganash, A.A., Alotaibi, M.M.: An electrochemical sensor for cadmium ion detection that is based on an AuNP-functionalized metal–organic framework-graphene nanocomposite. J. Mater. Res. **37**, 4368–4380 (2022). https://doi.org/10.1557/s43578-022-00808-7

63. Alencar, L.M., Silva, A.W.B.N., Trindade, M.A.G., Salvatierra, R.V., Martins, C.A., Souza, V.H.R.: One-step synthesis of crumpled graphene fully decorated by copper-based nanoparticles: application in H2O2 sensing. Sens. Actuators B: Chem. **360** (2022). https://doi.org/10.1016/j.snb.2022.131649
64. Durai, L., Badhulika, S.: One-pot synthesis of rGO supported Nb2O5 nanospheres for ultra-selective sensing of bisphenol A and hydrazine in water samples. IEEE Sens. J. **21**, 4152–4159 (2021). https://doi.org/10.1109/jsen.2020.3032028
65. Shen, H., Liu, J., Pan, P., Yang, X., Yang, Z., Li, P., Liu, G., Zhang, X., Zhou, J.: One-step synthesis of nanosilver embedding laser-induced graphene for H(2)O(2) sensor. Synth. Met. **293**, 117235 (2023). https://doi.org/10.1016/j.synthmet.2022.117235
66. Nagajyoti, P.C., Lee, K.D., Sreekanth, T.V.M.: Heavy metals, occurrence and toxicity for plants: a review. Environ. Chem. Lett. **8**, 199–216 (2010). https://doi.org/10.1007/s10311-010-0297-8
67. Lindqvist, O.: Environmental impact of mercury and other heavy metals. J. Power Sources **57**, 3–7 (1995). https://doi.org/10.1016/0378-7753(95)02229-5
68. Aragay, G., Pons, J., Merkoçi, A.: Recent trends in macro-, micro-, and nanomaterial-based tools and strategies for heavy-metal detection. Chem. Rev. **111**, 3433–3458 (2011). https://doi.org/10.1021/cr100383r
69. Xu, X., Zeng, J., Wu, Y., Wang, Q., Wu, S., Gu, H.: Preparation and application of graphene-based materials for heavy metal removal in tobacco industry: a review. Separations **9**, 401 (2022)
70. Vilela, D., Parmar, J., Zeng, Y., Zhao, Y., Sánchez, S.: Graphene-based microbots for toxic heavy metal removal and recovery from water. Nano Lett. **16**, 2860–2866 (2016). https://doi.org/10.1021/acs.nanolett.6b00768
71. Benjamin, S.R., Miranda Ribeiro Júnior, E.J.: Graphene-based electrochemical sensors for detection of environmental pollutants. Curr. Opin. Environ. Sci. Health **29**, 100381 (2022). https://doi.org/10.1016/j.coesh.2022.100381
72. Shan, Q., Tian, J., Ding, Q., Wu, W.: Electrochemical sensor based on metal-free materials composed of graphene and graphene oxide for sensitive detection of cadmium ions in water. Mater. Chem. Phys. **284** (2022). https://doi.org/10.1016/j.matchemphys.2022.126064
73. Baghayeri, M., Ghanei-Motlagh, M., Tayebee, R., Fayazi, M., Narenji, F.: Application of graphene/zinc-based metal-organic framework nanocomposite for electrochemical sensing of As(III) in water resources. Anal. Chim. Acta **1099**, 60–67 (2020). https://doi.org/10.1016/j.aca.2019.11.045
74. Sang, S., Li, D., Zhang, H., Sun, Y., Jian, A., Zhang, Q., Zhang, W.: Facile synthesis of AgNPs on reduced graphene oxide for highly sensitive simultaneous detection of heavy metal ions. RSC Adv. **7**, 21618–21624 (2017). https://doi.org/10.1039/c7ra02267k
75. Sahoo, S., Sahoo, P.K., Satpati, A.K.: Gold nano particle and reduced graphene oxide composite modified carbon paste electrode for the ultra trace detection of arsenic (III). Electroanalysis **29**, 1400–1409 (2017). https://doi.org/10.1002/elan.201600676
76. Zhu, Y., Pan, D., Hu, X., Han, H., Lin, M., Wang, C.: An electrochemical sensor based on reduced graphene oxide/gold nanoparticles modified electrode for determination of iron in coastal waters. Sens. Actuators, B Chem. **243**, 1–7 (2017). https://doi.org/10.1016/j.snb.2016.11.108
77. Rahman, M.T., Kabir, M.F., Gurung, A., Reza, K.M., Pathak, R., Ghimire, N., Baride, A., Wang, Z., Kumar, M., Qiao, Q.: Graphene oxide-silver nanowire nanocomposites for enhanced sensing of Hg2+. ACS Appl. Nano Mater. **2**, 4842–4851 (2019). https://doi.org/10.1021/acsanm.9b00789
78. Han, H., Pan, D., Wang, C., Zhu, R.: Controlled synthesis of dendritic gold nanostructures by graphene oxide and their morphology-dependent performance for iron detection in coastal waters. RSC Adv. **7**, 15833–15841 (2017). https://doi.org/10.1039/c6ra27075a
79. Liu, M., Pan, D., Pan, W., Zhu, Y., Hu, X., Han, H., Wang, C., Shen, D.: In-situ synthesis of reduced graphene oxide/gold nanoparticles modified electrode for speciation analysis of copper in seawater. Talanta **174**, 500–506 (2017). https://doi.org/10.1016/j.talanta.2017.06.054

80. Chimezie, A.B., Hajian, R., Yusof, N.A., Woi, P.M., Shams, N.: Fabrication of reduced graphene oxide-magnetic nanocomposite (rGO-Fe 3 O 4) as an electrochemical sensor for trace determination of As(III) in water resources. J. Electroanal. Chem. **796**, 33–42 (2017). https://doi.org/10.1016/j.jelechem.2017.04.061
81. Devi, P., Sharma, C., Kumar, P., Kumar, M., Bansod, B.K.S., Nayak, M.K., Singla, M.L.: Selective electrochemical sensing for arsenite using rGO/Fe(3)O(4) nanocomposites. J. Hazard. Mater. **322**, 85–94 (2017). https://doi.org/10.1016/j.jhazmat.2016.02.066
82. Zhang, W., Xu, Y., Zou, X.: A ZnO–RGO-modified electrode coupled to microwave digestion for the determination of trace cadmium and lead in six species fish. Anal. Methods **9**, 4418–4424 (2017). https://doi.org/10.1039/c7ay01574g
83. Zhou, S.-F., Han, X.-J., Fan, H.-L., Huang, J., Liu, Y.-Q.: Enhanced electrochemical performance for sensing Pb(II) based on graphene oxide incorporated mesoporous MnFe2O4 nanocomposites. J. Alloy. Compd. **747**, 447–454 (2018). https://doi.org/10.1016/j.jallcom.2018.03.037
84. Xu, J., Li, Z., Yue, X., Xie, F., Xiong, S.: Electrochemical detection of Cu(ii) using amino-functionalized MgFe2O4/Reduced graphene oxide composite. Anal. Methods **10**, 2026–2033 (2018). https://doi.org/10.1039/c8ay00452h
85. Karthik, R., Thambidurai, S.: Synthesis of cobalt doped ZnO/reduced graphene oxide nanorods as active material for heavy metal ions sensor and antibacterial activity. J. Alloy. Compd. **715**, 254–265 (2017). https://doi.org/10.1016/j.jallcom.2017.04.298
86. Mejri, A., Mars, A., Elfil, H., Hamzaoui, A.H.: Graphene nanosheets modified with curcumin-decorated manganese dioxide for ultrasensitive potentiometric sensing of mercury(II), fluoride and cyanide. Mikrochim. Acta **185**, 529 (2018). https://doi.org/10.1007/s00604-018-3064-3
87. Zuo, Y., Xu, J., Jiang, F., Duan, X., Lu, L., Xing, H., Yang, T., Zhang, Y., Ye, G., Yu, Y.: Voltammetric sensing of Pb(II) using a glassy carbon electrode modified with composites consisting of Co3O4 nanoparticles, reduced graphene oxide and chitosan. J. Electroanal. Chem. **801**, 146–152 (2017). https://doi.org/10.1016/j.jelechem.2017.07.046
88. Dahaghin, Z., Kilmartin, P.A., Mousavi, H.Z.: Simultaneous determination of lead(II) and cadmium(II) at a glassy carbon electrode modified with GO@Fe3O4 @benzothiazole-2-carboxaldehyde using square wave anodic stripping voltammetry. J. Mol. Liq. **249**, 1125–1132 (2018). https://doi.org/10.1016/j.molliq.2017.11.114
89. Wang, Y., Zhao, G., Zhang, G., Zhang, Y., Wang, H., Cao, W., Li, T., Wei, Q.: An electrochemical aptasensor based on gold-modified MoS2/rGO nanocomposite and gold-palladium-modified Fe-MOFs for sensitive detection of lead ions. Sens. Actuators B: Chem. **319** (2020). https://doi.org/10.1016/j.snb.2020.128313
90. Priya, T., Dhanalakshmi, N., Thennarasu, S., Karthikeyan, V., Thinakaran, N.: Ultra sensitive electrochemical detection of Cd2+ and Pb2+ using penetrable nature of graphene/gold nanoparticles/modified L-cysteine nanocomposite. Chem. Phys. Lett. **731** (2019). https://doi.org/10.1016/j.cplett.2019.136621
91. Ru, J., Wang, X., Zhao, J., Yang, J., Zhou, Z., Du, X., Lu, X.: Evaluation and development of GO/UiO-67@PtNPs nanohybrid-based electrochemical sensor for invisible arsenic (III) in water samples. Microchem. J. **181** (2022). https://doi.org/10.1016/j.microc.2022.107765
92. Muralikrishna, S., Nagaraju, D.H., Balakrishna, R.G., Surareungchai, W., Ramakrishnappa, T., Shivanandareddy, A.B.: Hydrogels of polyaniline with graphene oxide for highly sensitive electrochemical determination of lead ions. Anal. Chim. Acta **990**, 67–77 (2017). https://doi.org/10.1016/j.aca.2017.09.008
93. Jayaraman, N., Palani, Y., Jonnalagadda, R.R., Shanmugam, E.: Covalently dual functionalized graphene oxide-based multiplex electrochemical sensor for Hg(II) and Cr(VI) detection. Sens. Actuators B: Chem. **367** (2022). https://doi.org/10.1016/j.snb.2022.132165
94. Wei, P., Zhu, Z., Song, R., Li, Z., Chen, C.: An ion-imprinted sensor based on chitosan-graphene oxide composite polymer modified glassy carbon electrode for environmental sensing application. Electrochim. Acta **317**, 93–101 (2019). https://doi.org/10.1016/j.electacta.2019.05.136

95. Duoc, P.N.D., Binh, N.H., Hau, T.V., Thanh, C.T., Trinh, P.V., Tuyen, N.V., Quynh, N.V., Tu, N.V., Duc Chinh, V., Thi Thu, V., Thang, P.D., Minh, P.N., Chuc, N.V.: A novel electrochemical sensor based on double-walled carbon nanotubes and graphene hybrid thin film for arsenic(V) detection. J. Hazard Mater. **400**, 123185 (2020). https://doi.org/10.1016/j.jhazmat.2020.123185
96. Wang, L., Peng, X., Fu, H.: An electrochemical aptasensor for the sensitive detection of Pb2+ based on a chitosan/reduced graphene oxide/titanium dioxide. Microchem. J. **174** (2022). https://doi.org/10.1016/j.microc.2021.106977
97. Huang, Z., Song, H., Feng, L., Qin, J., Wang, Q., Guo, B., Wei, L., Lu, Y., Guo, H., Zhu, D., Ma, X., Guo, Y., Zheng, H., Li, M., Su, Z.: A novel ultrasensitive electrochemical sensor based on a hybrid of rGO/MWCNT/AuNP for the determination of lead(II) in tea drinks. Microchem. J. **186** (2023). https://doi.org/10.1016/j.microc.2022.108346
98. Chalupniak, A., Merkoci, A.: Graphene oxide-poly(dimethylsiloxane)-based lab-on-a-chip platform for heavy metals preconcentration and electrochemical detection. ACS Appl. Mater. Interfaces **9**, 44766–44775 (2017). https://doi.org/10.1021/acsami.7b12368
99. Guo, Z., Li, D.D., Luo, X.K., Li, Y.H., Zhao, Q.N., Li, M.M., Zhao, Y.T., Sun, T.S., Ma, C.: Simultaneous determination of trace Cd(II), Pb(II) and Cu(II) by differential pulse anodic stripping voltammetry using a reduced graphene oxide-chitosan/poly-l-lysine nanocomposite modified glassy carbon electrode. J. Colloid Interface Sci. **490**, 11–22 (2017). https://doi.org/10.1016/j.jcis.2016.11.006
100. Nguyen, L.D., Doan, T.C.D., Huynh, T.M., Nguyen, V.N.P., Dinh, H.H., Dang, D.M.T., Dang, C.M.: An electrochemical sensor based on polyvinyl alcohol/chitosan-thermally reduced graphene composite modified glassy carbon electrode for sensitive voltammetric detection of lead. Sens. Actuators B: Chem. **345** (2021). https://doi.org/10.1016/j.snb.2021.130443
101. Palanisamy, S., Thangavelu, K., Chen, S.-M., Velusamy, V., Chang, M.-H., Chen, T.-W., Al-Hemaid, F.M.A., Ali, M.A., Ramaraj, S.K.: Synthesis and characterization of polypyrrole decorated graphene/β-cyclodextrin composite for low level electrochemical detection of mercury (II) in water. Sens. Actuators, B Chem. **243**, 888–894 (2017). https://doi.org/10.1016/j.snb.2016.12.068
102. Teodoro, K.B.R., Migliorini, F.L., Facure, M.H.M., Correa, D.S.: Conductive electrospun nanofibers containing cellulose nanowhiskers and reduced graphene oxide for the electrochemical detection of mercury(II). Carbohydr. Polym. **207**, 747–754 (2019). https://doi.org/10.1016/j.carbpol.2018.12.022
103. Gumpu, M.B., Veerapandian, M., Krishnan, U.M., Rayappan, J.B.B.: Amperometric determination of As(III) and Cd(II) using a platinum electrode modified with acetylcholinesterase, ruthenium(II)-tris(bipyridine) and graphene oxide. Mikrochim. Acta **185**, 297 (2018). https://doi.org/10.1007/s00604-018-2822-6
104. Cui, X., Yang, B., Zhao, S., Li, X., Qiao, M., Mao, R., Wang, Y., Zhao, X.: Electrochemical sensor based on ZIF-8@dimethylglyoxime and β-cyclodextrin modified reduced graphene oxide for nickel (II) detection. Sens. Actuators B: Chem. **315** (2020). https://doi.org/10.1016/j.snb.2020.128091
105. Ruengpirasiri, P., Punrat, E., Chailapakul, O., Chuanuwatanakul, S.: Graphene oxide-modified electrode coated within-situAntimony film for the simultaneous determination of heavy metals by sequential injection-anodic stripping voltammetry. Electroanalysis **29**, 1022–1030 (2017). https://doi.org/10.1002/elan.201600568
106. Lin, X., Lu, Z., Zhang, Y., Liu, B., Mo, G., Li, J., Ye, J.: A glassy carbon electrode modified with a bismuth film and laser etched graphene for simultaneous voltammetric sensing of Cd(II) and Pb(II). Mikrochim. Acta **185**, 438 (2018). https://doi.org/10.1007/s00604-018-2966-4
107. Lee, C.-S., Yu, S.H., Kim, H.: One-step electrochemical fabrication of reduced graphene oxide/gold nanoparticles nanocomposite-modified electrode for simultaneous detection of dopamine, ascorbic acid, and uric acid. Nanomaterials **8**, 17 (2018)
108. Jane-Llopis, E., Matytsina, I.: Mental health and alcohol, drugs and tobacco: a review of the comorbidity between mental disorders and the use of alcohol, tobacco and illicit drugs. Drug Alcohol Rev. **25**, 515–536 (2006). https://doi.org/10.1080/09595230600944461

109. van Amsterdam, J., Opperhuizen, A., Koeter, M., van den Brink, W.: Ranking the harm of alcohol, tobacco and illicit drugs for the individual and the population. Eur. Addict. Res. **16**, 202–207 (2010). https://doi.org/10.1159/000317249
110. Mohamed, M.A., Atty, S.A., Salama, N.N., Banks, C.E.: Highly selective sensing platform utilizing graphene oxide and multiwalled carbon nanotubes for the sensitive determination of tramadol in the presence of co-formulated drugs. Electroanalysis **29**, 1038–1048 (2017). https://doi.org/10.1002/elan.201600668
111. Ahmed, M.J., Perveen, S., Hussain, S.G., Khan, A.A., Ejaz, S.M.W., Rizvi, S.M.A.: Design of a facile, green and efficient graphene oxide-based electrochemical sensor for analysis of acetaminophen drug. Chem Zvesti **77**, 2275–2294 (2023). https://doi.org/10.1007/s11696-022-02628-9
112. Yeh, S.-H., Huang, M.-S., Huang, C.-H.: Electrochemical sensors for sulfamethoxazole detection based on graphene oxide/graphene layered composite on indium tin oxide substrate. J. Taiwan Inst. Chem. Eng. **131** (2022). https://doi.org/10.1016/j.jtice.2021.11.022
113. Baghayeri, M., Nabavi, S., Hasheminejad, E., Ebrahimi, V.: Introducing an electrochemical sensor based on two layers of ag nanoparticles decorated graphene for rapid determination of methadone in human blood serum. Top. Catal. **65**, 623–632 (2021). https://doi.org/10.1007/s11244-021-01483-4
114. Veera Manohara Reddy, Y., Bathinapatla, S., Łuczak, T., Osińska, M., Maseed, H., Ragavendra, P., Subramanyam Sarma, L., Srikanth, V.V.S.S., Madhavi, G.: An ultra-sensitive electrochemical sensor for the detection of acetaminophen in the presence of etilefrine using bimetallic Pd–Ag/reduced graphene oxide nanocomposites. New J. Chem. **42**, 3137–3146 (2018). https://doi.org/10.1039/c7nj04775d
115. Sebastian, N., Yu, W.C., Hu, Y.C., Balram, D., Yu, Y.H.: Sonochemical synthesis of iron-graphene oxide/honeycomb-like ZnO ternary nanohybrids for sensitive electrochemical detection of antipsychotic drug chlorpromazine. Ultrason. Sonochem. **59**, 104696 (2019). https://doi.org/10.1016/j.ultsonch.2019.104696
116. Atta, N.F., Galal, A., El-Ads, E.H., Hassan, S.H.: Cobalt oxide nanoparticles/graphene/ionic liquid crystal modified carbon paste electrochemical sensor for ultra-sensitive determination of a narcotic drug. Adv. Pharm. Bull. **9**, 110–121 (2019). https://doi.org/10.15171/apb.2019.014
117. Aflatoonian, M.R., Tajik, S., Aflatoonian, B., Beitollahi, H., Zhang, K., Le, Q.V., Cha, J.H., Jang, H.W., Shokouhimehr, M., Peng, W.: A screen-printed electrode modified with graphene/Co(3)O(4) nanocomposite for electrochemical detection of tramadol. Front. Chem. **8**, 562308 (2020). https://doi.org/10.3389/fchem.2020.562308
118. He, Q., Wu, Y., Tian, Y., Li, G., Liu, J., Deng, P., Chen, D.: Facile electrochemical sensor for nanomolar rutin detection based on magnetite nanoparticles and reduced graphene oxide decorated electrode. Nanomaterials (Basel) **9** (2019). https://doi.org/10.3390/nano9010115
119. Mohamed, M.A., Yehia, A.M., Banks, C.E., Allam, N.K.: Novel MWCNTs/graphene oxide/pyrogallol composite with enhanced sensitivity for biosensing applications. Biosens. Bioelectron. **89**, 1034–1041 (2017). https://doi.org/10.1016/j.bios.2016.10.025
120. Hadi, B., Fariba Garkani, N.: Magnetic core–shell graphene oxide/Fe3O4@SiO2 nanocomposite for sensitive and selective electrochemical detection of morphine using modified graphite screen printed electrode. J. Anal. Chem. **75**, 127–134 (2020). https://doi.org/10.1134/s1061934820010049
121. Riahifar, V., Haghnazari, N., Keshavarzi, F., Ahmadi, E.: A sensitive voltammetric sensor for methamphetamine determination based on modified glassy carbon electrode using Fe3O4@poly pyrrole core-shell and graphene oxide. Microchem. J. **170** (2021). https://doi.org/10.1016/j.microc.2021.106748
122. Anvari, L., Ghoreishi, S.M., Faridbod, F., Ganjali, M.R.: Electrochemical determination of methamphetamine in human plasma on a nanoceria nanoparticle decorated reduced graphene oxide (rGO) glassy carbon electrode (GCE). Anal. Lett. **54**, 2509–2522 (2021). https://doi.org/10.1080/00032719.2021.1875229

123. Wang, H., Zhang, S., Li, S., Qu, J.: Electrochemical sensor based on palladium-reduced graphene oxide modified with gold nanoparticles for simultaneous determination of acetaminophen and 4-aminophenol. Talanta **178**, 188–194 (2018). https://doi.org/10.1016/j.talanta.2017.09.021
124. Garima;Sachdev, A., Matai, I.: An electrochemical sensor based on cobalt oxyhydroxide nanoflakes/reduced graphene oxide nanocomposite for detection of illicit drug-clonazepam. J. Electroanalytical Chem. **919** (2022). https://doi.org/10.1016/j.jelechem.2022.116537
125. Silva, M.K.L., Simões, R.P., Leão, A.L., Lapa, R.M.L., Rascón, J., Cesarino, I.: Competitive host-guest electrochemical detection of ivermectin drug using a β-cyclodextrin/graphene-based electrode. Electroanalysis **35** (2022). https://doi.org/10.1002/elan.202100649
126. Sohouli, E., Ghalkhani, M., Rostami, M., Rahimi-Nasrabadi, M., Ahmadi, F.: A noble electrochemical sensor based on TiO(2)@CuO-N-rGO and poly (L-cysteine) nanocomposite applicable for trace analysis of flunitrazepam. Mater. Sci. Eng. C Mater. Biol. Appl. **117**, 111300 (2020). https://doi.org/10.1016/j.msec.2020.111300
127. Baccarin, M., Santos, F.A., Vicentini, F.C., Zucolotto, V., Janegitz, B.C., Fatibello-Filho, O.: Electrochemical sensor based on reduced graphene oxide/carbon black/chitosan composite for the simultaneous determination of dopamine and paracetamol concentrations in urine samples. J. Electroanal. Chem. **799**, 436–443 (2017). https://doi.org/10.1016/j.jelechem.2017.06.052
128. Liu, Z., Jin, M., Cao, J., Wang, J., Wang, X., Zhou, G., van den Berg, A., Shui, L.: High-sensitive electrochemical sensor for determination of Norfloxacin and its metabolism using MWCNT-CPE/pRGO-ANSA/Au. Sens. Actuators, B Chem. **257**, 1065–1075 (2018). https://doi.org/10.1016/j.snb.2017.11.052
129. Kumary, V.A., Abraham, P., Renjini, S., Swamy, B.E.K., Nancy, T.E.M., Sreevalsan, A.: A novel heterogeneous catalyst based on reduced graphene oxide supported copper coordinated amino acid—a platform for morphine sensing. J. Electroanalytical Chem. **850** (2019). https://doi.org/10.1016/j.jelechem.2019.113367
130. Abraham, P., Renjini, S., Nancy, T.E.M., Kumary, V.A.: Electrochemical synthesis of thin-layered graphene oxide-poly(CTAB) composite for detection of morphine. J. Appl. Electrochem. **50**, 41–50 (2019). https://doi.org/10.1007/s10800-019-01367-2
131. Rokhsefid, N., Shishehbore, M.R.: Synthesis and characterization of an Au nanoparticles/graphene nanosheet nanocomposite and its application for the simultaneous determination of tramadol and acetaminophen. Anal. Methods **11**, 5150–5159 (2019). https://doi.org/10.1039/c9ay01497g
132. Nazari, Z., Es' haghi, Z.: A new electrochemical sensor for the simultaneous detection of morphine and methadone based on thioglycolic acid decorated CdSe doped graphene oxide multilayers. Anal. Bioanalytical Electrochem. **14**, 228–245 (2022)
133. Florea, A., Cowen, T., Piletsky, S., De Wael, K.: Electrochemical sensing of cocaine in real samples based on electrodeposited biomimetic affinity ligands. Analyst **144**, 4639–4646 (2019). https://doi.org/10.1039/c9an00618d
134. Bagheri, H., Shirzadmehr, A., Rezaei, M., Khoshsafar, H.: Determination of tramadol in pharmaceutical products and biological samples using a new nanocomposite carbon paste sensor based on decorated nanographene/tramadol-imprinted polymer nanoparticles/ionic liquid. Ionics **24**, 833–843 (2017). https://doi.org/10.1007/s11581-017-2252-1
135. Dehghani, M., Nasirizadeh, N., Yazdanshenas, M.E.: Determination of cefixime using a novel electrochemical sensor produced with gold nanowires/graphene oxide/electropolymerized molecular imprinted polymer. Mater. Sci. Eng. C Mater. Biol. Appl. **96**, 654–660 (2019). https://doi.org/10.1016/j.msec.2018.12.002
136. Fu, K., Zhang, R., He, J., Bai, H., Zhang, G.: Sensitive detection of ketamine with an electrochemical sensor based on UV-induced polymerized molecularly imprinted membranes at graphene and MOFs modified electrode. Biosens. Bioelectron. **143**, 111636 (2019). https://doi.org/10.1016/j.bios.2019.111636
137. Prakash, V., Prabhakar, N., Kaur, G., Diwan, A., Mehta, S.K., Sharma, S.: Structural insight on thiourea doped graphene: An efficient electrochemical sensor for voltammetric detection of morphine in alcoholic and non-alcoholic beverages. Curr. Res. Green Sustain. Chem. **5** (2022). https://doi.org/10.1016/j.crgsc.2022.100267

138. Mohamed, M.A., El-Gendy, D.M., Ahmed, N., Banks, C.E., Allam, N.K.: 3D spongy graphene-modified screen-printed sensors for the voltammetric determination of the narcotic drug codeine. Biosens. Bioelectron. **101**, 90–95 (2018). https://doi.org/10.1016/j.bios.2017.10.020
139. Magdy, N., Sobaih, A.E., Hussein, L.A., Mahmoud, A.M.: Graphene-based disposable electrochemical sensor for chlorhexidine determination. Electroanalysis **35** (2022). https://doi.org/10.1002/elan.202200119
140. Rocha, R.G., Ribeiro, J.S., Santana, M.H.P., Richter, E.M., Munoz, R.A.A.: 3D-printing for forensic chemistry: voltammetric determination of cocaine on additively manufactured graphene-polylactic acid electrodes. Anal. Methods **13**, 1788–1794 (2021). https://doi.org/10.1039/d1ay00181g
141. Saisahas, K., Soleh, A., Somsiri, S., Senglan, P., Promsuwan, K., Saichanapan, J., Kanatharana, P., Thavarungkul, P., Lee, K., Chang, K.H., Abdullah, A.F.L., Tayayuth, K., Limbut, W.: Electrochemical sensor for methamphetamine detection using laser-induced porous graphene electrode. Nanomaterials (Basel) **12** (2021). https://doi.org/10.3390/nano12010073
142. Thomas, R., Balachandran, M.: Heteroatom engineered graphene-based electrochemical assay for the quantification of high-risk abused drug oxytocin in edibles and biological samples. Food Chem. **400**, 134106 (2023). https://doi.org/10.1016/j.foodchem.2022.134106
143. Abd-Rabboh, H.S.M., El-Galil, E.A.A., Sayed, A.Y.A., Kamel, A.H.: Paper-based potentiometric sensing devices modified with chemically reduced graphene oxide (CRGO) for trace level determination of pholcodine (opiate derivative drug). RSC Adv. **11**, 12227–12234 (2021). https://doi.org/10.1039/d1ra00581b
144. Narang, J., Malhotra, N., Singhal, C., Mathur, A., Chakraborty, D., Anil, A., Ingle, A., Pundir, C.S.: Point of care with micro fluidic paper based device integrated with nano zeolite-graphene oxide nanoflakes for electrochemical sensing of ketamine. Biosens. Bioelectron. **88**, 249–257 (2017). https://doi.org/10.1016/j.bios.2016.08.043
145. Aktar, M.W., Sengupta, D., Chowdhury, A.: Impact of pesticides use in agriculture: their benefits and hazards. Interdiscip. Toxicol. **2**, 1–12 (2009). https://doi.org/10.2478/v10102-009-0001-7
146. Wang, W., Wang, X., Cheng, N., Luo, Y., Lin, Y., Xu, W., Du, D.: Recent advances in nanomaterials-based electrochemical (bio)sensors for pesticides detection. TrAC, Trends Anal. Chem. **132**, 116041 (2020). https://doi.org/10.1016/j.trac.2020.116041
147. Wei, P., Gan, T., Wu, K.: N-methyl-2-pyrrolidone exfoliated graphene as highly sensitive analytical platform for carbendazim. Sens. Actuators, B Chem. **274**, 551–559 (2018). https://doi.org/10.1016/j.snb.2018.07.174
148. Saritha Rani, K., Mutyala, S., Reddy, S.M.: Electrocatalytic oxidation of hydrazine at sulphur-doped graphene-modified glassy carbon electrode. Bull. Mater. Sci. **44** (2021). https://doi.org/10.1007/s12034-021-02498-z
149. Jahromi, M.N., Tayadon, F., Bagheri, H.: A new electrochemical sensor based on an Au-Pd/reduced graphene oxide nanocomposite for determination of Parathion. Int. J. Environ. Anal. Chem. **100**, 1101–1117 (2019). https://doi.org/10.1080/03067319.2019.1649398
150. Amin, H.M.A., El-Kady, M.F., Atta, N.F., Galal, A.: Gold nanoparticles decorated graphene as a high performance sensor for determination of trace hydrazine levels in water. Electroanalysis **30**, 1757–1766 (2018). https://doi.org/10.1002/elan.201800125
151. Yola, M.L.: Electrochemical activity enhancement of monodisperse boron nitride quantum dots on graphene oxide: its application for simultaneous detection of organophosphate pesticides in real samples. J. Mol. Liq. **277**, 50–57 (2019). https://doi.org/10.1016/j.molliq.2018.12.084
152. Rajaji, U., Murugan, K., Chen, S.-M., Govindasamy, M., Chen, T.-W., Lin, P.H., Lakshmi Prabha, P.: Graphene oxide encapsulated 3D porous chalcopyrite (CuFeS2) nanocomposite as an emerging electrocatalyst for agro-hazardous (methyl paraoxon) detection in vegetables. Compos. Part B: Eng. **160**, 268–276 (2019). https://doi.org/10.1016/j.compositesb.2018.10.042

153. Velusamy, V., Palanisamy, S., Chen, S.W., Balu, S., Yang, T.C.K., Banks, C.E.: Novel electrochemical synthesis of cellulose microfiber entrapped reduced graphene oxide: a sensitive electrochemical assay for detection of fenitrothion organophosphorus pesticide. Talanta **192**, 471–477 (2019). https://doi.org/10.1016/j.talanta.2018.09.055
154. Ensafi, A.A., Noroozi, R., Zandi—Atashbar, N., Rezaei, B.: Cerium(IV) oxide decorated on reduced graphene oxide, a selective and sensitive electrochemical sensor for fenitrothion determination. Sens. Actuators B: Chem. **245**, 980–987 (2017). https://doi.org/10.1016/j.snb.2017.01.186
155. Aghaie, A., Khanmohammadi, A., Hajian, A., Schmid, U., Bagheri, H.: Nonenzymatic electrochemical determination of paraoxon ethyl in water and fruits by graphene-based NiFe bimetallic phosphosulfide nanocomposite as a superior sensing layer. Food Anal. Methods **12**, 1545–1555 (2019). https://doi.org/10.1007/s12161-019-01486-8
156. Özcan, A., Topçuoğulları, D., Atılır Özcan, A.: Fenitrothion sensing with reduced graphene oxide decorated fumed silica nanocomposite modified glassy carbon electrode. Sens. Actuators, B Chem. **284**, 179–185 (2019). https://doi.org/10.1016/j.snb.2018.12.122
157. Kokulnathan, T., Wang, T.J., Duraisamy, N., Kumar, E.A., An Ni, S.: Hierarchical nanoarchitecture of zirconium phosphate/graphene oxide: robust electrochemical platform for detection of fenitrothion. J. Hazard Mater. **412**, 125257 (2021). https://doi.org/10.1016/j.jhazmat.2021.125257
158. Karimian, N., Fakhri, H., Amidi, S., Hajian, A., Arduini, F., Bagheri, H.: A novel sensing layer based on metal–organic framework UiO-66 modified with TiO2–graphene oxide: application to rapid, sensitive and simultaneous determination of paraoxon and chlorpyrifos. New J. Chem. **43**, 2600–2609 (2019). https://doi.org/10.1039/c8nj06208k
159. Yan, L., Yan, X., Li, H., Zhang, X., Wang, M., Fu, S., Zhang, G., Qian, C., Yang, H., Han, J., Xiao, F.: Reduced graphene oxide nanosheets and gold nanoparticles covalently linked to ferrocene-terminated dendrimer to construct electrochemical sensor with dual signal amplification strategy for ultra-sensitive detection of pesticide in vegetable. Microchemical J. **157** (2020). https://doi.org/10.1016/j.microc.2020.105016
160. Khalifa, M.E., Abdallah, A.B.: Molecular imprinted polymer based sensor for recognition and determination of profenofos organophosphorous insecticide. Biosens. Bioelectron.: X **2** (2019). https://doi.org/10.1016/j.biosx.2019.100027
161. Zhang, J., Hu, H., Wang, P., Zhang, C., Wu, M., Yang, L.: A stable biosensor for organophosphorus pesticide detection based on chitosan modified graphene. Biotechnol. Appl. Biochem. **69**, 567–575 (2022). https://doi.org/10.1002/bab.2133
162. Thet Tun, W.S., Saenchoopa, A., Daduang, S., Daduang, J., Kulchat, S., Patramanon, R.: Electrochemical biosensor based on cellulose nanofibers/graphene oxide and acetylcholinesterase for the detection of chlorpyrifos pesticide in water and fruit juice. RSC Adv. **13**, 9603–9614 (2023). https://doi.org/10.1039/d3ra00512g
163. Itsoponpan, T., Thanachayanont, C., Hasin, P.: Sponge-like CuInS2 microspheres on reduced graphene oxide as an electrocatalyst to construct an immobilized acetylcholinesterase electrochemical biosensor for chlorpyrifos detection in vegetables. Sens. Actuators B: Chem. **337** (2021). https://doi.org/10.1016/j.snb.2021.129775
164. Bao, J., Huang, T., Wang, Z., Yang, H., Geng, X., Xu, G., Samalo, M., Sakinati, M., Huo, D., Hou, C.: 3D graphene/copper oxide nano-flowers based acetylcholinesterase biosensor for sensitive detection of organophosphate pesticides. Sens. Actuators, B Chem. **279**, 95–101 (2019). https://doi.org/10.1016/j.snb.2018.09.118
165. Wang, B., Li, Y., Hu, H., Shu, W., Yang, L., Zhang, J.: Acetylcholinesterase electrochemical biosensors with graphene-transition metal carbides nanocomposites modified for detection of organophosphate pesticides. PLoS ONE **15**, e0231981 (2020). https://doi.org/10.1371/journal.pone.0231981
166. Zhang, P., Sun, T., Rong, S., Zeng, D., Yu, H., Zhang, Z., Chang, D., Pan, H.: A sensitive amperometric AChE-biosensor for organophosphate pesticides detection based on conjugated polymer and Ag-rGO-NH(2) nanocomposite. Bioelectrochemistry **127**, 163–170 (2019). https://doi.org/10.1016/j.bioelechem.2019.02.003

167. Miyazaki, C.M., Adriano, A.M., Rubira, R.J.G., Constantino, C.J.L., Ferreira, M.: Combining electrochemically reduced graphene oxide and Layer-by-Layer films of magnetite nanoparticles for carbofuran detection. J. Environ. Chem. Eng. **8** (2020). https://doi.org/10.1016/j.jece.2020.104294
168. Wang, W., Zhao, Z., Yang, H., Li, P., Yu, Z., Zhang, W., Shi, J., Hu, J., Chen, Y.: Synthesis and electrochemical performance of gold nanoparticles deposited onto a reduced graphene oxide/nickel foam hybrid structure for hydrazine detection. J. Mater. Sci. **55**, 9470–9482 (2020). https://doi.org/10.1007/s10853-020-04684-6
169. Qi, P., Wang, J., Wang, X., Wang, X., Wang, Z., Xu, H., Di, S., Wang, Q., Wang, X.: Sensitive determination of fenitrothion in water samples based on an electrochemical sensor layered reduced graphene oxide, molybdenum sulfide (MoS2)-Au and zirconia films. Electrochim. Acta **292**, 667–675 (2018). https://doi.org/10.1016/j.electacta.2018.09.187
170. Ding, L., Wei, J., Qiu, Y., Wang, Y., Wen, Z., Qian, J., Hao, N., Ding, C., Li, Y., Wang, K.: One-step hydrothermal synthesis of telluride molybdenum/reduced graphene oxide with Schottky barrier for fabricating label-free photoelectrochemical profenofos aptasensor. Chem. Eng. J. **407** (2021). https://doi.org/10.1016/j.cej.2020.127213
171. Zhang, R., Chen, W.: Recent advances in graphene-based nanomaterials for fabricating electrochemical hydrogen peroxide sensors. Biosens. Bioelectron. **89**, 249–268 (2017). https://doi.org/10.1016/j.bios.2016.01.080
172. Giorgio, M., Trinei, M., Migliaccio, E., Pelicci, P.G.: Hydrogen peroxide: a metabolic by-product or a common mediator of ageing signals? Nat. Rev. Mol. Cell Biol. **8**, 722–728 (2007). https://doi.org/10.1038/nrm2240
173. Lippert, A.R., Van de Bittner, G.C., Chang, C.J.: Boronate oxidation as a bioorthogonal reaction approach for studying the chemistry of hydrogen peroxide in living systems. Acc. Chem. Res. **44**, 793–804 (2011). https://doi.org/10.1021/ar200126t
174. Winterbourn, C.C.: Reconciling the chemistry and biology of reactive oxygen species. Nat. Chem. Biol. **4**, 278–286 (2008). https://doi.org/10.1038/nchembio.85
175. Zhong, L., Gan, S., Fu, X., Li, F., Han, D., Guo, L., Niu, L.: Electrochemically controlled growth of silver nanocrystals on graphene thin film and applications for efficient nonenzymatic H2O2 biosensor. Electrochim. Acta **89**, 222–228 (2013). https://doi.org/10.1016/j.electacta.2012.10.161
176. Zhang, P., Zhang, X., Zhang, S., Lu, X., Li, Q., Su, Z., Wei, G.: One-pot green synthesis, characterizations, and biosensor application of self-assembled reduced graphene oxide-gold nanoparticle hybrid membranes. J. Mater. Chem. B **1**, 6525–6531 (2013). https://doi.org/10.1039/c3tb21270j
177. Mahmoudian, M.R., Alias, Y., Basirun, W.J., Woi, P.M., Sookhakian, M.: Facile preparation of MnO2 nanotubes/reduced graphene oxide nanocomposite for electrochemical sensing of hydrogen peroxide. Sens. Actuators, B Chem. **201**, 526–534 (2014). https://doi.org/10.1016/j.snb.2014.05.030
178. Zhang, Y., Wang, Z., Ji, Y., Liu, S., Zhang, T.: Synthesis of Ag nanoparticle–carbon nanotube–reduced graphene oxide hybrids for highly sensitive non-enzymatic hydrogen peroxide detection. RSC Adv. **5**, 39037–39041 (2015). https://doi.org/10.1039/c5ra04246a
179. Liu, J., Bo, X., Zhao, Z., Guo, L.: Highly exposed Pt nanoparticles supported on porous graphene for electrochemical detection of hydrogen peroxide in living cells. Biosens. Bioelectron. **74**, 71–77 (2015). https://doi.org/10.1016/j.bios.2015.06.042
180. Zhang, M., Wang, Z.: Nanostructured silver nanowires-graphene hybrids for enhanced electrochemical detection of hydrogen peroxide. Appl. Phys. Lett. **102** (2013). https://doi.org/10.1063/1.4807921
181. Radhakrishnan, S., Kim, S.J.: An enzymatic biosensor for hydrogen peroxide based on one-pot preparation of CeO2-reduced graphene oxide nanocomposite. RSC Adv. **5**, 12937–12943 (2015). https://doi.org/10.1039/c4ra12841a
182. Karuppiah, C., Palanisamy, S., Chen, S.-M., Veeramani, V., Periakaruppan, P.: A novel enzymatic glucose biosensor and sensitive non-enzymatic hydrogen peroxide sensor based on graphene and cobalt oxide nanoparticles composite modified glassy carbon electrode. Sens. Actuators, B Chem. **196**, 450–456 (2014). https://doi.org/10.1016/j.snb.2014.02.034

183. Yang, Z., Qi, C., Zheng, X., Zheng, J.: Facile synthesis of silver nanoparticle-decorated graphene oxide nanocomposites and their application for electrochemical sensing. New J. Chem. **39**, 9358–9362 (2015). https://doi.org/10.1039/c5nj01621e
184. Palanisamy, S., Lee, H.F., Chen, S.-M., Thirumalraj, B.: An electrochemical facile fabrication of platinum nanoparticle decorated reduced graphene oxide; application for enhanced electrochemical sensing of H2O2. RSC Adv. **5**, 105567–105573 (2015). https://doi.org/10.1039/c5ra20512c
185. Liu, J., Zhang, M., Liu, J., Zheng, J.: Synthesis of Ag@Pt core–shell nanoparticles loaded onto reduced graphene oxide and investigation of its electrosensing properties. Anal. Methods **8**, 1084–1090 (2016). https://doi.org/10.1039/c5ay02672e
186. Wang, J., Sun, H.-B., Pan, H.-Y., Ding, Y.-Y., Wan, J.-G., Wang, G.-H., Han, M.: Detection of hydrogen peroxide at a palladium nanoparticle-bilayer graphene hybrid-modified electrode. Sens. Actuators B: Chem. **230**, 690–696 (2016). https://doi.org/10.1016/j.snb.2016.02.117
187. Ning, L., Liu, Y., Ma, J., Fan, X., Zhang, G., Zhang, F., Peng, W., Li, Y.: Synthesis of palladium, ZnFe2O4 functionalized reduced graphene oxide nanocomposites as H2O2 detector. Ind. Eng. Chem. Res. **56**, 4327–4333 (2017). https://doi.org/10.1021/acs.iecr.6b04964
188. Yusoff, N., Rameshkumar, P., Mehmood, M.S., Pandikumar, A., Lee, H.W., Huang, N.M.: Ternary nanohybrid of reduced graphene oxide-nafion@silver nanoparticles for boosting the sensor performance in non-enzymatic amperometric detection of hydrogen peroxide. Biosens. Bioelectron. **87**, 1020–1028 (2017). https://doi.org/10.1016/j.bios.2016.09.045
189. Karthikeyan, C., Ramachandran, K., Sheet, S., Yoo, D.J., Lee, Y.S., Satish kumar, Y., Kim, A.R., Gnana Kumar, G.: Pigeon-Excreta-Mediated synthesis of reduced graphene oxide (rGO)/CuFe2O4 nanocomposite and its catalytic activity toward sensitive and selective hydrogen peroxide detection. ACS Sustain. Chem. Eng. **5**, 4897–4905 (2017). https://doi.org/10.1021/acssuschemeng.7b00314
190. Cheng, C., Zhang, C., Gao, X., Zhuang, Z., Du, C., Chen, W.: 3D network and 2D paper of reduced graphene oxide/Cu(2)O composite for electrochemical sensing of hydrogen peroxide. Anal. Chem. **90**, 1983–1991 (2018). https://doi.org/10.1021/acs.analchem.7b04070
191. Kubendhiran, S., Thirumalraj, B., Chen, S.M., Karuppiah, C.: Electrochemical co-preparation of cobalt sulfide/reduced graphene oxide composite for electrocatalytic activity and determination of H(2)O(2) in biological samples. J. Colloid Interface Sci. **509**, 153–162 (2018). https://doi.org/10.1016/j.jcis.2017.08.087
192. Bai, Z., Dong, W., Ren, Y., Zhang, C., Chen, Q.: Preparation of nano Au and Pt alloy microspheres decorated with reduced graphene oxide for nonenzymatic hydrogen peroxide sensing. Langmuir **34**, 2235–2244 (2018). https://doi.org/10.1021/acs.langmuir.7b02626
193. Liu, X., Zhao, Z., Shen, T., Qin, Y.: Graphene/Gold nanoparticle composite-based paper sensor for electrochemical detection of hydrogen peroxide. Fullerenes, Nanotubes, Carbon Nanostruct. **27**, 23–27 (2018). https://doi.org/10.1080/1536383x.2018.1479695
194. Wang, M., Ma, J., Guan, X., Peng, W., Fan, X., Zhang, G., Zhang, F., Li, Y.: A novel H2O2 electrochemical sensor based on NiCo2S4 functionalized reduced graphene oxide. J. Alloy. Compd. **784**, 827–833 (2019). https://doi.org/10.1016/j.jallcom.2019.01.043
195. Salazar, P., Fernández, I., Rodríguez, M.C., Hernández-Creus, A., González-Mora, J.L.: One-step green synthesis of silver nanoparticle-modified reduced graphene oxide nanocomposite for H2O2 sensing applications. J. Electroanalytical Chem. **855** (2019). https://doi.org/10.1016/j.jelechem.2019.113638
196. Guan, J.F., Huang, Z.N., Zou, J., Jiang, X.Y., Peng, D.M., Yu, J.G.: A sensitive non-enzymatic electrochemical sensor based on acicular manganese dioxide modified graphene nanosheets composite for hydrogen peroxide detection. Ecotoxicol. Environ. Saf. **190**, 110123 (2020). https://doi.org/10.1016/j.ecoenv.2019.110123
197. Chen, X., Gao, J., Zhao, G., Wu, C.: In situ growth of FeOOH nanoparticles on physically-exfoliated graphene nanosheets as high performance H2O2 electrochemical sensor. Sens. Actuators B: Chem. **313** (2020). https://doi.org/10.1016/j.snb.2020.128038

198. Patella, B., Buscetta, M., Di Vincenzo, S., Ferraro, M., Aiello, G., Sunseri, C., Pace, E., Inguanta, R., Cipollina, C.: Electrochemical sensor based on rGO/Au nanoparticles for monitoring H2O2 released by human macrophages. Sens. Actuators B: Chem. **327** (2021). https://doi.org/10.1016/j.snb.2020.128901
199. Jiang, L., Zhao, Y., Zhao, P., Zhou, S., Ji, Z., Huo, D., Zhong, D., Hou, C.: Electrochemical sensor based on reduced graphene oxide supported dumbbell-shaped CuCo2O4 for real-time monitoring of H2O2 released from cells. Microchem. J. **160** (2021). https://doi.org/10.1016/j.microc.2020.105521
200. Verma, S., Mal, D.S., de Oliveira, P.R., Janegitz, B.C., Prakash, J., Gupta, R.K.: A facile synthesis of novel polyaniline/graphene nanocomposite thin films for enzyme-free electrochemical sensing of hydrogen peroxide. Mol. Syst. Des. Eng. **7**, 158–170 (2022). https://doi.org/10.1039/d1me00130b
201. Qian, Y., Wang, Y., Wang, L.: Preparation of cuprous oxide-supported silver-modified reduced graphene oxide nanocomposites for non-enzymatic electrochemical sensor. Rev. Anal. Chem. **41**, 189–197 (2022). https://doi.org/10.1515/revac-2022-0045
202. Zhang, S., Zhang, X., Yue, S., Ma, Y., Tang, Y.: Synthesis of Ag nanoparticles dispersed on Co3O4–3D porous reduced graphene oxide and their application for electrochemical sensing of hydrogen peroxide. Nano **18** (2023). https://doi.org/10.1142/s179329202350008x

Two-Dimensional (2D) Materials for Bio-sensing Applications

J. M. Rajwade, A. Padhye, and S. Kulkarni

Abstract Two-dimensional (2D) materials have attracted the attention of researchers all over the world since the discovery of graphene in 2004. Several routes of synthesis for graphene, transition metal dichalcogenides (TMDs), hexagonal boron nitride (hBN), black phosphorus (BP), metal oxides, MXenes, silicene, and other 2D materials were optimized. Due to their peculiar layered nature, 2D materials offer immense potential as sensors. Extremely high surface area to volume ratio, tunable band gap, the possibility of doping, and possibilities for surface modifications with specific biomolecules for bio-recognition are properties that allow the development of sensors. The mechanical properties and flexible nature of 2D materials have led to the inclusion of these in wearable electronics for biomonitoring. With the advancements in communication systems such sensors are set to revolutionize biomedicine. However, careful assessment of toxicity, biocompatibility, and design and implementation of strategies for safe disposal is required. Scalable production methods, material characterization protocols, and the establishment of uniform regulations and risk assessment guidelines would help the commercialization of 2D materials as sensor platforms.

Keywords Graphene · Biosensors · Wearable sensors · Biocompatibility · Toxicity

1 Introduction

Structures with nanoscale dimensions are categorized into zero- (0D), one- (1D), or two-dimensional (2D) materials. Fullerenes and quantum dots were among the first examples of 0D nanostructures. One-dimensional structures having confinement in two directions were introduced subsequently [1]. The concept of 2D materials goes

J. M. Rajwade (✉) · A. Padhye · S. Kulkarni
Nanobioscience Group, Agharkar Research Institute, G. G. Agarkar Road, Pune 411004, Maharashtra, India
e-mail: jrajwade@aripune.org

N. Talreja et al. (eds.), *Two-dimensional Hybrid Composites*, Engineering Materials,
https://doi.org/10.1007/978-981-99-8010-9_8

back to the early twentieth century when scientists first started exploring the properties of single-layer crystals [2]. In 1917, the French physicist Georges Urbach proposed that thin layers of materials, which he called "lamelles," could have unique electronic and optical properties. The Russian physicist Vladimir Fock further developed this idea in the 1930s, who suggested that 2D crystals could exist in nature. The first experimental evidence of 2D materials came in 1962 when the American physicist John C. Hannon observed the diffraction pattern of a single layer of molybdenum disulfide (MoS_2) under an electron microscope. Since the discovery of graphene in 2004 by Novoselov and Geim, researchers have discovered many other 2D materials, including transition metal dichalcogenides (TMDs) such as MoS_2, tungsten diselenide (WSe_2), and molybdenum diselenide ($MoSe_2$), as well as hexagonal boron nitride (hBN), black phosphorus (BP), and, and silicene, among others [3, 4]. 2D materials are materials that have a thickness of only a few atomic layers or less, typically in the range of 0.5–100 nm. They can be thought of as thin films that are just a few atoms thick [5, 6]. Thus, the discovery of 2D materials has a long history.

Two-dimensional (2D) materials differ from their bulk counterparts in several ways:

1. **Electronic properties**: In 2D materials, the electronic properties can be tuned by controlling the thickness of the material. For example, as the thickness of graphene is reduced, it exhibits unique electronic properties such as high electron mobility and linear dispersion relation near the Fermi level. These properties are not present in bulk graphite [7, 8].
2. **Mechanical properties**: The mechanical properties of 2D materials are different from those of bulk materials due to the absence of interlayer interactions. For example, graphene is an extremely strongest material with a tensile strength 200 times greater than steel. In contrast, graphite, which is a bulk form of graphene, is a relatively soft material [9].
3. **Optical properties**: 2D materials exhibit unique optical properties due to their thickness and atomic structure. For example, transition metal dichalcogenides (TMDs) exhibit strong light-matter interactions [10, 11].
4. **Surface area**: 2D materials have a high surface area to volume ratio due to their layered nature [12].
5. **Chemical reactivity**: The chemical reactivity of 2D materials is the result of the high surface area and unique electronic properties. For example, graphene oxide, a derivative of graphene, is more reactive than bulk graphite due to the presence of oxygen-containing functional groups [13, 14].

These materials have unique properties such as high mechanical strength, flexibility, high thermal conductivity, and excellent electronic and optical properties, making them potential candidates for a wide range of applications, from electronics and photonics to energy and biomedical devices [15]. Further, their properties can be tuned by controlling their thickness and composition. Research in this field is ongoing and has the potential to revolutionize many industries [16, 17].

2 Types of 2D Materials and Their Properties

2.1 Graphene Family

A six-membered cyclic arrangement of carbon atoms makes up graphene, a 2D allotrope of carbon. In air, several layers of graphene are stable, and doping graphene oxide changes the polarity of those layers. Additionally, the linear electronic structure of graphene makes it possible to greatly modify the work function using an external electric field, chemical alterations, metal configurations, or thermal annealing. Graphene displays exceptionally high carrier mobility approaching 106 $cm^2 V^{-1} s^{-1}$ at 1.8 K and 105 $cm^2 V^{-1} s^{-1}$ at ambient temperature [18]. With regard to manipulating infrared and visible light, graphene offers a great deal of versatility and opportunity. Theoretically, monolayer graphene absorbs 2.3% of transmitted white light as a collection of distinct crystal lattice configurations, and the opacity rises by 2.3% with each additional layer [19, 20]. A typical semiconductor has a band gap ranging from 1–2 eV, but graphene is an exception which does not have any band gap that is zero band gap [21] but it can be opened up to 250 eV by NHG–NHG heterojunctions [22]. The preparation, synthesis, properties, and characterization of graphene and graphene-based 2D nanomaterials for their potential applications in bioelectronics and biosensors have been a subject of several reviews. 2D nanomaterials such as graphene would require more studies that could substantiate its utility as an alternative to conventional materials that will be cost-effective [23].

2.2 Metal Oxides (MOXs)

2D MOXs are grouped into two categories viz., layered and non-layered, based on their crystal structures. There are several layered MOXs materials assembled from M–O (M: V, Mo, or Mn) octahedral, in which the in-plane atoms are connected via strong chemical bonding, and the stacking layers are combined via weak van der Waals (vdW) interaction [24]. 2D oxides include non-layered metal oxides such as ZnO, MoO_2, HfO_2, Y_2O_3, VO_2, TiO_2 and layered oxides such as h-TiO_2, RuO_2, Cr_2O_3, MnO_2, hFe_2O_3 etc. [25]. The band gap of metal oxides 2D materials are mostly in the range of 3–5 eV [26], in case of monolayer MOs, it can be 1.22–6.48 eV [27].

2.3 Chalcogenides

Two-dimensional metal chalcogenides display an extraordinary optical response and carrier transport ability. These are layered materials with strong in-plane bonding and weak out-of-plane interactions enabling layer-controlled production [28]. Single-layer transition metal dichalcogenides (TMDs) have optical and electronic properties and may show a transition from indirect to direct band gaps. Because of their layer-dependent opto-electrical characteristics and narrow band gaps, single-layer TMDs outperform graphene. They are compatible with common fabrication methods and have a high surface-to-volume ratio, planar structure, low operational voltage, and an ultra-fast response time [29]. The band gap of TMDs depends on the type of transition metal used for example, GaN has a band gap of 1.32 eV [30], $Na_4CdGe_2S_7$ has a band gap of 3.35 eV [31].

2.4 MXenes

MXenes, a fast growing families of 2D materials, include the carbides and nitrides from the transition metals. In a 2D flake of MXene, $n + 1$ ($n = 1$–3) layers of early transition metals (Ti, V, Cr, Y, Zr, Nb, Mo, Hf, Ta, and W) are interleaved with n layers of carbon or nitrogen (X elements) with a general formula of $M_{n+1}X_nT_x$ [32].

2.5 Black Phosphorus (BP)

Phosphorene, i.e. two-dimensional black phosphorus (2D BP), was discovered in 2014 and since then it has garnered the attention of several researchers. Phosphorene is present as a puckered monolayer structure, which confers interesting properties. The potential applications have been realized in catalysis, energy storage sensor materials, etc. The unique properties such as the large surface area, good electric conductivity, and high theoretical specific capacity, make 2D BP an extensively studied electrode material which can significantly enhance the performance of energy storage devices [33]. In its thin film form, BP has a band gap of 0.3 eV which can be a bridge between the zero band gap graphene and relatively large band gap of TMDs (1.5–2.5 eV). In 4 nm thin film of BP, the band gap is tunable upto 75 meV [34, 35].

Several unique features, such as excellent biocompatibility, extraordinarily high interlayer spacing and conductivity, and environment-friendliness, have attracted researchers to these 2D materials. Especially for BP, the possibility of structural alteration and its exceptional surface chemistry present an exciting interface for the creation of new biosensors [36]. The extremely high surface area to volume ratio offers a multitude of sites for the binding of biomolecules, receptors, and dyes. The very high mechanical strength combined with flexibility and biocompatibility permits their usage in wearable biosensors. Typical electron structure gives the attribute of conductivity making materials for use in electrochemical and impedimetric biosensors. MXenes represent new 2D structures with electroactivity, robustness, functionality-friendliness, metallic conductivity, biocompatibility, hydrophilic surface, and 2D layered atomic structure (Fig. 1).

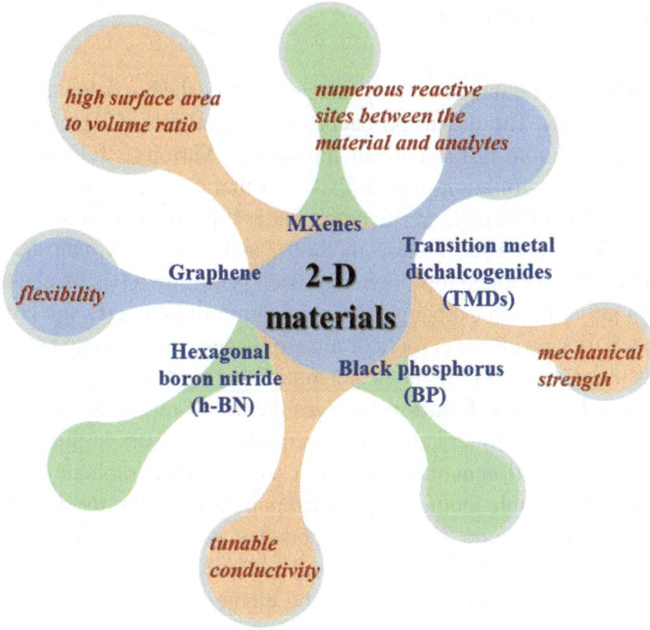

Fig. 1 2D material attributes beneficial for biosensing applications

A large surface area and numerous active sites make MXenes ideal and potential adsorbents for various small molecules [37]. Two-dimensional (2D) layered materials known as MXenes were first identified in 2011 (Ti_3C_2X). MXenes have excellent conductivity, they have been pursued as the material of choice in the field of electrochemical supercapacitors and research has been focused on enhancing interlayer spacing and the conversion of functional groups to manifest high energy and power density. Thus, the MXene family has been proven to be promising electrode material for ES applications [38].

3 Synthesis of 2D Material

Various methods have been reported for the synthesis of 2D materials out of which beam epitaxy, atomic layer deposition, and chemical vapor deposition are non-scalable, expensive processes for the synthesis of 2D materials. On the other hand, ultrafast liquid-phase laser processing offers the swift delivery of 2D quantum materials free from flaws [39]. A limited discussion on the methods is presented to make the reader aware of the various materials synthesis strategies. The synthesis approaches for 2D materials are broadly divided into two types viz., top-down approaches and bottom-up approaches.

3.1 Top-Down Approaches

Top-down approach of nanomaterial synthesis involves the breakdown of bulk material into the nanoscale. Top-down approaches to the synthesis of 2D materials involve chemical-based reactions or mechanical approaches. Various exfoliation techniques such as mechanical exfoliation, liquid-phase exfoliation, microwave-assisted exfoliation, photo exfoliation, etc., are included under top-down approach of 2D material synthesis. With the use of these methods, incredibly thin nanosheets can be obtained. While the chemical top-down approach primarily relies on chemical reactions brought about by ion exchange, the application of heat, etc., the physical top-down approach uses mechanical force or ultrasonic wave to exfoliate layered van der Waals solids into single and few-layer 2D materials [35, 40].

Physical/mechanical exfoliation involves the use of mechanical forces to break down the Wan Der Vaals interactions between adjacent layers in the bulk material. The mechanical exfoliation method was first used to obtain graphene from graphite. In recent years, this simple method has gained popularity and is used to obtain high-quality single-layer graphene. A universal, one-step, contamination-free Au-assisted mechanical exfoliation technique was developed and demonstrated for superconductors, magnets, metal-dichalcogenides, and elemental two-dimensional crystals. Effective exfoliation was possible by enhanced adhesion between the crystals and the substrates [41]. To obtain high-quality molybdenum disulfide (MoS_2) flakes, Siepi et al. employed lysozyme and the exfoliated material presented both hydrophobic and hydrophilic groups hindering re-aggregation [42]. The material was characterized for its physicochemical properties and was also reported to be biocompatible in studies involving cell lines. According to authors, the approach can be used to produce 2D graphene and the denatured lysozyme can be used in surface functionalization.

The **photo-exfoliation** synthesis involves the use of lasers generated from krypton fluoride (KrF) to breakdown the bulk material into nanoscale. Kumar et al. synthesized boron nitride (BN), molybdenum disulfide (MoS_2), and atomic graphene layers using powdered parent materials dissolved in N,N-dimethylformamide (DMF) and irradiated with a strong KrF laser were reported [39]. The researchers observed that with an increase in the laser irradiation period, the number of atomic layers and the lateral size of the sheets steadily decreased. The lateral size and the quantity of layers are also controlled critically by the laser fluence. The ratio of sheets with fewer layers increases while the average lateral dimension decreases from 400 nm at 1.5 J/cm^2 to 20–30 nm at 4 J/cm^2. The interactions at the scale of molecules and atoms correlate with the laser processing parameters with the sample size. The interlayer distance is stretched to 6.68 and the activation energy of exfoliation is lowered, according to simulation and DFT calculations, by the mild out-of-plane thermal expansion of atomic layers followed by solvent intercalation.

Microwave-assisted exfoliation is a new universal and rapid method to obtain ultrathin 2D structures using liquid-nitrogen and microwave treatments. This method involves the exfoliation of bulk-layered materials [43]. Common principles underlying the topochemical synthesis of 2D materials were documented by [44]. The method produces novel materials such as 2D transition metal carbides, nitrides, and carbonitrides by selective etching of metal from the MAX phases. Several 2D catalysts were prepared by microwave-assisted synthesis. These procedures can be classified as exfoliation, synthesis, doping, modification, and construction methods, based on which material properties can differ [30]. Thus, for next-generation electronic devices, photo-exfoliation synthesis, and microwave-assisted synthesis of pure crystals of 2D materials can be a promising approach.

3.2 *Bottom-Up Approaches*

Chemical vapor deposition (CVD) is one of the most crucial and dependable methods for creating 2D materials [45]. Common 2D materials such as graphene, boron nitride, and transition metal dichalcogenides (TMDs) are synthesized by CVD because of the quality of materials as well as the high process efficiency. The CVD method is also used for synthesizing high-quality 2D TMDs with unique and tunable electronic, optical, and chemical properties [46, 47]. Theoretical studies on the CVD process for obtaining traditional 2D nanomaterials (graphene, hexagonal boron nitride (hBN), and transition metal dichalcogenides, MoS_2/WSe_2) and new materials (silicene, phosphorene, and borophene) were performed and its utility was proved in understating and improving the process [48]. A multiscale/multiphysics model based on coupling continuum fluid mechanics and phase-field models was introduced for guiding the growth of CVD 2D materials, thus paving the way for synthesis-by-design [49].

Solvo-thermal and hydrothermal synthesis of 2D materials is a wet chemical synthesis method in which chemicals are solubilized in aqueous and organic solutions under high vapor pressure at high temperatures such as 300 °C [50]. Xu et al. reported a solvo-thermal process to create black phosphorus quantum dots (BPQDs), with an average size of 2.1 ± 0.9 nm. BPQDs displayed good nonlinear optical response, as confirmed by femtosecond laser Z-scan measurement [51].

Molecular beam epitaxy (MBE) is an epitaxial process, in which growth of the thin film takes place on a heated crystalline substrate through the interaction of adsorbed species supplied by atomic or molecular beams under ultrahigh vacuum (UHV) conditions. In this procedure, materials placed in ultrapure crucibles are evaporated or sublimed to produce beams with thermal energy. The deposition rate is typically less than 1000 nm/h, under vacuum which allows the films to grow epitaxially. The absence of carrier gases as well as the UHV environment results in the highest achievable purity in the grown films, making the material extra-pure [20, 52].

Photochemical synthesis is a recently developed technique for the synthesis of 2D metallic nanostructures. In this unique and efficient photochemical synthesis method, the metal precursors are reduced to two-dimensional metal nanostructures by using light radiation. Here, light radiation plays a vital role and acts as a powerful tool to reduce metal precursors into 2D nanostructures [53]. Vanadium diselenide

(VSe_2) was converted to Vo_2(r)[R] in a reductive atmosphere. It was produced in the form of porous nanosheets, increasing the surface area and introducing defects. The material containing holes was evaluated as catalyst for water-splitting applications [54].

Apart from these standard methods, newer approaches for synthesis of 2D material are emerging such as synthesis using salt crystals, topochemical synthesis, etching strategies, etc. Huang et al. described the synthesis of 2D nanomaterials using salt crystals for meeting the requirements of technologies for fabrication. They proposed a rational design of synthesis route [55]. Liu et al. review a few topochemical synthesis strategies such as the salt-templating method for non-layered 2D materials, the molten Lewis acid etching strategy for novel MXenes, and the chalcogen vapors etching and substitution strategy for phase-controlled 2D materials [56]. The use of phosphorene sheets as the phosphorus precursors and 2D templates as a general strategy for the bottom-up topochemical synthesis of Co_2P, $Ni_{12}P_5$, and $Co_xFe_{2-x}P$ was attempted [57].

For realizing large-scale applications and commercialization of 2D materials, scalable, energy-efficient methods will be required.

4 Characterization of 2D Materials

Characterization forms the basis to study the size, shape, and structural properties of synthesized 2D materials. In the case of multiple layers, hetero conjugations, and doping, it is necessary to confirm the presence of bonding and specific functional groups on the material. Figure 2 describes various methods used in the characterization of 2D materials along with the information on the material using the technique.

5 Applications of 2D Materials for Biosensing

The transduction mechanism and the 2D material used as the sensor are the main factors that affect sensing applications. Several attributes of 2D materials have contributed to research on applications in bio-sensing. 2D materials provide a large surface area-to-volume ratio, increasing the interaction with the analyte resulting in high sensitivity and low limit of detection. Doping with nanomaterials and other entities that enhance conductivity is possible with 2D materials especially relying on conductimetric measurements. The incorporation of electrocatalytic materials is also preferred for lowering the redox potential. 2D materials are used in sensors that measure strain, which could be used especially in flexible wearable electronics. Several biomolecules (antibodies, antigens, oligonucleotides, etc.) can be coupled to the 2D materials by appropriate strategies (covalent, electrostatic, van der Waal's,

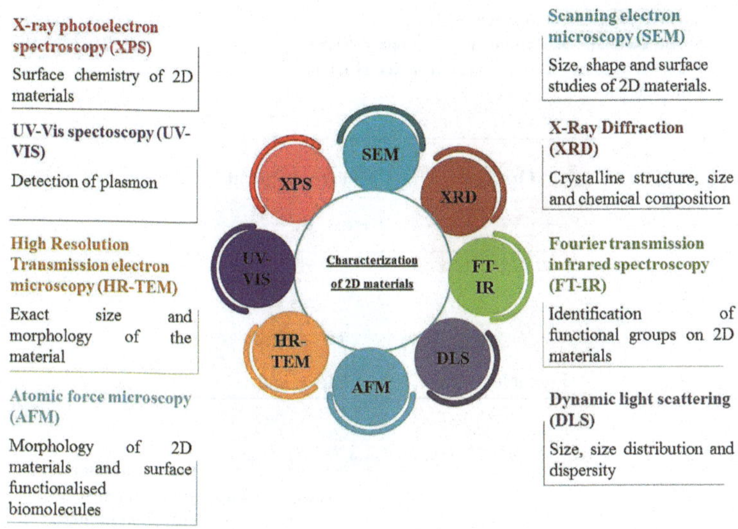

Fig. 2 Characterization techniques of 2D nanomaterials and their applicability

π–π interactions) resulting in surface modifications of 2D materials to increase selectivity and specificity for analyte detection.

The major transducing mechanisms are optical, electrical, piezoelectrical, and electrochemical and specific reviews could provide more insights into the transducing systems which are an integral part of biosensors [58]. Figure 3 is a schematic showing the transducing mechanisms. Materials-specific properties which are important for biosensing are also a subject of several recent reviews. According to Nangare and Patil, BP-based 2D materials are explored for highly sensitive and selective recognition of biomolecules and environmental pollutants because of optical properties, high carrier mobility, stronger immobilization of receptors and biomolecules, electronic bridging which play an important role in the highly selective and sensitive detection of analyte [59]. Garg and Pamme review graphene and other 2D nanostructures in microfluidic electrochemical and optical sensors [60]. Microfluidics has transformed measurement science by offering low sample consumption, instant readout, and high throughput. 2D carbon materials were reviewed for their application as a photo electrochemical (PEC) sensing platform. Characteristics of 2D carbon materials such as high surface area and ease of surface modifications for improved electrical transfer and attachment of bio recognition elements support their applications as biosensors [61]. According to Li et al., skin bioelectronics, biosensing technology, and neural interfaces will witness the most prominent uses of 2D materials [62].

The following section discusses specific applications related to biosensing.

From both a historical and analytical standpoint, the most significant developments in the use of 2D materials are outlined for four main categories of analytes, viz. gases, ions, volatiles, and biomolecules. The discussion of sensing performance

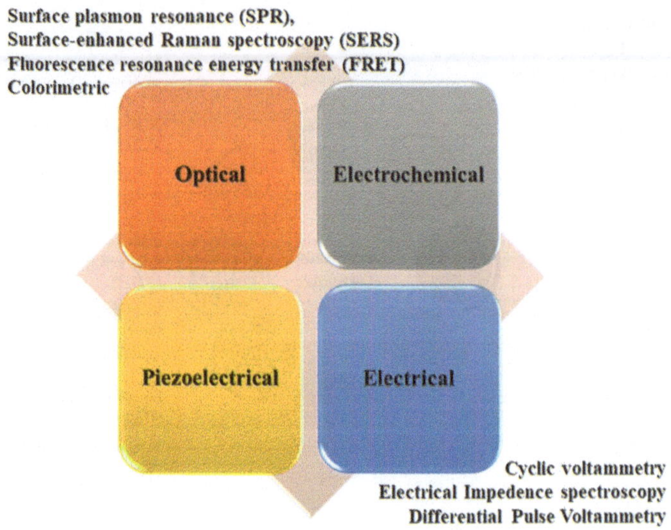

Fig. 3 Transduction sensing mechanisms

takes place within the framework of molecular design, structure–property relationships, and technology for device fabrication (Meng et al. 2019). Figure 4 provides a schematic of the diverse applications of 2D materials.

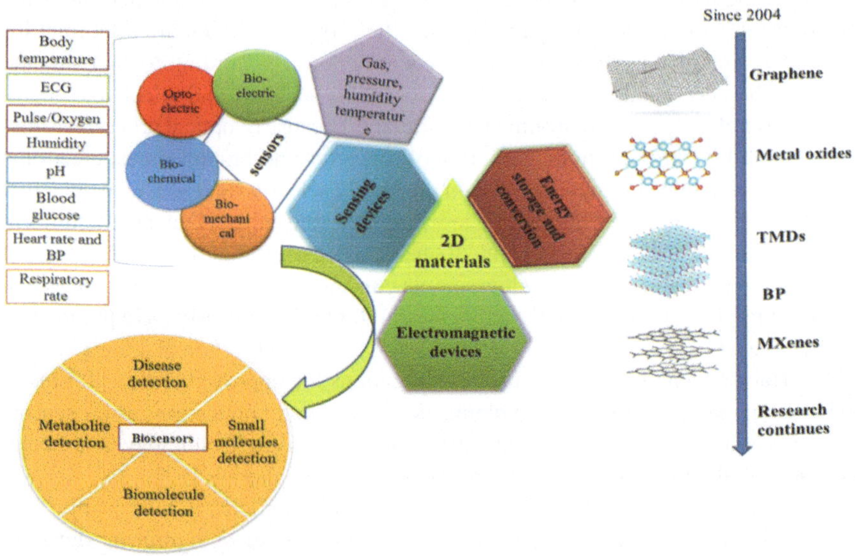

Fig. 4 Applications of 2D materials

5.1 Detection of Metabolites, Proteins, ROS, Small Molecules

Tao et al. have demonstrated a reliable sensing method for detecting a variety of targets, including metal ions, DNA, and tiny molecules, using single-stranded probe DNA and the hemin-graphene hybrid (GH) [63]. According to the researchers, this technique can be used for the quantitative detection of a wide range of analytes because of its benefits, such as straightforward operating process, affordable portable instrument, and user-friendly applications.

2D-MOF nanomaterials were modified with specific aptamers for application as enzyme mimic in the detection of carcinoembryonic antigen (CEA) with a linear range from 1 pg/mL to 1000 ng/mL and a limit of detection (LOD) of 0.742 pg/mL [64]. Wu et al. describe a SPR biosensor based on a Ti_3C_2-MXene-based sensing platform and multi-walled carbon nanotube (MWCNTs)-polydopamine (PDA)-Ag nanoparticle (AgNPs) signal enhancer for detecting carcinoembryonic antigen (CEA) [65]. A monoclonal antibody was the biorecognition element and in sandwich format, CEA determination of 2×10^{-16} to 2×10^{-8} M and a detection limit of 0.07 fM was obtained. In another study, an electrically neutral nanocomposite MXC-Fe_3O_4-Ru:Fe_3O_4 and $[Ru(NH_3)_6]^{3+}$ was synthesized to be used as a electrochemical sensor with antifouling characteristics. The sensor detected as low as 0.62 pg/mL carcinoembryonic antigen (CEA) with a linear response between 1 pg/mL and 1 μg/mL. The authors employed ratiometric strategy to achieve high selectivity, accuracy, and sensitivity. The biosensor could detect targets present in complex samples, such as FBS and clinical sera [66]. Kumar et al. reported an enhanced sensitivity in detection of CEA when a 2D material MoS and Mxene is sandwiched between the silver and graphene layer of the SPR sensor. Maximum sensitivity 144.72 deg./RIU and detection accuracy 2.24 was achieved [39]. Xu et al. used p-sulfonated calix 8 arene (SCX8)/polydopamine (PDA)/black phosphorene (BPene) nanocomposite as a receptor for cancer cells [67]. Through host–guest recognition, the SCX8 in BPene@PDA SCX8 binds to folic acid (FA) to create a stable BPene@PDA SCX8·FA conjugate. Through FA and the folate receptor (FR) on the cancer cells, cancer cells can bind to the BPene@PDA SCX8·FA modified electrode, generating a SCX8FA FR sandwich-type compound that causes a rise in impedance that is proportionate to the number of cells present. In the magnetoplasmonic biosensor, Faridi et al. observed minimal reflection, resonance angle shift, and transverse magneto-optical Kerr effect (TMOKE) responses only in the presence of the ss-DNA monolayer. In comparison to the standard Au/Co/Au trilayer, due to the presence of three-layer Gr and two-layer MoS_2 on top of the Au/Py bilayer, a significant boost in sensitivity (9- and 4-times, respectively) was observed. According to these findings, magnetoplasmonic devices have a high DNA sensitivity attributed to the coupling of light with 2D materials. Investigations on the comparative performance of nanoporous graphene, MoS_2, and titanium carbide MXene (Ti_3C_2) for their DNA detection performance and sensitivity reveal that nanoporous graphene was the most sensitive membrane for distinguishing the DNA bases [68]. However, the authors reported better distinction of A & T compared to C & G with MoS_2. The statistical measures, viz., the

Kolmogorov–Smirnov test and the absolute pairwise difference tests were employed in the simulation studies. Cai et al. used graphene functionalized black phosphorus composite (PG-BP) to create a unique, electrochemical sensor with high sensitivity for the detection of bisphenol A (BPA) in biological samples and everyday products [69]. BPA was quantified using differential pulse voltammetry under ideal circumstances. A good linear relationship was observed between BPA detection current in the concentration range (4.3×10^8 to 5.5×10^5 mol/L), with a detection limit of 7.8×10^9 mol/L (S/N = 3). Wu et al. used cysteine (CYS) to protect BP from oxidation and degradation. The CYS-functionalized BP and an aptamer were used to construct an silicon interdigital electrode to detect malachite green (MG), which is a toxic dye. With the lowest detection limit for MG of 0.3 ng/L, it becomes a selective biosensor for future on-site MG surveillance [70].

A new, highly sensitivity bismuth vanadate/two dimensional-carbon nitride/deoxyribonucleic acid ($BiVO_4$/2D-C_3N_4/DNA) aptamer photoelectrochemical (PEC) sensor for Microcystin-LR (MC-LR) is described [71]. Xiang et al. developed a new BP-modified nanosensor for detecting ochratoxin (OTA) [72]. With a good linear electrochemical response to OTA in the concentration range of 0.3–10 g/mL and a detection limit of 0.18 g/mL (S/N = 3), the sensor also exhibited strong anti-fouling capability, and good electrochemical stability. In real sample analysis, recoveries between 98.8 and 103.3% were obtained, which is a satisfactory practicability. Zhao et al. developed a novel electrochemical aptasensor based on gold nanoparticles decorated black phosphorus (AuNPs-BP) nanomaterial for the simultaneous detection of patulin (PAT) and OTA in apple juice [73]. To construct the aptasensor, ferrocene functionalized OTA aptamers (Fc-OTA-aptamers) and methylene blue functionalized PAT aptamers (Mb-PAT-aptamers) were used. According to the concentrations of OTA and PAT, two different electrochemical signal alterations were observed on Mb and Fc. The sensor was reported to have outstanding reproducibility, stability, and selectivity. The sensor was demonstrated for effective, simultaneous detection of PAT and OTA in apple juice. Xu et al. developed an aptamer based, impedimetric assay to detect mycotoxin PAT [74]. The biosensor consisted of black phosphorus nanosheets (BP NSs) and PAT aptamer on a glassy carbon electrode (GCE). The differences in electron transfer resistance at the changed electrode surface serve as the basis for detection. With a 0.3 nM detection limit, this assay can detect PAT over a linear range spanning from 1.0 nM to 1.0 M. The modification of the biosensor with gold nanoparticles and thiolated PAT aptamer resulted in enhancement of the sensor performance.

For early diagnosis and effective therapy of a disease, the development of a single test that detects miRNAs and other biomarkers with high selectively and sensitivity, in a simple manner yet rapidly is most desirable. Rahaie and Noroozi reported the early detection of miR-137 and miR-142, as two Alzheimer's biomarkers employing a sensitive fluorescence assay. The assay based on enzyme-free and isothermal hybridization chain reaction with SYBR Green and graphene oxide (GOX) achieved sensitive detection of the miRNAs from 0.05 to 5 nM [75]. The LOD for the newly constructed nanobiosensor was 82 pM. The method is promising as it will permit in detecting dual and multiple targets due to sensitivity, rapidity, low cost, and specificity for detection of miRNAs in blood which are present in nanomolar to femtomolar concentrations.

Sun et al. developed Mn-doped Ni-based metal–organic frameworks (Mn-MOF) with 3D hierarchical flower-like superstructures [76]. The aptamer/BPNSs/Mn-MOF/GCE was investigated as a smart aptasensor for the capture and selective and efficient detection of stress-induced phosphoprotein 1 (STIP1). With a low detection limit of 1 pg/mL, this aptasensor allowed for the efficient detection of STIP1 throughout a wide range (2×10^{-3}–1×10^{4} ng/mL). This aptasensor demonstrated good reliability and practicability by successfully determining STIP1 in real-world samples.

The electrochemical polymerization of conducting poly(3,4-ethylenedioxythiophene) nanorods (PEDOTNRs), negatively charged BPQDs, and template molecules vitamin C (VC) was proposed by Zhang et al. as a novel molecularly imprinted electrochemical sensor [77]. Peak currents measured by differential pulse voltammetry (DPV) under optimal conditions showed a linear relationship with VC concentrations ranging from 0.01 to 4 mM with a detection limit of 0.0033 mM. Additionally, the imprinted electrode was demonstrated to show repeatability, reproducibility, stability, and selectivity for the electrochemical analysis of VC. It was effectively employed to detect VC in commercial drink soft sample. The use of 2D nanomaterials for the optical and electrochemical methods for detecting oxidative stress is reviewed [78]. Fang et al. demonstrated an electrochemiluminescence (ECL) and photothermal dual-mode biosensor based on black phosphorous quantum dots (BPQDs) and MXenes (that serves as a signal amplifier), for the detection of exosomes [79]. The constructed self-enhanced Ru(dcbpy)32 + @BPQDs ECL system created a strong ECL signal by minimizing the energy loss and shortening the electron transfer distance. Xue et al. described preparation of a nanocomposite containing amino-functionalized multi-walled carbon nanotubes (NH_2-MWCNT) and in-situ deposited silver nanoparticles [80]. The material was used as an electrochemical sensor for simultaneous analysis of uric acid (UA), xanthine (XT), and hypoxanthine (HX); with an excellent linear range of 0.1–800 M. The concurrent electrochemical determination of clenbuterol (CLB) and ractopamine (RAC), was described by Ge et al. using several nanocomposites [81]. The BP nanohybrid was created by co-decorating with poly(3,4-ethylenedioxythiophene) and mono(6-mercapto-6-deoxy)-cyclodextrin nanoparticles. With relative limits of detection to be 0.14 and 0.12 M for CLB and RAC respectively, the sensor could detect the analytes from spiked beef, feed, and bovine serum samples.

Rutin is a common flavonoid glycoside that has been extensively employed in clinical chemistry and human health due to its good physiologic properties. Therefore, developing sensitive and trustworthy methods for rutin determination is crucial. Niu et al. [82] developed a poly(3,4-ethylenedioxythiophene)-poly(styrenesulfonate) (PEDOT:PSS) as the film and stabilizer to create a black phosphorene (BP) nanosheet modified glassy carbon electrode (GCE). The performance of the BP-PEDOT:PSS nanocomposite was assessed by SEM, EDS, and Raman spectroscopy. During analysis, high conductivity and stability were reported. A few-layer BP-based non-enzymatic hydrogen peroxide (H_2O_2) sensor was reported for the first time which utilized BP degradation under ambient conditions [83]. The fabricated H_2O_2 sensor exhibited a considerably lower detection limit (1×10^{-7} M) compared to other electrochemical methods (1×10^{-7}–5×10^{-5} M). In the work carried out by Ding et al., hydrogen peroxide (H_2O_2) was detected using a modified glassy carbon electrode (BPQDs@ZnO/GCE) enhanced with black phosphorus quantum dots (BPQDs) [84]. ZnO and BPQDs together raise the conductivity of various materials. The redesigned electrode works well for real sample detection and demonstrates long-term durability and good reproducibility. The limit of detection (LOD) for H_2O_2 was evaluated at 2.5 μmol/L with the signal-to-noise ratio of 3. Cai et al. described the use of an environmentally friendly immunosensor, based on porous graphene functionalized black phosphorus for the detection of leptin in biological samples [69]. Under the optimal conditions, the proposed immunosensor exhibited a wide linear range of 0.150–2500 pg/mL with a low detection limit of 0.036 pg/mL. The sensor exhibited excellent sensitivity and anti-interference capabilities. Jiang et al. created a flexible enzyme catalyzed, ratiometric biosensor using constructed nanohybrids of silver nanoclusters (AgNCs) and metal–organic frameworks (MOFs) doped with black phosphorus quantum dots (BPQDs) [85]. AgNCs adhering to MOF cause an increase in the red-FL (response signal), while BPQDs doped into MOF just slightly alter the blue-FL (reference signal). This biosensor has a detection limit of 3 ng/mL and can detect baicalin at concentrations between 0.01 and 500 g/mL. High sensitivity, selectivity, and stability make this biosensor ideal for detecting baicalin in real-world samples. This study investigated a quick and effective semi-quantitative technique for adaptable FL visual detection, which can encourage the creation of sophisticated chemo/bio-sensors and analytic techniques.

5.2 *Detection of Diseases*

For the early identification of diseases, the development of new sensors for the precise detection of biomarkers in biological fluids is crucial. Vilela et al. reported a sensor technology based on graphene oxide and upconversion nanoparticles (NPs) for the targeted detection of mRNA-related oligonucleotide markers in intricate biological fluids [86]. Low-background photon counting readout in conjunction with near-infrared light upconversion allowed for the accurate detection of femtomolar-sized amounts of short oligonucleotide sequences. The sensor was proved for detection of mRNAs associated with prostate cancer and Alzheimer's disease in human blood serum.

The high sensitive detection of tDNA, a biomarker associated with Alzheimer's disease (AD) was reported by Hua et al. [87]. In this electrochemical sensor, X-DNA building blocks were designed to be assembled by four DNA strands (known as S1, S2, S3, and S4) and expanded using DNA polymerase to form two X-DNA motifs.

Alkaline phosphatase (ALP) was also enclosed within a network of hydrogels to create a porous material of the type ALP@DNAhg. Gold nanoparticles (Au@rGO) functionalized reduced graphene oxide was used to modify the glassy carbon electrode GCE. If strand displacement successfully captures ALP@DNAhg, tDNA recycling assembly for signal amplification is started. upon the addition of pyrophosphate and molybdate (MoO_{42}), ALP catalyzes the reaction to create phosphomolybdate anions ($PMo_{12}O_40_3$) as a redox active product. The detection limit achieved was 3.4 $\times 10^3$ pM, and its amperometric signal is dependent on the concentration of tDNA in the range of 1.0×10^2 to 1.0×10^4 pM. With a detection limit of 36 cells/mL, the cytosensor demonstrated a linear relationship with LNCaP cell concentration in the range of 2×10^2 to 1×10^5 cells/mL. Due to particular supramolecular recognition, the developed cytosensing platform displayed good sensitivity, acceptable stability, and favorable reproducibility, exhibiting potential to be used in the detection and diagnosis of cancer. BP nanosheets with gold nanoparticle (AuNP) anchors were created by Liu et al. by in situ method [88]. The authors reported detection of circulating tumor cells (CTCs) after immunomagnetic separation by electrochemical method using BP@AuNPs@aptamer as a probe. A detection limit of 2 cells/mL was achieved and confirmed through the accurate detection of MCF-7 cells in human blood. In an assay based on the photoelectrochemical detection for the protein S100β (a biomarker for Alzheimer's disease located in the brain's astrocytes) Tabreizi et al. used rGO-Au/ITO electrode bound with anti-S100β antibody as primary antibody [89]. The anti-S100β antibody labeled with CdS quantum dots was used as a secondary antibody in a sandwich assay. The progressive changes in the electrochemical characteristics of electrode surface were verified using cyclic voltammetry and electrochemical impedance spectroscopy. The technique was successfully used to identify S100β in samples of human serum. Another study described a non-enzyme, target-triggered signal amplification method based on the entropy-driven strand displacement reaction (ESDR) and graphene oxide (GO) for detection of biomarkers for Alzheimer's disease [90]. According to this method, hairpin structure probes (H) open upon binding with the Aβ-42 oligomers, thus exposing the bases which are complementary to the FAM-labeled replacement probes R (R1 and R2). The replacement probes hybridize with H releasing the bound Aβ-42 oligomers. The subsequent cycle continues, magnifying the signal. With a low limit of detection (20 pM), high accuracy, outstanding precision, and convenience the sensor offers an excellent possibility for the early diagnosis of AD. The fabrication of magnetic hematite (Fe_2O_3) decorated electrochemically reduced graphene oxide (Fe_2O_3@erGO) nano-composite for the detection of Parkinson's disease biomarkers is presented by Mathew et al. [91]. The technique uses a glassy carbon electrode (GCE) for direct electrochemical reduction of self-assembled, ex-situ synthesized -Fe_2O_3 anchored GO to erGO (-Fe_2O_3@erGO) for the selective detection of dopamine (DA), a key biomarker for Parkinson's disease. The sensor (Fe_2O_3@erGO/GCE) was used to analyze DA in commercially available pharmaceutical formulations and human serum samples, with satisfactory results (LOD of 0.024 μM (S/N = 3)).

5.3 Detection of Bacteria and Viruses

Impedimetric methods are FDA approved methods useful in the detection of microorganisms. The sensitivity of these systems can be increased by employing 2D materials as the electrode surface. In these types of sensing devices the presence or absence of bacteria, leads to changes in the Faradaic or non-Faradaic interfacial impedance, which is then used to determine the concentration. Several researchers also describe the development of colorimetric sensors to detect bacteria and viruses. A wireless graphene sensor with several advantages such as portability, continuous, nondestructive analysis was described by Mannoor et al. [92]; which contained graphene functionalized with antimicrobial peptides (AMPs) and a coil microantenna. The sensor monitored changes in resistance and detected single *E. coli* cell in PBS and $\sim 10^2$ CFU/mL of *Helicobacter pylori* in saliva. The latter was detected after implantation on bovine tooth. According to the findings of Thakur et al., rGO-based FET sensor passivated with ultrathin layer of Al_2O_3 is a cost-effective, label-free sensor which can be produced in mass quantities for detection of pathogens in water [93]. A linear range (10^3–10^5 CFU/ mL) was demonstrated for *E. coli*. Another graphene biosensor functionalized with *E. coli* O157:H7 specific antibodies and with interdigitated microelectrodes as capacitors was fabricated [94]. The capture of pathogenic *E. coli* O157:H7 on the sensor surface causes (i) polarization of cell-surface charges, (ii) intrinsic bioactivity of the cells, and (iii) changes in the electronegativity or dipole moment of cell-wall and charge carrier mobility of graphene. The sensor is thus a label-free biosensor with good sensitivity (10–100 cells/mL), is fast, specific, and capable of detecting bacteria in situ. CTAB-MoS_2-NS (cetyltrimethyl ammonium bromide functionalized MoS_2 nanosheets) were used in the polydimethylsiloxane (PDMS) microfluidic device employing the electrochemical impedance spectroscopy (EIS) technique for detection of *S. typhimurium* with sensitivity of 1.79 kΩ/CFU/ mL/cm^2 and the detection limit of 1.56 CFU/ mL in the detection range of 10^1–10^7 CFU/mL [95]. For the sensitive and selective electrochemical detection of uropathogenic *Escherichia coli (E. coli)* UTI89 bacteria in aqueous and serum samples, Jijie et al. reported the fabrication of gold electrodes modified with thin active layers of reduced graphene oxide/polyethyleneimine (rGO/PEI) [96]. Electrophoretic deposition (EPD) technique was used to achieve selectivity for *E. coli*, anti-fimbrial antibodies were covalently attached to pyrene-polyethyleneglycol (pyrene-PEG) modified electrode surface and detection limit was 10 CFU/mL. A Paper immunosensor (a free-standing graphene paper), with greatly enhanced sensing performance was described for label-free detection of *E. coli*. A wide linear range (1.5×10^2–1.5×10^7 CFU/mL), low detection limit (1.5×10^2 CFU/mL), and excellent specificity was observed [97]. Using a multifunctional chemical and biochemical sensing platform, Li et al. detected p-aminophenol (PAP), -galactosidase (-Gal), and *Escherichia coli (E. coli)* [98]. The biosensing platform consisted of bimetallic platinum palladium (PtPd) nanoparticle on the BP surface. The biochemical reaction involved PAP production from the substrate PAP-galactopyranoside, which indicated β-galactosidase activity in the linear range of 0.2–0.8 mU/L. The enzyme is an indicator of total coliforms, which could be detected in the range 6×10^6–1.6×10^8 CFU/mL. These findings are in support of a multifunctional platform based on 2D layered nanomaterials with bimetallic nanoelectrocatalysts for chemical and biological sensing.

Gold nanoparticles (AuNPs) and anti-*Shewanella* antibodies (Ab) assembled on bovine serum albumin (BSA)-modified GO (Ab/AuNPs/BSA/GO) was described as a biosensor for detecting *Shewanella* which is particularly important in the process of bioremediation. Silver treatment enhanced the biosensor response [99] which was quantified using Anodic stripping voltammetry. Screen printed graphene probes immobilized with bacteriophage were used as a sensor for detection of *Staphylococcus arlettae*, a pathogenic, coagulase-negative staphylococcus in spiked water and apple juice samples. The sensor characteristics were fast response time (2 min), low limit of detection (2 CFU), specificity, and stability over a prolonged period (3 months) [100]. A facile, sensitive, and reliable impedimetric immunosensor doped with reduced graphene sheets (RGSs) and chitosan was developed for the selective detection of sulphate-reducing bacteria (SRB). Faradic impedance spectroscopy was used and a linear relationship was obtained between 1.8×10^1 and 1.8×10^7 CFU/mL [101]. Immunomagnetic beads (IMB, SiO_2/Fe_3O_4) conjugated with *Salmonella pullorum* antibody and reduced graphene oxide coated with gold nanoparticles (rGO/AuNPs) were used for sensitive detection of the food pathogen *S. pullorum*. The latter served as an electrochemical label for the assay developed on a four channel screen-printed carbon electrode. The assay was carried out using differential pulse voltammetry (DPV) for detection of 10^2–10^6 CFU/mL, with a LOD of 89 CFU/mL/ for *S. pullorum* [102]. For the rapid detection of the *Enterobacter sakazakii*, electrochemically reduced graphene oxide (ERGO) based immunosensor was developed. Horseradish peroxidase-labeled *Enterobacter sakazakii* specific antibody was used to enhance sensitivity. This was reported as the first screen-printed carbon electrode (SPCE) sensor [103]. A glassy carbon electrode coated with graphene decorated with a specific aptamer was used to quantify *Salmonella*, the food-borne pathogen. Electrochemical impedance was measured giving a detection limit as low as 3 CFU/mL [104]. Using two-photon polymerization and graphene an enhanced biosensing platform was fabricated for the detection of motile bacteria [105].

A colorimetric sensor based on $ZnFe_2O_4$–rGO hybrid nanostructures was used to detect *S. enterica* serovar typhimurium [106]. The $ZnFe_2O_4$–rGO hybrid nanostructures conjugated with specific aptamer was a nanozyme with peroxidase activity and another biotin-aptamer was used as the capture probe to form a sandwich structure in the presence of the bacterium. The sensor had a limit of detection of 11 CFU/mL. In another study, Kaushal et al. described the detection of *E. coli* and *S. typhimurium* based on use of a specific antibody conjugated to GO-AuNP 2D hybrid material. The material was also proved to have antibacterial effect, which was enhanced by near infrared light [107].

Similar to bacteria, for detecting viruses graphene-based hybrid materials have shown promise. The synthesis of hemin-modified graphene nanosheets by a wet chemical method was reported and the resultant 2D material was used for the label-free detection of Hepatitis B virus (HBV) [108]. A free-standing reduced graphene oxide film was prepared and used with $[Fe(CN)_6]^{3-/4-}$ redox system. The film was modified with pyrene derivatives and covalently linked with anti-rotavirus antibodies. The biosensor detected rotavirus using cyclic voltammetry [109]. Zhan et al. described colorimetric detection of respiratory virus (RSV) using a mercury-catalyzed AuNP–GO hybrid system [110]. The GO nanosheets were dispersed in tannic acid (TA) which served as a reductant for chloroauric acid ($HAuCl_4$), resulting in formation of AuNPs on the GO surface. The nanohybrid exhibited peroxidase-like activity. In presence of RSV, the peroxidase-like activity of AuNP–GO was restored by adding Hg^{2+} to the solution resulting in a colorimetric sensor. The synthesis of a composite comprising 3-Aminopropyltriethoxysilane (APTES) functionalized graphene oxide (APTES-GO) wrapped on SiO_2 particles (SiO_2@APTES-GO) was reported by Jin et al. [111]. This composite was used in a sensitive, specific impedimetric sensing device for dengue DNA and RNA with a 1-Femto Molar limit of detection. The norovirus-like particles (NOV-LP) were detected rapidly using antibody-modified graphene/AuNP hybrids as a colorimetric sensor [112]. Similar to the report of Zhan et al., the graphene/AuNP hybrids served as a peroxidase, changing the color of TMB solution to blue, upon addition of H_2O_2. The specificity originated from the antibody bound on the GO/AuNP hybrid [110].

The nanocomposite constituting gold@palladium nanoparticles (Au@Pd NPs) and molybdenum disulfide functionalized multiwalled carbon nanotubes (Au@Pd/MoS_2@MWCNTs) with large surface area and excellent electrocatalytic properties was used to fabricate an immunosensor for detecting hepatitis B e antigen (HBeAg). A low detection limit of 26 fg/mL was achieved and concentrations in the range of 0.1–500 pg/mL were detected successfully [113]. The development of a 2D material viz., semiconducting transition metal dichalcogenide (TMDC) WSe_2 conjugated with monoclonal antibody against the SARS-CoV-2 spike protein was reported as a FET-based biosensor for detection of SARS-CoV-2. It showed a detection limit of 25 fg/μL in 0.01X phosphate-buffered saline (PBS) [114]. Two-dimensional (2D) transition-metal carbides ($Ti_3C_2T_x$ MXene) were functionalized with probe DNA through noncovalent interactions and used as electrodes for sensitive detection of nucleocapsid (N) gene of SARS-CoV-2. Nucleic acid hybridization and chemoresistive transduction was the principle underlying the selective and sensitive detection (as low as 10(5) copies/mL in saliva). The authors state that the interlayer spacing between MXene serves as molecular sieve [115]. Chekin et al. described a potential point-of-care assay for detection of human papillomavirus (HPV). The glassy carbon (GC) electrodes were modified with porous reduced graphene oxide (prGO) and molybdenum sulfide (MoS_2) and detected the L1-major capsid protein. For specificity an RNA aptamer Sc5-c3 was used for recognition. DPV was performed and linear relationship was found in the range 0.2–2 ng/mL (3.5–35.3 pM) with a LOD of 0.1 ng/mL (1.75 pM) [116].

5.4 Disease Control

Shao et al. reported a thermosensitive hydrogel [poly(D,L-lactide)-poly(ethylene glycol)-poly(D,L-lactide) (PDLLA-PEG-PDLLA: PLEL)] and combined it with black phosphorus (BP) nanosheets to provide a new PTT system for the postoperative treatment of cancer. The BP@PLEL hydrogel has strong in vitro and in vivo biodegradability and biocompatibility, great near infrared (NIR) photothermal performance, and a quick NIR-induced sol–gel transition. An in vivo PTT postoperative therapy technique was developed in which the sprayed BP@PLEL hydrogel, when exposed to NIR irradiation, permits quick gelation, establishing a gelled membrane on wounds, and provides high PTT efficacy to remove any remaining tumor tissues after tumor removal surgery. Additionally, the effective and biodegradable PTT system is particularly promising in the postoperative treatment of cancer due to its strong photothermal antibacterial activity, which helps to prevent infection [117]. Ye et al. made a customized photothermal vaccination using the surgically excised tumor and PD-1 checkpoint blocking antibody to stop metastasis and recurrence of the tumor. The membrane of surgically excised tumor cells was coated on black phosphorus quantum dot nanovesicles (BPQD-CCNVs), which were then put into a thermosensitive hydrogel containing GM-CSF and LPS. Gel-BPQD-CCNVs were hypodermically injected, and the prolonged GM-CSF release from these injections successfully attracted dendritic cells to collect tumor antigen. DCs were expanded and activated by NIR radiation and LPS, and they then moved to the lymph nodes to deliver antigen to CD8+ T lymphocytes. Additionally, the addition of the PD-1 antibody greatly improved the eradication of the surgical residual and lung metastatic tumor by tumor-specific CD8+ T cells. This approach has a potential advantage of developing customized cancer vaccine [118]. With central nervous system (CNS) illnesses, delivery efficiency with gene transfection has been a significant problem in obtaining maximum therapeutic efficacy. Hsieh et al. in their study created a nanoplatform for gene therapy in CNS illness by combining an external NIR-laser-triggered assistance with neuro-specific targeting peptide [119]. A precisely controlled two-stage near-infrared (NIR)-laser photothermal treatment on reduced graphene oxide (rGO) nanoparticles to modify them with polyethyleneimine (PEI), a neuro-specific peptide that enhances the ability to target neurons and achieve high gene transfection in neurons, was designed. In comparison to natural release, at least twice as much pDNA can be released from rGO-PEI-NT/pDNA using an NIR laser trigger. Furthermore, it has been shown that the maximum transfection effectiveness occurs when NT alteration is combined with external multi-stage stimuli-responsive NIR laser treatment in mouse (C57BL/6) brain transfection tests and in vitro differentiated PC-12 cell experiments. Deepa et al. reviewed the interactions between some typical 2D materials (i.e. graphene, graphene oxide, nitrogen-doped graphene, molybdenum disulfide, phosphorene, silylene, and germanene) and biomacromolecules (i.e. silk protein, lysozyme, bovine serum albumin, bovine hemoglobin, ovalbumin, villin, bovine fibrinogen, DNA/RNA, glucose oxidase and chitosan) which are important for antibacterial and disease therapy applications [23].

5.5 *Wearable Sensors*

Specific applications of 2D materials for monitoring human body temperature, electrography, sweat detection, respiratory gas, and saliva sensors are reviewed by Hassan et al. [120]. 2D materials (graphene-based materials, transition metal dichalcogenides) have been used to make electrochemical sensors for monitoring small molecules such as H_2O_2, ·NO, glucose, etc. These could potentially be useful in applications especially for humans [121]. Wearable electrochemical sensors would permit real-time biomonitoring ensuring the best possible patient care. For wearable materials such as silk have been used in combination with 2D materials [122]. Wearable systems that can address both sensing and therapy are being developed. Hassan et al. present various aspects such as the technology needs, the parameters that can be monitored and the challenges [120]. Few examples of wearable sensors/devices based on 2D materials are presented in Table 1.

Several TMDs and their hybrids (MoS_2,WS_2; hybrid of MoS_2 with Ag nanoparticles, SnO_2, Cu_2S and graphene) were proved to be promising materials for the fabrication of gas sensors [139] because of remarkable semiconducting properties with a tunable bandgap, depending on their thickness and doping molecules. Kim et al. have demonstrated that $Ti_3C_2T_x$ exhibits both low electrical noise and a strong response signal for volatile organic compounds (VOCs) [140]. Graphene hybrids were used in disposable sensors for monitoring various analytes important in human health care [120]. The work of Wang et al. presents in-depth information on wearable sensors [141]. Along with wearable sensors, point-of-care analytical testing requires reliable product design, scalability, and low-cost manufacturing for 2D materials. Parameters such as mass production, low cost, stability, and ease of transport of the MOFs nanomaterials, as well as the ability for visual detection will make this sensor suitable for point-of-care (POC) testing in remote or resource-poor areas [142].

6 Toxicological Effects

Increasing applications of 2D materials in physical, chemical, and biological sciences especially for sensors has led to research on new materials. During production, transportation, applications, and destruction cycle several interactions between the biotic and abiotic components in the environment occur. A careful study of the toxicological effects of the 2D materials on humans and the environment is essential to ensure safety. Just like nanomaterials, in vitro and in vivo toxicity need to be established using high-throughput methods because mishandling materials without an understanding of the potential effects can result in severe, chronic health issues and even loss of lives [58, 143, 144]. Extensive evaluation of the inhalation toxicity, skin irritation, inflammatory responses, carcinogenicity, and genotoxicity are required across several model species and humans along with toxicogenomic studies.

Table 1 The 2D materials-based wearable sensors for human applications

Analyte/Application	2D material used in the application	Response detection/ Stimulation system	Reference
Glucose and uric acid (UA) in saliva	Uricase immobilized on graphene	Organic electrochemical transistors (OECTs)	[123]
Moisture (humidity) on skin	Reduced graphene oxide-polyurethane composite	Conductive elastomers	[124]
High-resolution neurophysiological recording	Graphene-based, carbon-layered electrode array	Optical interference	[125]
Electrocardiography (ECG)	Graphene films on polyethylene terephthalate (PET) substrates and graphene paper	Alternating current impedance spectroscopy	[126]
Electrocardiography (ECG)	Graphene-coated Ag/ AgCl electrode	Impedance	[127]
Electromyographical signals	Cell-sheet-graphene hybrid	Electrically and/or optically	[128]
Skin tactile sensor	MoS_2 and graphene	Piezoresistive effect	[129]
Electrooculography (EOG)	Graphene on PMMA	Electrical	[130]
Subtle and large human motions	Reduced graphene oxide on textile	Electrical resistance	[131]
Subtle pulse beat and movement in real-time	Wrinkled graphene film (WGF)/interconnected polyvinyl alcohol (PVA) nanowires/interdigital electrodes	Piezoresistive effect	[132]
Screening for arteriosclerotic disease and monitoring health	Janus graphene film	Pressure	[133]
Kinesthetic sensing	Graphene-based composite fiber sensor with a "compression spring" structure	Electrical conductivity	[134]
Monitoring respiration, joint movement and pulses in real-time	MXene-sponge	Piezoresistive sensor	[135]
Detection of human motions	Composite films of reduced graphene oxide	Strain sensor	[136]
Detection of pulse	MXene/reduced graphene oxide (MX/rGO) hybrid	Piezoresistive sensor	[137]

(continued)

Table 1 (continued)

Analyte/Application	2D material used in the application	Response detection/ Stimulation system	Reference
Monitoring diverse stimuli such as weak gas flow, acoustic sound, wrist pulse pressure, respiration, and foot pressure	GO/PVDF composite film	Piezoresistive pressure sensing	[138]
Monitoring glucose level in tear fluid and intraocular pressure	Graphene-AgNWs hybrid	Electrical resistance and capacitance	[128]

Several studies indicate antimicrobial characteristics of the 2D materials [145–147]. Although several 2D materials are biocompatible, differences in toxicity of materials are reported. The observed effects are highly dependent on the cell lines (primary or established) used for testing [148, 149]. The toxicity is highly influenced by size, shape, chemical composition, stabilizing/functionalizing material, surface charge, impurities/dopants, duration of exposure, etc. [144]. In a recent review, by Naikoo et al., the findings of studies on the toxicity/compatibility of graphene family, MXenes, chalcogenides, and 2D oxides as sensing materials are provided. The environmental toxicity and exposure toxicity of materials used for biosensors for clinical analysis and non-implantable applications are more important [150].

7 Challenges

Although new 2D materials are designed for several biological sensing applications and proof-of-the-concept studies are presented, it is only after achieving a large-scale synthesis at low production costs, and reliable product design, sensors can be commercialized [120]. Especially for 2D TMD materials to be applied in wearable and flexible biosensors, techniques for scalable synthesis of materials at industrial scale, device fabrication, and studies on the material-substrate interactions are extremely important [29, 151]. The latter may lead to inflammatory responses, toxicity, and ultimately rejection by the biological system. The 2D materials used for biosensing applications require surface modifications with biomolecules such as antibodies, aptamers, proteins, and antigens which increase their specificity and selectivity toward the analyte to be tested. Thus, assessment of the in vivo biocompatibility of these 2D materials is the main challenge for their clinical use [29, 144, 152, 153]. Further, implantable and wearable devices will also require studies with respect to biomimetic mechanical properties, long-term stability, and functionality before commercialization. Thus, devices/sensors employing robust materials with environmentally safe, reliable, and reproducible characteristics have to be developed for commercialization. Further, the establishment of uniform regulations and risk assessment guidelines would require participation of researchers and regulatory agencies.

8 Conclusion

The chapter presents information on the characteristics and routes of synthesis for 2D materials such as graphene, transition metal dichalcogenides (TMDs), hexagonal boron nitride (hBN), black phosphorus (BP), silicene, etc. Due to their peculiar layered nature, and materials characteristics such as extremely high surface area to volume ratio, tunable band gap, the possibility of doping, and possibilities for surface modifications with specific biomolecules for bio-recognition the development of sensors for biological applications is possible. The research on applications of these materials in the biosensing of small biomolecules, bacteria, viruses, disease biomarkers, etc., is summarized. The mechanical properties and flexible nature of 2D materials have led to the inclusion of these in wearable electronics for biomonitoring. However, careful assessment of toxicity, biocompatibility, and design and implementation of strategies for safe disposal is required. Scalable production methods, material characterization protocols, and the establishment of uniform regulations and risk assessment guidelines would help the commercialization of 2D materials as sensor platforms.

References

1. Dubertret, B., Heine, T., Terrones, M.: The rise of two-dimensional materials. Acc. Chem. Res. **48**, 1–2 (2015). https://doi.org/10.1021/ar5004434
2. Prasad, S.V.S., Mishra, R.K., Gupta, S., Prasad, S.B., Singh, S.: Introduction, history, and origin of two dimensional (2D) materials. In: Materials Horizons: From Nature to Nanomaterials, pp. 1–9 (2021). https://doi.org/10.1007/978-981-16-3322-5_1/COVER
3. Choudhary, K., Tavazza, F.: Discovery and characterization of 2D materials and their heterostructures. Inc (2020). https://doi.org/10.1016/B978-0-12-817678-8.00001-4
4. Mueller, T., Malic, E.: Exciton physics and device application of two-dimensional transition metal dichalcogenide semiconductors. Npj 2D Mater. Appl. **2**, 1–12 (2018). https://doi.org/10.1038/s41699-018-0074-2
5. Ahn, E.C.: 2D materials for spintronic devices. Npj 2D Mater. Appl. **4** (2020). https://doi.org/10.1038/s41699-020-0152-0
6. Zhang, H.: Ultrathin two-dimensional nanomaterials. ACS Nano **9**, 9451–9469 (2015). https://doi.org/10.1021/acsnano.5b05040
7. Dutta, T., Yadav, N., Wu, Y., Cheng, G.J., Liang, X., Ramakrishna, S., Sbai, A., Gupta, R., Mondal, A., Hongyu, Z., Yadav, A.: Electronic properties of 2D materials and their junctions. Nano Mater. Sci. (2023). https://doi.org/10.1016/j.nanoms.2023.05.003
8. Kanungo, S., Ahmad, G., Sahatiya, P., Mukhopadhyay, A., Chattopadhyay, S.: 2D materials-based nanoscale tunneling field effect transistors: current developments and future prospects. Npj 2D Mater. Appl. **6** (2022). https://doi.org/10.1038/s41699-022-00352-2
9. Akinwande, D., Brennan, C.J., Bunch, J.S., Egberts, P., Felts, J.R., Gao, H., Huang, R., Kim, J.S., Li, T., Li, Y., Liechti, K.M., Lu, N., Park, H.S., Reed, E.J., Wang, P., Yakobson, B.I.,

Zhang, T., Zhang, Y.W., Zhou, Y., Zhu, Y.: A review on mechanics and mechanical properties of 2D materials—graphene and beyond. Extrem. Mech. Lett. **13**, 42–77 (2017). https://doi.org/10.1016/j.eml.2017.01.008
10. Tan, T., Jiang, X., Wang, C., Yao, B., Zhang, H.: 2D material optoelectronics for information functional device applications: status and challenges. Adv. Sci. **7** (2020). https://doi.org/10.1002/advs.202000058
11. Choi, W., Choudhary, N., Han, G.H., Park, J., Akinwande, D., Lee, Y.H.: Recent development of two-dimensional transition metal dichalcogenides and their applications. Mater. Today **20**, 116–130 (2017). https://doi.org/10.1016/j.mattod.2016.10.002
12. Shanmugam, V., Mensah, R.A., Babu, K., Gawusu, S., Chanda, A., Tu, Y., Neisiany, R.E., Försth, M., Sas, G., Das, O.: A review of the synthesis, properties, and applications of 2D materials. Part. Part. Syst. Charact. **39** (2022). https://doi.org/10.1002/ppsc.202200031
13. Baig, N.: Two-dimensional nanomaterials: a critical review of recent progress, properties, applications, and future directions. Compos. Part A Appl. Sci. Manuf. **165**, 107362 (2023). https://doi.org/10.1016/j.compositesa.2022.107362
14. Sun, P.Z., Xiong, W.Q., Bera, A., Timokhin, I., Wu, Z.F., Mishchenko, A., Sellers, M.C., Liu, B.L., Cheng, H.M., Janzen, E., Edgar, J.H., Grigorieva, I.V., Yuan, S.J., Geim, A.K.: Unexpected catalytic activity of nanorippled graphene. Proc. Natl. Acad. Sci. **120**, 2017 (2023). https://doi.org/10.1073/pnas.2300481120
15. Ryu, B., Wang, L., Pu, H., Chan, M.K.Y., Chen, J.: Understanding, discovery, and synthesis of 2D materials enabled by machine learning. Chem. Soc. Rev. **51**, 1899–1925 (2022). https://doi.org/10.1039/d1cs00503k
16. Varghese, S.S., Varghese, S.H., Swaminathan, S., Singh, K.K., Mittal, V.: Two-dimensional materials for sensing: graphene and beyond. Electron. **4**, 651–687 (2015). https://doi.org/10.3390/electronics4030651
17. Kumbhakar, P., Jayan, J.S., Sreedevi Madhavikutty, A., Sreeram, P.R., Saritha, A., Ito, T., Tiwary, C.S.: Prospective applications of two-dimensional materials beyond laboratory frontiers: a review. IScience **26**, 106671 (2023). https://doi.org/10.1016/j.isci.2023.106671
18. Wang, Q.H., Kalantar-Zadeh, K., Kis, A., Coleman, J.N., Strano, M.S.: Electronics and optoelectronics of two-dimensional transition metal dichalcogenides. Nat. Nanotechnol. **7**, 699–712 (2012). https://doi.org/10.1038/nnano.2012.193
19. Cheng, J., Gao, L., Li, T., Mei, S., Wang, C., Wen, B., Huang, W., Li, C., Zheng, G., Wang, H., Zhang, H.: Two-dimensional black phosphorus nanomaterials: emerging advances in electrochemical energy storage science. Springer Singapore (2020). https://doi.org/10.1007/s40820-020-00510-5
20. Cheng, P., Zhang, W., Zhang, L., Gou, J., Wong, P.K.J., Chen, L.: Molecular beam epitaxy fabrication of two-dimensional materials. Elsevier Inc. (2019). https://doi.org/10.1016/B978-0-12-816187-6.00004-2
21. Nandee, R., Chowdhury, M.A., Shahid, A., Hossain, N., Rana, M.: Band gap formation of 2D materialin graphene: future prospect and challenges. Results Eng. **15**, 100474 (2022). https://doi.org/10.1016/j.rineng.2022.100474
22. Cao, X., Shi, J.J., Zhang, M., Jiang, X.H., Zhong, H.X., Huang, P., Ding, Y.M., Wu, M.: Band gap opening of graphene by forming heterojunctions with the 2D carbonitrides nitrogenated holey graphene, g-C_3N_4, and g-CN: electric field effect. J. Phys. Chem. C **120**, 11299–11305 (2016). https://doi.org/10.1021/acs.jpcc.6b03308
23. Deepa, C., Rajeshkumar, L., Ramesh, M.: Preparation, synthesis, properties and characterization of graphene-based 2D nano-materials for biosensors and bioelectronics. J. Mater. Res. Technol. **19**, 2657–2694 (2022). https://doi.org/10.1016/j.jmrt.2022.06.023
24. Xie, H., Li, Z., Cheng, L., Haidry, A.A., Tao, J., Xu, Y., Xu, K., Ou, J.Z.: Recent advances in the fabrication of 2D metal oxides. IScience **25**, 1–30 (2022). https://doi.org/10.1016/j.isci.2021.103598
25. Hu, X., Liu, K., Cai, Y., Zang, S.-Q., Zhai, T.: 2D oxides for electronics and optoelectronics. Small Sci. **2**, 2200008 (2022). https://doi.org/10.1002/smsc.202200008

26. Zhou, K., Shang, G., Hsu, H., Han, S., Roy, V.A.L., Zhou, Y.: Emerging two-dimensional metal oxides: from synthesis to device integration. Adv. Mater. **2207774**, 2207774 (2022). https://doi.org/10.1002/adma.202207774
27. Guo, Y., Ma, L., Mao, K., Ju, M., Bai, Y., Zhao, J., Zeng, X.C.: Eighteen functional monolayer metal oxides: wide bandgap semiconductors with superior oxidation resistance and ultrahigh carrier mobility. Nanoscale Horiz. **4**, 592–600 (2019). https://doi.org/10.1039/c8nh00273h
28. Yuan, H., Wang, H., Cui, Y.: Two-dimensional layered chalcogenides: from rational synthesis to property control via orbital occupation and electron filling. Acc. Chem. Res. **48**, 81–90 (2015). https://doi.org/10.1021/ar5003297
29. Mia, A.K., Meyyappan, M., Giri, P.K.: Two-dimensional transition metal dichalcogenide based biosensors: from fundamentals to healthcare applications. Biosensors **13**, 1–28 (2023). https://doi.org/10.3390/bios13020169
30. Wang, Z., Wang, G., Liu, X., Wang, S., Wang, T., Zhang, S., Yu, J., Zhao, G., Zhang, L.: Two-dimensional wide band-gap nitride semiconductor GaN and AlN materials: properties, fabrication and applications. J. Mater. Chem. C **9**, 17201–17232 (2021). https://doi.org/10.1039/d1tc04022g
31. Zhang, K., Lui, V.C.H., Chen, Y., Lok, C.N., Wong, K.K.Y.: Delayed application of silver nanoparticles reveals the role of early inflammation in burn wound healing. Sci. Rep. **10**, 1–12 (2020). https://doi.org/10.1038/s41598-020-63464-z
32. Gogotsi, Y., Anasori, B.: The rise of MXenes. ACS Nano **13**, 8491–8494 (2019). https://doi.org/10.1021/acsnano.9b06394
33. Tahir, M.B., Fatima, N., Fatima, U., Sagir, M.: A review on the 2D black phosphorus materials for energy applications. Inorg. Chem. Commun. **124**, 108242 (2021). https://doi.org/10.1016/j.inoche.2020.108242
34. Deng, B., Tran, V., Xie, Y., Jiang, H., Li, C., Guo, Q., Wang, X., Tian, H., Koester, S.J., Wang, H., Cha, J.J., Xia, Q., Yang, L., Xia, F.: Efficient electrical control of thin-film black phosphorus bandgap. Nat. Commun. **8** (2017). https://doi.org/10.1038/NCOMMS14474
35. Wang, J., Li, G., Li, L.: Synthesis strategies about 2D materials. In: Two-Dimensional Materials—Synthesis, Characterization and Potential Applications, pp. 1–20 (2016). https://doi.org/10.5772/63918
36. Alwarappan, S., Nesakumar, N., Sun, D., Hu, T.Y., Li, C.Z.: 2D metal carbides and nitrides (MXenes) for sensors and biosensors. Biosens. Bioelectron. **205**, 113943 (2022). https://doi.org/10.1016/j.bios.2021.113943
37. Zhu, J., Ha, E., Zhao, G., Zhou, Y., Huang, D., Yue, G., Hu, L., Sun, N., Wang, Y., Lee, L.Y.S., Xu, C., Wong, K.Y., Astruc, D., Zhao, P.: Recent advance in MXenes: a promising 2D material for catalysis, sensor and chemical adsorption. Coord. Chem. Rev. **352**, 306–327 (2017). https://doi.org/10.1016/j.ccr.2017.09.012
38. Ali Khan, A., Tahir, M.: Constructing S-scheme heterojunction of CoAlLa-LDH/g-C_3N_4 through monolayer Ti_3C_2-MXene to promote photocatalytic CO_2 re-forming of methane to solar fuels. ACS Appl. Energy Mater. **5**, 784–806 (2022). https://doi.org/10.1021/acsaem.1c03266
39. Kumar, P., Dey, A., Roques, J., Assaud, L., Franger, S., Parida, P., Biju, V.: Photoexfoliation synthesis of 2D materials. ACS Mater. Lett. **4**, 263–270 (2022). https://doi.org/10.1021/acsmaterialslett.1c00651
40. Dong, R., Zhang, T., Feng, X.: Interface-assisted synthesis of 2D materials: trend and challenges. Chem. Rev. **118**, 6189–6325 (2018). https://doi.org/10.1021/acs.chemrev.8b00056
41. Huang, L., Hu, Z., Jin, H., Wu, J., Liu, K., Xu, Z., Wan, J., Zhou, H., Duan, J., Hu, B., Zhou, J.: Salt-assisted synthesis of 2D materials. Adv. Funct. Mater. **30**, 1–27 (2020). https://doi.org/10.1002/adfm.201908486
42. Siepi, M., Morales-Narváez, E., Domingo, N., Monti, D.M., Notomista, E., Merkoçi, A.: Production of biofunctionalized MoS_2 flakes with rationally modified lysozyme: a biocompatible 2D hybrid material. 2D Mater. **4** (2017). https://doi.org/10.1088/2053-1583/aa7966

43. Zhu, X., Li, Y.: Review of two-dimensional materials for electrochemical CO_2 reduction from a theoretical perspective. Wiley Interdiscip. Rev. Comput. Mol. Sci. **9**, 1–13 (2019). https://doi.org/10.1002/wcms.1416
44. Xie, X., Zhou, Y., Huang, K.: Advances in microwave-assisted production of reduced graphene oxide. Front. Chem. **7**, 1–11 (2019). https://doi.org/10.3389/fchem.2019.00355
45. Yu, J., Li, J., Zhang, W., Chang, H.: Synthesis of high quality two-dimensional materials via chemical vapor deposition. Chem. Sci. **6**, 6705–6716 (2015). https://doi.org/10.1039/c5sc01941a
46. Deokar, G., Jin, J., Schwingenschlögl, U., Costa, P.M.F.J.: Chemical vapor deposition-grown nitrogen-doped graphene's synthesis, characterization and applications. Npj 2D Mater. Appl. **6**, 1–17 (2022). https://doi.org/10.1038/s41699-022-00287-8
47. You, W.X., Tsai, C.P., Su, P.: Short-channel effects in 2D negative-capacitance field-effect transistors. IEEE Trans. Electron Devices **65**, 1604–1610 (2018). https://doi.org/10.1109/TED.2018.2805716
48. Bhowmik, S., Govind Rajan, A.: Chemical vapor deposition of 2D materials: a review of modeling, simulation, and machine learning studies. IScience **25**, 103832 (2022). https://doi.org/10.1016/j.isci.2022.103832
49. Momeni, K., Ji, Y., Chen, L.Q.: Computational synthesis of 2D materials grown by chemical vapor deposition. J. Mater. Res. **37**, 114–123 (2022). https://doi.org/10.1557/s43578-021-00384-2
50. Choi, S.H., Yun, S.J., Won, Y.S., Oh, C.S., Kim, S.M., Kim, K.K., Lee, Y.H.: Large-scale synthesis of graphene and other 2D materials towards industrialization. Nat. Commun. **13**, 1–5 (2022). https://doi.org/10.1038/s41467-022-29182-y
51. Xu, Y., Wang, Z., Guo, Z., Huang, H., Xiao, Q., Zhang, H., Yu, X.F.: Solvothermal synthesis and ultrafast photonics of black phosphorus quantum dots. Adv. Opt. Mater. **4**, 1223–1229 (2016). https://doi.org/10.1002/adom.201600214
52. Dong, J., Zhang, L., Dai, X., Ding, F.: The epitaxy of 2D materials growth. Nat. Commun. **11** (2020). https://doi.org/10.1038/s41467-020-19752-3
53. Shelke, N.T., Late, D.J.: Synthesis and characterization of 2D materials. Inc (2021). https://doi.org/10.1016/B978-0-12-821993-5.00003-0
54. Najafi, M.D., Kowsari, E., Naderi, H.R., Chinnappan, A., Ramakrishna, S., Ehsani, A., Shokravi, A.: Functionalization of graphene oxide via chromium complexes coordinated on 5-aminopyridine-2-carboxylic acid as a symmetric supercapacitor electrode materials in energy storage devices. Compos. Sci. Technol. **211**, 108844 (2021). https://doi.org/10.1016/j.compscitech.2021.108844
55. Huang, Y., Guo, J., Li, Y., Li, H., Fan, D.E.: 2D-material-integrated micromachines: competing propulsion strategy and enhanced bacterial disinfection. Adv. Mater. **34**, 1–10 (2022). https://doi.org/10.1002/adma.202203082
56. Liu, N., Chang, L., Wu, J., Fang, J., Xiao, X.: Emerging topochemical strategies for designing two-dimensional energy materials. Micromachines **12**, 1–17 (2021). https://doi.org/10.3390/mi12080867
57. Yang, S., Chen, G., Ricciardulli, A.G., Zhang, P., Zhang, Z., Shi, H., Ma, J., Zhang, J., Blom, P.W.M., Feng, X.: Topochemical synthesis of two-dimensional transition-metal phosphides using phosphorene templates. Angew. Chem. - Int. Ed. **59**, 465–470 (2020). https://doi.org/10.1002/anie.201911428
58. Sulleiro, M.V., Dominguez-Alfaro, A., Alegret, N., Silvestri, A., Gómez, I.J.: 2D materials towards sensing technology: from fundamentals to applications. Sens. Bio-Sens. Res. **38** (2022). https://doi.org/10.1016/j.sbsr.2022.100540
59. Nangare, S., Patil, P.: Poly(allylamine) coated layer-by-layer assembly decorated 2D carbon backbone for highly sensitive and selective detection of Tau-441 using surface plasmon resonance biosensor. Anal. Chim. Acta **1271**, 341474 (2023). https://doi.org/10.1016/j.aca.2023.341474
60. Garg, M., Pamme, N.: Microfluidic (bio)-sensors based on 2-D layered materials. TrAC - Trends Anal. Chem. **158** (2023). https://doi.org/10.1016/j.trac.2022.116839

61. Tan, A.Y.S., Lo, N.W., Cheng, F., Zhang, M., Tan, M.T.T., Manickam, S., Muthoosamy, K.: 2D carbon materials based photoelectrochemical biosensors for detection of cancer antigens. Biosens. Bioelectron. **219**, 114811 (2023). https://doi.org/10.1016/j.bios.2022.114811
62. Li, S., Ma, L., Zhou, M., Li, Y., Xia, Y., Fan, X., Cheng, C., Luo, H.: New opportunities for emerging 2D materials in bioelectronics and biosensors. Curr. Opin. Biomed. Eng. **13**, 32–41 (2020). https://doi.org/10.1016/j.cobme.2019.08.016
63. Tao, Y., Lin, Y., Ren, J., Qu, X.: A dual fluorometric and colorimetric sensor for dopamine based on BSA-stabilized Au nanoclusters. Biosens. Bioelectron. **42**, 41–46 (2013). https://doi.org/10.1016/j.bios.2012.10.014
64. Zeng, Y., Wang, M., Sun, Z., Sha, L., Yang, J., Li, G.: Colorimetric immunosensor constructed using 2D metal–organic framework nanosheets as enzyme mimics for the detection of protein biomarkers. J. Mater. Chem. B **10**, 450–455 (2022). https://doi.org/10.1039/d1tb02192c
65. Wu, Q., Li, N., Wang, Y., Liu, Y., Xu, Y., Wei, S., Wu, J., Jia, G., Fang, X., Chen, F., Cui, X.: A 2D transition metal carbide MXene-based SPR biosensor for ultrasensitive carcinoembryonic antigen detection. Biosens. Bioelectron. **144**, 111697 (2019). https://doi.org/10.1016/j.bios.2019.111697
66. Yang, H., Xu, Y., Hou, Q., Xu, Q., Ding, C.: Magnetic antifouling material based ratiometric electrochemical biosensor for the accurate detection of CEA in clinical serum. Biosens. Bioelectron. **208**, 114216 (2022). https://doi.org/10.1016/j.bios.2022.114216
67. Xu, Y., Sprick, R.S., Brownbill, N.J., Blanc, F., Li, Q., Ward, J.W., Ren, S., Cooper, A.I.: Bottom-up wet-chemical synthesis of a two-dimensional porous carbon material with high supercapacitance using a cascade coupling/cyclization route. J. Mater. Chem. A **9**, 3303–3308 (2021). https://doi.org/10.1039/d0ta11649a
68. Cao, Z., Yadav, P., Barati Farimani, A.: Which 2D material is better for DNA detection: graphene, MoS_2, or MXene?, Nano Lett. **22**, 7874–7881 (2022). https://doi.org/10.1021/acs.nanolett.2c02603
69. Cai, J., Gou, X., Sun, B., Li, W., Li, D., Liu, J., Hu, F., Li, Y.: Porous graphene-black phosphorus nanocomposite modified electrode for detection of leptin. Biosens. Bioelectron. **137**, 88–95 (2019). https://doi.org/10.1016/j.bios.2019.04.045
70. Wu, L., Xu, Z., Meng, Q., Xiao, Y., Cao, Q., Rathi, B., Liu, H., Han, G., Zhang, J., Yan, J.: A new aptamer/black phosphorous interdigital electrode for malachite green detection. Anal. Chim. Acta **1099**, 39–45 (2020). https://doi.org/10.1016/J.ACA.2019.11.026
71. Li, Y., Bu, Y., Jiang, F., Dai, X., Ao, J.P.: Fabrication of ultra-sensitive photoelectrochemical aptamer biosensor: based on semiconductor/DNA interfacial multifunctional reconciliation via 2D–C_3N_4. Biosens. Bioelectron. **150**, 111903 (2020). https://doi.org/10.1016/J.BIOS.2019.111903
72. Xiang, Y., Camarada, M.B., Wen, Y., Wu, H., Chen, J., Li, M., Liao, X.: Simple voltammetric analyses of ochratoxin A in food samples using highly-stable and anti-fouling black phosphorene nanosensor. Electrochim. Acta **282**, 490–498 (2018). https://doi.org/10.1016/j.electacta.2018.06.055
73. Zhao, H., Qiao, X., Zhang, X., Niu, C., Yue, T., Sheng, Q.: Simultaneous electrochemical aptasensing of patulin and ochratoxin A in apple juice based on gold nanoparticles decorated black phosphorus nanomaterial. Anal. Bioanal. Chem. **413**, 3131–3140 (2021). https://doi.org/10.1007/S00216-021-03253-3/METRICS
74. Xu, J., Qiao, X., Wang, Y., Sheng, Q., Yue, T., Zheng, J., Zhou, M.: Electrostatic assembly of gold nanoparticles on black phosphorus nanosheets for electrochemical aptasensing of patulin. Microchim. Acta **186**, 1–8 (2019). https://doi.org/10.1007/S00604-019-3339-3/METRICS
75. Rahaie, M., Noroozi, S.K.: A nanobiosensor based on graphene oxide and DNA binding dye for multi-microRNAs detection. Biosci. Rep. **39**, 1–11 (2019). https://doi.org/10.1042/BSR20181404
76. Sun, Z., Jin, H., Sun, Y., Jiang, X., Gui, R.: Mn-doping-induced hierarchical petal growth of a flower-like 3D MOF assembled with black phosphorous nanosheets as an electrochemical aptasensor of human stress-induced phosphoprotein 1. Nanoscale **12**, 14538–14548 (2020). https://doi.org/10.1039/D0NR02342F

77. Zhang, Z., Li, Y., Xu, J., Wen, Y.: Electropolymerized molecularly imprinted polypyrrole decorated with black phosphorene quantum dots onto poly(3,4-ethylenedioxythiophene) nanorods and its voltammetric sensing of vitamin C. J. Electroanal. Chem. **814**, 153–160 (2018). https://doi.org/10.1016/J.JELECHEM.2018.02.059
78. Garg, M., Gupta, A., Sharma, A.L., Singh, S.: Advancements in 2D materials based biosensors for oxidative stress biomarkers. ACS Appl. Bio Mater. **4**, 5944–5960 (2021). https://doi.org/10.1021/acsabm.1c00625
79. Fang, D., Zhao, D., Zhang, S., Huang, Y., Dai, H., Lin, Y.: Black phosphorus quantum dots functionalized MXenes as the enhanced dual-mode probe for exosomes sensing. Sens. Actuators, B Chem. **305**, 127544 (2020). https://doi.org/10.1016/j.snb.2019.127544
80. Xue, T., Sheng, Y., Xu, J., Li, Y., Lu, X., Zhu, Y., Duan, X., Wen, Y.: In-situ reduction of Ag^+ on black phosphorene and its NH_2-MWCNT nanohybrid with high stability and dispersibility as nanozyme sensor for three ATP metabolites. Biosens. Bioelectron. **145**, 111716 (2019). https://doi.org/10.1016/j.bios.2019.111716
81. Ge, Y., Qu, M., Xu, L., Wang, X., Xin, J., Liao, X., Li, M., Li, M., Wen, Y.: Phosphorene nanocomposite with high environmental stability and antifouling capability for simultaneous sensing of clenbuterol and ractopamine. Microchim. Acta. **186** (2019). https://doi.org/10.1007/s00604-019-3908-5
82. Niu, X., Weng, W., Yin, C., Niu, Y., Li, G., Dong, R., Men, Y., Sun, W.: Black phosphorene modified glassy carbon electrode for the sensitive voltammetric detection of rutin. J. Electroanal. Chem. **811**, 78–83 (2018). https://doi.org/10.1016/j.jelechem.2018.01.038
83. Yan, S., Wang, B., Wang, Z., Hu, D., Xu, X., Wang, J., Shi, Y.: Supercritical carbon dioxide-assisted rapid synthesis of few-layer black phosphorus for hydrogen peroxide sensing. Biosens. Bioelectron. **80**, 34–38 (2016). https://doi.org/10.1016/j.bios.2016.01.043
84. Ding, H., Tang, Z., Zhang, L., Dong, Y.: Electrogenerated chemiluminescence of black phosphorus nanosheets and its application in the detection of H_2O_2. Analyst **144**, 1326–1333 (2019). https://doi.org/10.1039/c8an01838c
85. Jiang, X., Jin, H., Sun, Y., Sun, Z., Gui, R.: Assembly of black phosphorus quantum dots-doped MOF and silver nanoclusters as a versatile enzyme-catalyzed biosensor for solution, flexible substrate and latent fingerprint visual detection of baicalin. Biosens. Bioelectron. **152**, 112012 (2020). https://doi.org/10.1016/j.bios.2020.112012
86. Vilela, P., El-Sagheer, A., Millar, T.M., Brown, T., Muskens, O.L., Kanaras, A.G.: Graphene oxide-upconversion nanoparticle based optical sensors for targeted detection of mRNA biomarkers present in Alzheimer's disease and prostate cancer. ACS Sens. **2**, 52–56 (2017). https://doi.org/10.1021/acssensors.6b00651
87. Hua, X., Zhou, X., Guo, S., Zheng, T., Yuan, R., Xu, W.: Determination of Alzheimer biomarker DNA by using an electrode modified with in-situ precipitated molybdophosphate catalyzed by alkaline phosphatase-encapsulated DNA hydrogel and target recycling amplification. Microchim. Acta **186**, 2–9 (2019). https://doi.org/10.1007/s00604-019-3283-2
88. Liu, S., Luo, J., Jiang, X., Li, X., Yang, M.: Gold nanoparticle–modified black phosphorus nanosheets with improved stability for detection of circulating tumor cells. Microchim. Acta. **187** (2020). https://doi.org/10.1007/s00604-020-04367-8
89. Tabrizi, M.A., Ferré-Borrull, J., Kapruwan, P., Marsal, L.F.: A photoelectrochemical sandwich immunoassay for protein S100β, a biomarker for Alzheimer's disease, using an ITO electrode modified with a reduced graphene oxide-gold conjugate and CdS-labeled secondary antibody. Microchim. Acta **186** (2019). https://doi.org/10.1007/s00604-018-3159-x
90. Zhou, J., Meng, L., Ye, W., Wang, Q., Geng, S., Sun, C.: A sensitive detection assay based on signal amplification technology for Alzheimer's disease's early biomarker in exosome. Anal. Chim. Acta **1022**, 124–130 (2018). https://doi.org/10.1016/j.aca.2018.03.016
91. Mathew, G., Dey, P., Das, R., Chowdhury, S.D., Paul Das, M., Veluswamy, P., Neppolian, B., Das, J.: Direct electrochemical reduction of hematite decorated graphene oxide (α-Fe_2O_3@erGO) nanocomposite for selective detection of Parkinson's disease biomarker. Biosens. Bioelectron. **115**, 53–60 (2018). https://doi.org/10.1016/j.bios.2018.05.024

92. Mannoor, M.S., Tao, H., Clayton, J.D., Sengupta, A., Kaplan, D.L., Naik, R.R., Verma, N., Omenetto, F.G., McAlpine, M.C.: Graphene-based wireless bacteria detection on tooth enamel. Nat. Commun. **3** (2012). https://doi.org/10.1038/ncomms1767
93. Thakur, B., Zhou, G., Chang, J., Pu, H., Jin, B., Sui, X., Yuan, X., Yang, C.H., Magruder, M., Chen, J.: Rapid detection of single E. coli bacteria using a graphene-based field-effect transistor device. Biosens. Bioelectron. **110**, 16–22 (2018). https://doi.org/10.1016/j.bios.2018.03.014
94. Pandey, A., Gurbuz, Y., Ozguz, V., Niazi, J.H., Qureshi, A.: Graphene-interfaced electrical biosensor for label-free and sensitive detection of foodborne pathogenic E. coli O157:H7. Biosens. Bioelectron. **91**, 225–231 (2017). https://doi.org/10.1016/j.bios.2016.12.041
95. Singh, C., Ali, M.A., Kumar, V., Ahmad, R., Sumana, G.: Functionalized MoS_2 nanosheets assembled microfluidic immunosensor for highly sensitive detection of food pathogen. Sens. Actuators, B Chem. **259**, 1090–1098 (2018). https://doi.org/10.1016/j.snb.2017.12.094
96. Jijie, R., Kahlouche, K., Barras, A., Yamakawa, N., Gharbi, T., Szunerits, S., Boukherroub, R., Jijie, R., Kahlouche, K., Barras, A., Yamakawa, N., Bouckaert, J.: Reduced graphene oxide/polyethylenimine based immunosensor for the selective and sensitive electrochemical detection of uropathogenic Escherichia coli. To cite this version: HAL Id: hal-01925128 (2019)
97. Wang, Y., Ping, J., Ye, Z., Wu, J., Ying, Y.: Impedimetric immunosensor based on gold nanoparticles modified graphene paper for label-free detection of Escherichia coli O157: H7. Biosens. Bioelectron. **49**, 492–498 (2013). https://doi.org/10.1016/j.bios.2013.05.061
98. Li, Z., Fu, Y., Zhu, Q., Wei, S., Gao, J., Zhu, Y., Xue, T., Bai, L., Wen, Y.: High-stable phosphorene-supported bimetallic Pt-Pd nanoelectrocatalyst for p-Aminophenol, β-Galactosidase, and escherichia coli. Int. J. Electrochem. Sci. **15**, 3089–3103 (2020). https://doi.org/10.20964/2020.04.41
99. Wen, J., Zhou, S., Yuan, Y.: Graphene oxide as nanogold carrier for ultrasensitive electrochemical immunoassay of Shewanella oneidensis with silver enhancement strategy. Biosens. Bioelectron. **52**, 44–49 (2014). https://doi.org/10.1016/j.bios.2013.08.022
100. Bhardwaj, N., Bhardwaj, S.K., Mehta, J., Mohanta, G.C., Deep, A.: Bacteriophage immobilized graphene electrodes for impedimetric sensing of bacteria (Staphylococcus arlettae). Anal. Biochem. **505**, 18–25 (2016). https://doi.org/10.1016/j.ab.2016.04.008
101. Wan, Y., Lin, Z., Zhang, D., Wang, Y., Hou, B.: Impedimetric immunosensor doped with reduced graphene sheets fabricated by controllable electrodeposition for the non-labelled detection of bacteria. Biosens. Bioelectron. **26**, 1959–1964 (2011). https://doi.org/10.1016/j.bios.2010.08.008
102. Fei, J., Dou, W., Zhao, G.: Amperometric immunoassay for the detection of Salmonella pullorum using a screen-printed carbon electrode modified with gold nanoparticle-coated reduced graphene oxide and immunomagnetic beads. Microchim. Acta **183**, 757–764 (2016). https://doi.org/10.1007/s00604-015-1721-3
103. Hu, X., Dou, W., Fu, L., Zhao, G.: A disposable immunosensor for Enterobacter sakazakii based on an electrochemically reduced graphene oxide-modified electrode. Anal. Biochem. **434**, 218–220 (2013). https://doi.org/10.1016/j.ab.2012.11.013
104. Ma, X., Jiang, Y., Jia, F., Yu, Y., Chen, J., Wang, Z.: An aptamer-based electrochemical biosensor for the detection of Salmonella. J. Microbiol. Methods **98**, 94–98 (2014). https://doi.org/10.1016/j.mimet.2014.01.003
105. Li, B., Tan, H., Anastasova, S., Power, M., Seichepine, F., Yang, G.Z.: A bio-inspired 3D micro-structure for graphene-based bacteria sensing. Biosens. Bioelectron. **123**, 77–84 (2019). https://doi.org/10.1016/j.bios.2018.09.087
106. Wu, S., Duan, N., Qiu, Y., Li, J., Wang, Z.: Colorimetric aptasensor for the detection of Salmonella enterica serovar typhimurium using $ZnFe_2O_4$-reduced graphene oxide nanostructures as an effective peroxidase mimetics. Int. J. Food Microbiol. **261**, 42–48 (2017). https://doi.org/10.1016/j.ijfoodmicro.2017.09.002
107. Kaushal, S., Pinnaka, A.K., Soni, S., Singhal, N.K.: Antibody assisted graphene oxide coated gold nanoparticles for rapid bacterial detection and near infrared light enhanced antibacterial

activity. Sens. Actuators, B Chem. **329**, 129141 (2021). https://doi.org/10.1016/j.snb.2020.129141
108. Guo, Y., Deng, L., Li, J., Guo, S., Wang, E., Dong, S.: Hemin–graphene hybrid nanosheets with intrinsic peroxidase-like activity for label-free colorimetric detection of single-nucleotide polymorphism. ACS Nano **5**, 1282–1290 (2011). https://doi.org/10.1021/nn1029586
109. Liu, F., Choi, K.S., Park, T.J., Lee, S.Y., Seo, T.S.: Graphene-based electrochemical biosensor for pathogenic virus detection. Biochip J. **5**, 123–128 (2011). https://doi.org/10.1007/s13206-011-5204-2
110. Zhan, L., Li, C.M., Wu, W.B., Huang, C.Z.: A colorimetric immunoassay for respiratory syncytial virus detection based on gold nanoparticles-graphene oxide hybrids with mercury-enhanced peroxidase-like activity. Chem. Commun. **50**, 11526–11528 (2014). https://doi.org/10.1039/c4cc05155f
111. Jin, S.A., Poudyal, S., Marinero, E.E., Kuhn, R.J., Stanciu, L.A.: Impedimetric dengue biosensor based on functionalized graphene oxide wrapped silica particles. Electrochim. Acta **194**, 422–430 (2016). https://doi.org/10.1016/j.electacta.2016.02.116
112. Ahmed, S.R., Takemeura, K., Li, T.C., Kitamoto, N., Tanaka, T., Suzuki, T., Park, E.Y.: Size-controlled preparation of peroxidase-like graphene-gold nanoparticle hybrids for the visible detection of norovirus-like particles. Biosens. Bioelectron. **87**, 558–565 (2017). https://doi.org/10.1016/j.bios.2016.08.101
113. Gao, Z., Li, Y., Zhang, X., Feng, J., Kong, L., Wang, P., Chen, Z., Dong, Y., Wei, Q.: Ultrasensitive electrochemical immunosensor for quantitative detection of HBeAg using Au@Pd/MoS_2@MWCNTs nanocomposite as enzyme-mimetic labels. Biosens. Bioelectron. **102**, 189–195 (2018). https://doi.org/10.1016/j.bios.2017.11.032
114. Fathi-Hafshejani, P., Azam, N., Wang, L., Kuroda, M.A., Hamilton, M.C., Hasim, S., Mahjouri-Samani, M.: Two-dimensional-material-based field-effect transistor biosensor for detecting COVID-19 virus (SARS-CoV-2). ACS Nano **15**, 11461–11469 (2021). https://doi.org/10.1021/acsnano.1c01188
115. Chen, W.Y., Lin, H., Barui, A.K., Gomez, A.M.U., Wendt, M.K., Stanciu, L.A.: DNA-functionalized $Ti_3C_2T_x$ MXenes for selective and rapid detection of SARS-CoV-2 nucleocapsid gene. ACS Appl. Nano Mater. **5**, 1902–1910 (2022). https://doi.org/10.1021/acsanm.1c03520
116. Chekin, F., Bagga, K., Subramanian, P., Jijie, R., Singh, S.K., Kurungot, S., Boukherroub, R., Szunerits, S.: Nucleic aptamer modified porous reduced graphene oxide/MoS_2 based electrodes for viral detection: application to human papillomavirus (HPV). Sens. Actuators, B Chem. **262**, 991–1000 (2018). https://doi.org/10.1016/j.snb.2018.02.065
117. Shao, J., Ruan, C., Xie, H., Li, Z., Wang, H., Chu, P.K., Yu, X.F.: Black-phosphorus-incorporated hydrogel as a sprayable and biodegradable photothermal platform for postsurgical treatment of cancer. Adv. Sci. **5** (2018). https://doi.org/10.1002/advs.201700848
118. Ye, X., Liang, X., Chen, Q., Miao, Q., Chen, X., Zhang, X., Mei, L.: Surgical tumor-derived personalized photothermal vaccine formulation for cancer immunotherapy. ACS Nano **13**, 2956–2968 (2019). https://doi.org/10.1021/acsnano.8b07371
119. Hsieh, T.Y., Huang, W.C., Da Kang, Y., Chu, C.Y., Liao, W.L., Chen, Y.Y., Chen, S.Y.: Neurotensin-conjugated reduced graphene oxide with multi-stage near-infrared-triggered synergic targeted neuron gene transfection in vitro and in vivo for neurodegenerative disease therapy. Adv. Healthc. Mater. **5**, 3016–3026 (2016). https://doi.org/10.1002/adhm.201600647
120. Hassan, J.Z., Raza, A., Din Babar, Z.U., Qumar, U., Kaner, N.T., Cassinese, A.: 2D material-based sensing devices: an update. J. Mater. Chem. A 6016–6063 (2023). https://doi.org/10.1039/d2ta07653e
121. Munteanu, R.E., Moreno, P.S., Bramini, M., Gáspár, S.: 2D materials in electrochemical sensors for in vitro or in vivo use. Anal. Bioanal. Chem. (2020). https://doi.org/10.1007/s00216-020-02831-1
122. Lu, W., Yu, P., Jian, M., Wang, H., Wang, H., Liang, X., Zhang, Y.: Molybdenum disulfide nanosheets aligned vertically on carbonized silk fabric as smart textile for wearable pressure-sensing and energy devices. ACS Appl. Mater. Interfaces **12**, 11825–11832 (2020). https://doi.org/10.1021/acsami.9b21068

123. Liao, C., Mak, C., Zhang, M., Chan, H.L.W., Yan, F.: Flexible organic electrochemical transistors for highly selective enzyme biosensors and used for saliva testing. Adv. Mater. **27**, 676–681 (2015). https://doi.org/10.1002/adma.201404378
124. Trung, T.Q., Duy, L.T., Ramasundaram, S., Lee, N.E.: Transparent, stretchable, and rapid-response humidity sensor for body-attachable wearable electronics. Nano Res. **10**, 2021–2033 (2017). https://doi.org/10.1007/s12274-016-1389-y
125. Park, D.W., Schendel, A.A., Mikael, S., Brodnick, S.K., Richner, T.J., Ness, J.P., Hayat, M.R., Atry, F., Frye, S.T., Pashaie, R., Thongpang, S., Ma, Z., Williams, J.C.: Graphene-based carbon-layered electrode array technology for neural imaging and optogenetic applications. Nat. Commun. **5** (2014). https://doi.org/10.1038/ncomms6258
126. Lou, C., Li, R., Li, Z., Liang, T., Wei, Z., Run, M., Yan, X., Liu, X.: Flexible graphene electrodes for prolonged dynamic ECG monitoring. Sensors (Switzerland) **16**, 1–12 (2016). https://doi.org/10.3390/s16111833
127. Yapici, M.K., Alkhidir, T., Samad, Y.A., Liao, K.: Graphene-clad textile electrodes for electrocardiogram monitoring. Sens. Actuators B Chem. **221**, 1469–1474 (2015). https://doi.org/10.1016/j.snb.2015.07.111
128. Kim, J., Kim, M., Lee, M.S., Kim, K., Ji, S., Kim, Y.T., Park, J., Na, K., Bae, K.H., Kim, H.K., Bien, F., Lee, C.Y., Park, J.U.: Wearable smart sensor systems integrated on soft contact lenses for wireless ocular diagnostics. Nat. Commun. **8** (2017). https://doi.org/10.1038/ncomms14997
129. Park, M., Park, Y.J., Chen, X., Park, Y.K., Kim, M.S., Ahn, J.H.: MoS_2-based tactile sensor for electronic skin applications. Adv. Mater. **28**, 2556–2562 (2016). https://doi.org/10.1002/adma.201505124
130. Ameri, S.K., Kim, M., Kuang, I.A., Perera, W.K., Alshiekh, M., Jeong, H., Topcu, U., Akinwande, D., Lu, N.: Imperceptible electrooculography graphene sensor system for human–robot interface. Npj 2D Mater. Appl. **2**, 1–7 (2018). https://doi.org/10.1038/s41699-018-0064-4
131. Yang, Z., Pang, Y., Han, X.L., Yang, Y., Yang, Y., Ling, J., Jian, M., Zhang, Y., Ren, T.L.: Graphene textile strain sensor with negative resistance variation for human motion detection. ACS Nano **12**, 9134–9141 (2018). https://doi.org/10.1021/acsnano.8b03391
132. Liu, W., Liu, N., Yue, Y., Rao, J., Cheng, F., Su, J., Liu, Z., Gao, Y.: Piezoresistive pressure sensor based on synergistical innerconnect polyvinyl alcohol nanowires/wrinkled graphene film. Small **14**, 1–8 (2018). https://doi.org/10.1002/smll.201704149
133. Song, Z., Li, W., Bao, Y., Wang, W., Liu, Z., Han, F., Han, D., Niu, L.: Bioinspired microstructured pressure sensor based on a Janus graphene film for monitoring vital signs and cardiovascular assessment. Adv. Electron. Mater. **4**, 1–9 (2018). https://doi.org/10.1002/aelm.201800252
134. Cheng, Y., Wang, R., Sun, J., Gao, L.: A stretchable and highly sensitive graphene-based fiber for sensing tensile strain, bending, and torsion. Adv. Mater. **27**, 7365–7371 (2015). https://doi.org/10.1002/adma.201503558
135. Yue, Y., Liu, N., Liu, W., Li, M., Ma, Y., Luo, C., Wang, S., Rao, J., Hu, X., Su, J., Zhang, Z., Huang, Q., Gao, Y.: 3D hybrid porous Mxene-sponge network and its application in piezoresistive sensor. Nano Energy **50**, 79–87 (2018). https://doi.org/10.1016/j.nanoen.2018.05.020
136. Liu, Q., Chen, J., Li, Y., Shi, G.: High-performance strain sensors with fish-scale-like graphene-sensing layers for full-range detection of human motions. ACS Nano **10**, 7901–7906 (2016). https://doi.org/10.1021/acsnano.6b03813
137. Ma, Y., Yue, Y., Zhang, H., Cheng, F., Zhao, W., Rao, J., Luo, S., Wang, J., Jiang, X., Liu, Z., Liu, N., Gao, Y.: 3D synergistical MXene/reduced graphene oxide aerogel for a piezoresistive sensor. ACS Nano **12**, 3209–3216 (2018). https://doi.org/10.1021/acsnano.7b06909
138. Lee, Y., Park, J., Cho, S., Shin, Y.E., Lee, H., Kim, J., Myoung, J., Cho, S., Kang, S., Baig, C., Ko, H.: Flexible ferroelectric sensors with ultrahigh pressure sensitivity and linear response over exceptionally broad pressure range. ACS Nano **12**, 4045–4054 (2018). https://doi.org/10.1021/acsnano.8b01805

139. Abdel-Karim, R., Reda, Y., Abdel-Fattah, A.: Review—nanostructured materials-based nanosensors. J. Electrochem. Soc. **167**, 037554 (2020). https://doi.org/10.1149/1945-7111/ab67aa
140. Kim, S., Nam, S.N., Park, C.M., Jang, V, Taheri-Qazvini, N., Yoon, Y.: Effect of single and multilayered $Ti_3C_2T_X$ MXene as a catalyst and adsorbent on enhanced sonodegradation of diclofenac and verapamil. J. Hazard. Mater. **426** (2022). https://doi.org/10.1016/j.jhazmat.2021.128120
141. Wang, Y., Li, T., Li, Y., Yang, R., Zhang, G.: 2D-materials-based wearable biosensor systems. Biosensors **12**, 1–19 (2022). https://doi.org/10.3390/bios12110936
142. Zeng, R., Gong, H., Li, Y., Li, Y., Lin, W., Tang, D., Knopp, D.: CRISPR-Cas12a-derived photoelectrochemical biosensor for point-of-care diagnosis of nucleic acid. Anal. Chem. (2022). https://doi.org/10.1021/acs.analchem.2c01373
143. Arora, S., Peters, A.L., Agy, C., Menchine, M.: A mobile health intervention for inner city patients with poorly controlled diabetes: proof-of-concept of the TExT-MED program. Diabetes Technol. Ther. **14**, 492–496 (2012). https://doi.org/10.1089/dia.2011.0252
144. Rohaizad, N., Mayorga-Martinez, C.C., Fojtů, M., Latiff, N.M., Pumera, M.: Two-dimensional materials in biomedical, biosensing and sensing applications. Chem. Soc. Rev. **50**, 619–657 (2021). https://doi.org/10.1039/d0cs00150c
145. Pal, A., Jana, T.K., Roy, T., Pradhan, A., Maiti, R., Choudhury, S.M., Chatterjee, K.: MoS_2-TiO_2 nanocomposite with excellent adsorption performance and high antibacterial activity. ChemistrySelect **3**, 81–90 (2018). https://doi.org/10.1002/slct.201702618
146. Pandit, S., Karunakaran, S., Boda, S.K., Basu, B., De, M.: High antibacterial activity of functionalized chemically exfoliated MoS_2. ACS Appl. Mater. Interfaces **8**, 31567–31573 (2016). https://doi.org/10.1021/acsami.6b10916
147. Yang, X., Li, J., Liang, T., Ma, C., Zhang, Y., Chen, H., Hanagata, N., Su, H., Xu, M.: Antibacterial activity of two-dimensional MoS_2 sheets. Nanoscale **6**, 10126–10133 (2014). https://doi.org/10.1039/C4NR01965B
148. Appel, J.H., Li, D.O., Podlevsky, J.D., Debnath, A., Green, A.A., Wang, Q.H., Chae, J.: Low cytotoxicity and genotoxicity of two-dimensional MoS_2 and WS_2. ACS Biomater. Sci. Eng. **2**, 361–367 (2016). https://doi.org/10.1021/acsbiomaterials.5b00467
149. Bianco, A., Cheng, H.M., Enoki, T., Gogotsi, Y., Hurt, R.H., Koratkar, N., Kyotani, T., Monthioux, M., Park, C.R., Tascon, J.M.D., Zhang, J.: All in the graphene family—a recommended nomenclature for two-dimensional carbon materials. Carbon N. Y. **65**, 1–6 (2013). https://doi.org/10.1016/j.carbon.2013.08.038
150. Naikoo, G.A., Arshad, F., Almas, M., Hassan, I.U., Pedram, M.Z., Aljabali, A.A.A., Mishra, V., Serrano-Aroca, Á., Birkett, M., Charbe, N.B., Goyal, R., Negi, P., El-Tanani, M., Tambuwala, M.M.: 2D materials, synthesis, characterization and toxicity: a critical review. Chem. Biol. Interact. **365** (2022). https://doi.org/10.1016/j.cbi.2022.110081
151. Tyagi, D., Wang, H., Huang, W., Hu, L., Tang, Y., Guo, Z., Ouyang, Z., Zhang, H.: Recent advances in two-dimensional-material-based sensing technology toward health and environmental monitoring applications. Nanoscale **12**, 3535–3559 (2020). https://doi.org/10.1039/c9nr10178k
152. Fan, Q., Wang, L., Xu, D., Duo, Y., Gao, J., Zhang, L., Wang, X., Chen, X., Li, J., Zhang, H.: Solution-gated transistors of two-dimensional materials for chemical and biological sensors: status and challenges. Nanoscale **12**, 11364–11394 (2020). https://doi.org/10.1039/d0nr01125h
153. Sadighbayan, D., Minhas-Khan, A., Ghafar-Zadeh, E.: Laser-induced graphene-functionalized field-effect transistor-based biosensing: a potent candidate for COVID-19 detection. IEEE Trans. Nanobiosci. **21**, 232–245 (2022). https://doi.org/10.1109/TNB.2021.3119996

Two-Dimensional Material-Based Novel Drug Delivery System

Subhamoy Banerjee and Ruchira Mukherjee

Abstract The research on nanoscience and technology has unraveled many exciting properties of materials which are markedly different from its bulk phase. The invention of graphene not only gifted an interesting molecule with many new properties; but also it motivated the researchers to work on the whole new set of materials which are atomically thin in one direction. These materials are called 2-dimensional material or 2D material. In this book chapter, our primary area of discussion is on 2D material-based drug delivery. We have classified the 2D materials into six different types namely graphene, transition metal dichalcogenides, MXenes, black phosphorus, hexagonal boron nitride, and graphitic carbon nitirde and briefly discussed its properties. The unique properties like very high surface area, optical tunability, low or no photobleaching, etc., made them attractive theranostic agent. Most of the materials are good contrasting agent. Also, they can exert photodynamic therapy, generate reactive oxygen species, and take advantage of size modification to kill bacterial and cancer cells. Recent research works have been discussed in the current chapter.

Keywords 2D materials · Graphene · Graphene oxides · TMD · MXene · Black phosphorus · Cancer therapeutics

1 Introduction

Reduction in size of a material to the nanoscale unravels many new properties. Some of the unique properties of the nanoparticles which are different than their bulk counterparts are high surface-to-volume ratio, unique optical properties for metal and metal oxide nanoparticles, improved catalytic activities owing to higher surface area, quantum confinement effect which is more prominent in nanoscale. However the nanoparticles are three-dimensional materials. The reduction of size

S. Banerjee (✉) · R. Mukherjee
Department of Basic Science and Humanities, Institute of Engineering and Management, University of Engineering and Management, Management House, D1, Sector-V, Salt Lake Electronics Complex, Kolkata 700091, West Bengal, India
e-mail: subhamoy.banerjee@iem.edu.in

N. Talreja et al. (eds.), *Two-dimensional Hybrid Composites*, Engineering Materials,
https://doi.org/10.1007/978-981-99-8010-9_9

takes place in all three dimensions (e.g.—nanoparticles or nanocubes) or either of any dimensions (e.g.—nanotube). But reduction in dimensionality further offers many new possibilities. The discovery of graphene by Giem and Novoselov in 2004 opened the world of 2 dimensional (2D) materials [1]. Typically 2 dimensional materials are one or few molecules thick nanomaterials. Graphene is the best example of 2 dimensional materials which is one atom thick hexagonal lattice of carbon atoms. The properties of 2D materials are as follows.

1.1 Quantum Effect

The reduction of dimension leads to quantum confinement effect. Quantum confinement appears when the dimension of the material approaches de Broglie wavelength of the electron. The energy level becomes discrete and forms bandgap. As a result, bandgap exhibits size-dependent effects and optical and electronic properties can be modulated by controlling size. In 2D material like graphene nanoribbon, quantum confinement leads to transform its semimetallic property to semiconducting property. Also, the quantum tunneling effect is observed in 2D material. Excitons (the bound state of electron and hole) are enhanced due to Coulombic interaction in 2D materials which increases optical properties [2, 3].

1.2 Mechanical Strength

As the 2D sheets are characterized as one or few atoms thick structure, they show wonderful mechanical properties like intrinsic strength and stiffness. Some of 2D materials exhibit anisotropic mechanical property like differential response to stress depending on the direction or angle of applied force. The fracture toughness of the 2D materials is relatively low which means the structure may collapse if a critical crack length is reached. Although, they are highly flexible and bendable. This makes them attractive candidates for foldable electronic gadgets [4].

1.3 Optical Property

Due to the quantum confinement effect, 2D materials show tunable and diverse optical properties. Also, due to their similar refractive index with their surrounding, 2D materials show low reflectance. Some of the transition metal dichalcogenides show strong photoluminescence. Nonlinear optical property is also exhibited by different 2D materials like graphene, metal chalcogenides, black phosphorus, hexagonal boron nitride, etc. [2].

1.4 Electrical Property

Some of the 2D material like metal chalcogenides and graphene is conductive owing to the high electron mobility over its surface. Electrical conductivity of dichalcogenides depends on their finite bandgap, which confers both semiconducting and metallic properties. Being a 2D material, the defects and doping can be introduced and conduction property can be modified [5].

1.5 Thermal Conductivity

Thermal conductivity is the ability of a material to conduct heat. Being 2D material, graphene has superior thermal conductivity due to its sp^2 bonding. Phonon scattering, lattice vibration, etc., confers 2D material excellent thermal conductivity. Also, 2D-layered materials exhibit anisotropic thermal conductivity. Also, they exhibit thermoelectric property [6].

1.6 Chemical Reactivity

Owing to its high surface-to-volume ratio, 2D materials exhibit high chemical reactivity. As the 2D materials are prone to oxidation, surface modification with different functional groups is comparatively easier. As a result, 2D materials can have more than one reactive property. Due to this property, they can interact with varieties of substrates and can act as environmental sensor [7].

Due to its unique properties, 2D materials have been investigated to construct supercapacitor, ultra-sensitive sensors for nanolectronics, biosensors, drug delivery platform, and many more. In the current chapter, we shall briefly introduce different types of 2D materials (Fig. 1), and its application in biomedical field. Next we shall discuss the role of 2D materials as a drug delivery agent in detail.

2 Types of 2D Materials

Researchers are working on different types of 2D materials. Here some of the most prominent 2D materials will be discussed. An overview of properties and application of the 2D materials are discussed in Table 1.

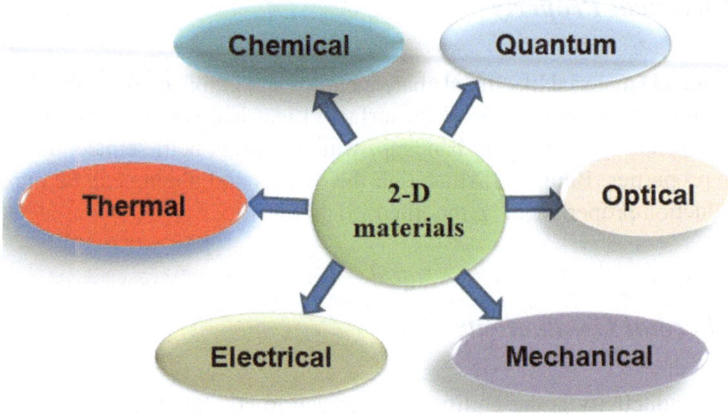

Fig. 1 Different properties of 2D materials which have promising applications in science engineering, technology, and medicine

2.1 Graphene

It is a hexagonal arrangement of carbon atoms with one or few atoms thickness. It has a honeycomb structure with sp^2 hybridization with inter-atomic distance of 0.142 nm and lattice constant of 0.246 nm. The presence of three σ bonds in each lattice stabilizes the structure. The presence of π bonds vertical to the lattice plane confers the electrical conductivity property of graphene. Graphene nanoribbon is a structure of interest as energy barrier changes with size. It has huge potential in the manufacturing of electronic gadgets. Graphene has nearly 98% transparency in visible light owing to its one atom thickness. Multilayer graphene has tunable optical property with diverse applications. Graphene is the most conductive material at room temperature given its excellent electron mobility. The conductivity of graphene is 10^6 S/m. The sp^2 hybridized bond shares an extra π electron which is mobile in room temperature. The partial overlapping of conduction band and valence band makes graphene a semimetal. The electrons present at the higher energy level of valence and move freely to conduction band without any external energy. This type of electron transport in graphene is called anomalous quantum Hall effect. The distribution of electrons at valence and conduction band makes a cone shape for both. The point at which both cones meet is called Dirac point. Due to these wonderful properties, graphene is applied to electronic devices, supercapacitors, drug delivery vehicle, highly efficient catalyst, etc. [8].

Table 1 Different key properties and proposed or potential applications of different 2D materials

Material	Properties	Potential application
Graphene	Monolayer of C atom, sp^2 bonded, bond length 0.142 nm, lattice constant 0.246 nm, tensile strength 130 GPa, Young's modulus 1 TPa, conductivity 10^6 S/m	Smart window, solar panel, touch panel, bioengineering, filtration, composite material, energy storage
Transition metal dichalcogenides	Thickness varies depending on composites, e.g. MoS_2 0.65 nm thick, spin–orbit coupling results in spin–orbit splitting of few hundred meV valence band and few meV conduction band, e.g.—MoS_2 148 meV (VB) and 3 meV (CB), composed of 3 atomic planes, hexagonal lattice, 2nd harmonic generation	Nonlinear optics, valleytronics (manipulation of K valley for quantum computation), sensor
MXenes	Higher X-ray attenuation, dark color, reflect IR, ceramic photothermal agent, although properties vary with type of MXene composite	Energy storage, super-capacitor, photocatalysis, water purification, SERS substrate, photonic diode, triboelectric generator, imaging agent
Black phosphorus	Heat of formation −39 kJ/mol, orthorhombic planed, honeycomb structure of phosphorus, $3s^2 3p^3$ electronic configuration	Sensor technology, imaging, drug delivery
Hexagonal boron nitride (hBN)	Low density (2.27 g cm^{-3}), high melting point (2600 °C), band gap −5.9 eV, nonreactive	Sensor, photodetector, piezoelectric device
Graphitic carbon nitride	Low band gap energy (2.7 eV), low electrical conductivity (0.9×10^{-9} Sm^{-1}), tunable band gap energy	Dye degradation, hydrogen generation, LED, filtration, and air purification

2.2 *Transition Metal Dichalcogenides (TMDs)*

The semiconductor property of TMDs is of special interest as 2D material. The common formula of TMD is MX_2 where M is the metal such as Molybdenum (Mo) or Tungsten (W) and X denotes chalcogenides which is salt of chalcogens (Group 16 elements of the periodic table such as sulfur (S), selenium (Se), and tellurium (Te)). TMDs have energetically most stable structures are 2H (trigonal prismatic) or 1T phase (octahedral). The stacking configuration and distortion further attribute to the structural variability of TMDs. MoS_2 and WS_2 are two TMDs naturally found in nature in layered crystal form [9]. TMDs are superconductive in nature. The superconducting state of the 2D TMDs is of the Kosterlitz–Thouless type, which was proven by the voltage–current characteristic measured at the critical temperature (3 K) [9].

2.3 MXenes

Another important 2D material with huge potential for application in biomedical field is known as MXenes. It is a group of material consisting of early transition metals (M) with intermittent stacking of various layers of C and N (X) and terminated with a surface functional group T_x/T_z. The general formula is $M_{n+1}X_nT_x$, where $n = 1–3$ [10]. First 2D MXene was synthesized by hydrofluoric acid (HF) mediated exfoliation of bulk titanium aluminum carbide (Ti_3AlC_2) (MAX phase) to 2D titanium carbide (Ti_3C_2) layers. The precursor MAX phase has a general formula $M_{n+1}AX_n$ with ternary carbide or nitride. In MAX phase, M is early transition metal, A is group IIIA or IVA elements and X is either carbon or nitrogen with $n = 1–3$. The exfoliation of A happens as M-X bond is stronger than M-A bond. The early transition metals are group III Scandium (Sc) group IV Titanium (Ti), Zirconium (Zr) and Hafnium (Hf) and group V Chromium (Cr) and Molybdenum (Mb). As the MXenes are high atomic number materials, they have better X-ray attenuation capacity and used in CT imaging.

2.4 Black Phosphorus (BP)

BP is a new member of layered material family. It is stable form of phosphorus at room temperature. It is a layered structure of phosphorus atoms like graphene. Phosphorus has $3s^2 3p^3$ electronic configuration in BP and is bound with each other via sp^3 hybridization. As a result, each P atom is bound with four adjacent P atoms and creates a pucker structure. As a result of puckering, each single layer of BP has two layers and two different types of interatomic binding. It has a band gap energy of 0.3–1.5 eV. Although the band gap strongly depends on number of layers. The integration ability of BP with silicon has made it an interesting candidate for optoelectronic devices. The narrow band gap of BP is advantageous for infrared detection, and fiber optics-based communication. Atomic force microscopy (AFM) study revealed the zigzag and arm chair conformation for BP. Direct band gap, high ON–OFF current ratio, surface electron mobility, etc., contribute to its excellent electrical property. BP nanosheet has been used as field effect transistor (FET) for sensing application. BP nanosheet coupled with gold nanoparticles and antibodies used to determine IgG [11]. BP is also explored as a potential electrochemical biosensors. Aptamer functionalized BP is used to determine cardiac myoglobin as a biomarker [12]. BP nanoparticles (BPNP) were also used for fluorescent-based detection. The tunable size, composition, etc., confer Black phosphorus quantum dots (BPQDs) higher quantum yield, photostability, and wide range of desired fluorescence [13]. The instability of BP in aqueous environment is exploited in colorimetric biosensing assay.

2.5 *Hexagonal Boron Nitride (hBN)*

Hexagonal boron nitride is a layered material similar to graphene and BP. hBN is single atom thin with alternating arrangement of B and N. The band gap in hBN is almost −5.9 eV, thus it acts as an efficient insulator. It is of special interest due to its high thermal conductivity, non-reactive nature, and tribological properties make it a lubricant and ceramic material. It can be deposited on graphene as monolayer and the heterostructure of graphene and hBN nanosheet exhibit superior electrical properties and exact planar structure. hBN is an excellent dielectric substrate which can be used in FET. The planar and compact structure is used as a spacer in tunneling devices. The hBN stacked graphene heterostructure is used as a pressure sensor, spin-valve devices, efficient photo detectors, nano-generators, etc. 2D hBN nanosheets are not symmetric crystals. As a result, it exhibits piezoelectric effect [14]. This property is exploited in fabrication of actuators, more efficient sensors, and nanoelectronic devices. hBN exhibits 99% transparency in near UV to far red spectrum (250–900 nm). The optical band gap of monolayer hBN is 6.07 eV [14]. The wide optical band gap and luminescent property in UV region make hBN as attractive candidate for photon emission, UV laser source, and deep UV detectors. hBN monolayer has much improved mechanical and thermal properties. It is used with polymers to improve different physical properties. Thermal expansion coefficient decreases and elastic modulus increases for polymethyl methacrylate (PMMA). The elastic modulus and strength of polyvinyl alcohol-hBN nanocomposite increase with increment of hBN amount [15].

2.6 *Graphitic Carbon Nitride (gC_3N_4)*

Graphitic carbon nitride is an organic semiconductor which can be used as photocatalyst and optoelectronic device. Visible light-induced hydrogen evolution reaction (HER) and efficient conversion of solar energy has turned researcher's interest toward gC_3N_4. The simple synthesis process, low band gap energy (2.7 eV), ability to absorb light energy in visible spectrum, etc., makes it an interesting component for light emitting diode, phototransistor, photodetector, etc. Different gC_3N_4 layered composite materials have been synthesized with metal, metal organic framework, conducting polymer, graphene, etc., which has expanded its use [16].

3 Application of 2D Material in Biomedical Field

The unique structural features of 2D materials make it an interesting candidate for biomedical applications. In this section, we shall briefly discuss different facets of biomedical engineering where 2D materials are applied (Fig. 2).

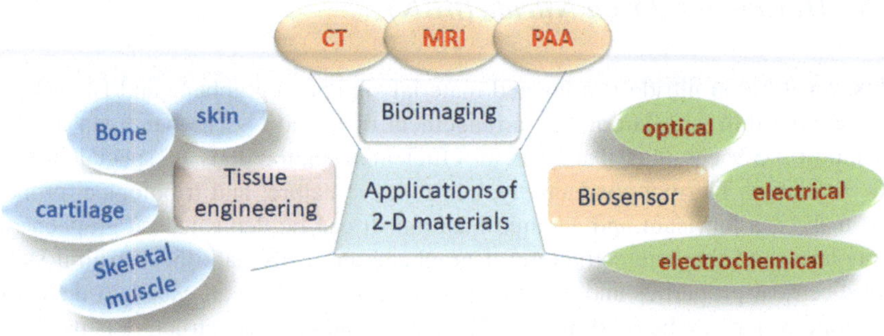

Fig. 2 Biomedical applications of 2D materials which are broadly classified into 3 sections—bioimaging, biosensor, and tissue engineering

3.1 Bioimaging

The advancement in imaging techniques has developed the field of theranostics. Generally, fluorescent dyes are used to visualize different cellular events. But, due to their photobleaching event when interrogated with laser light, the imaging process has a limited life span. Hence, the search for alternative material has led to 2D materials which are good contrasting agents. Hence it is used in different imaging applications like magnetic resonance imaging (MRI), photoacoustic imaging (PA), surface-enhanced Raman scattering (SERS), computed tomography (CT), positron emission-computed tomography imaging (PET), etc.

3.1.1 Computed Tomography (CT)

It is a very important medical imaging technique routinely used for diagnostic process. TMDs and MXenes are considered as ideal candidate for CT imaging. Being high atomic number element, they can absorb X-rays through their inner shell electrons and attenuate the energy. As a result, MXenes can be used as a contrasting agent. PEG-functionalized WS_2 was used as contrasting agent. Hounsfield unit (HU) values are used to measure CT, and PEG-WS_2 showed 22.01 HU L/g as compared to ipromide, an iodine-based commercial contrasting agent (15.9 HU L/g) [17]. In different other theranostic application (some of which are discussed in Sect. 4), CT-based in vivo detection has been addressed.

3.1.2 Magnetic Resonance Imaging (MRI)

Another important and widely applied diagnostic imaging technique is MRI. It has several advantages viz. noninvasive, large spatial resolution, and nice contrast in 3D tissue. 2D materials have been explored as contrasting agent in MRI. Ti_3C_2T flakes have been covalently functionalized with a chelating agent, diethylenetriaminepentaacetic acid (DTPA) and complexed with gadolinium ion (Gd^{3+}). Functionalization conferred paramagnetic response to the intrinsically diamagnetic Ti_3C_2T flakes for MRI. Flake concentration-dependent magnetic relaxation time helped to identify spatial distribution of flakes [18]. In different theranostic applications, MXenes have been used. MnO_x/Ti_3C_2 composite MXenes have been designed as a theranostic agent and cellular location tracking synthesized compound was done using T1-MRI [19]. Mn^{2+} ion bound on Ti_3C_2 was synthesized and PEG was functionalized on the surface. The synthesized material acted as contrasting material in tumor cells and detected by T1-MRI [20].

3.1.3 Photoacoustic Imaging (PAI)

PAI requires contrasting agents and 2D material-based materials have strong absorbance in NIR range. As a result, these materials can be used for PAI. Very brief laser pulse applied on 2D materials which are associated with acoustic wave generation (PA wave) and ultrasonic transducer-based signal receiving and computer-based image construction [21]. Different theranostic applications of 2D materials are discussed in Sect. 4, one of the dual-mode imaging is PAI [22, 23].

3.2 *Biosensing*

Easy surface modification, functional group addition, and high surface-to-volume ratio have made 2D materials as attractive agent for biosensing application. Based on the mode of detection, three different types of biosensors are prepared—optical, electrical, and electrochemical. Optical plasmonic biosensors made of 2D TMDs are highly sensitive and have stronger response than conventional noble metal-based biosensors [24]. 2D MoS_2 FET-based biosensor has been able to detect prostate specific antigen (PSA), a biomarker for prostate cancer at the nM range [25]. N-doped graphene functionalized with Cu–Ni nanoparticles is used to nonenzymatically detect glucose and H_2O_2 [26]. In order to detect circulating DNA, poly-xanthurenic acid (PXA) film functionalized MoS_2 nanosheets-based electrochemical sensor has been constructed and circulating DNA was detected in femto molar to atto molar range [27]. Cardiac troponin T (cTnT) detection was made using anti-cTnT covered C dot functionalized on MoS_2 nanosheets using FRET principle. Typical detection range was 01–50 ng/mL [28]. There are many such applications are published which is out of the scope of this chapter.

4 2D Material-Based Drug Delivery System

Smart drug delivery system is the next big challenge in pharmaceutical industry. A plethora of drug has been discovered and its pharmacokinetics and pharmacodynamics is well studied. But for most of the drugs, the efficiency obtained in lab condition is not found when the drug is administered or prescribed to the patients. One of the major reasons is bulk fraction of the drug is lost before reaching the target. As a result, higher amount of drug dosage is prescribed which causes unintended side effect. As a result, precise, in situ, and on-demand drug delivery is under intense research. Besides nanoparticle-mediated drug delivery system, 2D materials have gained attention as a drug delivery agent. 2D nanomaterial-based nanomedicines hold some advantages over conventional drugs like it is more biocompatible, target specific; limited side effects, lower photobleaching effect, etc.

4.1 Graphene and Its Oxides

Graphene is the most studied 2D material in the field of biomedicine. Its antibacterial action is exerted via different modes like physical damage by its sharp edges, wrapping, and photothermal ablation. Also, it exerts oxidative stress via reactive oxygen species (ROS) and charge transfer [29]. Dental caries is a result of biofilm formation by *Streptococcus mutans* and some other types of bacteria. Graphene oxide (GO) has been tested for its antibacterial effect on *S. mutans* and as a result, it may be used as anticaries agent. GO with more oxygen carrying functional groups show higher bactericidal effect at lower concentration which was ascertained by different techniques like 3-(4,5-dimethylthiazol-2-yl)-2,5-diphenyl tetrazolium bromide (MTT) reduction assay, colony forming unit (CFU) counting, etc. [30]. In another study, polydopamine-curcumin (PDA-Cur) composite is functionalized on GO to prepare GO/PDA-Cur nanocomposite and tested on *Staphylococcus aureus* which is causative agent for different infectious diseases. The nanocomposite was irradiated with white light and curcumin acts as photosensitizer. The photodynamic effect induced by the nanocomposite was able to reduce the *S. aureus* count to the 4th order of magnitude (99.99%). ROS generated by white light and mechanical damage of bacterial cell wall caused the reduction of bacterial number [31]. Tetracycline is an important antibiotic. Polyethyleneimine (PEI) modified GO (p-GO) was used as carrier for tetracycline hydrochloride (TCH) to exert synergistic effect on Gram-positive *S. aureus* and Gram-negative *E. coli.* pGO acted as a carrier for TCH and helps in delivery of TCH [32]. Copper oxide nanoparticles or nanowires have antibacterial activities. 3-dimesional nanocomposite of GO and CuO nanowire has been synthesized. Light induced antibacterial mechanism had been observed against Gram negative *E. coli.* The authors proposed 3 steps for light induced antibacterial activity which are as follows: Transfer of electron from CuO to rGO followed by localization of electrons on excess electrons on rGO functional groups and release of bactericidal ROS [33].

Anticancer effect of graphene and its oxide along with other components has been investigated. Different physical properties like size, shape, and charge on the surface play important role in the induction of apoptosis in cancer cells. The shape is widely regarded as an important parameter for nanomaterial. The lamellar structure of the graphene confers an inherent advantage as a nanotherapeutic platform as it has a large surface area. The interaction of the nanoparticles with surface receptor depends on the shape of the nanoparticle. Also, the phagocytosis of particles by cells depends on size, shape, and surface charge [34]. Still the toxicity of graphene oxide sheets is a debatable topic. It has been observed that smaller size of the sheets has lower toxicity. Graphene oxide nanoplatelets with dimension of 50 nm^2 were synthesized from stacked graphite nanofiber. It was administered with cisplatin, a common water soluble anticancer drug to nonsmall cell lung cancer cell line A549 for synergistic action. It was found that co-administration potentiated the effect of cisplatin on A549 [35]. In another study, to enhance the anticancer effect of daunorubicin, magnetic nanoparticle-based core–shell system was synthesized. Daunorubicin-loaded fluorescence resonance energy transfer (FRET) system $MnFe_2O_4$@SiO_2@graphene quantum dots were prepared as a pH-sensitive nanoplatform. In the FRET system, daunorubicin acted as acceptor and graphene-based platform as donor. pH-sensitive drug release was tested on breast cancer cell line MCF-7. It was estimated that at pH 5.5, 60% drug was released in 48 h. The cytotoxic efficiency of the system reached 95% [36]. Graphene oxide (GO) itself inhibits tubulin assembly. It disrupts the structural integrity of the protein and blocks the cell cycle at the S phase. It showed higher ROS mediated cytotoxicity toward colon cancer cell line HCT116 over noncancerous human embryonic kidney cell line HEK239 [37]. Reduced graphene oxide (rGO) nanosheet was synthesized using thiourea (T). The anticancer efficacy of T-rGO nanosheet was investigated against human colon cancer cell line HT29. Cell viability was decreased along with DNA fragmentation was observed [38]. In another study, GO nanoplatelet was incorporated into alginate emulsion. Then doxorubicin (DOX) was encapsulated via electrostatic interaction in alginate-GO nanogel. pH change accelerated the release of DOX. The higher accumulation of DOX inside the A549 was observed and showed higher cytotoxic ability when assisted by photothermal therapy [39]. The efficacy of GO and vanillin functionalized GO (V-rGO) was tested on human acute monocytic leukemia cell line (THP-1). GO and V-rGO induce oxidative stress-mediated apoptosis of the cells. The charge transfer on the GO or rGO surface caused inflammatory response in the cell [40]. Osteosarcoma is another type of cancer which is recalcitrant to many chemotherapeutic agents. In one study, cell penetrating peptide (CPP) modified and polyethylene-glycol- (PEG-) grafted GO (pGO) functionalized with photosensitive agents HPPH (2-(1-hexyloxyethyl)-2-devinyl pyropheophorbide-alpha) and epirubicin tested on osteosarcoma cell line MG-63 and osteosarcoma xenograft mice. Due to the elevated amount of reactive singlet oxygen, the composite system has higher cytotoxic effect on MG-63. Similarly, in mice, the size of the tumor was reduced [41]. In another study, anticancer efficacy was checked for pristine graphene, reduced graphene oxide, and graphene oxide on glioblastoma cell line U87MG. Microarray and qRT-PCR assay revealed that GO causes more downregulation of oxidative phosphorylation genes of complex I, III, IV, and V which results in decreased ATP level [42]. Many more such experimental studies have concluded that graphene and its derivatives may act as potent nanomedicine and drug delivery platform.

4.2 Materials Beyond Graphene

Besides graphene, MXenes are other candidates of special interest as drug delivery platform. Ti_3C_2 and Ti_3C_2-SP (soyabean phospholipid) nanosheets have been synthesized and used to load DOX to it. It has found high DOX loading capacity, of 211%. pH-responsive drug release was observed. Also, photothermal ablation was observed when near-infrared (NIR) laser (wavelength 880 nm) was irradiated for 5 min. The efficacy was tested on murein breast cancer cell line 4T1 and BALB/c xenograft mice model. Tumor eradication was observed and multimodal activity of Ti_3C_2 MXenes was shown in the study [43]. In another study, polyvinyl pyrrolidone (PVP) based 2D Ti_3C_2 was synthesized and two anti-cancer drugs namely DOX which are topoisomerase II inhibitor and deferasirox, an iron chelator was loaded on it. The dual drug promoted apoptosis in colorectal and hepatocellular carcinoma cell lines and downregulation of transferrin receptor. Tumor regression was also observed in BALB/c nude mice [22]. Another strategy used to kill cancer cells is called enzyme dynamic therapy (EDT). Glucose oxidase (GOX) and chloroperoxide (CPO) were conjugated with 2D Ti_3C_2 and acted as enzyme nanoreactor (as dubbed by the authors). Hypoxic cytotoxin drug tirapazamine (TRZ) was conjugated on it. The system was loaded in a membrane vesicle expressing CD47. Due to camouflaging, the Ti_3C_2-GOX-CPO/TPZ vesicle was target-specific and selectively engulfed by cancer cells. Once the membrane was internalized, Ti_3C_2-GOX-CPO/TPZ system was released. GOX and CPO generated HClO through cascade reaction inside the cell which further activated TPZ and killed the cancer cell [44]. Other than cancer therapeutics, multiple stimul-responsive hydrogel with MXenes have been used for wound healing purpose. MXene-wrapped magnetic nanoparticle was entrapped in dual polymeric poly (*N*-isopropyl acrylamide)-alginate meshwork. Thus synthesized MXene-based hydrogel was used to deliver wound healing drug in photo and magneto responsive manner [45]. Many more researches have been conducted using MXenes as stimuli responsive drug delivery platform.

Black phosphorus nanosheet (BPNS) is another 2D material used as drug delivery platform in cancer therapy owing to its stimuli-responsive nature. BPNS is a good light absorbing material. This property is used for photothermal therapy and ROS generation upon absorbing light. Dual responsive BPNS-based nanosystem has been synthesized as carrier for photosensitizer. The lack of oxygen in tumor microenvironment is one of the major obstacles faced by photosensitizers. BPNS-based dual responsive self-oxygen supporting system was synthesized to kill the cancer cells.

The platform was functionalized with folate and blocker DNA duplex of 5′ Cy5-aptamer-heme and 3′ heme labeled oligonucleotides. The resulting heme dimer would attenuate its peroxidase activity. After the aptamer target recognition, fluorescence is turned on. The peroxidase became active and converted H_2O_2 to reactive oxygen thus continued the supply of oxygen in hypoxic environment. As a result, photodynamic therapy efficiency increased and killed cancer cells in vitro and in vivo [46]. In the extension of the previous study, dual mode monitoring oxygen supply system was developed using Rhodamine B (RhB) encapsulated manganese dioxide (R-MnO_2) as O_2 supplier. Peptide functionalized black phosphorus labeled by fluorescein isothiocyanate (FITC) was used as theranostic agent. After reaching in low pH environment of cancer cells, R-MnO_2 was dissociated and generated oxygen from H_2O_2. As a result, hypoxia-mediated resistance to PDT was circumvented. The presence of Mn^{2+} and RhB facilitated dual mode monitoring via MRI and fluorescence imaging. This theranostic agent was able to kill the cancer cells [23]. In another study, incorporation of cuprous ion (Cu^{2+}) on BP (Cu^{2+}-BP nanosheet) enabled chemodynamic therapy enhanced PTT. Also, Cu^{2+} can be tracked through PET imaging. In this study, both rapid degradation and photothermal stability were achieved at the same time. Cu^{2+} incorporation accelerated BP degradation and cytotoxic hydroxyl radical production [47]. In another study, polydopamine (PDA) functionalized BP NS was used to target mitochondria. PDA-BPNS was functionalized with chlorine e6 (Ce6) and tripolyphosphate (TPP) and formed unique BP@PDA–Ce6&TPP NSs. The presence of Ce6-generated ROS and TPP-guided mitochondria targeting enhanced cytotoxic ability of the nanosystem. The nanosystem was used on both HeLa cell line and mouse model. NIR light mediated PTT and dual mode imaging guided theranostic system had been developed [48]. Platinum-based anticancer drugs namely DACHPt and $Pt(NH_3)_2$ were complexed with BP NS to form BP/DACHPt and BP/$Pt(NH_3)_2$ complexes. The drugs were loaded onto BP NS and released in lower pH and NIR stimulation. The complexes were tested on HeLa, HepG2, and A549 cell lines and killed the cells [49]. MnO_x/Ti_3C_2 nanosheet had been designed as theranostic agent and used to kill tumor cells using hyperthermia via photothermal agent. Dual mode tracking was achieved by MRI and photoacoustic imaging and PTT was initiated by NIR laser when synthesized material reached the tumor microenvironment [19].

Another important drug delivery application of 2D material is protein delivery for therapeutic purpose. Nanomaterial-based protein delivery is an established field of research. Different methodologies have been applied to deliver functional form of protein to the cells, e.g., liposome-mediated delivery [50, 51]; hydrogel-mediated delivery [52, 53]; nanocomposite based delivery [54–56]; aptamer-based delivery [57, 58], etc. Usage of 2D materials like GO, MXene, etc., has advantage like capture and release of protein in its functional form. In one study, catalase is used as a model protein functionalized on $Ti_3C_2T_x$. It was tested on macrophage cell line RAW264.7 to check its cytotoxic effect [59]. In another study, NIR-mediated release of hepatic growth factor (HGF) from Ti_3C_2@Agarose hydrogel inside the cell to remotely activate the c-met mediated pathway to promote proliferation, migration, etc. Also, authors have loaded tumor necrosis factor-α (TNF-α) in the composite hydrogel and released in the cell upon NIR irradiation [60]. GO nanosheets can adsorb

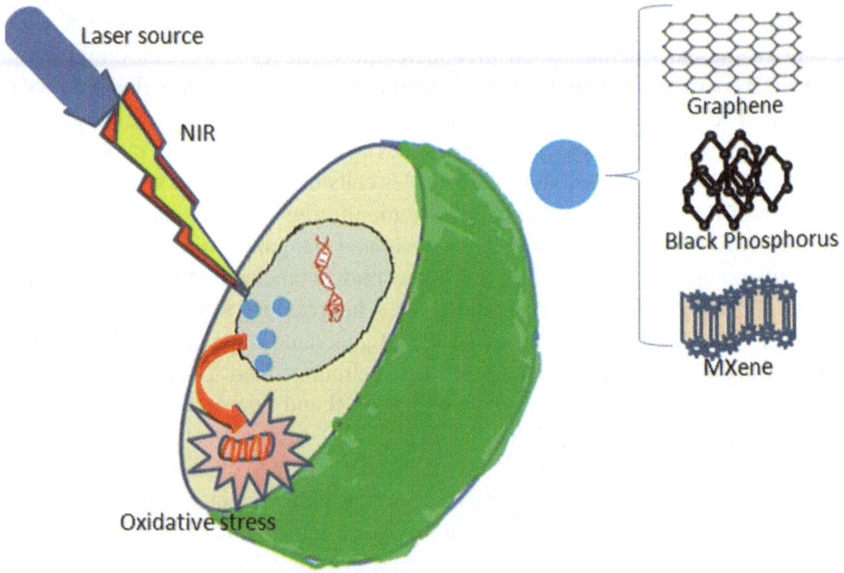

Fig. 3 Overview of the working mechanism of 2D material-based drug delivery against cancer cells. Different types of 2D material-based drug delivery agents are activated using NIR laser rays inside the cells and induce oxidative stress

protein on its surface. Ovalbumin adsorbed on GO nanosheet has successfully crossed dendritic cell membrane and elicits immune response through $CD8^+$ cells [61]. Bone morphogenetic protein-2 (BMP-2) is required in high dose for bone regeneration but it has side effects. For controlled delivery of BMP-2, it is encapsulated in GO nanoflakes suspended in fibrin gels. It showed significant osteogenic differentiation of human bone marrow-derived mesenchymal stem cells in in vitro experiment. In vivo experiment also showed promising result with bone regeneration with a lower dose [62]. An overview on the working mechanism of 2-D materials is graphically represented in Fig. 3.

5 Conclusion and Future Prospect

In this chapter, the application of 2D material in different biomedical fields has been discussed with strong focus on drug delivery system. Different properties of 2D material have been discussed. 2D materials are characterized into 6 different types in which some properties like electrical and thermal conductivity; ease of functionalization due to large surface area, etc., are common. Also, its electronic structure and unique characteristics were also discussed. The diverse range of source materials used for synthesis gave rise to the varieties of 2D material. Graphene is the hexagonal arrangement of C atoms with one or few atoms thickness. TMDs are next type of material with empirical formula MX_2 where M is the metal (Mo and W) and X is S, Se, and Te. Another important 2D material is MXenes where M is early transition metal and X is C and N. Black phosphorus is another type of 2D material where phosphorus is stable at room temperature. Phosphorus is arranged on BP similar to the way C is arranged on graphene. Another material of interest is hexagonal boron nitride. It is a layered material similar to graphene and BP. The last type discussed in this chapter is graphitic carbon nitride. These structural features and varieties of materials have been used in various applications ranging from supercapacitors to nanoelectronic devices. Also, due to electron absorbing ability of high molecular weight transition metals, 2D materials like TMDs and MXenes are used in CT, MRI, PAI, etc. Also, ease of surface functionalization and electronic properties are explored in the field of biosensor. Various 2D materials like GO, rGO, TMDs, etc., have been extensively used in chemical [63, 64], electrochemical [65–67], and electronic biosensing using FET [68, 69].

The main focus of the current chapter is 2D material-based drug delivery. We have subdivided the section into two parts: graphene and its oxide (GO and rGO) based delivery and 2D materials other than graphene like TMDs, MXenes, BP, etc. Various applications of graphene have been further subdivided into antibacterial and anticancer therapeutics. The role of graphene, GO, and rGO as drug delivery agent and therapeutic agent is explored in this section. The application of graphene-based drug delivery and mode of antibacterial mechanism for recalcitrant and pathogenic strains like *S. mutans*, *S. aureus*, etc., have been discussed. Unlike conventional drugs, these platforms offer light-induced controlled drug delivery. Graphene-based nanotherapeutic platforms have been extensively tested as anticancer therapeutic module. Besides piggybacking chemotherapeutic agents like cisplatin, dauxorubicin, daunorubicin, etc., it acted as pH-sensitive nanoplatform and killed cancer cells using photothermal effect. MXenes have also been investigated as a drug delivery platform for cancer therapeutics. High drug loading capacity is an added advantage for MXenes. Also PDT and PTT can be achieved when irradiated with NIR. Also, enzyme dynamic therapy was used to kill the cancer cells by generating toxic HClO ions. BP is a good light absorbing material and generates ROS. Using suitable functional groups on it, BP was able to generate oxidative stress even in low oxygen condition in tumor microenvironment. Also, it was used as a theranostic agent where the nanoplatform was induced to exert PDT when it reaches the desired site. One of the major obstacles of the 2D materials to be used inside the body is its probable toxicity due to nonspecific interaction with biological components. Due to very high surface, it may interact different cells in the blood or metabolites or ions or circulating DNA or RNA. Any material used as therapeutic component must be well studied for its interaction with different biological components, toxicity, absorption, circulation half-life, and excretion [70]. Hence, long road is ahead before 2D based drug delivery comes from bench to bedside.

References

1. Novoselov, K.S., Geim, A.K., Morozov, S.V., Jiang, D., Zhang, Y., Dubonos, S.V., et al.: Electric field effect in atomically thin carbon films. Science **306**(5696), 666–669 (2004). https://doi.org/10.1126/science.1102896
2. Ranasinghe, J.C., Jain, A., Wu, W., Zhang, K., Wang, Z., Huang, S.: Engineered 2D materials for optical bioimaging and path toward therapy and tissue engineering. J. Mater. Res. **37**(10), 1689–1713 (2022). https://doi.org/10.1557/s43578-022-00591-5
3. Yi, Y., Chen, Z., Yu, X.-F., Zhou, Z.-K., Li, J.: Recent advances in quantum effects of 2D materials. Adv. Quantum Technol. **2**(5–6), 1800111 (2019). https://doi.org/10.1002/qute.201800111
4. Wei, C., Wu, C.: Nonlinear fracture of two-dimensional transition metal carbides (MXenes). Eng. Fract. Mech. **230**, 106978 (2020). https://doi.org/10.1016/j.engfracmech.2020.106978
5. Naumis, G.G.: 5 - Electronic properties of two-dimensional materials. In: Yang, E.-H., Datta, D., Ding, J., Hader, G. (eds.) Synthesis, Modeling, and Characterization of 2D Materials, and Their Heterostructures, pp. 77–109. Elsevier (2020)
6. Lin, Y.-C., Torsi, R., Younas, R., Hinkle, C.L., Rigosi, A.F., Hill, H.M., et al.: Recent advances in 2D material theory, synthesis, properties, and applications. ACS Nano **17**(11), 9694–9747 (2023). https://doi.org/10.1021/acsnano.2c12759
7. Kang, M.-H., Lee, D., Sung, J., Kim, J., Kim, B.H., Park, J.: 2.04 - Structure and chemistry of 2D materials. In: Andrews, D.L., Lipson, R.H., Nann, T. (eds.) Comprehensive Nanoscience and Nanotechnology, 2nd edn., pp. 55–90. Academic Press, Oxford (2019)
8. Zhen, Z., Zhu, H.: 1 - Structure and properties of graphene. In: Zhu, H., Xu, Z., Xie, D., Fang, Y. (eds.) Graphene, pp. 1–12. Academic Press (2018)
9. Manzeli, S., Ovchinnikov, D., Pasquier, D., Yazyev, O.V., Kis, A.: 2D transition metal dichalcogenides. Nat. Rev. Mater. **2**(8), 17033 (2017). https://doi.org/10.1038/natrevmats.2017.33
10. Zamhuri, A., Lim, G.P., Ma, N.L., Tee, K.S., Soon, C.F.: MXene in the lens of biomedical engineering: synthesis, applications and future outlook. Biomed. Eng. Online **20**(1), 33 (2021). https://doi.org/10.1186/s12938-021-00873-9
11. Chen, Y., Ren, R., Pu, H., Chang, J., Mao, S., Chen, J.: Field-effect transistor biosensors with two-dimensional black phosphorus nanosheets. Biosens. Bioelectron. **89**(Pt 1), 505–510 (2017). https://doi.org/10.1016/j.bios.2016.03.059
12. Kumar, V., Brent, J.R., Shorie, M., Kaur, H., Chadha, G., Thomas, A.G., et al.: Nanostructured aptamer-functionalized black phosphorus sensing platform for label-free detection of myoglobin, a cardiovascular disease biomarker. ACS Appl. Mater. Interfaces **8**(35), 22860–22868 (2016). https://doi.org/10.1021/acsami.6b06488
13. Niu, X., Li, Y., Shu, H., Wang, J.: Anomalous size dependence of optical properties in black phosphorus quantum dots. J. Phys. Chem. Lett. **7**(3), 370–375 (2016). https://doi.org/10.1021/acs.jpclett.5b02457
14. Zhang, K., Feng, Y., Wang, F., Yang, Z., Wang, J.: Two dimensional hexagonal boron nitride (2D-hBN): synthesis, properties and applications. J. Mater. Chem. C **5**(46), 11992–12022 (2017). https://doi.org/10.1039/C7TC04300G
15. Zhi, C., Bando, Y., Tang, C., Kuwahara, H., Golberg, D.: Large-scale fabrication of boron nitride nanosheets and their utilization in polymeric composites with improved thermal and mechanical properties. Adv. Mater. **21**(28), 2889–2893 (2009). https://doi.org/10.1002/adma.200900323
16. Zhu, J., Xiao, P., Li, H., Carabineiro, S.A.C.: Graphitic carbon nitride: synthesis, properties, and applications in catalysis. ACS Appl. Mater. Interfaces **6**(19), 16449–16465 (2014). https://doi.org/10.1021/am502925j
17. Cheng, L., Liu, J., Gu, X., Gong, H., Shi, X., Liu, T., et al.: PEGylated WS(2) nanosheets as a multifunctional theranostic agent for in vivo dual-modal CT/photoacoustic imaging guided photothermal therapy. Adv. Mater. **26**(12), 1886–1893 (2014). https://doi.org/10.1002/adma.201304497

18. Neubertova, V., Guselnikova, O., Yamauchi, Y., Olshtrem, A., Rimpelova, S., Čižmár, E., et al.: Covalent functionalization of Ti_3C_2T MXene flakes with Gd-DTPA complex for stable and biocompatible MRI contrast agent. Chem. Eng. J. **446**, 136939 (2022). https://doi.org/10.1016/j.cej.2022.136939
19. Dai, C., Lin, H., Xu, G., Liu, Z., Wu, R., Chen, Y.: Biocompatible 2D titanium carbide (MXenes) composite nanosheets for pH-responsive MRI-guided tumor hyperthermia. Chem. Mater. **29**(20), 8637–8652 (2017). https://doi.org/10.1021/acs.chemmater.7b02441
20. An, D., Wu, X., Gong, Y., Li, W., Dai, G., Lu, X., et al.: Manganese-functionalized MXene theranostic nanoplatform for MRI-guided synergetic photothermal/chemodynamic therapy of cancer. Nanophotonics **11**(22), 5177–5188 (2022). https://doi.org/10.1515/nanoph-2022-0533
21. Jeong, W.Y., Kang, M.S., Lee, H., Lee, J.H., Kim, J., Han, D.W., et al.: Recent trends in photoacoustic imaging techniques for 2D nanomaterial-based phototherapy. Biomedicines **9**(1) (2021). https://doi.org/10.3390/biomedicines9010080
22. Xu, Y., Wang, Y., An, J., Sedgwick, A.C., Li, M., Xie, J., et al.: 2D-ultrathin MXene/DOXjade platform for iron chelation chemo-photothermal therapy. Bioact. Mater. **14**, 76–85 (2022). https://doi.org/10.1016/j.bioactmat.2021.12.011
23. Liu, J., Du, P., Liu, T., Cordova Wong, B.J., Wang, W., Ju, H., et al.: A black phosphorus/manganese dioxide nanoplatform: oxygen self-supply monitoring, photodynamic therapy enhancement and feedback. Biomaterials **192**, 179–188 (2019). https://doi.org/10.1016/j.biomaterials.2018.10.018
24. Ouyang, Q., Zeng, S., Jiang, L., Qu, J., Dinh, X.-Q., Qian, J., et al.: Two-dimensional transition metal dichalcogenide enhanced phase-sensitive plasmonic biosensors: theoretical insight. J. Phys. Chem. C **121**(11), 6282–6289 (2017). https://doi.org/10.1021/acs.jpcc.6b12858
25. Lee, J., Dak, P., Lee, Y., Park, H., Choi, W., Alam, M.A., et al.: Two-dimensional layered MoS(2) biosensors enable highly sensitive detection of biomolecules. Sci. Rep. **4**, 7352 (2014). https://doi.org/10.1038/srep07352
26. Shabnam, L., Faisal, S.N., Roy, A.K., Minett, A.I., Gomes, V.G.: Nonenzymatic multispecies sensor based on Cu-Ni nanoparticle dispersion on doped graphene. Electrochim. Acta **224**, 295–305 (2017). https://doi.org/10.1016/j.electacta.2016.12.056
27. Zhang, W., Dai, Z., Liu, X., Yang, J.: High-performance electrochemical sensing of circulating tumor DNA in peripheral blood based on poly-xanthurenic acid functionalized MoS_2 nanosheets. Biosens. Bioelectron. **105**, 116–120 (2018). https://doi.org/10.1016/j.bios.2018.01.038
28. Gogoi, S., Khan, R.: Fluorescence immunosensor for cardiac troponin T based on Förster resonance energy transfer (FRET) between carbon dot and MoS_2 nano-couple. Phys. Chem. Chem. Phys. **20**(24), 16501–16509 (2018). https://doi.org/10.1039/C8CP02433B
29. Ji, H., Sun, H., Qu, X.: Antibacterial applications of graphene-based nanomaterials: recent achievements and challenges. Adv. Drug Deliv. Rev. **105**(Pt B), 176–189 (2016). https://doi.org/10.1016/j.addr.2016.04.009
30. Zhao, M., Shan, T., Wu, Q., Gu, L.: The antibacterial effect of graphene oxide on Streptococcus mutans. J. Nanosci. Nanotechnol. **20**(4), 2095–2103 (2020). https://doi.org/10.1166/jnn.2020.17319
31. Yan, H., Li, P., Jiang, X., Wang, X., Hu, Y., Zhang, Y., et al.: Preparation of graphene oxide/polydopamine-curcumin composite nanomaterials and its antibacterial effect against Staphylococcus aureus induced by white light. Biomater. Adv. **139**, 213040 (2022). https://doi.org/10.1016/j.bioadv.2022.213040
32. Jiang, L., Su, C., Ye, S., Wu, J., Zhu, Z., Wen, Y., et al.: Synergistic antibacterial effect of tetracycline hydrochloride loaded functionalized graphene oxide nanostructures. Nanotechnology **29**(50), 505102 (2018). https://doi.org/10.1088/1361-6528/aae424
33. Kiani, F., Astani, N.A., Rahighi, R., Tayyebi, A., Tayebi, M., Khezri, J., et al.: Effect of graphene oxide nanosheets on visible light-assisted antibacterial activity of vertically-aligned copper oxide nanowire arrays. J. Colloid Interface Sci. **521**, 119–131 (2018). https://doi.org/10.1016/j.jcis.2018.03.013

34. Yang, W., Deng, X., Huang, W., Qing, X., Shao, Z.: The physicochemical properties of graphene nanocomposites influence the anticancer effect. J. Oncol. **2019**, 7254534 (2019). https://doi.org/10.1155/2019/7254534
35. Rosli, N.F., Fojtu, M., Fisher, A.C., Pumera, M.: Graphene oxide nanoplatelets potentiate anticancer effect of cisplatin in human lung cancer cells. Langmuir: ACS J. Surf. Colloids **35**(8), 3176–3182 (2019). https://doi.org/10.1021/acs.langmuir.8b03086
36. Kiani Nejad, Z., Akbar Khandar, A., Khatamian, M.: Graphene quantum dots based MnFe(2)O(4)@SiO(2) magnetic nanostructure as a pH-sensitive fluorescence resonance energy transfer (FRET) system to enhance the anticancer effect of the drug. Int. J. Pharm. **628**, 122254 (2022). https://doi.org/10.1016/j.ijpharm.2022.122254
37. Bera, S., Ghosh, S., Ali, A., Pal, M., Chakrabarti, P.: Inhibition of microtubule assembly and cytotoxic effect of graphene oxide on human colorectal carcinoma cell HCT116. Arch. Biochem. Biophys. **708**, 108940 (2021). https://doi.org/10.1016/j.abb.2021.108940
38. Vimalanathan, B., Vijaya, J.J., Mary, B.C.J., Ignacimuthu, S., Daniel, M., Jayavel, R., et al.: The anticancer efficacy of thiourea-mediated reduced graphene oxide nanosheets against human colon cancer cells (HT-29). J. Funct. Biomater. **13**(3) (2022). https://doi.org/10.3390/jfb130 30130
39. Xu, X., Wang, J., Wang, Y., Zhao, L., Li, Y., Liu, C.: Formation of graphene oxide-hybridized nanogels for combinative anticancer therapy. Nanomed. Nanotechnol. Biol. Med. **14**(7), 2387–2395 (2018). https://doi.org/10.1016/j.nano.2017.05.007
40. Gurunathan, S., Kang, M.H., Jeyaraj, M., Kim, J.H.: Differential immunomodulatory effect of graphene oxide and vanillin-functionalized graphene oxide nanoparticles in human acute monocytic leukemia cell line (THP-1). Int. J. Mol. Sci. **20**(2) (2019). https://doi.org/10.3390/ijms20020247
41. Zhang, Y.F., Wu, Y.F., Lan, T.J., Chen, Y., Su, S.H.: Codelivery of anticancer drug and photosensitizer by PEGylated graphene oxide and cell penetrating peptide enhanced tumor-suppressing effect on osteosarcoma. Front. Mol. Biosci. **7**, 618896 (2020). https://doi.org/10.3389/fmolb.2020.618896
42. Szmidt, M., Stankiewicz, A., Urbanska, K., Jaworski, S., Kutwin, M., Wierzbicki, M., et al.: Graphene oxide down-regulates genes of the oxidative phosphorylation complexes in a glioblastoma. BMC Mol. Biol. **20**(1), 2 (2019). https://doi.org/10.1186/s12867-018-0119-2
43. Han, X., Huang, J., Lin, H., Wang, Z., Li, P., Chen, Y.: 2D ultrathin MXene-based drug-delivery nanoplatform for synergistic photothermal ablation and chemotherapy of cancer. Adv. Healthcare Mater. **7**(9), e1701394 (2018). https://doi.org/10.1002/adhm.201701394
44. Zhang, X., Cheng, L., Lu, Y., Tang, J., Lv, Q., Chen, X., et al.: A MXene-based bionic cascaded-enzyme nanoreactor for tumor phototherapy/enzyme dynamic therapy and hypoxia-activated chemotherapy. Nano-Micro Lett. **14**(1), 22 (2021). https://doi.org/10.1007/s40820-021-007 61-w
45. Yang, X., Zhang, C., Deng, D., Gu, Y., Wang, H., Zhong, Q.: Multiple stimuli-responsive MXene-based hydrogel as intelligent drug delivery carriers for deep chronic wound healing. Small **18**(5), e2104368 (2022). https://doi.org/10.1002/smll.202104368
46. Liu, J., Du, P., Mao, H., Zhang, L., Ju, H., Lei, J.: Dual-triggered oxygen self-supply black phosphorus nanosystem for enhanced photodynamic therapy. Biomaterials **172**, 83–91 (2018). https://doi.org/10.1016/j.biomaterials.2018.04.051
47. Hu, K., Xie, L., Zhang, Y., Hanyu, M., Yang, Z., Nagatsu, K., et al.: Marriage of black phosphorus and Cu(2+) as effective photothermal agents for PET-guided combination cancer therapy. Nat. Commun. **11**(1), 2778 (2020). https://doi.org/10.1038/s41467-020-16513-0
48. Yang, X., Wang, D., Zhu, J., Xue, L., Ou, C., Wang, W., et al.: Functional black phosphorus nanosheets for mitochondria-targeting photothermal/photodynamic synergistic cancer therapy. Chem. Sci. **10**(13), 3779–3785 (2019). https://doi.org/10.1039/C8SC04844D
49. Liu, G., Tsai, H.-I., Zeng, X., Qi, J., Luo, M., Wang, X., et al.: Black phosphorus nanosheets-based stable drug delivery system via drug-self-stabilization for combined photothermal and chemo cancer therapy. Chem. Eng. J. **375**, 121917 (2019). https://doi.org/10.1016/j.cej.2019.121917

50. Koide, H., Ochiai, H., Suzuki, H., Hirata, S., Watanabe, M., Yonezawa, S., et al.: Easy preparation of a liposome-mediated protein delivery system by freeze–thawing a liposome–protein complex. J. Mater. Chem. B **10**(35), 6768–6776 (2022). https://doi.org/10.1039/D2TB00271J
51. Jiang, T., Mo, R., Bellotti, A., Zhou, J., Gu, Z.: Gel–liposome-mediated co-delivery of anticancer membrane-associated proteins and small-molecule drugs for enhanced therapeutic efficacy. Adv. Func. Mater. **24**(16), 2295–2304 (2014). https://doi.org/10.1002/adfm.201303222
52. Townsend, K.L., Pritchard, E., Coburn, J.M., Kwon, Y.M., Blaszkiewicz, M., Lynes, M.D., et al.: Silk hydrogel-mediated delivery of bone morphogenetic protein 7 directly to subcutaneous white adipose tissue increases browning and energy expenditure. Front. Bioeng. Biotechnol. **10**, 884601 (2022). https://doi.org/10.3389/fbioe.2022.884601
53. Banerjee, S., Sahoo, A.K., Chattopadhyay, A., Ghosh, S.S.: Hydrogel nanocarrier encapsulated recombinant IκBα as a novel anticancer protein therapeutics. RSC Adv. **3**(33), 14123–14131 (2013). https://doi.org/10.1039/C3RA23181J
54. Banerjee, S., Sahoo, A.K., Chattopadhyay, A., Ghosh, S.S.: Recombinant IkappaBalpha-loaded curcumin nanoparticles for improved cancer therapeutics. Nanotechnology **25**(34), 345102 (2014). https://doi.org/10.1088/0957-4484/25/34/345102
55. Song, J.G., Lee, S.H., Han, H.K.: Development of an M cell targeted nanocomposite system for effective oral protein delivery: preparation, in vitro and in vivo characterization. J. Nanobiotechnol. **19**(1), 15 (2021). https://doi.org/10.1186/s12951-020-00750-y
56. Luther, D.C., Lee, Y.W., Nagaraj, H., Clark, V., Jeon, T., Goswami, R., et al.: Cytosolic protein delivery using modular biotin-streptavidin assembly of nanocomposites. ACS Nano **16**(5), 7323–7330 (2022). https://doi.org/10.1021/acsnano.1c06768
57. Jiang, Y., Pan, X., Chang, J., Niu, W., Hou, W., Kuai, H., et al.: Supramolecularly engineered circular bivalent aptamer for enhanced functional protein delivery. J. Am. Chem. Soc. **140**(22), 6780–6784 (2018). https://doi.org/10.1021/jacs.8b03442
58. Abune, L., Davis, B., Wang, Y.: Aptamer-functionalized hydrogels: an emerging class of biomaterials for protein delivery, cell capture, regenerative medicine, and molecular biosensing. Wiley Interdiscip. Rev. Nanomed. Nanobiotechnol. **13**(6), e1731 (2021). https://doi.org/10.1002/wnan.1731
59. Seitak, A., Shanti, A., Al Adem, K., Farid, N., Luo, S., Iskandarov, J., et al.: 2D MXenes for controlled releases of therapeutic proteins. J. Biomed. Mater. Res., Part A **111**(4), 514–526 (2023). https://doi.org/10.1002/jbm.a.37469
60. Wang, S., Zhang, Z., Wei, S., He, F., Li, Z., Wang, H.H., et al.: Near-infrared light-controllable MXene hydrogel for tunable on-demand release of therapeutic proteins. Acta Biomater. **130**, 138–148 (2021). https://doi.org/10.1016/j.actbio.2021.05.027
61. Li, H., Fierens, K., Zhang, Z., Vanparijs, N., Schuijs, M.J., Van Steendam, K., et al.: Spontaneous protein adsorption on graphene oxide nanosheets allowing efficient intracellular vaccine protein delivery. ACS Appl. Mater. Interfaces **8**(2), 1147–1155 (2016). https://doi.org/10.1021/acsami.5b08963
62. La, W.-G., Jung, M.-J., Yoon, J.-K., Bhang, S.H., Jang, H.-K., Lee, T.-J., et al.: Bone morphogenetic protein-2 for bone regeneration—dose reduction through graphene oxide-based delivery. Carbon **78**, 428–438 (2014). https://doi.org/10.1016/j.carbon.2014.07.023
63. Tuz Johra, F., Jung, W.-G.: Low temperature synthesis of RGO-Au nanocomposite with apparently reduced time and its application as a chemical sensor. Appl. Surf. Sci. **362**, 169–175 (2016). https://doi.org/10.1016/j.apsusc.2015.11.145
64. Lee, K., Yoo, Y.K., Chae, M.S., Hwang, K.S., Lee, J., Kim, H., et al.: Highly selective reduced graphene oxide (rGO) sensor based on a peptide aptamer receptor for detecting explosives. Sci. Rep. **9**(1), 10297 (2019). https://doi.org/10.1038/s41598-019-45936-z
65. Devi, P., Sharma, C., Kumar, P., Kumar, M., Bansod, B.K.S., Nayak, M.K., et al.: Selective electrochemical sensing for arsenite using rGO/Fe(3)O(4) nanocomposites. J. Hazard. Mater. **322**(Pt A), 85–94 (2017). https://doi.org/10.1016/j.jhazmat.2016.02.066
66. Li, Y., Cheng, C., Yang, Y., Dun, X., Gao, J., Jin, X.-J.: A novel electrochemical sensor based on CuO/H-C_3N_4/rGO nanocomposite for efficient electrochemical sensing nitrite. J. Alloy. Compd. **798**, 764–772 (2019). https://doi.org/10.1016/j.jallcom.2019.05.137

67. Patella, B., Buscetta, M., Di Vincenzo, S., Ferraro, M., Aiello, G., Sunseri, C., et al.: Electrochemical sensor based on rGO/Au nanoparticles for monitoring H_2O_2 released by human macrophages. Sens. Actuators, B Chem. **327**, 128901 (2021). https://doi.org/10.1016/j.snb.2020.128901
68. Aspermair, P., Mishyn, V., Bintinger, J., Happy, H., Bagga, K., Subramanian, P., et al.: Reduced graphene oxide–based field effect transistors for the detection of E7 protein of human papillomavirus in saliva. Anal. Bioanal. Chem. **413**(3), 779–787 (2021). https://doi.org/10.1007/s00216-020-02879-z
69. Thakur, B., Zhou, G., Chang, J., Pu, H., Jin, B., Sui, X., et al.: Rapid detection of single E. coli bacteria using a graphene-based field-effect transistor device. Biosens. Bioelectron. **110**, 16–22 (2018). https://doi.org/10.1016/j.bios.2018.03.014
70. Rajakumari, R., Tharayil, A., Thomas, S., Kalarikkal, N.: Chapter twelve—Toxicity of graphene based nanomaterials—a general overview of origin, exposure and mechanisms. In: Hussain, C.M. (ed.) Comprehensive Analytical Chemistry, pp. 281–325. Elsevier (2020)

Two-Dimensional Graphene Quantum Dots in Drug Delivery Applications

Mohd Jahir Khan, Agung Wibowo, and Chularat Sakdaronnarong

Abstract Graphene quantum dots (GQDs) possess properties like a large surface area, photostability, and biocompatibility, and they can be tailored simply over in-situ synthesis and post-synthesis. GQDs can be altered with biomolecules such as polysaccharides, proteins, DNA, and polymers to generate a hybrid QD system. GQDs and other molecules in hybrid systems serve as carriers for drug delivery of several anticancer treatments. The use of these substances to modify GQDs reduces their cytotoxicity and increases effectiveness as carriers. Because they are less toxic and more biocompatible, the GQDs are potential candidates for biological purposes such as bioimaging, delivering therapeutic agents, and theranostics. This chapter discusses recent breakthroughs in the synthesis of GQDs and their drug delivery applications. Physicochemical, optical, and biological characteristics such as size, chemical composition-dependent fluorescence, therapies, biocompatibility, and cellular toxicity are extensively investigated and summarized. It also provides vital insight into the fact that the performance of QDs as a drug delivery carrier is dependent on a combination of particle formulation factors and the level of cellular absorption.

1 Introduction

Graphene is a two-dimensional (2D) hexagonal structure composed of a flat monolayer densely packed with carbon atoms [1]. The graphene exhibit sp^2 bonded carbon networks with a zero band gap property [2]. Therefore, it has no photoluminescence (PL) in virgin form, and it owns limitations of optoelectronic possibilities. Eda et al. [3] reported that introducing the defect into the structural graphite can be efficient strategic to produce graphene oxide (GO), resulting in emitting PL performance [3]. The structural defect can be achieved either by inducing oxygen-containing groups or

M. J. Khan (✉) · A. Wibowo · C. Sakdaronnarong
Department of Chemical Engineering, Faculty of Engineering, Mahidol University, Nakhon Pathom 73170, Thailand
e-mail: khanmohdjahir@gmail.com

N. Talreja et al. (eds.), *Two-dimensional Hybrid Composites*, Engineering Materials,
https://doi.org/10.1007/978-981-99-8010-9_10

by the removal of carbon from the graphene structure. The other profound properties of graphene are good transparency, light strength, and thermal conductivity [4].

The latest discovery of GQDs offers unique properties, including PL, water solubility, high photostability, excellent biocompatibility, and low cost. GQDs are one or a few functionalized graphene nanofragments with multiple atomic layers and few functional groups (e.g. oxygen and hydrogen), with a lateral size below 100 nm [5]. GQDs, unlike graphene, display band gaps due to quantum confinement and exhibit photoluminescence which mostly emerges from delocalized excitations [6, 7]. GQDs received a lot of interest because of its unique electrical, thermal, and mechanical properties in a range of areas: biomedicine, sensors, material composites, energy-related materials, and ultra-thin materials [8–10]. The PL properties of GQDs are extremely valuable in biomedical applications that include drug delivery [11], bioimaging [9, 12], photothermal therapy [13], and diagnostic [14]. GQDs are frequently confused with carbon dots (CDs) because both exhibit PL characteristics, however there are some differences. GQDs are made up of sp^2-hybridized carbon with strong crystallinity, whereas CDs are made up of sp^3-hybridized carbon with a spherical shape [10, 15, 16]. Fluorescence is a type of photoluminescence that is commonly used in modern research and practical applications. Fluorescent compounds are referred to either fluorochromes or fluorophores. The compounds have spectral properties that allow them to absorb light in particular wavelengths (photon excitation) and then emit light at higher wavelengths (photon emission) [17, 18]. The photon emission of GQDs during PL can be used as an optical signal interpretation for disease detection [19], thermal generating for photodynamic [20], and monitoring the metabolism of drugs in the body for drug delivery [21] etc.

2 Graphene Quantum Dots Synthesis

The known approaches for synthesizing GQDs can be broadly classified into two types (top-down and bottom-up) (Fig. 1). Bottom-up approaches require complex reaction pathways along with particular organic components, making optimization difficult. As a result, the top-down strategy, which involves cutting big blocks of carbon materials into little bits, is preferable. The inexpensive plentiful carbon sources were required for this process. On the other hand, relatively simple, and easy synthesis is another beneficial factor.

2.1 Top-Down Strategy

A number of approaches for synthesizing GQDs are included within the top-down approach. They include electrochemical oxidation, chemical oxidation, chemical vapor deposition, hydrothermal oxidation, ultrasonic-aided oxidation, and pulsed

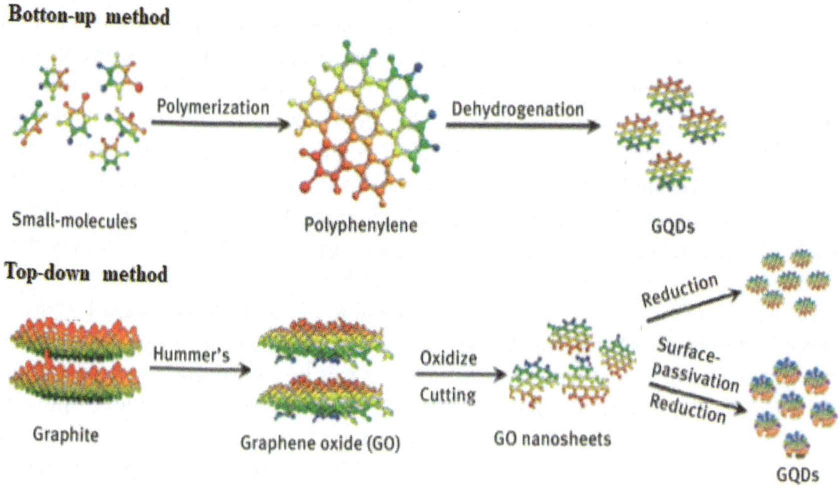

Fig. 1 Schematic diagram showing GQDs synthesis via bottom-up and top-down methods (reprinted/reproduced with the permission of Ref. [22], copyright 2012, Chemical Communications)

laser ablation. Each of these approaches could possibly be combined in a top-down method for GQD synthesis.

2.2 *Bottom-Up Strategy*

The starting materials in the bottom-up technique of GQDs synthesis include small molecules including amino acids, citric acid, small sugar molecules, and phenyl compounds. Bottom-up strategies often include molecular carbonization, electron beam irradiation, and microwave methods.

3 Properties of GQDs

3.1 *Optical Property*

Generally, GQDs, like carbon quantum dots (CQDs), provide strong optical absorption in the ultraviolet region with a long tail extending to the visible region. Mostly, it has an absorption band ranging 230–270 nm and an absorption peak around 320 nm which associate with π–π transition of C=C bonds and n–π^* transition of C=O respectively [23, 24]. The surface passivation of various elements (e.g., sulfur, nitrogen, boron) has been reported to change the absorption and emission of GQDs. Other important factors i.e., surface functional groups, size, and edge defect, affects the optical property of GQDs (Fig. 2). Wang et al. noted that PL is the most outstanding feature of GQDs [25] in which the performance of PL relies on emission wavelength and intensity [26]. Photoluminescence is a process in which a molecule absorbs a photon at a certain wavelength. This light absorption makes molecules get excitation from one exited state to higher energy exited state. At the excited state molecules are unstable, therefore, they return to the initial state by emitting light (light emission). The PL of GQDs can be influenced by various factors, including size [27], pH [7], solvent [28] surface functional group [29–31], heteroatom doping [32], etc.

Both the intensity and wavelength are crucial to assess fluorescence (FL) performance. Therefore, several approaches are conducted to improve these properties of GQDs [10]. Jin et al. reported PL property can be improved by increasing the bandgap [7]. Liu et al. reported that particle size plays an important role to optimize fluorescence properties since decreasing the particle size leads to a growing role of emission centers at their edges [27], thus reducing their band gap. The photoluminescence characteristics and quantum yield of GQDs were investigated in a previous report before and after doping with boron. The band gap was reduced and the absorption and photoluminescence/fluorescence spectra exhibited significantly

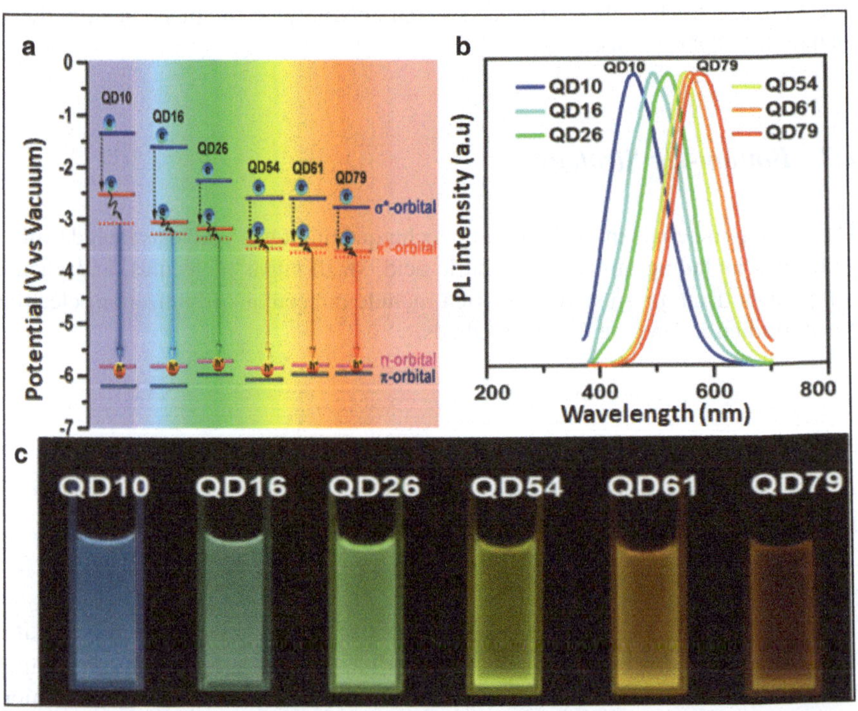

Fig. 2 Effect of particle size: **a** bandgap energy and **b** PL emission at 350 nm of excited wavelength; **c** appearance of GQDs under 365 nm of irradiation (reprinted/reproduced with the permission of Ref. [33] copyright 2012, Journal of Physical Chemistry Letters)

red shifted phenomenon as a result of boron doping and surface modification [34]. In a study, Jin et al. reported that charge transfers between functional groups and GQDs are responsible for PL shifts. They studied GQDs synthesized from graphene oxide (GOs), by a two-step cutting technique and the obtained GQDs were functionalized with amine groups. The results showed that PL emission of functionalized GQDs displayed a redshift (ca. <5 nm) when compared to un-functionalized GQDs [7]. Introduction of a dopant (e.g., sulfur, nitrogen, and zinc) was found to changes the surface properties and prevents π–π stacking which can reduce photoluminescence, subsequently, enhancing photon emission [9]. The effect of pH against fluorescence has been studied and reported that a pH 7.0 generates the best fluorescent intensity [7, 35].

3.2 *Quantum Confinement Effect*

Quantum confinement effect is the most favorable model for describing PL properties in GQDs. This effect occurs when GQDs' size (sp^2 conjugated domain) is smaller than the exciton Bohr radius [36]. Consequently, electron energy level changes from quasi-continuous energy level to discrete energy level. As the sp^2 domain size grows up, the effective band gap enhances moderately resulting in a redshifted emission absorption [37–39]. The observed quantum confinement is described as increasing size (sp^2 domains), leading to π^*–π gap of low energy, but unchanging n-state levels (Fig. 2a). There are 2 typical wavelength emission generations ($\pi^* \rightarrow n$ and $\sigma^* \rightarrow \pi^*$). The relaxation of electrons from pi antibonding to nonbonding, is responsible for the excitation wavelength-independent PL, induced by an excitation wavelength and electron scattering whereas the $\sigma^* \rightarrow \pi^*$ recombination, crucial for PL emissions, generated by short-wavelength excitation. The bandgap level declining when the size of GQDs increased from QD10 to QD79 then causes extending wavelength emission (Fig. 2b) and hence tuning the appearance of GQD (Fig. 2c) with various PL colors from blue (QD10) to orange-red for (QD79). QD10 and QD16, presented in Fig. 2, represented 1.0 nm and 1.6 nm, respectively [33].

3.3 *The Edge States*

At the armchair edges of GQDs structure, triple carbon bonds are found, and the edge has a carbyne-like structure (Fig. 3a), but at the zigzag edges, two unshared valence electrons are situated on each carbon atom at the edges (Fig. 3b) [40]. Localized states in zigzag-edged are pushed to the edge sites, whereas in armchair-edged are dispersed throughout the center. As a result, localized states at zigzag edge sites reduce bandgap energy [17], which is responsible for the PL characteristics [7, 40]. When the GQD is hexagonal in shape, the same type of edges is observed, and six 120° angles make the hexagonal shape (Fig. 3a, b) while both types of edges (armchair

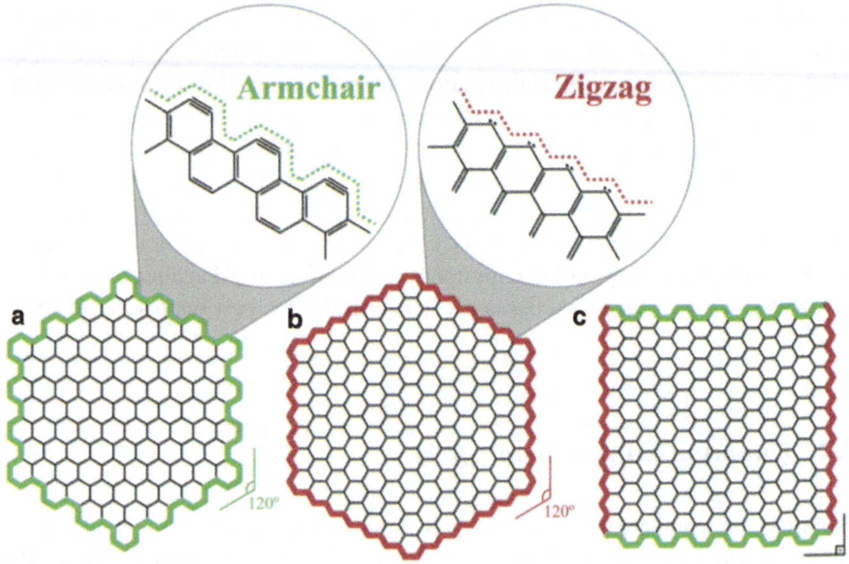

Fig. 3 The hexagonal shapes with **a** armchair, **b** zigzag, and **c** hybrid armchair-zigzag edges of GQDs (reprinted/reproduced with the permission of Ref. [45] copyright 2020, Environmental Science Nano)

and zigzag) are present in GQDs with rectangular shapes, generating four 90° angles (Fig. 3c). Kittiratanawasin et al. [41] discovered that the same shapes of GQDs can have various band gaps if the edge types are different. Lin et al. [42] examined the impact of the edge state on FL and showed that the carbene structures (zigzag edges) are more inclined to cause the induction of GQD strong luminescence than quantum confinement effects Zhu et al., [43] modified edge with functional groups can tune the PL properties, such as carboxyl and amide groups for green emission and hydroxyl groups for blue emission. The quantum confinement effect is an important model for explaining the PL behavior of GQDs. When the size of QGDs is smaller than their exciton Bohr radius, this phenomenon happens [44].

3.4 Biocompatibility

The fundamental requirements for materials employed for biological imaging and drug delivery are non-cytotoxicity and biocompatibility. GQDs have been demonstrated to be more biocompatible over traditional semiconductor quantum dots and CDs in in-vivo and in-vitro studies [10, 22]. The capability of a material to interact with its intended role in biological systems without triggering undesired biological reactions is referred to as biocompatibility [46]. Although, carbon-based nanomaterials like carbon nanotubes are less toxic than carbon fibers, and carbon nanoparticles. The toxicity of carbon nanotubes enhanced significantly when surfaces contains carboxyl, hydroxyl, and carbonyl functional groups [47]. GQDs are made solely of carbon, thus, they are non-toxic and have a high solubility in water, making them an excellent alternative to metal-based nanoparticles [46]. Cytotoxicity tests (in-vivo or in-vivo) are another approach used to assess biocompatibility. However, in-vitro testing employing cell culture is more recommended because of the reduced number of experiments with animals required, lower cost, and shorter testing time [48].

Zhou et al. [49] used an MTT assay in the dark and under photo-irradiation to assess the toxicity of virgin and functionalized GQDs (GQDs-PH; GQDs functionalized with phenylhydrazine, GQDs-BA; functionalized with benzoic anhydride, and GQDs-BrPE; functionalized with 2-bromo-1-phenylethanone) (Fig. 4) against the Hela cells. They observed that GQDs and their derivatives had substantially the same vitality as control cells in dark conditions, showing strong biocompatibility (Fig. 4a). When exposed to light, the cells treated with different functionalized GQDs had a relatively low viability (Fig. 4b). This was attributed to GQDs generating reactive oxygen species (ROS) upon excitation. The ROS produced by GQDs were responsible for the photoinduced cytotoxicity. Derivatives of GQD reduced photoinduced cytotoxicity that might be attributed to their lower ROS production activity, particularly GQDs-PH produced less ROS, resulting in the least cytotoxicity. In another report, modification of GQDs surface gave different toxicities such as GQDs modified with -OH displayed toxicity at concentration above 100 μg mL^{-1}, whereas –NH_2, and –COOH showed relatively low toxicity at concentration up to 200 μg mL^{-1} [50]. MTT test was used to evaluate the cytotoxicity of pristine GQDs generated from carbon fibers in two distinct human breast cancer cell lines, MDA-MB-231 and T47D. The GQDs posed no cytotoxicity to the treated cells at 50 mg mL^{-1} concentrations. For long time, GQDs functionalized with phthalhydrazide-like groups and hydrazide groups at the edge displayed low cytotoxicity. The cell viability of three types of stem cells i.e., neurosphere cells, pancreas progenitor cells, and cardiac progenitor cells, was essentially unaffected after 72 h of incubation with 25 mg mL^{-1} of GQDs. Cell viabilities of 80% for neurosphere cells and 65% for cardiac progenitor cells were attained even at a concentration of 100 mg mL^{-1} for GQDs [51, 52].

3.5 *Physical Property*

GQDs exhibits both graphene's property as well as quantum confinement in three-dimensional quantum dot [53]. This kind of nanoparticle is commonly produced from graphene and graphene oxide (GO). Graphene and graphene oxide consist of a single layer sp^2-hybridized carbon atoms [54] which play a key role in photoluminescence properties [55]. Graphene itself has no functional group [56], while graphene oxide contains an oxygen functional group [57]. GQDs might be embellished with a range of functional groups including amide, amine, hydroxyl, and groups doped with certain

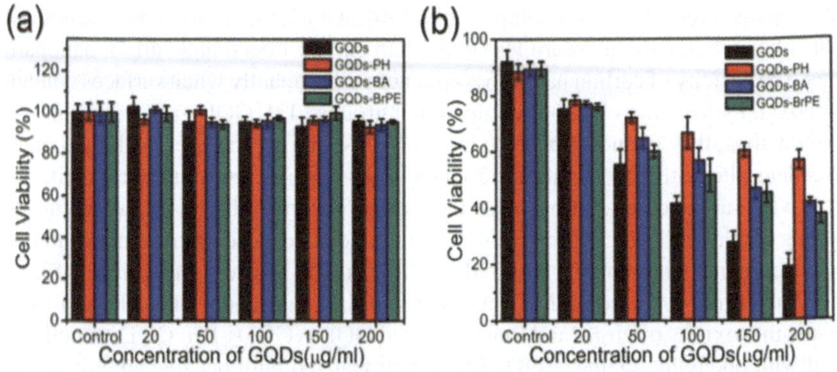

Fig. 4 Viability cells at different concentrations of GQDs, GQDs-PH, GQDs-BA, and GQDs-BrPE in the dark (**a**) or in irradiation (**b**) (reprinted/reproduced with the permission of Ref. [49], copyright 2017, Chemical Communications)

substances like sulfur and nitrogen to change their molecular structure and to improve their characteristics [58] (e.g., FL and antimicrobial properties). Strelko et al. [59] reported that the presence of heteroatoms, such as doping agents or functional groups, adds electron-donating atoms to the structures of carbon, enabling electron transfer. As a result, it leads to higher antibacterial ability. On the other hand, the smaller size which is owned by GQDs compared to graphene and graphene oxide, make GQD have better prospect in biomedical application [60].

4 Applications of GQDs in Drug Delivery

In recent years, GQDs applications have attained major success in drug delivery [61–63], bio-imaging [64, 65], magnetic hyperthermia [66–68], photothermal therapy [69, 70], antibacterial activity [71–73], and environmental protection [74]. Several studies employed the density functional theory (DFT) calculations, molecular dynamics (MD) simulations, or other methods to theoretically explore the properties of GQDs in order to optimize their application in drug administration. Vatanparast et al. [75] applied DFT simulations to study the interaction of 5-fluorouracil (FU) drug with doped/undoped GQDs. According to the findings, AlN and AlP-doped GQDs could be used as prospective FU drug carriers in the realm of nanomedicine. In another study, DFT simulations assisted to evaluate the suitability of GQDs and doped GQDs as prospective isoniazid (Iso) carriers [76]. The findings demonstrated that AlN- and AlP-doped GQDs might be used as drug delivery vehicles. In addition, the DFT calculations and MD simulations have been implemented to analyze the effects of different N-functionalized groups on their drug delivery performance. The drug release performance of center N-GQDs is obviously thought to be superior than that of pristine GQDs and edge N-GQDs [77].

Figure 5 depicts anti-cancer drug delivery, launch, and response. The near-infrared fluorescent molecule Cy5.5 (Cy) dye was covalently bonded to GQDs using a cathepsin D tuned peptide (Phe-AlaAla-Phe-Phe-Val-Leu-Cys, FAAFFVLC, P), and subsequently the functionalized GQDs were loaded with the anti-cancer drug doxorubicin (DOX) via a π–π interaction mechanism and the in-vitro study on cancer cells was monitored. There are several strategies for drug delivery however, focusing solely on delivery and ignoring drug release is not improving the therapeutic impact of drugs. As a result, researchers are focusing on the closed interaction between drug delivery and release in order to optimize drug's therapeutic impact by increasing delivery and release efficiency. Drug delivery and release using graphene or graphene-based nanostructures has been shown to improve delivery efficiency and therapeutic benefits.

GQDs exhibit huge surface/volume ratio, high water solubility and reduced cytotoxicity, as compared to graphene, making them more effective drug molecular loading cores [73, 79–81]. Additionally, high functionalization options of GQDs choices render these particles superior to other nanoparticles in drug delivery applications. The high surface/volume ratio and surface functionalization make GQDs an efficient carrier for drug loading and distribution. Furthermore, GQDs outperform quantum dots (QDs) of comparable size in terms of quantum confinement effects, simultaneous tracking potential because of programmable photoluminescence (PL) and substantially lesser toxicity due to the absence of heavy metal components. However, the distinctive π-orbitals in the sp^2-hybridized GQDs lattice can be exploited to link medicines with aromatic ring structures via π–π stacking, which broadens the potential of GQDs for drug delivery [67, 82, 83].

The drugs entered the target cells via the targeting ligand's enhanced permeability and retention action. GQDs, like various non-degradable nanoparticle drug carriers, released drugs through diffusion. Adsorbed drugs on GQDs are released

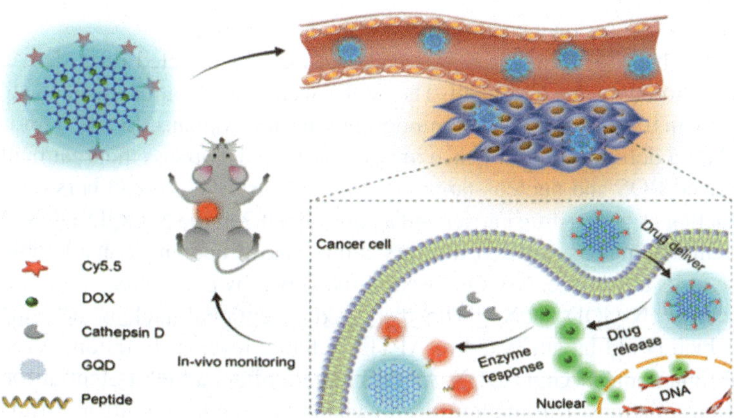

Fig. 5 Anti-cancer delivery, release, and response of GQD-based theranostic agent. (reproduced with the permission of Ref. [78], copyright 2017, ACS Applied Materials & Interfaces)

into the cytoplasm by desorption and diffusion. It has been reported that the GQDs can pass the blood–brain barrier (BBB) and inhibit α-synucleinopathy in Parkinson's disease. The size and charge of the nanoparticles have an important effect on their BBB permeability. Nanoparticles of small size can easily pass the BBB and diffuse more effectively across the brain. Gold nanoparticles as small as 15 nm can easily conquer the BBB without being functionalized. Nevertheless, gold nanoparticles of 50 nm and larger size are unable to pass the BBB and are not detected in the brain. The tiny size is believed to be the primary reason for GQDs that can cross BBB, and also most likely cross biological barriers via the transmembrane or paracellular pathway [84, 85]. Biomedical experts have long been concerned about the integration of cancer detection and treatment. Despite significant advances in targeted drug delivery systems to deliver anticancer treatments to specific locations of interest, new nanomaterials are frequently synthesized and examined for improved drug delivery efficacy. While developing novel nanomaterials, scientists explored new drug delivery-release modalities. EPR-pH delivery-release mode [86], ligand-pH delivery-release mode [87, 88], EPR-photothermal delivery-release mode [61], and core/shell-photothermal/magnetic thermal delivery release modes are the most common [89, 90].

Drug delivery schemes are uniquely developed for selective and regulated drug release which is critical to meet the requirement of current biomedical therapies. The development of nanocarriers for in-vivo drug storage along with controlled release with high specificity in targeted infected and/or malignant cells has taken significant time and effort. To enhance the drug selectivity and half-life, researcher modified GQDs with PEG, which increased therapeutic drug selectivity toward cancerous cells. Indeed, the PEG molecule dramatically enhances the half-life of GQDs by preventing white blood cells from recognizing it as a particle. Certainly, scientists are working to develop a more efficient GQDs-based vehicle for controlled drug delivery directly to infected cells. It has been demonstrated utilizing two types of nanoparticles, the first of which recognizes tumor cells and the second of which (containing therapeutic medications) focuses on a signal produced by the first [91, 92].

A GQDs-based nanomaterial was used for nucleus targeted drug delivery and regulated photodynamic treatment. The GQDs were doped with a nitrogen atom (N) to allow them to interact with (3-Aminopropyl) triethoxysilane (APTES) for nucleus targeted drug delivery. Fluorescence imaging was used to assess the treatment effect of generated ROS and nucleus-targeting drug delivery. The researchers discovered that by acting as both a drug carrier and a photosensitizer, the N-GQD-DOX-APTES could achieve nucleus-targeted delivery while also producing considerable ROS. Because of its small size, N-GQD is easily absorbed by cells. More importantly, as a drug carrier, N-GQD-DOXAPTES displayed exceptional nucleus targeting capability, which could be attributed to APTES' charge reversal function. As a result, the multifunctional N-GQD-DOX-APTES system offers a high potential for multimodal cancer therapy. This method offers a viable way for developing multifunctional therapy in a single nano platform for biomedical applications [93]. Iannazzo et al. created a GQD-based cancer-targeted medication delivery system. The introduction of an anticancer medicine via a PEG linker and the targeted ligand riboflavin through contact with a pyrene linker became possible by multimodal nanomaterial conjugation. The nanocarrier was chemically connected to the drug through a cleavable PEG linker, while the targeted ligand riboflavin was attached to the GQDs using a pyrene linker. On three cancer cell lines, the cytotoxic activity of the manufactured drug delivery method GQD-PEG-BFG@Pyr-RF was studied and compared to that of the same nano-vector lacking the RF ligand (GQD-PEG-BFG) or the anticancer agent (GQD@Pyr-RF). Biological studies demonstrated that the GQDs sample had weak cytotoxicity, however the drug deliver system (DDS) had cytotoxic activity against the examined cancer cell lines with greater or equivalent potency to that of the benzofuran structure (BFG) alone. It opens up new possibilities for the use of cytotoxic anticancer drugs with substantial adverse effects [94].

5 Conclusions

In conclusion, the ongoing substantial amount of GQDs research in the disciplines of nanomedicine and nano-theranostics is significant and promising, and clinical applications of GQDs are expected to be reality within the near future. It is believed that the future research may effectively address the current issues associated with the deployment of GQDs in order to build more safe and simple synthesis routes as well as efficient targeted drug delivery in the treatment of numerous life-threatening diseases.

References

1. Geim, A.K., Novoselov, K.S.: The rise of graphene. Nat. Mater. **6**, 183–191 (2007)
2. Pierrat, P., Gaumet, J.-J.: Graphene quantum dots: emerging organic materials with remarkable and tunable luminescence features. Tetrahedron Lett. **61**(49), 52554 (2020)
3. Eda, G., Lin, Y.Y., Mattevi, C., Yamaguchi, H., Chen, H.A., Chen, I.S., et al.: Blue photoluminescence from chemically derived graphene oxide. Adv. Mater. **22**(4), 505–509 (2010)
4. Loh, K.P., Bao, Q., Eda, G., Chhowalla, M.: Graphene oxide as a chemically tunable platform for optical applications. Nat. Chem. **2**(12), 1015–1024 (2010)
5. Zhao, C., Song, X., Liu, Y., Fu, Y., Ye, L., Wang, N., et al.: Synthesis of graphene quantum dots and their applications in drug delivery. J. Nanobiotechnol. **18**, 1–32 (2020)
6. Wang, L., Zhu, S.-J., Wang, H.-Y., Qu, S.-N., Zhang, Y.-L., Zhang, J.-H., et al.: Common origin of green luminescence in carbon nanodots and graphene quantum dots. ACS Nano **8**(3), 2541–2547 (2014)
7. Jin, S.H., Kim, D.H., Jun, G.H., Hong, S.H., Jeon, S.: Tuning the photoluminescence of graphene quantum dots through the charge transfer effect of functional groups. ACS Nano **7**(2), 1239–1245 (2013)
8. Tian, P., Tang, L., Teng, K., Lau, S.: Graphene quantum dots from chemistry to applications. Mater. Today Chem. **10**, 221–258 (2018)
9. Chen, W., Lv, G., Hu, W., Li, D., Chen, S., Dai, Z.: Synthesis and applications of graphene quantum dots: a review. Nanotechnol. Rev. **7**(2), 157–185 (2018)

10. Henna, T.K., Pramod, K.: Graphene quantum dots redefine nanobiomedicine. Mater. Sci. Eng. C **110**, 110651 (2020)
11. Zhao, C., Song, X., Liu, Y., Fu, Y., Ye, L., Wang, N., et al.: Synthesis of graphene quantum dots and their applications in drug delivery. J. Nanobiotechnol. **18**(1), 142 (2020)
12. Zheng, X.T., Ananthanarayanan, A., Luo, K.Q., Chen, P.: Glowing graphene quantum dots and carbon dots: properties, syntheses, and biological applications. Small **11**(14), 1620–1636 (2015)
13. Fan, J., Li, D., Wang, X.J.D.: Effect of modified graphene quantum dots on photocatalytic degradation property. Diam. Relat. Mater. **69**, 81–85 (2016)
14. Cai, N., Tan, L., Li, Y., Xia, T., Hu, T., Su, X.: Biosensing platform for the detection of uric acid based on graphene quantum dots and G-quadruplex/hemin DNAzyme. Anal. Chim. Acta **965**, 96–102 (2017)
15. Gulati, S., Baul, A., Amar, A., Wadhwa, R., Kumar, S., Varma, R.S.: Eco-friendly and sustainable pathways to photoluminescent carbon quantum dots (CQDs). Nanomaterials **13**(3), 554 (2023)
16. Tong, G., Wang, J., Wang, R., Guo, X., He, L., Qiu, F., et al.: Amorphous carbon dots with high two-photon fluorescence for cellular imaging passivated by hyperbranched poly(amino amine). J. Mater. Chem. B **3**(4), 700–706 (2015)
17. Sk, M.A., Ananthanarayanan, A., Huang, L., Lim, K.H., Chen, P.: Revealing the tunable photoluminescence properties of graphene quantum dots. J. Mater. Chem. C **2**(34), 6954–6960 (2014)
18. Valeur, B., Berberan-Santos, M.N.: A brief history of fluorescence and phosphorescence before the emergence of quantum theory. J. Chem. Educ. **88**(6), 731–738 (2011)
19. Mansuriya, B.D., Altintas, Z.: Applications of graphene quantum dots in biomedical sensors. Sensors (Basel) **20**(4) (2020)
20. Han, H.S., Choi, K.Y.: Advances in nanomaterial-mediated photothermal cancer therapies: toward clinical applications. Biomedicines **9**(3), 305 (2021)
21. Badıllı, U., Mollarasouli, F., Bakirhan, N.K., Ozkan, Y., Ozkan, S.A.: Role of quantum dots in pharmaceutical and biomedical analysis, and its application in drug delivery. TrAC - Trends Anal. Chem. **131**, 116013 (2020)
22. Shen, J., Zhu, Y., Yang, X., Li, C.: Graphene quantum dots: emergent nanolights for bioimaging, sensors, catalysis and photovoltaic devices. ChemComm **48**(31), 3686–3699 (2012)
23. Song, Y., Wang, J., Wan, H., Zhang, Y., Ningcong, Y., Yang, B.: Photoluminescence mechanism in graphene quantum dots: quantum confinement effect and surface/edge state. Nano Today **13**, 10–14 (2016)
24. Eda, G., Lin, Y.-Y., Mattevi, C., Yamaguchi, H., Chen, H.-A., Chen, I.S., et al.: Blue photoluminescence from chemically derived graphene oxide. Adv. Mater. **22**, 505–509 (2010)
25. Wang, Z., Zeng, H., Sun, L.: Graphene quantum dots: versatile photoluminescence for energy, biomedical, and environmental applications. J. Mater. Chem. C **3**, 1157–1165 (2014)
26. Wang, Y., Hu, A.: Carbon quantum dots: synthesis, properties and applications. J. Mater. Chem. C **2**, 6921–6939 (2014)
27. Liu, Z., Li, F., Luo, Y., Li, M., Hu, G., Pu, X., et al.: Size effect of graphene quantum dots on photoluminescence. Molecules **26**, 3922 (2021)
28. Yang, P.-C., Ting, Y.-X., Gu, S., Ashraf Gandomi, Y., Li, J., Hsieh, C.-T.: Effect of solvent on fluorescence emission from polyethylene glycol-coated graphene quantum dots under blue light illumination. Nanomaterials **11**(6), 1383 (2021)
29. Son, I., Son, S.-R., An, J., Choi, J.-W., Kim, S., Lee, W.Y., et al.: Photoluminescent surface-functionalized graphene quantum dots for spontaneous interfacial homeotropic orientation of liquid crystals. J. Mol. Liq. **332**, 115901 (2021)
30. Chhabra, V., Kaur, R., Kumar, N., Deep, A., Ch, R., Kim, K.-H.: Synthesis and spectroscopic studies of functionalized graphene quantum dots with diverse fluorescence characteristics. RSC Adv. **8**, 11446–11454 (2018)
31. Timofeeva, T., Egorova, M., Tomskaya, A., Smagulova, S.: Ab initio calculations of energy and IR spectra of edge functionalized graphene quantum dots. AIP Conf. Proc. **2041**, 020019 (2018)

32. Rani, P., Dalal, R., Srivastava, S.: Effect of surface modification on optical and electronic properties of graphene quantum dots. Appl. Surf. Sci. **609**, 155379 (2023)
33. Yeh, T.-F., Huang, W.-L., Chung, C.-J., Chiang, I.T., Chen, L.-C., Chang, H.-Y., et al.: Elucidating quantum confinement in graphene oxide dots based on excitation-wavelength-independent photoluminescence. J. Phys. Chem. Lett. **7**(11), 2087–2092 (2016)
34. Rani, P., Dalal, R., Srivastava, S.: Enhanced NIR fluorescence quantum yield of graphene quantum dots using dopants. Phys. Scr. **98**(6), 064004 (2023)
35. Jiang, D., Chen, Y., Li, N., Li, W., Wang, Z., Zhu, J., et al.: Synthesis of luminescent graphene quantum dots with high quantum yield and their toxicity study. PLoS ONE **10**(12), e0144906 (2015)
36. Kim, R., Perebeinos, V., Avouris, P.: Relaxation of optically excited carriers in graphene. Phys. Rev. B **84**(7), 075449 (2011)
37. Zhao, B., Ma, H., Zheng, M., Xu, K., Zou, C., Qu, S., et al.: Narrow-bandwidth emissive carbon dots: a rising star in the fluorescent material family. Carbon Energy **4**(1), 88–114 (2022)
38. Valappil, M.O., Pillai, V.K., Alwarappan, S.: Spotlighting graphene quantum dots and beyond: synthesis, properties and sensing applications. Appl. Mater. Today **9**, 350–371 (2017)
39. Kwon, W., Kim, Y.-H., Lee, C.-L., Lee, M., Choi, H.C., Lee, T.-W., et al.: Electroluminescence from graphene quantum dots prepared by amidative cutting of tattered graphite. Nano Lett. **14**(3), 1306–1311 (2014)
40. Facure, M.H., Schneider, R., Mercante, L.A., Correa, D.S.: A review on graphene quantum dots and their nanocomposites: from laboratory synthesis towards agricultural and environmental applications. Environ. Sci. Nano **7**(12), 3710–3734 (2020)
41. Kittiratanawasin, L., Hannongbua, S.: The effect of edges and shapes on band gap energy in graphene quantum dots. Integr. Ferroelectr. **175**(1), 211–219 (2016)
42. Lin, L., Zhang, S.: Creating high yield water soluble luminescent graphene quantum dots via exfoliating and disintegrating carbon nanotubes and graphite flakes. ChemComm **48**(82), 10177–10179 (2012)
43. Zhu, S., Song, Y., Wang, J., Wan, H., Zhang, Y., Ning, Y., et al.: Photoluminescence mechanism in graphene quantum dots: quantum confinement effect and surface/edge state. Nano Today **13**, 10–14 (2017)
44. Zhu, S., Shao, J., Song, Y., Zhao, X., Du, J., Wang, L., et al.: Investigating the surface state of graphene quantum dots. Nanoscale **7**(17), 7927–7933 (2015)
45. Schneider, R., Mercante, L., Correa, D.: A review on graphene quantum dots and their nanocomposites: from laboratory synthesis towards agricultural and environmental applications. Environ. Sci. Nano **7**, 3710–3734 (2020)
46. Chung, S., Revia, R.A., Zhang, M.: Graphene quantum dots and their applications in bioimaging, biosensing, and therapy. Adv. Mater. **33**(22), e1904362 (2021)
47. Magrez, A., Kasas, S., Salicio, V., Pasquier, N., Seo, J.W., Celio, M., et al.: Cellular toxicity of carbon-based nanomaterials. Nano Lett. **6**(6), 1121–1125 (2006)
48. Brown, R.P., Fowler, B.A., Fustinoni, S., Costa, M., Nordberg, M.: Toxicity of metals released from implanted medical devices. In: Nordberg, G.F., Costa, M.: Handbook on the Toxicology of Metals, pp. 127–36. Elsevier (2022)
49. Zhou, Y., Sun, H., Wang, F., Ren, J., Qu, X.: How functional groups influence the ROS generation and cytotoxicity of graphene quantum dots. ChemComm **53**(76), 10588–10591 (2017)
50. Xie, Y., Wan, B., Yang, Y., Cui, X., Xin, Y., Guo, L.-H.: Cytotoxicity and autophagy induction by graphene quantum dots with different functional groups. J. Environ. Sci. **77**, 98–209 (2019)
51. Peng, J., Gao, W., Gupta, B.K., Liu, Z., Romero-Aburto, R., Ge, L., et al.: Graphene quantum dots derived from carbon fibers. Nano Lett. **12**(2), 844–849 (2012)
52. Zhang, M., Bai, L., Shang, W., Xie, W., Ma, H., Fu, Y., et al.: Facile synthesis of water-soluble, highly fluorescent graphene quantum dots as a robust biological label for stem cells. J. Mater. Chem. **22**(15), 7461–7467 (2012)
53. Preeti Nigam, J., Subir, K., Sunil, K.S., Dhiman, S.: Graphene quantum dots-from emergence to nanotheranostic applications. In: Ali Demir, S.: Smart Drug Delivery Dystem, chap. 7. IntechOpen (2016)

54. Jiříčková, A., Jankovský, O., Sofer, Z., Sedmidubský, D.: Synthesis and applications of graphene oxide. Materials **15**(3), 920 (2022)
55. Tang, D., Liu, J., Yan, X., Kang, L.: The graphene oxide derived graphene quantum dots with different photoluminescence properties and peroxidase-like catalytic activity. RSC Adv. **6**, 50609–50617 (2016)
56. Zhen, Z., Zhu, H.: Structure and properties of graphene. In: Zhu, H., Xu, Z., Xie, D., Fang, Y.: Graphene, pp. 1–12. Academic Press (2018)
57. Ray, S.C.: Application and uses of graphene oxide and reduced graphene oxide. In: Ray, S.C.: Applications of Graphene and Graphene-Oxide Based Nanomaterials, pp. 39–55. William Andrew Publishing, Oxford (2015)
58. Schroer, Z.S., Wu, Y., Xing, Y., Wu, X., Liu, X., Wang, X., et al.: Nitrogen–sulfur-doped graphene quantum dots with metal ion-resistance for bioimaging. ACS Appl. Nano Mater. **2**(11), 6858–6865 (2019)
59. Strelko, V.V., Kuts, V.S., Thrower, P.A.: On the mechanism of possible influence of heteroatoms of nitrogen, boron and phosphorus in a carbon matrix on the catalytic activity of carbons in electron transfer reactions. Carbon **38**(10), 1499–1503 (2000)
60. Du, Y., Guo, S.: Chemically doped fluorescent carbon and graphene quantum dots for bioimaging, sensor, catalytic and photoelectronic applications. Nanoscale **8**, 2532–2543 (2016)
61. Xu, Y., Hu, X., Guan, P., Du, C., Tian, Y., Ding, S., et al.: A novel controllable molecularly imprinted drug delivery system based on the photothermal effect of graphene oxide quantum dots. J. Mater. Sci. **54**(12), 9124–9139 (2019)
62. Yue, J., He, L., Tang, Y., Yang, L., Wu, B., Ni, J.: Facile design and development of photoluminescent graphene quantum dots grafted dextran/glycol-polymeric hydrogel for thermoresponsive triggered delivery of buprenorphine on pain management in tissue implantation. J. Photochem. Photobiol. B: Biol. **197**, 111530 (2019)
63. Havanur, S., Batish, I., Cheruku, S.P., Gourishetti, K., JagadeeshBabu, P., Kumar, N.: Poly (N, N-diethyl acrylamide)/functionalized graphene quantum dots hydrogels loaded with doxorubicin as a nano-drug carrier for metastatic lung cancer in mice. Mater. Sci. Eng. C **105**, 110094 (2019)
64. Rajender, G., Goswami, U., Giri, P.: Solvent dependent synthesis of edge-controlled graphene quantum dots with high photoluminescence quantum yield and their application in confocal imaging of cancer cells. J. Colloid Interface Sci. **541**, 387–398 (2019)
65. Zhang, Q., Deng, S., Liu, J., Zhong, X., He, J., Chen, X., et al.: Cancer-targeting graphene quantum dots: fluorescence quantum yields, stability, and cell selectivity. Adv. Funct. Mater. **29**(5), 1805860 (2019)
66. Yao, X., Niu, X., Ma, K., Huang, P., Grothe, J., Kaskel, S., et al.: Graphene quantum dots-capped magnetic mesoporous silica nanoparticles as a multifunctional platform for controlled drug delivery, magnetic hyperthermia, and photothermal therapy. Small **13**(2), 1602225 (2017)
67. Chung, S., Revia, R.A., Zhang, M.: Graphene quantum dots and their applications in bioimaging, biosensing, and therapy. Adv. Mater. **33**(22), 1904362 (2021)
68. Alegret, N., Criado, A., Prato, M.: Recent advances of graphene-based hybrids with magnetic nanoparticles for biomedical applications. Curr. Med. Chem. **24**(5), 529–536 (2017)
69. Justin, R., Tao, K., Román, S., Chen, D., Xu, Y., Geng, X., et al.: Photoluminescent and superparamagnetic reduced graphene oxide–iron oxide quantum dots for dual-modality imaging, drug delivery and photothermal therapy. Carbon **97**, 54–70 (2016)
70. Guo, M., Xiang, H.-J., Wang, Y., Zhang, Q.-L., An, L., Yang, S.-P., et al.: Ruthenium nitrosyl functionalized graphene quantum dots as an efficient nanoplatform for NIR-light-controlled and mitochondria-targeted delivery of nitric oxide combined with photothermal therapy. ChemComm **53**(22), 3253–3256 (2017)
71. Kuo, W.-S., Chen, H.-H., Chen, S.-Y., Chang, C.-Y., Chen, P.-C., Hou, Y.-I., et al.: Graphene quantum dots with nitrogen-doped content dependence for highly efficient dual-modality photodynamic antimicrobial therapy and bioimaging. Biomaterials **120**, 185–194 (2017)
72. Sun, H., Gao, N., Dong, K., Ren, J., Qu, X.: Graphene quantum dots-band-aids used for wound disinfection. ACS Nano **8**(6), 6202–6210 (2014)

73. Biswas, A., Khandelwal, P., Das, R., Salunke, G., Alam, A., Ghorai, S., et al.: Oxidant mediated one-step complete conversion of multi-walled carbon nanotubes to graphene quantum dots and their bioactivity against mammalian and bacterial cells. J. Mater. Chem. B **5**(4), 785–796 (2017)
74. Xu, T., Wang, D., Dong, L., Shen, H., Lu, W., Chen, W.: Graphitic carbon nitride co-modified by zinc phthalocyanine and graphene quantum dots for the efficient photocatalytic degradation of refractory contaminants. Appl. Catal. B: Environ. **244**, 96–106 (2019)
75. Vatanparast, M., Shariatinia, Z.: AlN and AlP doped graphene quantum dots as novel drug delivery systems for 5-fluorouracil drug: theoretical studies. J. Fluor. Chem. **211**, 81–93 (2018)
76. Vatanparast, M., Shariatinia, Z.: Computational studies on the doped graphene quantum dots as potential carriers in drug delivery systems for isoniazid drug. Struct. Chem. **29**, 1427–1448 (2018)
77. Vatanparast, M., Shariatinia, Z.: Revealing the role of different nitrogen functionalities in the drug delivery performance of graphene quantum dots: a combined density functional theory and molecular dynamics approach. J. Mater. Chem. B **7**(40), 6156–6171 (2019)
78. Ding, H., Zhang, F., Zhao, C., Lv, Y., Ma, G., Wei, W., et al.: Beyond a carrier: graphene quantum dots as a probe for programmatically monitoring anti-cancer drug delivery, release, and response. ACS Appl. Mater. Interfaces **9**(33), 27396–27401 (2017)
79. Nasrollahi, F., Koh, Y.R., Chen, P., Varshosaz, J., Khodadadi, A.A., Lim, S.: Targeting graphene quantum dots to epidermal growth factor receptor for delivery of cisplatin and cellular imaging. Mater. Sci. Eng. C **94**, 247–257 (2019)
80. Xue, Z., Sun, Q., Zhang, L., Kang, Z., Liang, L., Wang, Q., et al.: Graphene quantum dot assisted translocation of drugs into a cell membrane. Nanoscale **11**(10), 4503–4514 (2019)
81. De, S., Patra, K., Ghosh, D., Dutta, K., Dey, A., Sarkar, G., et al.: Tailoring the efficacy of multifunctional biopolymeric graphene oxide quantum dot-based nanomaterial as nanocargo in cancer therapeutic application. ACS Biomater. Sci. Eng. **4**(2), 514–531 (2018)
82. Jegannathan, P., Yousefi, A.T., Abd Karim, M.S., Kadri, N.A.: Enhancement of graphene quantum dots based applications via optimum physical chemistry: a review. Biocybern. Biomed. Eng. **38**(3), 481–497 (2018)
83. Pardo, J., Peng, Z., Leblanc, R.M.: Cancer targeting and drug delivery using carbon-based quantum dots and nanotubes. Molecules **23**(2), 378 (2018)
84. Kim, D., Yoo, J.M., Hwang, H., Lee, J., Lee, S.H., Yun, S.P., et al.: Graphene quantum dots prevent α-synucleinopathy in Parkinson's disease. Nat. Nanotechnol. **13**(9), 812–818 (2018)
85. Khlebtsov, N., Dykman, L.: Biodistribution and toxicity of engineered gold nanoparticles: a review of in vitro and in vivo studies. Chem. Soc. Rev. **40**(3), 1647–1671 (2011)
86. Khodadadei, F., Safarian, S., Ghanbari, N.: Methotrexate-loaded nitrogen-doped graphene quantum dots nanocarriers as an efficient anticancer drug delivery system. Mater. Sci. Eng. C **79**, 280–285 (2017)
87. Ko, N., Nafiujjaman, M., Lee, J., Lim, H.-N., Lee, Y.-K., Kwon, I.: Graphene quantum dot-based theranostic agents for active targeting of breast cancer. RSC Adv. **7**(19), 11420–11427 (2017)
88. Dong, J., Wang, K., Sun, L., Sun, B., Yang, M., Chen, H., et al.: Application of graphene quantum dots for simultaneous fluorescence imaging and tumor-targeted drug delivery. Sens. Actuators B Chem. **256**, 616–623 (2018)
89. Sung, S.-Y., Su, Y.-L., Cheng, W., Hu, P.-F., Chiang, C.-S., Chen, W.-T., et al.: Graphene quantum dots-mediated theranostic penetrative delivery of drug and photolytics in deep tumors by targeted biomimetic nanosponges. Nano Lett. **19**(1), 69–81 (2018)
90. Gao, Y., Zhong, S., Xu, L., He, S., Dou, Y., Zhao, S., et al.: Mesoporous silica nanoparticles capped with graphene quantum dots as multifunctional drug carriers for photo-thermal and redox-responsive release. Microporous Mesoporous Mater. **278**, 130–137 (2019)
91. Iannazzo, D., Pistone, A., Salamò, M., Galvagno, S., Romeo, R., Giofré, S.V., et al.: Graphene quantum dots for cancer targeted drug delivery. Int. J. Pharm. **518**(1–2), 185–192 (2017)
92. Cui, W., Li, J., Decher, G.: Self-assembled smart nanocarriers for targeted drug delivery. Adv. Mater. **28**(6), 1302–1311 (2016)

93. Ju, J., Regmi, S., Fu, A., Lim, S., Liu, Q.: Graphene quantum dot based charge-reversal nanomaterial for nucleus-targeted drug delivery and efficiency controllable photodynamic therapy. J. Biophotonics **12**(6), e201800367 (2019)
94. Iannazzo, D., Pistone, A., Celesti, C., Triolo, C., Patané, S., Giofré, S.V., et al.: A smart nanovector for cancer targeted drug delivery based on graphene quantum dots. Nanomaterials **9**(2), 282 (2019)

Two-Dimensional (2D)-Based Hybrid Composites for Cancer Diagnosis and Therapy

Mehdi Abedi, Mahnaz Asadi, Marzieh Mehrzadeh, Zahra Rahiminezhad, Younes Ghasemi, and Vahid Alimardani

Abstract Cancer is the deadliest disease that has affected human health all over the world. Despite the much research that has been done in this field so far, there is an urgent need for improved techniques for diagnosis and treatment of cancer. In recent years, innovative nanomaterial-assisted therapies and diagnostics have led to significant enhancements in cancer treatment. In this context, two-dimensional nanomaterials (2DNMs) and their nanocomposites have been explored for cancer diagnostics and therapy due to their unique shape, size, chemical composition, biodegradability, and biocompatibility. The studies indicate that 2DNMs have significant potential in biomedicine, particularly in multimodal imaging, biosensors, drug/gene delivery, and cancer therapy. Due to high specific surface area, 2DNMs can efficiently adsorb molecules via covalent or non-covalent interactions and usage as carriers in controlled release systems in response to external stimuli. In addition, unique sheet-like nanostructures and photothermal converting ability led to 2DNMs being promising candidates for optical therapies such as photothermal therapy (PTT), photodynamic therapy (PDT), and theranostic. Furthermore, integrating 2DNMs in

M. Abedi · Y. Ghasemi
Pharmaceutical Sciences Research Center, Shiraz University of Medical Sciences, Shiraz, Iran

M. Abedi
Research and Development Department, Danesh Salamat Kowsar Co, Shiraz, Iran

M. Asadi
Department of Chemistry, Borujerd Branch, Islamic Azad University, Borujerd, Iran

M. Mehrzadeh
Department of Pharmacognosy, School of Pharmacy, Shiraz University of Medical Sciences, Shiraz, Iran

Z. Rahiminezhad
Research and Innovation Division, Kani Godakht Tahmin Co, Marvdasht, Iran

Y. Ghasemi
Department of Pharmaceutical Biotechnology, School of Pharmacy, Shiraz University of Medical Sciences, Shiraz, Iran

V. Alimardani (✉)
Student Research Committee, Shiraz University of Medical Sciences, Shiraz, Iran
e-mail: vahidalimardani@ymail.com

N. Talreja et al. (eds.), *Two-dimensional Hybrid Composites*, Engineering Materials,
https://doi.org/10.1007/978-981-99-8010-9_11

nanocomposites with further functional moieties become a new class of therapeutic agents in biomedicine substantially improving their features for synergistic cancer diagnosis and therapy. This chapter highlights the current state and benefits of using 2DNMs in cancer diagnosis and therapy and discusses the obstacles and prospects of their future development. Then we focus on 2DNMs applications in cancer treatment including smart drug delivery systems, PTT, and PDT. Lastly, the 2DNMs biocompatibility is also discussed to provide a unique overview of the topic.

1 Introduction

Cancer's diversity, complexity, and heterogeneity make it one of the most deadly diseases that affect human health over the centuries [1]. Surgery, chemotherapy, and radiation are the gold standards in cancer treatment, yet they all have significant drawbacks that lead to a high recurrence and fatality rate [2, 3]. As a result, there is an urgent need for improved techniques for identifying and treating cancer early to enhance the quality of life, raise survival rates, and, most significantly, prolong people's lives [4]. Since the beginning of nanotechnology and its subsequent explosion in scope, innovative cancer nanomaterial-assisted therapies and diagnostics have made significant strides forward in recent years [5]. The field of cancer treatment has seen the emergence of various nanomaterials [6]. However, new evidence suggests that two-dimensional nanomaterials (2DNMs) and their nanocomposites have significant benefits over zero-, one-, and three-dimensional nanomaterials for application in cancer diagnostics and therapy. Reasons for this include their adaptability in terms of structure and surface chemistry as well as their ability to take on arbitrary or desirable physical qualities; in particular, their high aspect ratio, ultra-high surface area, and ultra-high therapeutic loading capacity [7, 8]. In 2004, Novoselov et al. successfully exfoliated graphene (G) from graphite, marking the beginning of the research of 2DNMs [9–11]. Graphene's achievements have rekindled research into similar ultrathin 2DNMs. Numerous 2DNMs, such as transition metal carbides and/or nitrides (MXenes), transition metal dichalcogenide (TMDC), black phosphorus (BP), graphene oxide (GO), manganese dioxide (MnO_2), layered double hydroxides (LDHs), and palladium (Pd), have garnered considerable interest in the realm of cancer diagnosis and treatment (Table 1) [12]. The two most common methods for synthesizing 2DNMs are the bottom-up and top-down methods. The bottom-up techniques include crystal phase transition, biological synthesis, hydro-/solvothermal approaches, seeded growth, small molecule and ion-mediated synthesis, and nanoparticle assembly. Top-down procedures include mechanical compression, exfoliation, and nanolithography [13].

Several types of 2DNMs have been explored, each with its own unique shape, size, chemical composition, biodegradability, and biocompatibility [14]. Nanomaterials with these varied qualities and features have significant potential in biomedicine, particularly in fields like multimodal imaging, biosensors, drug/gene delivery, and cancer therapy [15]. 2DNMs stand out because of their enormous specific surface

Table 1 Pros and cons of 2DNMs

2DNMs	Type	Pros	Cons	Refs.
Metallic	MXenes	High specific surface area. high NIR absorption, high photothermal conversion efficiency	Not easily synthesized	[19]
	Gold nanosheets	Optical property that can be tuned, size controllable characteristic, simple to produce in the lab	Having a low tolerance for heat, insufficient photothermal conversion, potential toxicity, poor drugs loading capacity	[20]
	Pd nanosheets	Extreme photothermal conversion efficiency	Long-term toxicity is unclear, and they are not biodegradable	[21]
	Layered double hydroxides (LDHs)	Superior charge density, with high biocompatibility, large potential for drug loading	Potential toxicity	[22]
	TMDCs	High thermal efficiency, large surface area, excellent NIR absorbance	Complicated preparation methods, size and thickness are hard to control	[23]
Metal-free	GO, rGO	Outstanding optical, mechanical, and electrical characteristics, the tendency to a large variety of wavelengths, powerful heat transfer. huge specific surface area, a high capacity for loading drugs	The probability of reactive oxygen species (ROS) generation, the dispersion process exhibits inefficiency; the data exhibits a significant variation in size	[24]
	2D boron-based NMs	With high biocompatibility, high thermal conductivity, and a wide band gap, great PTCE, superior capacity for drugs loading	Low solubility in living organisms, preparation is challenging	[25]
	Black phosphorus (BP)	photo-stable property, layer-dependent bandgap, high PTCE. non-toxic after decomposition, high drug load	Air–water instability, hard to mass-prepare	[26]

(continued)

Table 1 (continued)

2DNMs	Type	Pros	Cons	Refs.
	2D organic polypyrrole (PPy) nanosheets	PTCE is high in NIR-I but low in NIR-II, affordable price, excellent biocompatibility	Not very stable	[27]
	Graphdiyne (GDY)	High surface area to volume ratio, high PTCE, ideal for use in drugs and metal ion loading, capacity to load drugs excessively	Single-layer GDY preparation is challenging, low level of biocompatibility, possible biotoxicity over time	[28]

area, which allows them to efficiently adsorb molecules via covalent or non-covalent interactions, resulting in enhanced controlled release in response to external stimuli. 2DNMs, with their unique sheet-like nanostructures and possible reactivity to near-infrared (NIR) light, hold great promise for usage in optical therapies including photothermal therapy (PTT) and photodynamic therapy (PDT) [16, 17]. These benefits have led to 2DNMs being the most popular nanoplatforms for theranostic studies. Recently, most researchers have focused on the fabrication or optimization of monotherapy, despite the fact that lots of preclinical and clinical studies have shown that monotherapy is not as effective as expected, due to the recurrence and metastasis of cancer. 2D nanocomposites have the potential to become a new class of therapeutic agents in biomedicine if they are combined with other functional moieties in order to impart substantially enhanced features for synergistic cancer diagnosis and therapy on the 2DNMs [18].

This chapter focuses on the current state and benefits of using 2D nanomaterials in cancer detection and therapy, as well as discussing the obstacles and prospects of their future development. Here we focus on three of the most promising uses of 2D nanomaterials in cancer treatment: (1) smart drug delivery; (2) PTT; and (3) PDT. The biocompatibility of 2DNMs is also discussed in this chapter, which provides a unique overview of the topic.

2 Diagnosis of Cancer Using 2D-Based Hybrid Composite

Cancer's high incidence and mortality rate make it a major public health problem worldwide [29]. The death rate from cancer is expected to decrease as more cases are caught at an earlier stage [30]. Magnetic resonance imaging (MRI), computed tomography (CT), endoscopy, mammography, X-rays, ultrasound, and morphological study of cells or tissue (cytology or histopathology) are the methods that are now utilized in the diagnosis of many types of cancer patients. These methods are used to discriminate between healthy and unhealthy cells in the patient's body. However, cancer diagnosis with MRI and CT scanning is standard practice [31, 32]. There is still a major issue with the inability to identify cancer at an early stage due to low sensitivity and inefficiency. Consequently, it is clear that there is an urgent need to create cutting-edge technology that can rapidly and accurately detect the complete eradication of malignant cells from the body's tissues. Incorporating different contrast agents for tumor and metastasis detection into 2DNMs might open up new possibilities, allowing for rapid fulfillment of multimodal imaging needs [33]. On the other hand, cancer cells can be identified by a wide variety of biomarkers, including carbohydrate antigens, enzymes, embryonic antigens, isoenzymes, proteins, oncogenes, hormones, etc. [12]. The markers need to be examined both quantitatively and qualitatively for cancer detection at an early stage. So, obviously, it is evident that there is an immediate requirement for the development of cutting-edge technology that is able to identify, in a speedy and precise manner, the total malignant cells from the tissues of the body. This has led to the development of 2DNM-based biosensors as a diagnostic tool for cancer screening. These biosensors include surface field effect transistors (FET), plasmon resonance (SPR), electrochemical, and fluorescent sensors.

2.1 Fluorescence Biosensors

Monitoring alterations in cellular and subcellular ions, signaling molecules, and metabolites has been proposed as a useful tool in the diagnosis and treatment of a variety of illnesses and conditions, including cancer [34]. Fluorescent biosensors are able to quantify or detect target molecules due to the versatility of fluorescent proteins or dyes in real-time applications such as protein localization or gene expression [35]. To this end, a number of fluorescent biosensors based on 2DNMs have been developed. GO is a 2D nanomaterial with remarkable optical and electrical characteristics, making it a popular component of biosensors [36]. Changes in fluorescence intensity or lifespan attributed to biomolecule attachment to the surface of GO can be detected using GO biosensors. For instance, an aptasensor based on GO as a quencher, which could quell the fluorescence of a single-stranded Cy5-labeled-Mucin 1 (MUC1)-specific aptamer, could be able to detect mucin1, a glycoprotein tumor biomarker that appears in epithelial malignancies. This sensor possessed a wide range detection (0.04–10 μM) with a detection limit of 28 nM [37]. More recently, a nano biosensor based on a DNA-GO nanohybrid has been developed to detect deletion mutations causing lung cancer. In this method, mutations were detected using a Fluorescein amidite-labeled DNA probe with fluorescence spectrometry. Adsorption of the DNA probe onto the GO surface was followed by a rise in fluorescence intensity with the addition of target (healthy) DNA. Lung cancer was diagnosed despite the fact that the addition of (mutated) mDNA had no effect on the fluorescence intensity [38].

Transition metal dichalcogenides (TMDs) represent another type of 2DNMs that have exhibited significant promise in the realm of biosensing applications, primarily attributed to their robust photoluminescence and substantial surface area [39, 40]. The detection of biomolecules can be achieved through the utilization of TMDs biosensors, which operate by quantifying alterations in fluorescence intensity or energy transfer that occur between TMDs and fluorescent dyes upon binding to the surface of TMDs. In this setting, a number of different fluorescent biosensors based on 2DNMs have been created for the detection of cancer. Specifically, a molybdenum disulfide (MoS_2)-based biosensor has been developed for the detection of miR-21, a breast cancer biomarker. The fluorescence biosensor was made by combining a fluorescent dye-labeled DNA probe with the MoS_2 2DNMs, and then introducing non-complementary miRNA, one-base mismatched miRNA, and complementary miR-21. Following DNA-miRNA hybridization, miR-21 can be identified by comparing the fluorescence signal before and after the reaction [41]. Table 2 provides a brief overview of the several 2DNMs employed in the development of sensitive biosensors for cancer diagnosis. In summary, the 2DNMs-based platform demonstrated its ability to detect diverse cancer cells *in-vivo* and *in-vitro*, providing new insights into cancer diagnosis.

2.2 *Field-Effect Transistor (FET) Biosensors*

One of the most capable, efficient, and potentially fruitful biomedical devices or technologies is the FET biosensor, which, unlike the conventional method, does not necessitate the use of fluorescent or electrochemical tags [42]. FET biosensors are transducers that employ FETs to detect and analyze biological substances or events. There are three main parts to a FET biosensor: the gate electrode, the source and drain electrodes, and the sensing layer that is connected to the gate electrode [43]. 2DNMs-based FET biosensors are one FET biosensor that uses nanoscale materials and structures to improve the device's sensitivity and specificity [44]. As the sensing layer, 2DNMs like carbon nanotubes (CNTs), G, or nanowires are commonly used in these biosensors because of the large surface area they offer for target molecule binding and fast electron transport. Target molecule binding to the sensing layer, modified with biological material such as an enzyme or antibody that preferentially attaches to a target molecule or analyte, changes the FET's electrical properties, which may be monitored to determine the target molecule's amount or presence. When 2DNMs are used, numerous sensing components may be integrated into a single device, allowing for the simultaneous detection of many analytes. As a 2D nanomaterial with high sensitivity to surface charge changes and good compatibility with standard microfabrication techniques, MoS_2 has found extensive use in FET biosensors. For instance, miRNA-155 indicator for breast cancer was detected using a MoS_2-based FET biosensor in human serum and cell-line samples. MoS_2 flacks were drop-cast onto the FET surface during device fabrication, and the miRNA-155 probe was then immobilized on the MoS_2-FET. The manufactured miRNA155 probe-MoS_2-FET apparatus has a sensitivity of 0.03 fM in a range of 0.1 Fm to 10 nM [45]. Additionally, rGO was utilized in the detection of miRNA-21(1 fM) in human plasma through FET biosensor technology for the purpose of diagnosing breast cancer. It can detect precisely matched, single-base mismatched, and entirely

Table 2 Summary of 2DNMs-based nanocomposites in cancer diagnosis

Type	2DNMs	Application	Description	Refs.
Fluorescence	MoS_2	Detection of carcinoembryonic antigen (CEA), a tumor biomarker, with high sensitivity	Van der Waals force deposited CEA aptamer probe on MoS_2 nanosheets in close proximity, initiating fluorescence resonance energy transfer and quenching its fluorescence signal. The fluorescence signal was recovered when CEA protein bound to MoS_2 nanosheets and caused the aptamer probe to disengage	[50]
	MoS_2	miRNA-21	The strategy relies on folic acid (FA)-poly(ethylene glycol)-functionalized MoS2 nanosheets with adsorbed dye-labeled single-stranded DNA (ssDNA), which are then internalized into cancer cells and undergo hybridization with target miRNA, resulting in the release of green fluorescence from the nanoprobes	[51]
	GO	miRNA	The GO surface physically adsorbed fluorophore-labeled DNA probes, dimming fluorescence. The duplex-specific nuclease (DSN) released target miRNA from the DNA–RNA hybrid duplex by cleaving probe DNA into tiny pieces. Thus, target miRNA recycling produced many tiny DNA pieces	[52]

(continued)

Table 2 (continued)

Type	2DNMs	Application	Description	Refs.
FET	rGO	DNA	The rGO FET biosensor detected DNA using peptide nucleic acid-DNA hybridization instead of DNA as the capture probe	[53]
Electrochemical	AgNPs-rGO modified-screen-printed carbon electrode	Carcinoembryonic antigen (CEA)	Using a primary antibody and a secondary antibody coupled with horseradish peroxidase (HRP), CEA was immobilized on AgNPs-rGO modified-SPEs to create an electrochemical immunosensor	[54]

mismatched sequences, and it can do so without the need for miRNA-21 extraction, amplification, and tagging, making it useful for analyzing clinically complicated matrix materials, as shown in Fig. 1 [46]. Taken together, the developed 2D-based FET biosensors can be utilized for the detection of cancer biomarkers. A variety of FET biosensors based on 2DNMs has been developed for the detection of cancer biomarkers, as shown in the Table 2.

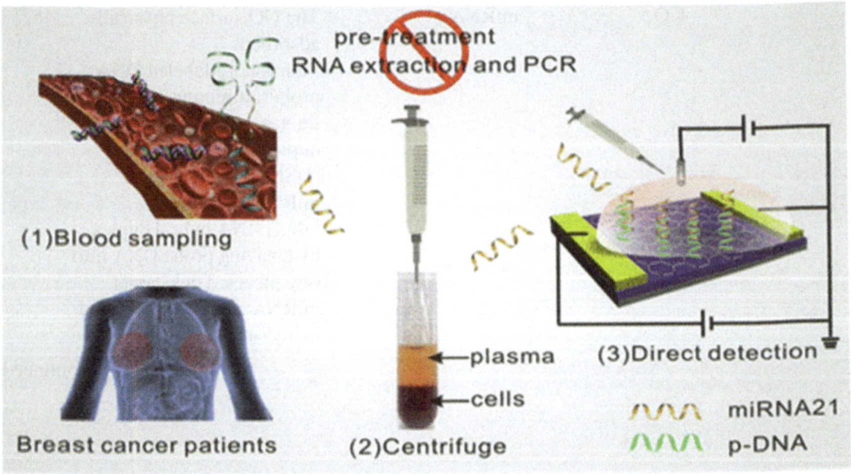

Fig. 1. rGO-based FET biosensor for sensitive detection of miRNA21 [46]

2.3 Electrochemical Biosensors

The electrochemical sensors are now being marketed as a very sensitive technology that may be used in a number of chemical and biological research, including the detection of cancer biomarkers [47]. The two main components of these biosensors are the receptor, which binds selectively to the target analyte, and the transducer, which turns the biological signal into an electrical signal. The electrochemical biosensor's selectivity, sensitivity, high accuracy, simplicity of operation, and compatibility with the downsizing of electrochemical devices make it an attractive candidate for the next analytical system. This technology offers novel perspectives and ample opportunities for early cancer diagnosis. 2DNMs, such as carbon nanotubes, graphene, and metal nanoparticles (NPs), have been used extensively in recent years due to their exceptional electrical and chemical characteristics in the development of very sensitive electrochemical biosensors, particularly for the diagnosis of cancer. Biomolecules like enzymes, antibodies, or DNA strands are commonly detected by these biosensors with acceptable specificity and selectivity, and little background noise, (Table 2). For example, more recently, neuron-specific enolase (NSE), a lung cancer biomarker, was detected using an anti-NSE-chitosan-functionalized MoS_2 electrochemical biosensor. The manufactured immunoelectrode has a sensitivity of 3.35 $\mu A\ ng^{-1}\ mL\ cm^{-2}$ and a wide linear detection range of 0.1–100 $ng\ mL^{-1}$ [48]. Nucleic acids (DNA or RNA) are another sensitive biomarker useful for detecting cancer at an early stage. After tumor cells die, their DNA and cellular contents may be released into the bloodstream. Moreover, abnormal expression of microRNAs has been associated with numerous types of malignancy. Electrochemical sensors based on 2DNMs have been also explored to detect nucleic acid biomarkers with a relatively cheap cost, a quick reaction, and a high degree of selectivity. Cytokeratin 19 fragment 21–1 (CYFRA21-1) is a biomarker for non-small cell lung cancer, particularly squamous cell carcinoma. In this context, early diagnosis of CYFRA 21–1 (DNA target probe) requires a novel approach. More recently, a reduced graphene oxide (rGO)-based label-free electrochemical DNA biosensor was developed for the diagnosis of non-small cell lung cancer. The glassy carbon electrode was modified by a single-strand DNA (ssDNA as capture probe), resulting in the determination of CYFRA21-1 with a wide range of 1.0×10^{-14}–1.0×10^{-10} M with a detection limit of 2.4 fM [49]. Table 2 summarizes the different 2DNMs used in the development of various biosensors. Taken together, 2D-NMs-based biosensors can be used for the sensitive detection of various cancer biomarkers.

3 2D-Based Hybrid Composites for Drug Delivery

3.1 Smart Drug Delivery

The phrase "smart drug delivery system" (SDDS) refers to a form of drug delivery system that may automatically transmit a signal, respond with the required action, provide the drugs, and then stop the administration process [55]. All smart drug delivery systems try to make sure that the medicine is given in the right amount, at the right time, and in the right place [56, 57]. SDDSs can be controlled by both internal and exterior signals. Internal signals include redox, pH, concentration of specific biomolecules, and enzyme activity. Ultrasound, electric fields, magnetic fields, and other forms of light are all examples of possible external signals [58]. As nanotechnology has progressed, a wide range of SDDSs built on this foundation have been used for the treatment of a wide range of disorders, most notably cancer [59]. In the years after graphene's introduction as the first 2D nanomaterial, several other 2DNMs-based SDDSs have been developed thanks to their remarkable features [15].

Light is one of the most popular stimuli for 2DNMs-based SDDSs because of its practical benefits and the ease with which it may be manipulated in vivo and in vitro [60]. The near infrared (NIR) spectral region is more advantageous than the visible region due to its lower toxicity and greater ability to penetrate tissue. The NIR light spectrum ranging from 700 to 1000 nm has been found to be highly advantageous in the field of biology. By converting the energy of the absorbed photons into heat, NIR-responsive nano devices can trigger the release of chemotherapeutic medicines. The primary benefits of NIR radiation treatment include lower toxicity to healthy cells, greater penetration, and reduced injury to living tissue [61, 62]. More recently, on the base of WS_2 for NIR light controlling cancer therapy, a hollow structured WS_2 was developed for the controlled DOX release under 808 nm laser irradiation. 1-tetradecane (TA) was used as the phase change material to seal the DOX-loaded PEG-modified WS_2 NPs, preventing the DOX leakage during the route to the tumor locations. The release of DOX and its transport to the tumor site were made possible by irradiation with an 808 nm laser and the conversion of NIR light into heat [63].

The pH-responsive SDDSs are gaining popularity due to their ability to precisely administer medications at certain times and sites [64]. Several organelles, including the lysosome and the endosome, as well as extracellular areas (including cancer cells), are capable of undergoing pH alterations, including the vagina, the small intestine, and the large intestine. For instance, the pH of the extracellular environment in tumors is acidic, in contrast to the pH of the internal environment, which is more gently basic. Because of this, pH has been thought to be a useful physiological property for SDDSs in their delivery to the targeted tumor locations [65]. There have been many reports of 2DNMs pH-responsive SDDSs, which stabilize and store drugs at physiological pH but only rapidly release them in the acidic environment of cancer cells, ensuring that the concentration of drugs reaches a peak in the cell. More recently, polyethyleneimine (PEI)-conjugated GQDs were used for the development of a pH-responsive DOX delivery system for colon cancer suppression. Drug release is triggered and cancer cells are inhibited when GQDs-polymer-DOX conjugates (GECD) are injected into the tumor lesion, where the acidic tumor microenvironment protonates the tertiary amine of GECD from neutral to mildly positive [66]. Table 3 summarizes the different 2DNMs SDDSs that have been used in cancer therapy.

3.2 PTT

PTT is increasingly being recognized as a feasible approach for treating cancer due to its cost-effectiveness, high selectivity, and minimal impact on healthy tissues [67]. Upon exposure to laser radiation, the photothermal transducing agents (PTAs) that have accumulated at the tumor site generate heat, leading to the destruction of cancer cells through thermal ablation [68]. The effectiveness of photothermal therapy for treating cancer is contingent upon various factors, such as the photothermal conversion efficiency (PTCE) of photothermal agents (PTAs), the power intensity of the laser, and the duration of laser exposure. The PTT treatment settings facilitate the regulation of two discrete forms of cellular demise: necrosis in response to high-energy irradiation and apoptosis in response to low-energy irradiation [69]. Tumor cells would be ablated due to the intense photothermal impact, while PTAs' moderate local heat would cause the rupture of endo/lysosomal membranes [70]. PEG-BPEI-rGO, synthesized from unreduced PEG-BPEI-GO, showed significantly improved DOX loading efficiency and good water stability due to the recovery of aromatic structures of GO [71]. The results of in vitro cellular experiments indicate that a nanocarrier loaded with DOX exhibits the capacity to effectively evade the endosome upon exposure to near-infrared (NIR) light. The nanocarrier was able to induce a significant amount of cancer cell apoptosis through the glutathione-mediated and light-triggered release of DOX. Through a combination of direct cell killing and the release of tumor-associated antigens, the local photothermal impact can facilitate the controlled drug release, activate antigen-presenting cells, and generate an anticancer immune response [72]. PEG- and FA-functionalized rGO was used to improve tumor accumulation on the nanoplatform, activate photothermal death of primary tumors, and release tumor-associated antigens; this is only one example of the many uses for rGO that have been developed into multifunctional nanoplatforms, as shown in Fig. 2 [73]. Clinical translation of the PTT synergistic antitumor immunity shows remarkable potential because of its many benefits over monotherapy in cancer treatment. Table 3 summarizes the different 2D-NMs PTT systems that have been used in cancer therapy.

3.3 PDT

PDT is a type of procedure used to destroy tumor cells. It works by using light of a certain wavelength to activate molecules called photosensitizers (PS). These

Table 3 Recent 2DNMs that have been used for cancer therapy and diagnosis

Composite formulation	Photosensitizer	Therapy	Result	Refs.
MoS_2-LA-PEG[a]	Chlorin e6 (Ce6)	PDT	Enhance in vitro cellular uptake for Ce6, enhance PDT of Ce6, synergistic cancer cell killing in vivo and in vitro	[94]
MoS_2-PEG- DMA peptide	Toluidine blue O (TBO)	PDT/PTT	High growth inhibition effect on cancer cell in vitro, enhance the tumor inhibition efficacy in vivo, significantly synergistic therapeutic efficiency in vivo in mice model	[95]
MoS_2/indocyanine green/curcumin	MoS_2	PDT/PTT	Excellent photothermal effect, relatively low toxicity, inhibit the P-gp effectively	[96]
MoS_2/BSA/ Cy5	MoS_2	PAT[C]/ PDT/PTT	A nanoplatform with four functions in one device fluorescent imaging/PAT/ PDT/PTT, excellent imaging capability, successful tumor elimination in vivo	[97]
MoS_2/Au nanorod	MoS_2	PDT/PTT	Synergistic effect of PTT and PDT, excellent antibacterial activity	[98]
MSNR/MoS_2-HSA/ Ce6	Ce6	Imaging/ PDT/PTT	Hyperthermia generated by MoS_2 could enhance the PDT effect with increasing blood flow, which would result in an increase in oxygen content in tumor tissue, successful in vivo imaging, effectively eliminate tumor from 4T1 tumor-bearing mouse	[99]
WS_2/Fe_3O_4/MSN/ PEG/DOX	WS_2	PDT/PTT	Synergistic therapeutic effect superior to the respective monotherapies, imaging, and therapeutic capability in one system	[100]
BSA/WS_2/Methylene blue	Methylene blue	PTT/PDT	Synergistic therapeutic effect superior to the respective monotherapies, an effective nano platform	[101]

(continued)

Table 3 (continued)

Composite formulation	Photosensitizer	Therapy	Result	Refs.
MSN/WS_2QDs/DOX	WS_2	PTT	NIR and pH-responsive DOX release system, an effective nano theranostics platform	[102]
Ti_3C_2/MXene/DOX/ hyaluronic acid	Ti_3C_2	PTT/PDT	Excellent biocompatibility, stimuli-responsive drug release, tumor-specific accumulation, effective cancer cell killing, synergistic therapy in vivo and in vitro	[93]
Ti_3C_2Tx/MXene/ indocyanine green	Ti_3C_2Tx	PTT/PDT	Synergistic antibacterial effect superior to the respective monotherapies under NIR irradiation	[103]
GQDs[d]/GOQDs[e]	GQDs and GOQDs	PDT	90% of cancer cells treated with GQDs/ GOQDs were killed after 5 min of UV exposure	[104]
PTX@GO-PEG-OSA	GO	PTT/PDT	Excellent ROS generation, effective therapeutic efficacy against PTX resistance gastric cancer	[105]

[a] lipoic acid-terminated polyethylene glycol
[b] chlorin e6
[c] photoacoustic imaging tomography (PAT)
[d] graphene quantum dots (GQDs)
[e] graphene oxide quantum dots (GOQDs)

molecules then create ROSs which are capable of killing the tumor cells. To make sure the PS reaches the tumor effectively, nanocarriers are needed. Common PS used in PDT include porphyrins, chlorins, and dyes [75]. However, these substances have some disadvantages. They are usually not soluble in water, require high amounts to be effective, are not very selective in targeting cancer cells, take a long time to enter cells, need a lot of time to be eliminated from the body, can make the skin sensitive to light for a long time, tend to aggregate together in solutions, and lose their ability to react to light [76]. After being given, PSs can be taken inside by both cancer and normal cells. Healthy tissues can get rid of PS over time, but tumor cells can't because the lymphatic system isn't working well enough. The way PDT works is that it helps keep the medicine in the tumor tissues and only activates it in that specific area when it's hit with radiation. This makes PDT a treatment that targets cancer cells specifically [77]. Right now, different treatments and tests are used together to achieve a theranostic effect. This can be made even better with the use of NM-based theranostic nanoplatforms [78, 79]. Additionally, as oxygen is an important factor in

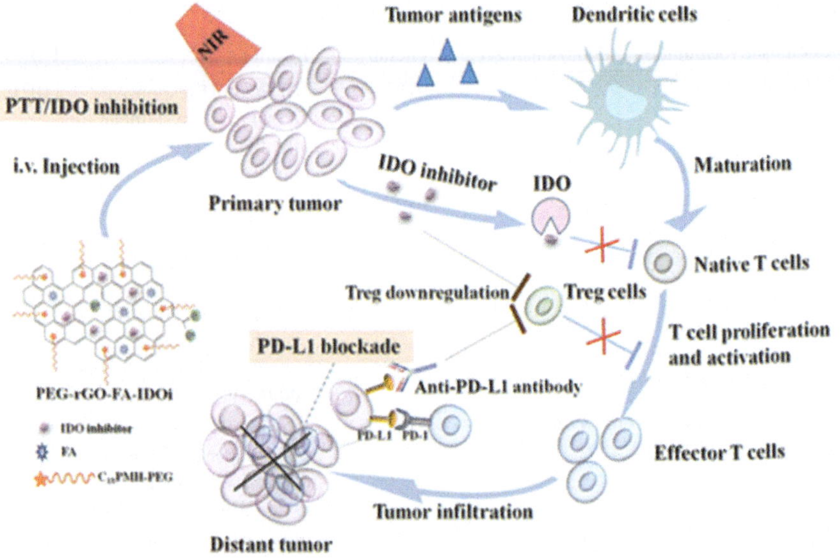

Fig. 2 Mechanism of PTT treatment mediated by PEG-rGO-FA- indoleamine-2,3 dioxygenase (IDO) inhibition (IDOi) nanosheets in combination with IDO inhibition and programmed cell death-ligand 1 (PD-L1) blockage for enhanced antitumor immunity [74]

successful PDT, it is essential to overcome low oxygen levels in most solid tumors to ensure the best possible result [80]. Several combinations of PS and 2DNMs have been used for cancer therapy and diagnosis.

Wang and his team made a new Fe_3O_4@MoS_2@SDS nanocomposite. They used it to fight against many different types of bacteria including *Escherichia coli, Methicillin-resistant, Staphylococcus aureus, and Pseudomonas aeruginosa.* This nanocomposite, when exposed to NIR radiation, produces certain radical substances that can kill bacteria. Furthermore, Fe_3O_4@MoS_2@SDS can effectively prevent the spread of antibiotic resistance genes through conjugative plasmids [81]. Cao and his team made a special system using a MoS_2 QDs together with PEG and FA. This system is able to respond to changes in the body and carry DOX for both light treatment and standard chemotherapy. This tiny platform can change NIR light into Visible light. This light activates MoS_2QDs and makes them produce ROS by using a special type of energy transfer that kills cancer cells. Under the use of NIR light, they found that combining different treatments was more effective than using only one treatment [82].

Recently, researchers have been studying MXenes and graphene for their optical loading properties use in the medical field, specifically as important components in tiny technology platforms for healthcare [83, 84]. Research in the field of theranostics began with using graphene-based material as carriers for both a treatment substance and imaging agent [78, 85]. This led to further studies on using nanotechnology-based PDT for more specific treatment of cancer. Graphene can capture light in the NIR region. This makes it possible to study its potential use for treating cancer using light therapy, both inside and outside the body [86, 87]. GO and MXenes can be used for imaging. GO-based nanoplatforms have a lot of potential for use in imaging because they can help make fluorescent materials better. GO can make these materials more stable, easier to spread out, safe for living things, and improve how well they work with light [88].

Scientists are looking into new and effective ways to treat cancer. Researchers are studying how these nanoplatforms can work together with other therapies to improve cancer treatment. Tian and his team combined PEG-GO with the Ce6 as PS using π-π stacking. The cervical cancer cells absorbed the substance and it caused the production of ROS when exposed to light. The use of GO-PEG-Ce6 in the treatment of cancer was more effective than using Ce6 on its own [79]. Huang and Collagenous suggested using GO as a carrier for Ce6. Similar to previous studies, Ce6 was attached to folic acid-functionalized GO by π-π stacking binding. The study found that the system can destroy gastric cancer cells when exposed to radiation [89]. In the same way, Zhou and colleagues effectively incorporated hypocrellin B (HB) as PS into GO by using π-π stacking interaction. They demonstrated that the substance could produce 1O2 when exposed to radiation [90, 91]. In this research, two substances called hypocrellin A and 7-ethyl-10-hydroxycamptothecin were put together on GO. When this combination was exposed to light, it caused more lung cancer cells to die. This shows that chemotherapy and PDT can work together to be more effective [90].

Recently, Zhang and colleagues suggested using Mo_2C/MXene platform to create tiny particles that can be used for treating cancer. These particles can be used for different types of cancer treatment such as PTT and PDT as well as for imaging using sound waves and computerized scans. Furthermore, in this study, the ability to produce ROS was studied by observing changes in the absorbance of a substance called DPBF at a specific wavelength of light (420 nm) when exposed to NIR light at a wavelength of 1064 nm. This ROS generation was also confirmed by observing that the degradation of DPBF was prevented in the presence of NaN_3, which can stop the activity of ROS. Additionally, a combination of PTT and PDT treatments on liver cancer cells (HepG2) in a lab setting showed that over 80% of the cells underwent apoptosis depending on the amount of treatment given. This confirms that the production of ROS played a key role in the effectiveness of the Mo_2C-mediated PTT/PDT therapy. The effectiveness of treating liver cancer was tested by observing the tumor growth in mice. After 14 days of treatment, mice that had received a 10-min NIR exposure with Mo_2C injected into the tumor showed no regrowth of the tumor and it was completely destroyed. On the other hand, mice in the control group, who did not receive any radiation or received radiation without Mo_2C, had a tumor volume that was 4 times larger. They studied the effects of Mo_2C/MXene, which is a type of injection used in cancer treatment. They looked at blood-related factors, body weight, and examined tissues after death to see if it was safe. Their findings showed that Mo_2C/MXene is a safe option for cancer treatment [92].

Liu and colleagues showed that tiny sheets of Ti_3C_2 MXene can be used as light-sensitive substances for a PDT system. This method combines light therapy, heat therapy, and chemotherapy to work together and be more effective. The researchers studied the production of ROS when titanium carbide nanosheets (Ti_3C_2 NSs) were in the water. They used 1,3-diphenyli-sobenzofuran (DPBF) to detect a specific type of reactive oxygen species called singlet $^1O_2$. When Ti_3C_2 NSs are exposed to NIR light at a wavelength of 808 nm for 10 min, DPBF absorbs less light at 420 nm, which indicates the presence of $^1O_2$. They found that Ti_3C_2 functionalized with DOX created similar levels of ROS when exposed to the same treatment, allowing the use of both PDT and chemotherapy together [93]. Some 2DNMs that applied for PDT are summarized in Table 3.

4 Biocompatibility of 2D-Based Hybrid Composite

Due to their distinctive mechanical, optical, and physicochemical properties, 2DNMs have emerged as promising materials for upcoming biomedical applications. Although 2DNMs have demonstrated significant promise in state-of-the-art biomedical devices, significant challenges still exist for the clinical application of such nanomaterials in drug delivery and therapy [106]. Indeed, the majority of them never have a chance to pass clinical trials as a significant portion of the research is devoted to the general strategy and greater effectiveness of their nanoscale item [106]. In other words, the risk of potential health problems limits the commonsense biomedical applications of nanomaterials. When the human body is exposed to the health risks of nanomaterials through various channels (e.g., ingestion, infusion, inhalation, skin contact, etc.). Then, these materials may undergo several variations including disintegration or agglomeration that may contribute to various levels of poisonous quality [107].

One of the biggest obstacles to the successful development of these highly advantageous materials in medical applications is actually health issues, which rank among the most important challenges facing nanomaterials. As a result, it is urgent to design and adopt methods for evaluating the potential danger of these engineered nanomaterials [108]. A few characteristics, such as a high ratio of the surface-to-volume (expansive dynamic surface zone), can uncover cells to these fabricated nanomaterials as more often as possible, expanding cellular intuitive and thus raising the chance of poisonous quality [109]. Additionally, it is well known that cytotoxicity strongly relates with various physicochemical characteristics of nanomaterials, including size, shape, thickness, chemical composition, and toxicity of the nanomaterials [110]. Therefore, investigation on the risk of 2DNMs shows their great potential for support in the biomedical field due to their low level of cytotoxicity. However, modification of some properties (such as surface property, chemistry, etc.) can improve their biocompatibility while minimizing the antagonistic natural impact induced by 2DNMs [111]. Recent studies have concentrated on the short-term cytotoxicity linked to these 2DNM applications, but they have also looked into other biosafety issues such as long-term cytotoxicity, genotoxicity, immunotoxicity, biodegradability, and solubility. We haven't yet looked into other variables that affect sex [112, 113].

4.1 Graphene Family

The concept of biocompatibility pertains to the capacity of materials to interact with biological cells, tissues, or the anatomical system without eliciting deleterious effects [114]. The biocompatibility of biological cells toward graphene-based nanomaterials is contingent upon their inherent properties, which may render them either suitable or hazardous for cellular interactions [115]. The reactivity of living cells toward nanomaterials is contingent upon an array of factors including their layer count, lateral dimensions, degree of purity, dosage, surface chemical properties, and hydrophilic nature [116]. The heterogeneity in the surface chemistries of graphene nanomaterials is attributable to the diverse methodological approaches employed in their synthesis, as well as the range of molecules and polymers accessible for surface functionalization [117]. In vitro assessment of nanomaterial toxicity commonly involves the utilization of multiple prominent cellular lineages, including phagocytic cell types (such as macrophages) and non-phagocytic counterparts (such as endothelial and epithelial cells, cancer cells, and erythrocytes) [118, 119].

An essential aspect of the utilization of graphene nanomaterials for medical purposes is the precise comprehension of their mechanism of interaction with cells [120, 121]. The interactions between graphene-based nanomaterials and cell membranes possess the potential to result in detrimental effects such as cellular membrane impairment and cytotoxicity [116, 122]. Cell membrane phospholipids are composed of a phosphate moiety attached to two fatty acyl chains. The main head groups comprise choline, serine, glycerol, ethanolamine, inositol, and phosphatic acid. The diverse head groups present in phospholipids impart unique and discernable characteristics to these molecules. In addition, cellular membranes possess cholesterol molecules that perform significant functions in stabilizing the structure of the membrane, upholding its fluid nature, and regulating the operations of proteins associated with it [123, 124]. Untainted graphene that lacks charges on its basal plane is unable to engage in electrostatic interactions with phospholipids [125]. Nevertheless, it exhibits a propensity to partake in hydrophobic interactions with the lipid tails. Furthermore, graphene backbone interactions with the cholesterol residue have the potential to extract or eliminate cholesterol molecules from the cell membrane, thereby resulting in membrane damage [116].

Recently, researchers exhibited that GO was able to extract phospholipids from the cell membranes and induce surface pores. Cells became less viable as a result of this effect, eventually dying. They suggest that this is due to the visual solidity between the hydrophobic gaps of GO and the lipid tail carbons [122]. The cellular interactions are significantly influenced by the surface charge as well as the surface chemistry of GO. GO shows a notable negative charge density due to the presence of oxygenated functional groups. As a result, plausible electrostatic couplings between GO and the membrane lipids may take place. Within this particular context, the examination completed by Li and colleagues revealed that the surface functionalize group had significant effects on cellular interaction. A collection of five lipids was synthesized that harbored an identical 18-carbon alkyl chain, yet their head groups had varying charges. Utilizing the Langmuir monolayer technique, investigations were conducted to examine the interactions between these lipids and GO [126]. The researchers discovered that GO can engage in electrostatic interactions with lipid head groups possessing positive charges. However, no such interaction was observed with lipid head groups that are either neutral or negatively charged. It is a well-established fact that the electrical charges of phospholipids present in mammalian cell membranes lean toward negative or neutral polarity. It is improbable that negatively charged lipids would exhibit an attractive force toward GO possessing equivalent negative charges [127]. The findings suggest that GO with a negative charge experiences electrostatic repulsion when interacting with lipids bearing a similar charge [127].

Furthermore, the adsorption of GO onto said lipids is also facilitated by hydrophobic interactions between them, particularly in situations where the GO has a negative charge. The evidence supporting this inference is documented in the reference. The present study reveals that GOs possess the ability to cause perturbations in the cell membrane through hydrophobic interactions, even in the absence of cell penetration [127]. Li and colleagues have recently explored the impact of the surface chemistry of GO on its interactions with lipid membranes. Pristine GO, hydrated GO (hGO), and rGO were utilized in this investigation. The authors have reported that hGO can trigger lipid peroxidation of the surface membrane, thereby causing membrane lysis and compromising cellular integrity [128]. The surface of rGO harbors free radicals endowed with reactivity as a consequence of the existence of unpaired electrons, which promote facile interaction with oxygen, ultimately resulting in the formation of superoxide radicals. These radicals bestow the capacity to instigate oxidation processes that lead to the generation of lipid peroxides by means of their interaction with unsaturated lipids and thiol groups of proteins. Consequently, there was a breakdown of the integrity of the cellular membrane resulting in the demise of the cell. The epoxy moiety present in pristine GO is known to generate carbon radicals, albeit to a relatively lesser degree as compared to hydroxylated GO.

In addition to the interactions between lipid membranes and nanomaterials, it has been observed that graphene-based nanomaterials have the capability to penetrate the cytoplasm owing to their small size and sharp edges. It has been observed that these agents have the ability to readily traverse the cell membrane, resulting in deleterious effects on the membrane and consequent release of intracellular content. Upon inhabiting living cells, they have the potential to instigate toxicity by generating ROS [116]. These ROS can trigger mitochondrial dysfunction by decreasing mitochondrial membrane potential (MMP), and also cause damage to the cell membrane by facilitating the release of lactate dehydrogenase (LDH). ROS has the ability to initiate the process of lipid peroxidation by reacting with unsaturated fatty acids of membrane lipids, resulting in the formation of lipid peroxides, including malondialdehyde (MDA). The outcomes of the study indicate that GO possesses the ability to stimulate both extracellular and intracellular ROS production, even at minimal dosages, in a fashion that is reliant on both dose and time. Hence, the generation of ROS, impairment of mitochondrial function, and release of LDH are the primary mechanisms associated with cell death [129].

The potential for nanomaterials to infiltrate the nucleus has implications for direct interaction with DNA and the potential for genotoxic effects to occur. Moreover, the biocompatibility and toxic effects of graphene nanomaterials can be influenced by various factors, including but not limited to the dosage of graphene administered, duration of exposure, type of cell or animal utilized in experimentation, and the method employed to evaluate cell viability, in both in vitro and in vivo contexts. Consequently, an accurate comprehension of the interplay between graphene-derived nanomaterials and biological entities, as well as their deleterious implications, is imperative for the advancement and secure application of such nanomaterials [116].

Several studies have demonstrated the biodistribution and in vivo toxicology of graphene and its functionalized derivatives following varied routes of animal administration. Liu et al. conducted a study investigating the toxicity and biodistribution of GO, which was dependent on its size and dosage [130]. The researchers observed that S-GO demonstrated primary deposition in the liver, while minimal aggregate formations were found in the lung and spleen. When examining the comparative behavior of GO particles of differing sizes, it was observed that intermediate-sized GO demonstrated a higher degree of accumulative behavior in the pulmonary system. Moreover, upon the passage of 180 min, intermediate-sized GO maintained a residual amount of approximately 19%, highlighting its potential for prolonged residence and sustained impact [130]. Analysis using transmission electron microscopy revealed that s-GO was found to accumulate within phagocytes, while GO particles above a size threshold of 1 μm were localized in the cellular interstitial space within the lungs [116].

In a separate investigation, Kurantowicz and colleagues conducted an inquiry into the intraperitoneal toxicity of GO, graphite, and nanodiamonds among female Wistar rats of six weeks. The administration of NP suspension commenced at a dosage of 4 mg/kg after a period of three days and was continued for either four or twelve weeks. Following a period of 1–3 months, the mice were subjected to euthanasia, and both their livers and blood samples were extracted for subsequent analyses [131]. The investigators identified clusters of NPs in the peritoneal cavity proximal to the injection site. The present study observes the presence of minute aggregates in the mesenteric and liver serosal membranes, thereby providing evidence for the efficient transportation and accumulation of NPs in the liver. During the observation period of either 4 weeks or 12 weeks, no significant deterioration in health was observed in relation to any of the tested NPs, namely GO, graphite, or nanodiamond. The conventional practice involves conducting blood tests and evaluating the levels of liver enzymes to ascertain normal liver functional ability and sound biocompatibility. When observed through a microscope, the lungs exhibit granulomatous reactions alongside interstitial fibrosis, necrosis, and notable vascular sclerosis [132]. Occasionally, extensive necrosis of tissues may lead to the formation of sizable cavities that contain a thick, dark fluid composed of fatty material. Objects that do not contain asbestos and contain black graphite cores exhibit similarities to the iron-containing materials that are typically encountered following exposure to asbestos. These materials are dispersed in the lungs and are occasionally discernible in sputum [115, 133].

4.2 MXenes

Currently, there exists a body of research investigating the potential biological applications of various MXenes. The biomedical utilization of Titanium Carbide MXene is constrained. The various carbide MXenes, including Titanium Carbide MXene (Ti_2CT_x), Tantalum Carbide MXene ($Ta_4C_3T_x$), and Niobium Carbide MXene (Nb2CTx) have been applied in biomedical devices [134–136]. Among the members of the MXene family, the extensively researched one is Ti3C2Tx, primarily attributed to its comparatively simple synthesis procedures [137]. One of the appealing characteristics of MXene is its notable proficiency in near-infrared absorption, which is conducive to theranostic applications, PTT, and the collaborative treatment of cancer [138, 139].

Furthermore, MXenes have been utilized as biosensors and for bioimaging following their conversion into quantum dots [140–142]. Despite the growing scientific inquiry into the potential applications of MXene, there is a paucity of research on its cytotoxicity. Broadly speaking, the employment of 2D MXenes in the domain of biomedicine remains at a nascent stage of development. In order to facilitate the translation of lab-scale experimentation with MXenes into practical applications within the biomedical industry, a rigorous evaluation pertaining to the biocompatibility and cytotoxicity of MXenes ought to be conducted [143].

The majority of scholarly investigations pertaining to the cytotoxicity of MXene have been concentrated on the subject of in vivo and in vitro toxicity [144–147]. The objective of these studies is to evaluate the interplay between cancerous and non-cancerous epithelial cell lines with several types of MXenes. Jastrzębska and colleagues conducted an inquiry into the in vitro cytotoxic potential of delaminated Ti_3C_2Tx MXene on four distinct cell lines, namely A549 and A375 (cancerous) and MRC-5 and HaCaT (normal), utilizing the MTT assay. The findings indicate that the Ti_3C_2Tx MXene, when present in a concentration below 62.5 mg/L, exhibits negligible cytotoxic effects. The findings from the microscopic images of the Ti3CTx MXenes revealed a notable degree of agglomeration on the cell surface, attributed to the pronounced attraction between MXene and the cell membrane [148].

The cytotoxicity of Ti_3C_2MXene was observed to be contingent upon the specific cell lines evaluated. The cell viability of cancerous A549 cells was found to be significantly lower (ranging from 20 to 90%) in comparison to various other normal cell types (with a range of 70–100%), and this disparity was contingent upon the concentration of Ti_3C_2 MXene. In addition, it was observed that the Ti_3C_2Tx MXenes exhibit toxicity toward cancer cells. The normal cells were found to have experienced a minimal reduction in their viability. It has been demonstrated that the mechanism of response underlying the toxicity exhibited on cancer cells is attributable to the generation of ROS, the levels of which surpass the established threshold of oxidative stress for such cells. This phenomenon ultimately results in the dyeing of malignant cells [148]. The Ti_3C_2 material exhibits a negative zeta potential, which can be attributed to terminal functional groups such as –F, –OH, and = O, as well as the surface passivation of TiO_2. The robust electrostatic interaction between the negatively-charged surface of MXene and the positively-charged phosphatidylcholine lipid results in a compromised cellular membrane integrity [120]. Scheibe et al. conducted an investigation into the in vitro cytotoxic impacts of various $Ti_3C_2T_x$ MXenes and corresponding precursors on human fibroblasts and HeLa cells. The study findings revealed that the exposure of cells to increased concentrations of TiC triggered significant harmful impacts on cell viability, originating from mechanical stress and the generation of oxide radicals. According to their findings, $Ti_3C_2T_x$ MXene appears to exhibit a cytocompatibility rate of over 80% when administered to non-malignant cells within the concentration range of 10–400 μg/mL. The findings indicate that MXenes exhibit cytotoxicity effect that is influenced by the dose and chemical composition.

The prior research has provided insight into the mechanism by which the cytotoxicity of MXene originates. The primary mechanism of toxicity attributed to MXene involves the production of ROS and physical interaction. Intracellular ROS can induce protein and DNA damage ultimately culminating in cell death upon the infiltration of MXene through cellular membranes in the presence of water, triggering ROS generation. The hydroxyl radical (·OH) and its reactive oxidative species (ROS) are of significant interest to researchers in various fields. This highly reactive and unstable molecule has been implicated in a range of biological processes, including aging, inflammation, and cancer. Studies have shown that ·OH can damage lipids, proteins, and DNA, leading to cellular dysfunction and ultimately, disease. However, the intricate mechanisms by which ·OH mediates these effects are not fully understood. This highlights the need for further investigation into the biology of OH and ROS, as well as their potential therapeutic implications.

An additional mechanism underlying the toxicity of MXene pertains to its robust adherence to the cellular membrane. The direct interaction between MXene and the cell membrane is facilitated through an array of fundamental physicochemical forces, including ionic interaction, hydrophobic interactions, van der Waals forces, as well as receptor-ligand binding. Consequently, membrane destabilization ensued along with the impairment of cellular integrity. The interaction between MXene and the cell membrane leads to the accrual of MXene, culminating in cell demise [149].

Furthermore, besides considering the oxidative state of MXenes, another important factor to take into account when studying them is their size. There is a suggestion that these substances may exhibit cytotoxic properties toward living organisms, including microbes. In a recent study, ArabiShamsabadi et al. systematically investigated the impact of varying lateral dimensions of Ti_3C_2 MXene nanosheets, ranging from 0.09 to 4.40 μm, on the inhibition of Bacillus subtilis and Escherichia coli. Their findings suggest Ti_3C_2 MXene nanosheets with reduced lateral dimensions (0.09 μm) exhibit superior antimicrobial efficacy against bacterial species. This phenomenon can be attributed to the diminutive lateral dimensions of MXene, which facilitate its facile penetration into the cytosol, resulting in damage to the cytoplasmic constituents in the bacterial DNA [144]. Wu and colleagues endeavored to investigate the cytotoxic properties of Ti_3C_2 nanosheets utilizing a select methodology. The transmission electron microscopy (TEM) findings reveal that the incorporation of Ti_3C_2 nanosheets with a width of 2 μm occurred within the neural stem cells at concentrations exceeding 25 μg/mL. This study indicates that the cellular viability subsequent to the application of MXene was contingent upon the lateral dimensions of the MXene sheet [150].

4.3 Chalcogenides

In contemporary scholarly discourse, considerable attention has been directed toward the investigation of 2D TMDs among the diverse range of chalcogenides. Amongst these, there exist TMDs that have received considerable attention with regard to their in vitro assessments, namely MoS_2, WS_2, and WSe_2 [151]. Primarily, the MoS_2 nanomaterials from the TDG cohort have received the greatest amount of research attention. The biocompatibility of the materials was further examined utilizing cell lines including leukemia, lung carcinoma, and human embryonic epidermal fibroblasts cells [152]. Upon exposure of the human embryonic epidermal fibroblasts cell line to the highest concentration of 3.52 mg of substance 1 for a duration of 48 h, it was observed that the administered particles did not produce any considerable effect on the cell growth. However, the present study revealed that the administration of 3.52 mg/l of MoS_2 particles solely resulted in roughly 10% of cellular demise, thereby reflecting a considerably elevated level of inherent toxicity.

Of the various TMDs, those containing sulfur (S), tellurium (Te), polonium (Po), or selenium (Se) are prevalent in their availability. Upon decomposition, the chalcogenides liberate unbound Te or Se moieties which are responsible for inducing pronounced toxic effects in both human organisms and the surrounding environment [153, 154]. It has been established that elevated levels of Selenium (Se) pose significant deleterious health effects in animals, including but not limited to alopecia, vision impairment, and premature mortality [155]. Moreover, the exposure of individuals to acute Te has been observed to result in the suppression of sebaceous and salivary glands, in conjunction with the paralysis of the secretory nerves, and a significant occurrence of inactivity [156]. Furthermore, introduction of either of these two moieties, including chalcogenides, can lead to substitution of sulfur iota within proteins and peptides [157].

Furthermore, while pursuing the advancement of two-dimensional materials based on Te or Se, noxious gases, such as H_2Se or H_2Te, are emitted, posing a detrimental impact on both the natural environment and the ecosystem [158]. In addition to the chemical composition of the materials, it has been demonstrated that the exfoliation technique can profoundly impact the in vitro and in vivo behavior of transition metal dichalcogenide (TMD) materials, as well as their potential for toxicity. As an illustration, when compared with MoS_2, the lithographically-exfoliated printing plate of MoS_2 (Lit-MoS_2) and Pluronic F87 (PF87-MoS_2) dispersed the material to a greater extent. This dispersion had detrimental effects on cells, including significant inflammatory and pro-fibrotic responses [159]. In comparison to Agg-MoS_2, it is advocated (reported) that 2D-MoS_2 nanomaterials exhibit a higher level of safety in relation to lung injury.

Moreover, in a separate investigation, Yin et al. substituted the surface of MoS with chitosan, resulting in an augmentation of biocompatibility. Consequently, a decline was noted in the KB and Panc-1 cell lines, which respectively belong to human epithelial cancer [160]. In vitro findings suggest that surface modification of 2D nanosheets can effectively mitigate the toxicities associated with these materials, rendering them suitable for biomedical applications with minimal risks. In addition, atomically thin TMDCs exhibit a diverse range of advantageous characteristics, including robust photothermal response, fluorescence, and modest conductivity, which hold significant promise for numerous biomedical applications. Nonetheless, these materials lack inherent biocompatibility and necessitate supplementary functionalization for their successful integration into biological systems. Additionally, it may be necessary to implement supplementary functionalization to augment the sensitivity of diagnostic utilities or to bestow additional functionalities upon TMDCs, such as drug delivery, DNA sensing, or contrast imaging. The functionalization process of TMDCs is deemed relatively straightforward owing to their remarkable surface-to-volume ratio. This process can be achieved through non-covalent or covalent methods.

Non-covalent functionalization is frequently applied to TMDCs during experimentation. Maintaining the intrinsic characteristics of 2D TMDC is of crucial significance, particularly in scenarios where extraction of their functional groups is an integral part of the diagnostic or delivery mechanism. The functionalization process typically entails the physisorption of the intended probe or drug carriers onto the comparatively expansive basal plane of the TMDC, thereby enabling optimal loading and engagement with the neighboring milieu. The process of covalent modification entails the creation of chemical bonds between the targeted chemical group and the surface of the TMDC with the purpose of functionalization. The occurrence of this phenomenon can be attributed to particular reaction mechanisms, such as thiol chemistry in the context of sulfur-containing TMDCs [161, 162].

Moreover, in spite of the fact that the cytotoxicity of TMDCs in organic frameworks has appeared to be generally low, the instruments taking put some time recently and after its cellular take-up or utilize as a restorative specialist has not been totally caught on however, particularly its bio-distribution, debasement within the body and its ensuing excretion. Harmfulness to non-targeted zones (sits) of the organic framework has too not been thoroughly considered however and will be fundamental some time recently any clinical trials can take put. 2DNMs, aside from TMDCs, namely boron nitride (BN) and graphene, possess the potential for functionalization and may provide practical benefits when applied in conjunction with TMDCs in biological systems. The functionalization of TMDCs for biomedical applications remains in its early developmental phase, yet it has demonstrated substantial potential as a novel nanomaterial capable of facilitating synergistic therapy, multi-modal imaging, highly sensitive biosensing, and externally regulated in vivo drug release. This unique capability has proven difficult due to present biodegradable materials such as liposomes, nanocrystals, micelles, and hydrogels [163–165]. Consequently, there are significant opportunities and requisite efforts for the functionalization of

TMDCs in order to render them genuinely practicable nanomaterials in the arena of biomedical applications.

4.4 2D Oxides

Numerous 2D oxides exhibit pronounced reactivity and toxicity, particularly when they are observed in the form of NPs. This is attributed to their manifestation of augmented catalytic performance and elevated dissolution rates with a reduction in the size of the 2D metal oxide. An example of a metal oxide exhibiting two-dimensional properties is V_2O_5. This compound possesses significant toxicity and is well-recognized for its status as a highly poisonous substance [166]. The present 2D material is abundantly released in the atmosphere through natural phenomena such as volcanic eruptions and forest fires, thereby posing a significant threat to the environment. Additionally, it is known to exhibit genotoxic effects [167]. Research conducted by Uche et al. exhibited that exposure to V_2O_5 has deleterious effects on the liver and testes of guinea pigs, as well as inhibits the developmental stage of mice [168]. Similarly, MoO_3, a 2D oxide material, exhibits notable toxicological characteristics. Whilst this 2D material serves as the building block for other 2D materials, such as MoS2, its impact on human and animal health has also been documented. An illustration of this scenario can be observed in the case of MoO_3 solids, whereby the ingestion of doses of 1. 2–6 g has been demonstrated to result in mortality in both guinea pigs and rats [158]. Nonetheless, the manifestation of toxicity is contingent solely upon the solid state of MoO_3, whereas the existence of gaseous MoO_3 does not elicit any lethal effects.

Numerous investigations have indicated that CuO NPs exhibit cytotoxic features across various cell lines. The aforementioned phenomenon has been observed in the study conducted by Morris et al. wherein it was exhibited that the size of NPs elicits a toxic response in Beas-2B cells that is sensitive to serum [169]. The study revealed that the implementation of CuO NPs served as stimuli and prompted the squamous differentiation of Beas-2B cells in the presence of serum. Furthermore, the present study investigated the cellular internalization of CuO NPs of differing dimensions. The findings revealed a reduction in the hydrodynamic diameter upon exposure to serum within the cell culture medium. Additionally, it was observed that the serum augments the aggregate cell area and induces dose-dependent cytotoxicity in CuO NPs of greater diameters. As such, the present investigation has demonstrated that the dimensions of CuO NPs, coupled with the associated serum levels in cellular culture, exert a discernible influence on the efficacy of Beas-2B cells. It can be inferred that the toxicity of CuO NPs is influenced by various factors such as particle size, shape, surface properties, chemical composition, and concentration [170]. Numerous studies have demonstrated that the toxicity of NPs is significantly influenced by the surface coatings they possess. The application of chitosan as a coating material for copper NPs has been demonstrated to mitigate the adverse impact of in vitro toxicity, while concurrently limiting the occurrence of inflammatory response in pulmonary cells [171]. Further investigation into the physical and chemical characteristics of NPs, encompassing attributes such as dissolution kinetics, surface modifications, and route of exposure, is essential to gain an improved understanding and mitigate the toxicity associated with 2DNMs [172].

5 Conclusion

The unique physicochemical properties of 2DNMs, such as their large surface area, simplicity of functionalization, photothermal ability, luminous ability, and high ion mobility, make them attractive for application in cancer diagnosis and therapy. The 2DNMs offer a number of benefits over traditional pharmaceuticals, including excellent stability, biocompatibility, increased permeability, and precise targeted administration. A greater capacity for loading drugs is afforded by the larger surface area. Oxygen-containing groups that are found in 2DNMs are easily functionalized with a wide variety of functional groups, including polymers, ligands, and other NMs. Particularly noteworthy is the possibility that functionalizing 2DNMs with PEG may improve their biocompatibility, drug loading capacity, and stability. An additional discovery that is worthy of notice is that the PTT may be enhanced by the straightforward synchronization of 2DNMs, such as functionalization, changes in therapeutic state, and prolonged exposure to NIR. 2DNMs may also help achieve further imaging and integrated therapeutic features. It is hypothesized that 2DNMs' particular or targeted drug delivery capacity is improved by tumor ligands that connect with them. Surface-functionalized or coated 2DNMs have the potential to serve as a nanoplatform for the detection and treatment of cancer. Since exposure to 2DNMs can cause serious health problems, studying their biocompatibility is an important area of study. According to the available research, 2DNMs may generate oxidative stress and ultimately lead to cell death when exposed to larger doses over longer periods of time. In addition, 2DNMs' biocompatibility may be improved by functionalization. However, other investigations showed that 2DNMs were not hazardous to healthy cells at all, but they were very toxic to cancer cell lines. Toxic effects of 2DNMs can differ between cell lines and materials utilized to make them. The study of 2DNMs' interactions with different cell lines is of the utmost importance. To summarize, 2D-NMs are distinct and promising nanomaterials for improved cancer diagnostics as well as cancer treatment, and the therapeutic limits of these nanomaterials may be circumvented with the assistance of various pharmacological and physicochemical approaches.

References

1. Abedi, M., Abolmaali, S.S., Heidari, R., Samani S.M.,Tamaddon, A.M.: Hierarchical mesoporous zinc-imidazole dicarboxylic acid MOFs: surfactant-directed synthesis, pH-responsive degradation, and drug delivery. Int. J. Pharm. **602**,120685 (2021)
2. Alimardani, V., Abolmaali, S.S., Tamaddon, A.M., Ashfaq, M.: Recent advances on microneedle arrays-mediated technology in cancer diagnosis and therapy. Drug Deliv. Transl. Res. **11**, 788–816 (2021)
3. Alipour, S., Shirazi, H.C., Kazemi, M., Dehshahri, A., Ahmadi, F.: Synthesis and cytotoxicity evaluation of doxorubicin-polyethyleneimine conjugate as a potential carrier for dual delivery of drug and gene. J. Drug Deliv. Sci. Technol. **68**,102994 (2022)
4. Alimardani, V., Farahavar, G., Salehi, S., Taghizadeh, S., Ghiasi, M.R., Abolmaali, S.S.: Gold nanocages in cancer diagnosis, therapy, and theranostics: a brief review. Front. Mater. Sci. **15**, 494–511 (2021)
5. Taghizadeh, S., Alimardani, V., Roudbali, P.L., Ghasemi, Y., Kaviani, E.: Gold nanoparticles application in liver cancer. Photodiagnosis Photodyn. Ther. **25**, 389–400 (2019)
6. Alimardani, V., et al.: Nanotechnology-based cell-mediated delivery systems for cancer therapy and diagnosis. Drug Deliv. Transl. Res. **13**, 189–221 (2023)
7. Chen, W., et al.: Black phosphorus nanosheet-based drug delivery system for synergistic photodynamic/photothermal/chemotherapy of cancer. Adv. Mater. (Deerfield Beach, Fla) 29 (2017)
8. Peng, L., et al.: Monolayer nanosheets with an extremely high drug loading toward controlled delivery and cancer. Theranostics Adv. Mater. (Deerfield Beach, Fla) **30**, e1707389 (2018)
9. Alimardani, V, Abolmaali, S.S., Borandeh, S.: Antifungal and antibacterial properties of graphene-based nanomaterials: a mini-review. J. Nanostructures **9**, 402–413 (2019)
10. Borandeh, S., Alimardani, V., Abolmaali, S.S., Seppala, J.: Graphene family nanomaterials in ocular applications: physicochemical properties and toxicity. Chem. Res. Toxicol. **34**, 1386–1402 (2021)
11. Novoselov, K.S., et al.: Electric field effect in atomically thin carbon films science **306**, 666–669 (2004)
12. Ashfaq, M., Talreja, N., Chauhan, D., Afreen, S., Sultana, A., Srituravanich, W.: Two-dimensional (2D) hybrid nanomaterials for diagnosis and treatment of cancer. J. Drug Deliv. Sci. Technol. 103268 (2022)
13. Chen, Y., et al.: Two-dimensional metal nanomaterials: synthesis. Prop., Appl. Chem. Rev. **118**, 6409–6455 (2018)
14. Wang, X., Cheng, L.: Multifunctional two-dimensional nanocomposites for photothermal-based combined cancer therapy. Nanoscale **11**, 15685–15708 (2019)
15. Zhang, H., Fan, T., Chen, W., Li, Y., Wang, B.: Recent advances of two-dimensional materials in smart drug delivery nano-systems. Bioact. Mater. **5**, 1071–1086 (2020)
16. Gong, L., Yan, L., Zhou, R., Xie, J., Wu, W., Gu, Z.: Two-dimensional transition metal dichalcogenide nanomaterials for combination cancer therapy. J. Mater. Chem. B **5**, 1873–1895 (2017)
17. Luo, S., et al.: Multifunctional photosensitizer grafted on polyethylene glycol and polyethylenimine dual-functionalized nanographene oxide for cancer-targeted near-infrared imaging and synergistic phototherapy. ACS Appl. Mater. Interfaces **8**, 17176–17186 (2016)
18. Yang, G., et al.: Manganese dioxide coated WS2@ Fe_3O_4/sSiO_2 nanocomposites for pH-responsive MR imaging and oxygen-elevated synergetic therapy. Small **14**, 1702664 (2018)
19. Li, H., Fan, R., Zou, B., Yan, J., Shi, Q., Guo, G.: Roles of MXenes in biomedical applications: recent developments and prospects. Journal of Nanobiotechnology **21**, 1–39 (2023)
20. Hong, X., Tan, C., Chen, J., Xu, Z., Zhang, H.: Synthesis, properties and applications of one-and two-dimensional gold nanostructures. Nano Res. **8**, 40–55 (2015)
21. Nosheen, F., et al.: Ultrathin Pd-based nanosheets: syntheses, properties and applications. Nanoscale **12**, 4219–4237 (2020)

22. Mishra, G., Dash, B., Pandey, S.: Layered double hydroxides: a brief review from fundamentals to application as evolving biomaterials. Appl. Clay Sci. **153**, 172–186 (2018)
23. Fan, J., Sun, M.: Transition metal dichalcogenides (tmdcs) heterostructures: synthesis, excitons and photoelectric properties. Chem. Rec. **22**, e202100313 (2022)
24. Tamang, S., Rai, S., Bhujel, R., Bhattacharyya, N.K., Swain, B.P., Biswas, J.: A concise review on GO, rGO and metal oxide/rGO composites: Fabrication and their supercapacitor and catalytic applications. J. Alloy. Compd. **169588** (2023)
25. Tian, Y., et al.: Inorganic boron-based nanostructures: synthesis, optoelectronic properties, and prospective applications. Nanomaterials **9**, 538 (2019)
26. Luo, M., Fan, T., Zhou, Y., Zhang, H., Mei, L.: 2D black phosphorus–based biomedical applications. Adv. Funct. Mater. **29**, 1808306 (2019)
27. Qin, Y., Wan, Y., Guo, J., Zhao, M.: Two-dimensional metal-organic framework nanosheet composites: preparations and applications. Chin. Chem. Lett. **33**, 693–702 (2022)
28. Fang, Y., Liu, Y., Qi, L., Xue, Y., Li, Y.: 2D graphdiyne: an emerging carbon material. Chem. Soc. Rev. (2022)
29. Soltani, A., Faramarzi, M., Farjadian, F., Parsa, S.A.M., Panahi, H.A.: pH-responsive glycodendrimer as a new active targeting agent for doxorubicin delivery. Int. J. Biol. Macromol. **221**, 508–522 (2022)
30. Farjadian, F., Ghasemi, S., Akbarian, M., Hoseini-Ghahfarokhi, M., Moghoofei, M., Doroudian, M.: Physically stimulus-responsive nanoparticles for therapy and diagnosis. Front. Chem. **10** (2022)
31. Chakraborty, D., Ghosh, D., Kumar, S., Jenkins, D., Chandrasekaran, N., Mukherjee, A.: Nano-diagnostics as an emerging platform for oral cancer detection: Current and emerging trends. Wiley Interdiscip. Rev. Nanomedicine Nanobiotechnology **15**, e1830 (2023)
32. Wen, W., et al.: Recent advances in emerging 2D nanomaterials for biosensing and bioimaging applications. Mater. Today **21**, 164–177 (2018)
33. Wang, L., Xiong, Q., Xiao, F., Duan, H.: 2D nanomaterials based electrochemical biosensors for cancer diagnosis. Biosens. Bioelectron. **89**, 136–151 (2017)
34. Nawrot, W., Drzozga, K., Baluta, S., Cabaj, J., Malecha, K.: A fluorescent biosensors for detection vital body fluids' agents. Sensors **18**, 2357 (2018)
35. Son, M.H., Park, S.W., Sagong, H.Y., Jung, Y.K.: Recent advances in electrochemical and optical biosensors for cancer biomarker detection. BioChip J. **17**, 44–67 (2023)
36. Mohammadpour-Haratbar, A., Boraei, S.B.A., Zare, Y., Rhee, K.Y., Park, S.-J.: Graphene-based electrochemical biosensors for breast cancer detection. Biosensors **13**, 80 (2023)
37. He, Y., Lin, Y., Tang, H., Pang, D.: A graphene oxide-based fluorescent aptasensor for the turn-on detection of epithelial tumor marker mucin 1 Nanoscale **4**, 2054–2059 (2012)
38. Kadhim, M.M., Rheima, A.M., Abbas, Z.S., Jlood, H.H., Hachim, S.K., Kadhum, W.R.: Evaluation of a biosensor-based graphene oxide-DNA nanohybrid for lung cancer RSC. Advances **13**, 2487–2500 (2023)
39. Hu, Y., et al.: Two-dimensional transition metal dichalcogenide nanomaterials for biosensing applications. Mater. Chem. Front. **1**, 24–36 (2017)
40. Presutti, D., et al.: Transition metal dichalcogenides (TMDC)-based nanozymes for biosensing and therapeutic applications. Materials **15**, 337 (2022)
41. Cai, B., Guo, S., Li, Y.: MoS 2-based sensor for the detection of miRNA in serum samples related to breast cancer. Anal. Methods **10**, 230–236 (2018)
42. Sadighbayan, D., Hasanzadeh, M., Ghafar-Zadeh, E.: Biosensing based on field-effect transistors (FET): recent progress and challenges TrAC. Trends Anal. Chem. **133**, 116067 (2020)
43. Syu, Y.-C., Hsu, W.-E., Lin, C.-T.: Field-effect transistor biosensing: Devices and clinical applications ECS. J. Solid State Sci. Technol. **7**, Q3196 (2018)
44. Mao, S., Chang, J., Pu, H., Lu, G., He, Q., Zhang, H., Chen, J.: Two-dimensional nanomaterial-based field-effect transistors for chemical and biological sensing. Chem. Soc. Rev. **46**, 6872–6904 (2017)

45. Majd, S.M., Salimi, A., Ghasemi, F.: An ultrasensitive detection of miRNA-155 in breast cancer via direct hybridization assay using two-dimensional molybdenum disulfide field-effect transistor biosensor. Biosens. Bioelectron. **105**, 6–13 (2018)
46. Cai, B., Xia, Z., Wang, J., Wu, S., Jin, X.: Reduced graphene oxide-based field effect transistor biosensors for high-sensitivity miRNA21 detection. ACS Appl. Nano Mater. **5**, 12035–12044 (2022)
47. Aziz, A., et al.: Advancements in electrochemical sensing of hydrogen peroxide, glucose and dopamine by using 2D nanoarchitectures of layered double hydroxides or metal dichalcogenides. A review Microchimica Acta **186**, 671 (2019)
48. Khatri, R., Puri, N.K.: Electrochemical studies of biofunctionalized MoS 2 matrix for highly stable immobilization of antibodies and detection of lung cancer protein biomarker. New J. Chem. **46**, 7477–7489 (2022)
49. Jafari-Kashi, A., Rafiee-Pour, H.-A., Shabani-Nooshabadi, M.: A new strategy to design label-free electrochemical biosensor for ultrasensitive diagnosis of CYFRA 21–1 as a biomarker for detection of non-small cell lung cancer. Chemosphere **301**, 134636 (2022)
50. Zhao, L., et al.: A fluorescent biosensor based on molybdenum disulfide nanosheets and protein aptamer for sensitive detection of carcinoembryonic antigen. Sens. Actuators, B Chem. **273**, 185–190 (2018)
51. Oudeng, G., Au, M., Shi, J., Wen, C., Yang, M.: One-step in situ detection of miRNA-21 expression in single cancer cells based on biofunctionalized MoS_2 Nanosheets. ACS Appl. Mater. Interfaces **10**, 350–360 (2018)
52. Guo, S., Yang, F., Zhang, Y., Ning, Y., Yao, Q., Zhang, G.-J.: Amplified fluorescence sensing of miRNA by combination of graphene oxide with duplex-specific nuclease. Anal. Methods **6**, 3598–3603 (2014)
53. Cai, B., Wang, S., Huang, L., Ning, Y., Zhang, Z., Zhang, G.J.: Ultrasensitive label-free detection of PNA-DNA hybridization by reduced graphene oxide field-effect transistor biosensor. ACS Nano **8**, 2632–2638 (2014)
54. Lee, S.X., Lim, H.N., Ibrahim, I., Jamil, A., Pandikumar, A., Huang, N.M.: Horseradish peroxidase-labeled silver/reduced graphene oxide thin film-modified screen-printed electrode for detection of carcinoembryonic antigen. Biosens. Bioelectron. **89**, 673–680 (2017)
55. Liu, D., Yang, F., Xiong, F., Gu, N.: The smart drug delivery system and its clinical potential. Theranostics **6**, 1306 (2016)
56. Alimardani, V., Sadat Abolmaali, S., Yousefi, G., Hossein Nowroozzadeh, M., Mohammad Tamaddon, A.: In-situ nanomicelle forming microneedles of poly NIPAAm-b-poly glutamic acid for trans-scleral delivery of dexamethasone. J. Ind. Eng. Chem. **119**, 485–498 (2023)
57. Ghasemi, S., Ahmadi, L., Farjadian, F.: Thermo-responsive PNIPAAm-b-PLA amphiphilic block copolymer micelle as nanoplatform for docetaxel drug release. J. Mater. Sci. **57**, 17433–17447 (2022)
58. Ahmadi, S., et al.: Stimulus-responsive sequential release systems for drug and gene delivery. Nano Today **34**, 100914 (2020)
59. Farjadian, F., Ghasemi, S., Andami, Z., Tamami, B.: Thermo-responsive nanocarrier based on poly (N-isopropylacrylamide) serving as a smart doxorubicin delivery system. Iran. Polym. J. **29**, 197–207 (2020)
60. Chu, Z., Liu, J., Guo, Z., Zhang, H.: 2 μm passively Q-switched laser based on black phosphorus. Opt. Mater. Express **6**, 2374–2379 (2016)
61. Fan, T., Zhou, Y., Qiu, M., Zhang, H.: Black phosphorus: a novel nanoplatform with potential in the field of bio-photonic nanomedicine. J. Innov. Opt. Health Sci. **11**, 1830003 (2018)
62. Zhao, Y., et al.: Stable and multifunctional dye-modified black phosphorus nanosheets for near infrared imaging-guided photothermal therapy. Chem. Mater. **29**, 7131–7139 (2017)
63. Liu, N., Fan, F., Xu, W., Zhang, H., Zhou, Q., Li, X.: Synthesis of functional hollow WS2 particles with large surface area for Near-Infrared (NIR) triggered drug delivery. J. Alloy. Compd. **875**, 160034 (2021)
64. Farzanfar, J., Farjadian, F., Roointan, A., Mohammadi-Samani, S., Tayebi, L.: Assessment of pH responsive delivery of methotrexate based on PHEMA-st-PEG-DA nanohydrogels. Macromol. Res. **29**, 54–61 (2021)

65. Sawant, R.M., Hurley, J., Salmaso, S., Kale, A., Tolcheva, E., Levchenko, T., Torchilin, V.: "SMART" drug delivery systems: double-targeted pH-responsive pharmaceutical nanocarriers. Bioconjugate Chem. **17**, 943–949 (2006)
66. Lee, G.-Y., Lo, P.-Y., Cho, E.-C., Zheng, J.-H., Li, M., Huang, J.-H., Lee, K.-C.: Integration of PEG and PEI with graphene quantum dots to fabricate pH-responsive nanostars for colon cancer suppression in vitro and in vivo. FlatChem **31**, 100320 (2022)
67. Jung, H.S., Verwilst, P., Sharma, A., Shin, J., Sessler, J.L., Kim, J.S.: Organic molecule-based photothermal agents: an expanding photothermal therapy universe. Chem. Soc. Rev. **47**, 2280–2297 (2018)
68. Melamed, J.R., Edelstein, R.S., Day, E.S.: Elucidating the fundamental mechanisms of cell death triggered by photothermal therapy. ACS Nano **9**, 6–11 (2015)
69. Lin, C., Hao, H., Mei, L., Wu, M.: Metal-free two-dimensional nanomaterial-mediated photothermal tumor therapy. Smart Mater. Med. **1**, 150–167 (2020)
70. Qin, Z., Bischof, J.C.: Thermophysical and biological responses of gold nanoparticle laser heating. Chem. Soc. Rev. **41**, 1191–1217 (2012)
71. Kim, H., Lee, D., Kim, J., Kim, T.-I., Kim, W.J.: Photothermally triggered cytosolic drug delivery via endosome disruption using a functionalized reduced graphene oxide. ACS Nano **7**, 6735–6746 (2013)
72. Liu, Y., Bhattarai, P., Dai, Z., Chen, X.: Photothermal therapy and photoacoustic imaging via nanotheranostics in fighting cancer. Chem. Soc. Rev. **48**, 2053–2108 (2019)
73. Yan, M., et al.: Nanoscale reduced graphene oxide-mediated photothermal therapy together with IDO inhibition and PD-L1 blockade synergistically promote antitumor immunity ACS Appl. Mater. Interfaces **11**, 1876–1885 (2018)
74. Yan, M., et al.: nanoscale reduced graphene oxide-mediated photothermal therapy together with IDO inhibition and PD-L1 blockade synergistically promote antitumor immunity. ACS Appl. Mater. Interfaces **11**, 1876–1885 (2019)
75. Spikes, J.D., Bommer, J.C.: Photosensitizing properties of mono-l-aspartyl chlorin e6 (NPe6): a candidate sensitizer for the photodynamic therapy of tumors. J. Photochem. Photobiol. B: Biol. **17**, 135–143 (1993)
76. Master, A., Livingston, M., Sen Gupta, A.: Photodynamic nanomedicine in the treatment of solid tumors: perspectives and challenges. J. Control. Release **168**, 88–102 (2013)
77. Gazzi, A., et al.: Photodynamic therapy based on graphene and MXene in cancer theranostics. Front. Bioeng. Biotechnol. **7** (2019)
78. Sahu, A., Choi, W.I., Lee, J.H., Tae, G.: Graphene oxide mediated delivery of methylene blue for combined photodynamic and photothermal therapy. Biomaterials **34**, 6239–6248 (2013)
79. Tian, B., Wang, C., Zhang, S., Feng, L., Liu, Z.: Photothermally enhanced photodynamic therapy delivered by nano-graphene oxide. ACS Nano **5**, 7000–7009 (2011)
80. Yang, G., et al.: Smart nanoreactors for pH-responsive tumor homing, mitochondria-targeting, and enhanced photodynamic-immunotherapy of cancer. Nano Lett. **18**, 2475–2484 (2018)
81. Wang, H., et al.: SDS coated Fe_3O_4@MoS_2 with NIR-enhanced photothermal-photodynamic therapy and antibiotic resistance gene dissemination inhibition functions. Colloids Surf.S B: Biointerfaces **214**,112457 (2022)
82. Cao, Y., et al.: A near-infrared triggered upconversion/MoS2 nanoplatform for tumour-targeted chemo-photodynamic combination therapy colloids and Surfaces B: Biointerfaces **213**, 112393 (2022)
83. Avitabile, E., Bedognetti, D., Ciofani, G., Bianco, A., Delogu, L.G.: How can nanotechnology help the fight against breast cancer? Nanoscale **10**, 11719–11731 (2018)
84. Viseu, T., Lopes, C.M., Fernandes, E., Real Oliveira, M.E.C., Lucio, M.: A systematic review and critical analysis of the role of graphene-based nanomaterials in cancer theranostics. Pharmaceutics **10**, 282 (2018)
85. Gollavelli, G., Ling, Y.-C.: Magnetic and fluorescent graphene for dual modal imaging and single light induced photothermal and photodynamic therapy of cancer cells. Biomaterials **35**, 4499–4507 (2014)

86. Gulzar, A., Xu, J., Yang, D., Xu, L., He, F., Gai, S., Yang, P.: Nano-graphene oxide-UCNP-Ce6 covalently constructed nanocomposites for NIR-mediated bioimaging and PTT/PDT combinatorial therapy. Dalton Trans. **47**, 3931–3939 (2018)
87. Wu, C., Zhu, A., Li, D., Wang, L., Yang, H., Zeng, H., Liu, Y.: Photosensitizer-assembled PEGylated graphene-copper sulfide nanohybrids as a synergistic near-infrared phototherapeutic agent. Expert Opin. Drug Deliv. **13**, 155–165 (2016)
88. Yang, K., Feng, L., Hong, H., Cai, W., Liu, Z.: Preparation and functionalization of graphene nanocomposites for biomedical applications. Nat. Protoc. **8**, 2392–2403 (2013)
89. Huang, P., et al.: Folic acid-conjugated graphene oxide loaded with photosensitizers for targeting photodynamic therapy. Theranostics **1**, 240 (2011)
90. Zhou, L., Jiang, H., Wei, S., Ge, X., Zhou, J., Shen, J.: High-efficiency loading of hypocrellin B on graphene oxide for photodynamic therapy. Carbon **50**, 5594–5604 (2012)
91. Zhou, L., et al.: Combination of chemotherapy and photodynamic therapy using graphene oxide as drug delivery system. J. Photochem. Photobiol. B. **135**, 7–16 (2014)
92. Zhang, Q., et al.: The theranostic nanoagent Mo_2C for multi-modal imaging-guided cancer synergistic phototherapy. Biomater. Sci. **7**, 2729–2739 (2019)
93. Liu, G., et al.: Surface modified Ti_3C_2 MXene nanosheets for tumor targeting photothermal/photodynamic/chemo synergistic therapy ACS. Appl. Mater. Interfaces **9**, 40077–40086 (2017)
94. Liu, T., et al.: Combined photothermal and photodynamic therapy delivered by PEGylated MoS_2 nanosheets. Nanoscale **6**, 11219–11225 (2014)
95. Peng, M.Y., Zheng, D.W., Wang, S.B., Cheng, S.X., Zhang, X.Z.: Multifunctional nanosystem for synergistic tumor therapy delivered by two-dimensional MoS(2). ACS Appl. Mater Interfaces **9**, 13965–13975 (2017)
96. Li, S., et al.: Enhanced photothermal-photodynamic therapy by indocyanine green and curcumin-loaded layered MoS(2) Hollow Spheres via Inhibition of P-Glycoprotein. Int. J. Nanomedicine **16**, 433–442 (2021)
97. Song, C., et al.: MoS2-Based multipurpose theranostic nanoplatform: realizing dual-imaging-guided combination phototherapy to eliminate solid tumor via a liquefaction necrosis process. J. Mater. Chem. B **5**, 9015–9024 (2017)
98. Yougbaré, S., et al.: Gold nanorod-decorated metallic MoS_2 nanosheets for synergistic photothermal and photodynamic antibacterial therapy. Nanomaterials **11**, 3064 (2021)
99. Yang, L., Wang, J., Yang, S., Lu, Q., Li, P., Li, N.: Rod-shape MSN@MoS(2) Nanoplatform for FL/MSOT/CT imaging-guided photothermal and photodynamic therapy. Theranostics **9**, 3992–4005 (2019)
100. Yang, G., Gong, H., Liu, T., Sun, X., Cheng, L., Liu, Z.: Two-dimensional magnetic WS2@Fe_3O_4 nanocomposite with mesoporous silica coating for drug delivery and imaging-guided therapy of cancer. Biomaterials **60**, 62–71 (2015)
101. Yong, Y, et al.: WS2 nanosheet as a new photosensitizer carrier for combined photodynamic and photothermal therapy of cancer cells. Nanoscale **6**, 10394–10403 (2014)
102. Liao, W., Zhang, L., Zhong, Y., Shen, Y., Li, C., An, N.: Fabrication of ultrasmall WS2 quantum dots-coated periodic mesoporous organosilica nanoparticles for intracellular drug delivery and synergistic chemo-photothermal therapy. OncoTargets Ther. **11**, 1949–1960 (2018)
103. Yu, C., et al.: $Ti_3C_2T_x$ MXene loaded with indocyanine green for synergistic photothermal and photodynamic therapy for drug-resistant bacterium. Colloids Surf. B. Biointerfaces **217**, 112663 (2022)
104. Ahirwar, S., Mallick, S., Bahadur, D.: Photodynamic therapy using graphene quantum dot derivatives. J. Solid State Chem. **282**, 121107 (2020)
105. Guo, W., et al.: Graphene oxide (GO)-based nanosheets with combined chemo/photothermal/photodynamic therapy to overcome gastric cancer (GC) paclitaxel resistance by reducing mitochondria-derived adenosine-triphosphate (ATP). J. Nanobiotechnology **19**, 146 (2021)
106. Ashkarran, A.A., Swann, J., Hollis, L., Mahmoudi, M.: The file drawer problem in nanomedicine. Trends Biotechnol. **39**, 425–427 (2021)

107. Liu, Y., Zhu, S., Gu, Z., Chen, C., Zhao, Y.: Toxicity of manufactured nanomaterials. Particuology **69**, 31–48 (2022)
108. Derakhshi, M., Daemi, S., Shahini, P., Habibzadeh, A., Mostafavi, E., Ashkarran, A.A.: Two-Dimensional Nanomaterials beyond graphene for biomedical applications. J. Funct. Biomater. **13** (2022)
109. Duch, M.C., et al.: Minimizing oxidation and stable nanoscale dispersion improves the biocompatibility of graphene in the lung. Nano Lett. **11**, 5201–5207 (2011)
110. Bussy, C., Ali-Boucetta, H., Kostarelos, K.: Safety considerations for graphene: lessons learnt from carbon nanotubes. Acc. Chem. Res. **46**, 692–701 (2013)
111. Jayakumar, A., Surendranath, A., Mohanan, P.: 2D materials for next generation healthcare applications. Int. J. Pharm. **551**, 309–321 (2018)
112. Li, M., Luo, Z., Zhao, Y.: Recent advancements in 2D nanomaterials for cancer therapy. Sci. China Chem. **61**, 1214–1226 (2018)
113. Lin, H., Chen, Y., Shi, J.: Insights into 2D MXenes for versatile biomedical applications: current advances and challenges ahead. Adv. Sci. **5**, 1800518 (2018)
114. Kyriakides, T.R., et al.: Biocompatibility of nanomaterials and their immunological properties. Biomed. Mater. **16** (2021)
115. Bullock, C.J., Bussy, C.: Biocompatibility considerations in the design of graphene biomedical materials advanced materials. Interfaces **6**, 1900229 (2019)
116. Liao, C., Li, Y., Tjong, S.C.: Graphene nanomaterials: synthesis, biocompatibility, and cytotoxicity. Int. J. Mol. Sci. 19 (2018).
117. Maio, A., Pibiri, I., Morreale, M., Mantia, F.P.L., Scaffaro, R.: An overview of functionalized graphene nanomaterials for advanced. Appl. Nanomater. **11**, 1717 (2021)
118. Jones, C.F., Grainger, D.W.: In vitro assessments of nanomaterial toxicity. Adv. Drug Deliv. Rev. **61**, 438–456 (2009)
119. Liao, C., Li, Y., Tjong, S.C.: Graphene nanomaterials: synthesis, biocompatibility, and cytotoxicity. Int. J. Mol. Sci. **19**, 3564 (2018)
120. Ou, L., et al.: Toxicity of graphene-family nanoparticles: a general review of the origins and mechanisms. Part. Fibre Toxicol. **13**, 57 (2016)
121. Zhang, B., Wei, P., Zhou, Z., Wei, T.: Interactions of graphene with mammalian cells: molecular mechanisms and biomedical insights. Adv. Drug Deliv. Rev. **105**, 145–162 (2016)
122. Duan, G., et al.: Graphene-induced pore formation on cell membranes. Sci. Rep. **7**, 42767 (2017)
123. Bach, D., Wachtel, E.: Phospholipid/cholesterol model membranes: formation of cholesterol crystallites. Biochimica et Biophysica Acta (BBA) - Biomembranes **1610**, 187–197 (2003)
124. Ermilova, I., Lyubartsev, A.P.: Cholesterol in phospholipid bilayers: positions and orientations inside membranes with different unsaturation degrees. Soft Matter **15**, 78–93 (2019)
125. Zhan, J., Lei, Z., Zhang, Y.: Non-covalent interactions of graphene surface: mechanisms and applications. Chem **8**, 947–979 (2022)
126. Li, S., Stein, A.J., Kruger, A., Leblanc, R.M.: Head groups of lipids govern the interaction and orientation between graphene oxide and lipids. J. Phys. Chem. C **117**, 16150–16158 (2013)
127. Hu, X., Lei, H., Zhang, X., Zhang, Y.: Strong hydrophobic interaction between graphene oxide and supported lipid bilayers revealed by AFM. Microsc. Res. Tech. **79**, 721–726 (2016)
128. Li, R.: et al.: Surface oxidation of graphene oxide determines membrane damage, lipid peroxidation, and cytotoxicity in macrophages in a pulmonary toxicity model. ACS Nano **12**, 1390–1402 (2018)
129. Pelin, M. et al.: Graphene and graphene oxide induce ROS production in human HaCaT skin keratinocytes: the role of xanthine oxidase and NADH dehydrogenase. Nanoscale **10**, 11820–11830 (2018)
130. Liu, J.H., Yang, S.T., Wang, H., Chang, Y., Cao, A., Liu, Y.: Effect of size and dose on the biodistribution of graphene oxide in mice. Nanomedicine (Lond.) **7**, 1801–1812 (2012)
131. Kurantowicz, N., et al.: Biodistribution of a high dose of diamond, graphite, and graphene oxide nanoparticles after multiple intraperitoneal injections in rats. Nanoscale Res. Lett. **10**, 398 (2015)

132. Strojny, B., et al.: Long term influence of carbon nanoparticles on health and liver status in rats. PloS one **10**, e0144821 (2015)
133. Baur, X., Sanyal, S., Abraham, J.L.: Mixed-dust pneumoconiosis: review of diagnostic and classification problems with presentation of a work-related case. Sci. Total. Environ. **652**, 413–421 (2019)
134. Lin, H., Gao, S., Dai, C., Chen, Y., Shi, J.: Correction to A two-dimensional biodegradable niobium carbide (MXene) for photothermal tumor eradication in NIR-I and NIR-II biowindows. J. Am. Chem. Soc. **142**, 10567–10567 (2020)
135. Lin, H., Wang, X., Yu, L., Chen, Y., Shi, J.: Two-dimensional ultrathin MXene ceramic nanosheets for photothermal conversion. Nano Lett. **17**, 384–391 (2017)
136. Lin, H., Wang, Y., Gao, S., Chen, Y., Shi, J.: Theranostic 2D tantalum carbide (MXene). Adv. Mater. **30**, 1703284 (2018)
137. Yang, C., Xu, D., Peng, W., Li, Y., Zhang, G., Zhang, F., Fan, X.: Ti_2C_3Tx nanosheets as photothermal agents for near-infrared responsive hydrogels. Nanoscale **10**, 15387–15392 (2018)
138. Dai, C. Lin, H., Xu, G., Liu, Z., Wu, R., Chen, Y.: Biocompatible 2D titanium carbide (MXenes) composite nanosheets for pH-responsive MRI-guided tumor hyperthermia. Chem. Mater. **29**, 8637–8652 (2017)
139. Xing, C., et al.: Two-dimensional MXene (Ti3C2)-integrated cellulose hydrogels: toward smart three-dimensional network nanoplatforms exhibiting light-induced swelling and bimodal photothermal/chemotherapy anticancer activity. ACS Appl. Mater. Interfaces **10**, 27631–27643 (2018)
140. Pandey, R.P., Rasool, K., Madhavan, V.E., Aïssa, B., Gogotsi, Y., Mahmoud, K.A.: Ultrahigh-flux and fouling-resistant membranes based on layered silver/MXene (Ti3C2Tx) nanosheets. J. Mater. Chem. A **6**, 3522–3533 (2018)
141. Rakhi, R., Nayak, P., Xia, C., Alshareef, H.N.: Novel amperometric glucose biosensor based on MXene nanocomposite. Sci. Rep. **6**, 1–10 (2016)
142. Wu, L., et al.: 2D transition metal carbide MXene as a robust biosensing platform for enzyme immobilization and ultrasensitive detection of phenol. Biosens. Bioelectron. **107**, 69–75 (2018)
143. Lim, G.P., Soon, C.F., Ma, N.L., Morsin, M., Nayan, N., Ahmad, M.K., Tee, K.S.: Cytotoxicity of MXene-based nanomaterials for biomedical applications: A mini review. Environ. Res. **201**, 111592 (2021)
144. Arabi Shamsabadi, A., Sharifian Gh, M., Anasori, B., Soroush, M.: Antimicrobial mode-of-action of colloidal $Ti_3C_2T_x$ MXene nanosheets. ACS Sustain. Chem. Eng. **6**, 16586–16596 (2018)
145. Jastrzębska, A.M., et al.: On the rapid in situ oxidation of two-dimensional V2CTz MXene in culture cell media and their cytotoxicity. Mater. Sci. Eng. C **119**, 111431 (2021)
146. Jastrzębska, A.M., et al.: On tuning the cytotoxicity of Ti_3C_2 (MXene) flakes to cancerous and benign cells by post-delamination surface modifications 2D. Materials **7**, 025018 (2020)
147. Liu, J., Pope, C.N.: Chapter 21 - Intrinsic and extrinsic factors that can modify toxicity. In: Pope, C.N., Liu, J. (eds.). An Introduction to Interdisciplinary Toxicology, pp. 285–293. Academic Press (2020). https://doi.org/10.1016/B978-0-12-813602-7.00021-1
148. Jastrzębska, A.M., et al.: In vitro studies on cytotoxicity of delaminated Ti_3C_2 MXene. J. Hazard. Mater. **339**, 1–8 (2017)
149. Ganguly, P., Breen, A., Pillai, S.C.: Toxicity of nanomaterials: exposure, pathways, assessment, and recent advances ACS. Biomater. Sci. Eng. **4**, 2237–2275 (2018)
150. Wu, W., Ge, H., Zhang, L., Lei, X., Yang, Y., Fu, Y., Feng, H.: Evaluating the Cytotoxicity of Ti3C2 MXene to neural stem. Cells Chem. Res. Toxicol. **33**, 2953–2962 (2020)
151. Teo, W.Z., Chng, E.L.K., Sofer, Z., Pumera, M.: Cytotoxicity of exfoliated transition-metal dichalcogenides (MoS2, WS2, and WSe2) is lower than that of graphene and its analogues. Chem. A Eur. J. **20**, 9627–9632 (2014)
152. Wu, H., et al.: Biocompatible inorganic fullerene-like molybdenum disulfide nanoparticles produced by pulsed laser ablation in water. ACS Nano **5**, 1276–1281 (2011)

153. Gupta, M., Gupta, S.: An overview of selenium uptake, metabolism, and toxicity in plants. Front. Plant Sci. **7**, 2074 (2017)
154. Pereira, B.B., Limongi, J.E., Campos Júnior, E.Od., Luiz, D.P., Kerr, W.E.: Effects of piperonyl butoxide on the toxicity of the organophosphate temephos and the role of esterases in the insecticide resistance of Aedes aegypti Revista da Sociedade Brasileira de Medicina Tropical **47**, 579–582 (2014)
155. Saha, U., Fayiga, A., Hancock, D., Sonon, L.: Selenium in animal nutrition: deficiencies in soils and forages, requirements, supplementation and toxicity. Int. J. Appl. Agric. Sci. **2**, 112–125 (2016)
156. Nurunnabi, M., McCarthy, J.: Biomedical applications of graphene and 2D nanomaterials. Elsevier (2019)
157. Moroder, L.: Isosteric replacement of sulfur with other chalcogens in peptides and proteins. J. Pept. Sci.: Off. Publ. Eur. Pept. Soc. **11**, 187–214 (2005)
158. Nath, N.C.D., Debnath, T., Nurunnabi, M., Kim, E.-K.: In vitro toxicity of 2D materials. In: Biomedical Applications of Graphene and 2D Nanomaterials. Elsevier, pp 165–186 (2019)
159. Wang, X., et al.: Differences in the toxicological potential of 2D versus aggregated molybdenum disulfide in the lung. Small **11**, 5079–5087 (2015)
160. Yin, W., et al.: High-throughput synthesis of single-layer MoS_2 nanosheets as a near-infrared photothermal-triggered drug delivery for effective cancer therapy. ACS Nano **8**, 6922–6933 (2014)
161. Chou, S.S., et al.: Ligand conjugation of chemically exfoliated MoS_2. J. Am. Chem. Soc. **135**, 4584–4587 (2013)
162. Kim, J.S., Yoo, H.W., Choi, H.O., Jung, H.T.: Tunable volatile organic compounds sensor by using thiolated ligand conjugation on MoS2. Nano Lett. **14**, 5941–5947 (2014)
163. Anielle Christine Almeida, S., et al.: Biocompatibility of doped semiconductors nanocrystals and nanocomposites. In: Tülay Aşkin, Ç. (ed.) Cytotoxicity. IntechOpen, Rijeka, p Ch. 9. (2018). https://doi.org/10.5772/intechopen.77197
164. Dymek, M., Sikora, E.: Liposomes as biocompatible and smart delivery systems—the current state. Adv. Colloid Interface Sci. **309**, 102757 (2022)
165. Saroia, J., Yanen, W., Wei, Q., Zhang, K., Lu, T., Zhang, B.: A review on biocompatibility nature of hydrogels with 3D printing techniques, tissue engineering application and its future prospective. Bio-Des. Manuf. **1**, 265–279 (2018)
166. Areecheewakul, S., et al.: Toxicity assessment of metal oxide nanomaterials using in vitro screening and murine acute inhalation studies. NanoImpact **18**, 100214 (2020)
167. Park, E.-J., Lee, G.-H., Yoon, C., Kim, D.-W.: Comparison of distribution and toxicity following repeated oral dosing of different vanadium oxide nanoparticles in mice. Environ. Res. **150**, 154–165 (2016)
168. Uche, F., Obianime, A., Gogo-Abite, M.: Effects of vanadium pentoxide on the histological and sperm parameters of male guinea pigs. Chem. A Eur. J. **12** (2008)
169. Morris, A.S., Givens, B.E., Silva, A., Salem, A.K.: Copper oxide nanoparticle diameter mediates serum-sensitive toxicity in BEAS-2B cells. Adv. NanoBiomed Res. **1**, 2000062 (2021)
170. Naz, S., Gul, A., Zia, M.: Toxicity of copper oxide nanoparticles: a review study. IET Nanobiotechnol. **14**, 1–13 (2020)
171. Worthington, K.L., et al.: Chitosan coating of copper nanoparticles reduces in vitro toxicity and increases inflammation in the lung. Nanotechnology **24**, 395101 (2013)
172. Sánchez-López, E., et al.: Metal-based nanoparticles as antimicrobial agents: an overview. Nanomaterials **10**, 292 (2020)

Two-Dimensional (2D) Based Hybrid Polymeric Nanoparticles as Novel Potential Therapeutics in the Treatment of Hepatocellular Carcinoma

Alok Raghav and Goo-Bo-Jeong

Abstract Liver cancer, or primary hepatic malignancy, accounts for the sixth most common form of human cancer worldwide, and among this, 90% of liver cancer cases exhibit hepatocellular carcinoma (HCC). Due to several associated limitations, current drug-based therapies for treating HCC are insufficient for an effective and efficient approach. Combination therapies with two-dimensional (2D) hybrid nanomaterials enhance the performance of the biocompatible nanomaterials in HCC treatment. 2D hybrid nanomaterials delivered an effective potential in ablating tumours. Many researchers exploited the 2D hybrid nanomaterials in the treatment and diagnostic of various diseases including graphene oxide, reduced graphene oxide, transition metal dichalcogenides (MoS_2, WS_2, $MoSe_2$, $NbSe_2$, TiS_2, ZrS_2, TaS_2, and WSe_2), g-C_3N_4, transition metal oxides (TiO_2 and MnO_2), layered double hydroxides, hexagonal boron nitride, black phosphorus, boron nitride, bismuth selenide, and MXenes, but here we explored their potential in effective treatment of HCC. The market potential of nanomaterials is widely expanding due to their potential to overcome several limitations associated with pharmacological-based treatment approaches commercially available, including side effects, bioavailability, stability, and efficiency.

1 Introduction

Liver cancer, or primary hepatic malignancy, accounts for the sixth most common form of human Cancer worldwide and among this, 90% of the liver cancer cases exhibit hepatocellular carcinoma (HCC) [1, 2]. The major risk factors for HCC include hepatitis B and C infections, fatty liver disease, and excess alcohol intake [3]. Among these risk factors, hepatitis B virus infection is among the prominent

A. Raghav (✉) · Goo-Bo-Jeong
Department of Anatomy and Cell Biology, College of Medicine, Gachon University, 155 Getbeol-Ro Yeonsu-Gu, Incheon 21999, South Korea
e-mail: alokalig@gachon.ac.kr

N. Talreja et al. (eds.), *Two-dimensional Hybrid Composites*, Engineering Materials,
https://doi.org/10.1007/978-981-99-8010-9_12

risk factors for developing HCC, which alone accounts for 50% of cases [4]. Unrelenting virological response (UVR) with the usage of antiviral drugs has significantly diminished the risk of HCC attributed to hepatitis C virus infection [5]. However, in the West, nonalcoholic steatohepatitis (NASH) attributed to metabolic disorders, including diabetes mellitus and obesity, is increasing at an alarming rate, contributing to the aetiology of HCC [6, 7]. Age is also considered to be the contributory risk factor in the progression of nonalcoholic fatty liver disease (NAFLD) related HCC. One of the previously published studies observed that patients with NAFLD-attributed HCC were more aged compared to virus-associated HCC [8]. Age-associated gut microbiota modulation in patients presenting NAFLD is also considered to be at high risk of developing HCC [9]. The global age-standardized rate of primary liver cancer per 100,000 populations in 2019 by country and territory is shown in Fig. 1.

World Health Organization (WHO) considers liver cancer the prime cause of cancer-related mortalities worldwide. It is also estimated that in the year 2020, about 0.83 million people died due to it [10]. In Asia, HCC is among the most common forms of liver cancer, accounting for 0.5 million deaths with 0.6 million new cases in 2020 [11, 12]. Asian men demonstrated higher incidence and mortality compared to Asian women, making it fourth highest incidence and second highest mortality. Moreover, among Asian women, liver cancer accounted for the seventh-highest incidence and sixth-highest mortality in 2020 [10], as shown in Fig. 2.

A sharp decline in the average annual percent change (AAPC) in incidence rates was observed in cases of liver cancer in South Korea (−2.2%), Japan, China (−1.6%), and the Philippines (−1.7%) as documented by a previously published study [12]. On the contrary, South-Western Asian countries, including Israel, showed a hike in AAPC [12]. In another report published by GLOBOCAN 2020, countries including Iran, Afghanistan, Qatar, Azerbaijan, Iraq, and Nepal also showed a worrying trend [13]. Studies demonstrated that eastern Asia, northern Africa, and Micronesia had the highest incidence rates, while the highest mortality rates were shown by eastern Asia, northern Africa, and south-eastern Asia [14, 15]. With the increased incidence and mortality rates of HCC in several parts of the world, it is necessary to focus on such issues with newer technologies and interventions.

Several staging approaches were fabricated to classify HCC, including Hong Kong Liver Cancer (HKLC), Cancer of the Liver Italian Program (CLIP), Okuda, Barcelona Clinic Liver Cancer (BCLC), American Association for the Study of Liver Diseases (AASLD) [16, 17]. Among these staging approaches, the latter two were widely used worldwide. Some other staging regimes involve the classification of HCC using molecular genetics, metabolism, immunological properties, and chromosomal arrangements [18].

Current interventional approaches for HCC include liver resection, transplantation, transarterial therapy, the implication of tyrosine kinase inhibitors (systemic therapy), and local ablative therapy. Moreover, along with conventional therapies, several drugs, including sorafenib, lenvatinib, atezolizumab, and bevacizumab, along with chemotherapeutic agents such as doxorubicin (DOX), showed restricted effects along with associated extrinsic and intrinsic drug resistance [19]. Table 1 demonstrates the clinical trials of a combination of drugs.

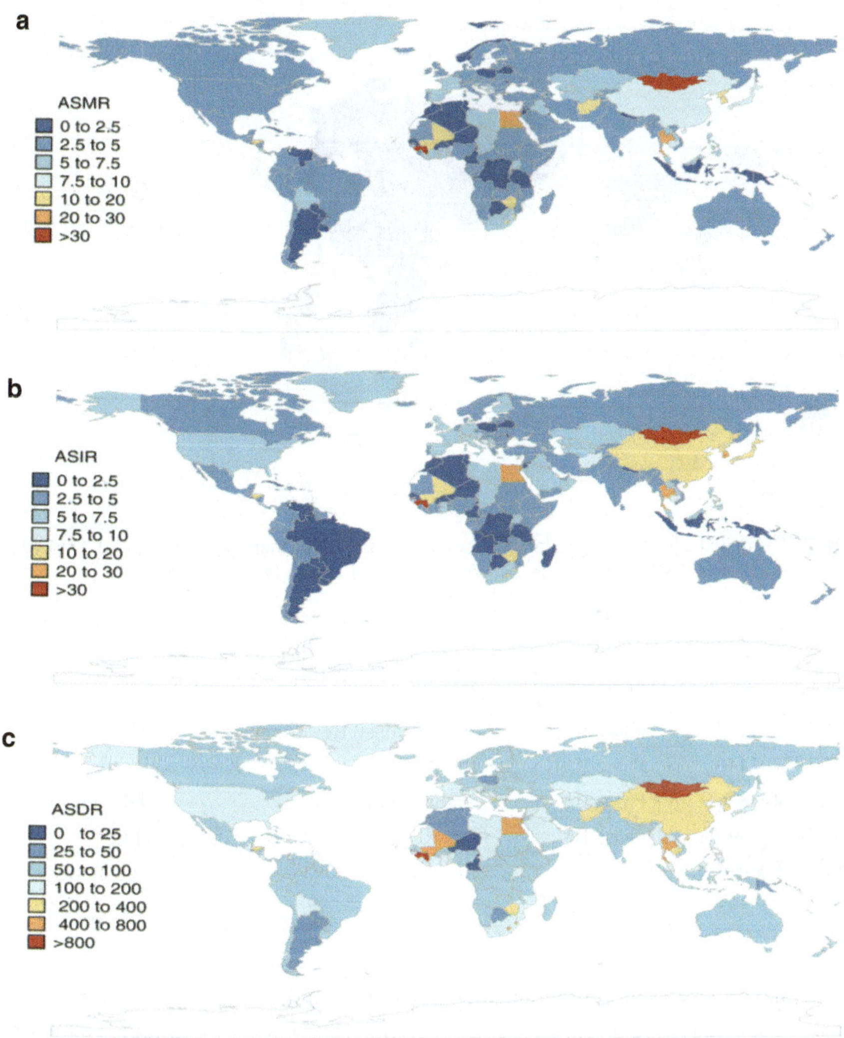

Fig. 1 The global age-standardized rate of primary liver cancer per 100,000 populations in 2019 by country and territory. (A) ASMR in 2019; (B) ASIR in 2019; and (C) ASDR in 2019. ASMR, age-standardized mortality rate. ASIR, age-standardized incidence rate. ASDR, age-standardized DALYs rate (Adopted from Reference No. 3 under the terms of the common creative license 4.0)

Despite multiple pharmacological interventions, treatment of advanced-stage HCC does not fulfill the standard health outcomes. Such unsatisfied outcomes may be attributed to several reasons, including drug-associated adverse effects, low bioavailability, high toxicity, non-specific delivery of pharmacological agents, high cost at large-scale production, immune complications, and anaphylactic responses [20].

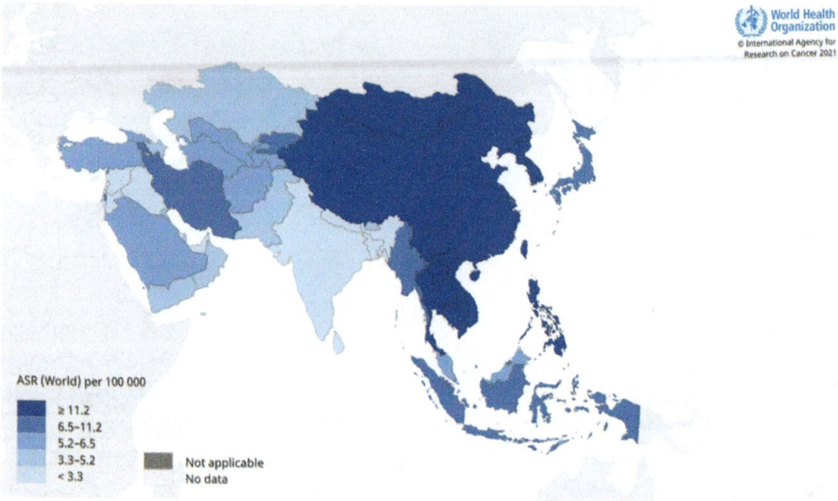

Fig. 2 Estimated age-standardized incidence rates (ASIR) of primary liver cancer in Asia in 2020. Graph production: IARC (http://gco.iarc.fr/today), World Health Organization

Table 1 Clinical trials of combination therapy for advanced HCC

Name	Study type	Drugs	Target
NCT03434379	Phase II	Atezolizumab Bevacizumab	PDL-1
NCT03794440	Phase II/III	Sintilizumab	VEGF
NCT03463876	Phase II	Camerlizumab	PDL-1
NCT03764293	Phase III		VEGF
NCT03006926	Phase Ib	Lenvatinib	Multikinase
NCT03713593	Phase III		PD1

Novel nanomaterial-based theranostics is widely implicated in drug delivery, diagnosis of severe to rare diseases, wound dressings, and cancer treatment. The market potential of nanomaterials is widely expanding due to their potential to overcome several limitations associated with pharmacological-based treatment approaches commercially available, including side effects, bioavailability, stability, and efficiency. Several published literature exploited the potential of 2D-nanomaterials because these exhibit ultra-thin architectures (thickness of at least one atomic layer) with good physical characteristics (high surface-area-to-volume ratio) and stability (high atoms presence). Some of the potential 2D nanomaterials include graphene, graphene oxide (GO), Transition metal dichalcogenide (TMDs), and Transition metal oxides (TMOs) [21–23]. To overcome such limitations, researchers are now exploring the possibility of using two-dimensional (2D) hybrid polymeric nanoparticles as novel theranostics for treating HCC.

Liver cancer or primary hepatic malignancy accounts for the sixth most common form of human Cancer worldwide, and among this, 90% of liver cancer cases exhibit hepatocellular carcinoma (HCC) [2, 24]. The major risk factors for HCC include hepatitis B and C infections, fatty liver disease, and excess alcohol intake [25]. Among these risk factors, hepatitis B virus infection is among the prominent risk factors for developing HCC, which alone accounts for 50% of cases [26]. Unrelenting virological response (UVR) using antiviral drugs has significantly diminished the risk of HCC attributed to hepatitis C virus infection [5]. However, in the West, nonalcoholic steatohepatitis (NASH) attributed to metabolic disorders, including diabetes mellitus and obesity, is increasing at an alarming rate, contributing to the aetiology of HCC [6, 7]. Age is also considered to be the contributory risk factor in the progression of nonalcoholic fatty liver disease (NAFLD) related HCC. One of the previously published studies observed that patients with NAFLD attributed HCC were more aged compared to virus-associated HCC [27]. Age-associated gut microbiota modulation in patients presenting NAFLD is also considered to be at high risk of developing HCC [9].

World Health Organization (WHO) considers liver cancer the prime cause of cancer-related mortalities worldwide. It is also estimated that in 2020, about 0.83 million people died due to it [28]. In Asia, HCC is among the most common forms of liver cancer, accounting for 0.5 million deaths, with 0.6 million new cases in 2020 [12, 29]. Asian men demonstrated higher incidence and mortality compared to Asian women, making it fourth highest incidence and second highest mortality. Moreover, among Asian women, liver cancer accounted for the seventh-highest incidence and sixth-highest mortality in 2020 [28]. A sharp decline in the average annual percent change (AAPC) in incidence rates was observed in cases of liver cancer in South Korea (−2.2%), Japan, China (−1.6%), and the Philippines (−1.7%) as documented by a previously published study [12]. On the contrary, South-Western Asian countries, including Israel, showed a hike in AAPC [12]. In another report published by GLOBOCAN 2020, countries including Iran, Afghanistan, Qatar, Azerbaijan, Iraq, and Nepal also showed a worrying trend [13]. Studies demonstrated that eastern Asia, northern Africa, and Micronesia have the highest incidence rates, while the highest mortality rates were shown by eastern Asia, northern Africa, and south-eastern Asia [15, 30] (Figs. 1, 2, and 3). With the increased incidence and mortality rates of HCC in several parts of the world, it is necessary to focus on such issues with newer technologies and interventions.

Several staging approaches were fabricated to classify HCC, including Hong Kong Liver Cancer (HKLC), Cancer of the Liver Italian Program (CLIP), Okuda, Barcelona Clinic Liver Cancer (BCLC), American Association for the Study of Liver Diseases (AASLD) [16, 31]. Among these staging approaches, the latter two were widely used worldwide. Some other staging regimes involve the classification of HCC using molecular genetics, metabolism, immunological properties, and chromosomal arrangements [18].

Current interventional approaches for HCC include liver resection, transplantation, transarterial therapy, the implication of tyrosine kinase inhibitors (systemic therapy), and local ablative therapy. Moreover, along with conventional therapies,

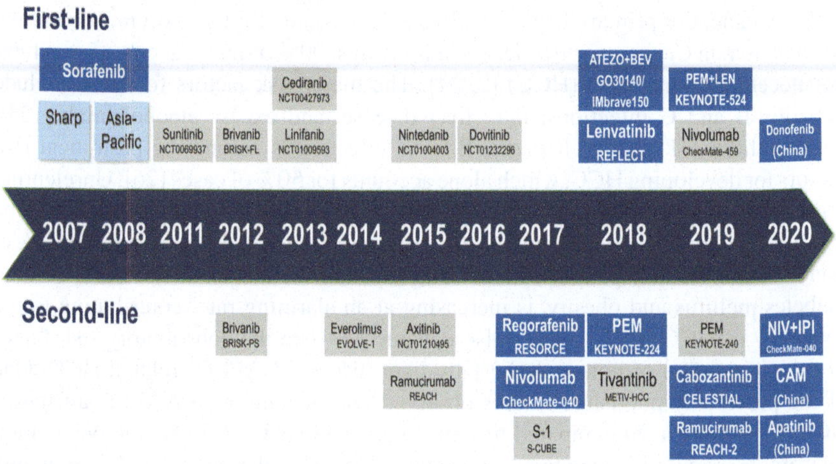

Fig. 3 Overview of the targeted agents approved for HCC. ATEZO atezolizumab, BEV bevacizumab, CAM camrelizumab, LEN lenvatinib, PEM pembrolizumab, NIV nivolumab, IPI ipilimumab ((Adopted from Reference No. 19 under the terms of the common creative license 4.0)

several drugs, including sorafenib, lenvatinib, atezolizumab, and bevacizumab, along with chemotherapeutic agents such as doxorubicin (DOX), showed restricted effects along with associated extrinsic and intrinsic drug resistance [32].

2 Epidemiology of Hepatocellular Carcinoma

Hepatocellular carcinoma is one of the primary causes of mortality among patients with chronic liver diseases. Despite available data, there is a great disparity among the regions reflecting India's prevalence and etiological considerations. National Cancer Registry Program (NCRP), which coordinates with the Indian Council of Medical Research (ICMR) and the National Centre for Disease Informatics and Research (NCDIR), Bengaluru, from 28 Population-Based Cancer Registries (PBCR) and 58 Hospital-Based Cancer Registries (HBCR) for the period between 2012 and 2016, does not specifically mention about liver cancer. In India, the first Cancer Registry was conducted by the Indian Cancer Society (ICS), covering a large population size from Mumbai, Pune, Aurangabad, and Nagpur. Hence, due to these limitations, it doesn't represent the entire country. According to India, the mortality due to HCC in the male population was estimated to be 6.8/100000, according to the ICMR consensus document [33]. In the Asian population, hepatitis B Virus and Hepatitis C virus are among the significant causes of HCC. However, NAFLD is also among the contributory cause of HCC [34]. Tertiary care centres in India demonstrate a large number of HCC cases [35–39]. One of the previously published studies reported that HCC incidence was 2.8/100000 in the population in the year 2015; among them, males contributed 3.9 while females contributed 1.6 [39]. Another study found that the incidence of HCC in patients with liver cirrhosis was 1.6/100 persons [36]. Another prospective study from Kerala over the period of 2 years (2018 to 2020) was accepted as the largest survey practiced in India [40]. Liver Imaging Reporting and Data System (LI-RADS) and computerized tomography/magnetic resonance imaging (CT/MRI) scan along with liver biopsy are some of the diagnostic tools of HCC according to the Indian National Association for Study of the Liver (INASL) consensus [41]. It is comforting to note that hepatitis B virus (HBV) as an etiological factor has contributed to only 7% of the entire cohort, indicating the reaping benefit of successful HBV vaccination in the state and high literacy rate. A recently published cohort study from 2007–2009 to 2013–2015 found that NAFLD is among the most common etiological factors for HCC [42].

3 Pathophysiology of Hepatocellular Carcinoma

The pathophysiology of HCC exhibits complex multifactor mechanisms. HCC progression and hepatocytes' malignant transformation depend on the interplay between various factors, including genetic predisposition, viral and non-viral elements, the severity of liver disease, and the cellular microenvironment at its early stage. It was seen that nearly 80% of liver cirrhotic patients develop HCC attributed to molecular alterations [43]. Viral elements include etiological infections associated with HCV and HBV, while the non-viral elements include alcohol consumption, NASH, and use of aflatoxin, tobacco, and aristolochic acid, which have been identified as a trigger of cancer mechanisms in the liver [44]. In addition to the aforementioned factors, some specific immune and molecular causes were identified as an initiator of HCC [44]. In this respect, studies of such molecular and immunological checkpoints are necessary to understand the onset, progression, and treatment using biopharmaceuticals for these targets. Some of the major checkpoints were extensively studied elsewhere [45–47].

4 Potential of 2D Nanomaterials as an Anticancer Candidate for Treatment of HCC

2D nanomaterials have been exploited well for the treatment of almost all types of cancers and, therefore, emerging as novel anticancer therapeutics candidates. Such nanomaterials exhibit unique features such as high surface area, ultra-thin thickness, planar structure, and good physicochemical properties. Such 2D nanomaterials are comprised of atoms bound in a single layer in nanosheet form [48]. Previous studies claimed that 2D nanomaterials, including graphite carbon nitride (GCN), hexagonal boron nitride (HBN), group-VA semiconductors, transition metal carbides (TMCs), and transition metal dichalcogenides (TMDs) showed improved dispersibility within the aqueous medium and can have the potential for effective cancer treatment [49–52]. It has been shown that some of the 2D nanomaterials including graphene oxide, reduced graphene oxide, transition metal dichalcogenides (MoS_2, WS_2, $MoSe_2$, $NbSe_2$, TiS_2, ZrS_2, TaS_2, and WSe_2), g-C_3N_4, transition metal oxides (TiO_2 and MnO_2), layered double hydroxides, hexagonal boron nitride, black phosphorus, boron nitride, bismuth selenide, and MXenes have been recently developed and explored for the application in various therapeutics to treat diseases.

Studies explored the potential of 2D nanomaterials for the treatment and diagnosis of Cancer due to improved transportability, efficient drug distribution, controlled drug release, and increased chemotherapy efficacy [53]. Among the most explored nature of 2D nanomaterials, their controlled release approaches with the drugs were most appreciated by researchers worldwide. Such stimuli (commonly pH and temperature) activated 2D nanomaterials were fabricated to release the anticancer drugs to the tumour niche to enhance chemotherapeutic agents' effect further or increase systematic toxicity. Therefore, such materials provide a strong synergistic effect to treat Cancer effectively with good outcomes.

In one of the previous studies, ultra-small graphene oxide nanomaterials conjugated with polyethylene glycol (PEG) loaded with C6 ceramide were used to assess the antitumor effect against HCC in combination with sorafenib [54]. The authors found that GO 2D nanomaterials significantly enhanced the C6 ceramide cellular uptake within the tumour cells primarily via a clathrin-mediated mechanism [54]. Such an approach produced a synergistic effect when used with sorafenib against HCC. Furthermore, the authors reported that this approach was effective in subverting multidrug resistance (MDR) in HCC cells mediated through Akt mechanism [54]. This 2D GO nanomaterial significantly inhibits tumour growth and improves the survival outcome in HCC, thereby proposing a potential therapeutic target in drug-resistant HCC.

In another study, it was found that graphene oxide plane/nanofilm (nfGO) with chicken embryo liver extract (CELE) effectively reduced the population of HepG2 cells in the G0/G1 phase and an increased the population in G2/M and altered expression of proto-oncogenes (*focal adhesion kinase* (*fak*), *e-cadherin, n-cadherin, β-catenin*) [29]. This 2D GO nanomaterial approach was found to increase the secretion of integrins proteins responsible for suppressing the HepG2 cancer cell growth [55]. The author of another study showed that 2D GO nanomaterials exhibit reduced expression of laminin-binding $\alpha 3$ integrin and fibronectin binding $\alpha 5$ subunits in HepG2 cancer cells [56, 57].

Ultrahigh photothermal conversion efficiency with significant computed tomography performance was seen with the combination of Au nanorods with Bi_2S_3 film. This approach was proven to be a promising nanotheranostic agent for PT/PA/CT imaging. Subsequently, it was found that poly (N-vinylpyrrolidone)-modified Au@Bi_2S_3 NBs (Au@Bi_2S_3-PVP NBs) loaded with doxorubicin were extremely effective in delivery to cancer cells and increasing the outcome during chemotherapy [58]. Gold nanoparticles (Aunps) were evaluated with some phytochemical extract for antitumor activity and biosafety in HepG2 cells [59]. The authors found that the combination of Aunps with the phytochemical extract is significantly efficient compared to plant extract alone against the HepG2 cells and thereby concluded that gold nanoparticles can be used as two-dimensional nanomaterials for liver cancer treatment [60]. In another study, the functionalized carbon nanotubes with carboxylation and esterification in the presence of polyethylene glycol (PEG) were loaded with doxorubicin and evaluated against human liver cancer cells (HepG2) and found effective against liver cancer through activation of caspases 3, 8 and 9 along with apoptosis pathway proteins Bcl-2 and BAX [61]. The synthesized DOX-loaded nanomaterial exhibited increased cytotoxicity and apoptosis in liver HepG2 cells and suggests that the DOX-loaded nanocarrier possesses strong anticancer properties and could be an applicable and potential drug carrier for liver cancer chemotherapy [61]. In another study, multiwalled functional carbon nanotubes (MWCNTs) with human serum albumin (HSA) further irradiated with 2W 808 nm laser beam showed effective internalization of HAS-MWCNTs mediated through albondin (aka Gp60) receptors present on the membrane of the HepG2 cells and mediate the caveolin mediated endocytosis [62]. The study showed that within 30 min of the cellular uptake, the apoptosis rate increased from 88.24 to 92.34% [62]. 166Ho-labeled hydroxyapatite particles were seen for their anticancer effects against liver cancer [59]. The authors concluded that 166Ho-HA exhibited promising features as an agent for liver cancer therapy in preliminary studies and warrants further investigation [63]. Naturally occurring hydroxyapatite can scavenge the sodium from the cytoplasm, thereby disrupting the internal homeostasis and activating apoptosis pathways [64]. The authors investigated the hydroxyapatite against liver cancer HepG2 cells and observed that as the concentration of sodium decreased by one-third due to the presence of hydroxyapatite, the rate of HepG2 cells apoptosis was increased and can be used as a potential anticancer candidate for the treatment of liver cancer [64]. In one previously published study, authors explored the 2D nanomaterial potential of hyaluronic acid-coated copper–aluminum layered double hydroxide (LDH) incorporated with doxorubicin for the treatment of HCC [65]. With the significant cellular uptake by the HepG2 cells, it produced a synergistic effect against the HCC [65]. In another study, the authors developed pH-sensitive layered double hydroxides incorporated with siRNA against hepatocellular carcinoma mediating via NR2F6 and PD-L1 checkpoint synergistically [66].

5 Biocompatibility of 2D Nanomaterial

As increasing research focuses on exploring the potential of 2D nanomaterials in biomedical therapeutics, nanotoxicity, and biocompatibility must be addressed thoroughly. All 2D nanomaterials were not explored for safety information extensively except graphene. Unfortunately, the biocompatibility and nanotoxicity of layered 2D nanomaterials cannot be inferred directly from their bulk counterparts because nano-bio interactions between 2D nanomaterials and biological entities are unique and heavily dependent on numerous nanomaterial physicochemical parameters. Such parameters include surface area, concentration, shape, lateral size, distribution, number of layers, chemical composition, and surface charge. Despite the importance of these physicochemical parameters in influencing nanomaterial biocompatibility and nanotoxicity, we find that there have been few studies that look at the individual effects of certain factors, such as shape, stability, surface charge, and purity. In one of the previously published studies, biocompatibility and nanotoxicology concentration-dependent behaviour of graphene, rGO, and GO were studied in hMSCs and other cells, including human erythrocytes, U87, U118 glioblastoma cells, and skin fibroblasts [67, 68]. It was also found that fluorinated graphene showed concentration-dependent toxicological effects on A549 cells due to the presence of fluorinated atoms [68]. In a separate study, graphene at a concentration of 100 μg/mL showed significant toxicity on neural phaeochromocytoma-derived PC12 and A549 cells due to the generation of reactive oxygen species (ROS) [69]. It was also suggested that graphene-based nanomaterials were among the potent toxicants for phospholipid bilayer [69]. Another study reported that a few layers of MoS2 nanosheets with small dimensions of 1 μM were safe and showed no cytotoxic effects [69]. Moreover, higher concentrations showed the toxicity. In another study, human lung cancer epithelial cells (A549) were found alive even after prolonged nanomaterial exposures when interacting with these TMD nanomaterials. Interestingly, ultrasmall MoS_2 nanodots functionalized with glutathione (MoS_2-GSH nanodots) are extremely biocompatible, have no detectable in vitro cytotoxicity, and can be rapidly removed from the body via urine [69]. The cellular viability of 4T1 cells was found to be high even after 24-h exposure to MoS2-GSH nanodots at a high concentration of 200 g/mL. In the presence of these nanomaterials, oxidative stress was not induced, and cellular membrane integrity was preserved. Furthermore, because of their ultrasmall size, the intravenously administered MoS_2-GSH nanodots achieved efficient body clearance via the renal route, unlike bigger MoS_2 nanoflakes [70].

6 Molecular Operators of Hepatocellular Carcinoma

In patients with liver cirrhosis, the neoplasm advances through a sequential cascade of histopathological modulations, ultimately initiating HCC. Histomorphological characteristics of HCC include highly vascularized tumours with prominent acinar and wide trabeculae and loss of Kupffer cells and reticulin network [71]. In advanced HCC, tumours were found to be encapsulated with positive septae for CD34 and α-smooth muscle actin (SMA). Studies have found that mature hepatocytes are the primary cells responsible for HCC origin and progression in addition to liver stromal cells [71, 72]. Studies in the past demonstrated that repetitive stress to regenerating hepatocytes triggers genetic lesions that initiate transformation and oncogenesis progression [73]. The study observed that alterations in cyclin-A2 or E1 proteins of the cell cycle favour the progression of HCC, especially in non-cirrhotic patients, which is further mediated by activation of E2F and ATR transcriptional pathways along with inactivation of RB1 and PTEN [74]. In patients with NASH, CD8 + PD1 + T cells promote hepatocyte death, favouring the micro-environment for HCC pathogenesis and progression [75]. On the contrary, somatic, genomic, and epigenetic modulations also trigger the HCC. A study showed the single nucleotide polymorphisms (SNP) of PNPLA3 (rs738409), TM6SF2 (rs585542926), and HSD17B13 (rs72613567) predispose to liver carcinogenesis that increases the probability of HCC [76]. Genotoxic compounds, including aflatoxin B1 and aristolochic acid (promote inversion of T to A), were known to trigger somatic mutations that again increased the risk of HCC progression [72].

7 Checkpoint Targets of Hepatocellular Carcinoma

Hepatocellular carcinoma pathogenesis is triggered by several mechanistic pathways that involve numerous checkpoints and can be explored as targeted therapies in HCC. The following checkpoints were considered to play a pivotal role in HCC.

7.1 *Wnt–β-Catenin Signalling*

7.1.1 Ctnnb1

A CTNNB1-related active mutation is a major canonical component of the Wnt signalling pathway and is exhibited in nearly 11–41% of patients with liver cancer [74–76]. CTNNB1 is actively involved in synthesizing actin cytoskeleton responsible for halting cell division [77]. Indeed, mutations of CTNNB1 were reported to be significantly correlated with TERT promoter, NFE2L2, MLL2, ARID2, and APOB [78, 79]. Studies related to human HCC found that CTNNB1 mutations concurrently occurred with the upregulation of Met, Myc, or Nrf2 [79–81]. Drugs, including sorafenib and gamma-secretase inhibitors, were also studied as effective targets indulging the CTNNB1 mechanism [82, 83].

7.1.2 Adenomatous Polyposis Coli (APC)

Human APC mutations originated within the central core region of the open reading frame (ORF), commonly known as MCR (mutation cluster region), that produces truncated proteins [83, 84]. Moreover, this event triggers the loss of several factors, including β-catenin binding sites (20R), nuclear localization sequences (NLS), axin binding sites (ABS), and C-terminal basic domain (CTBD), which are responsible for cytoskeletal interfaces. Sporadic APC mutations are considered to be the contributory factor for tumorigenesis. Mutations in APC significantly modulate the Wnt–β-catenin signalling, which in turn initiates the origin and progression of HCC.

7.1.3 AXIN1

AXIN1 mutations were found to be associated with nearly 5–19% of patients with liver cancer [53, 54]. AXIN1 negatively regulates the Wnt/β-catenin signalling by modulating the expression of β-catenin [85]. A study found that upregulated expression of wild-type AXIN1 intimidated the cellular proliferation in HCC along with induction of programmed cell death and thereby can be used as a molecular target to treat HCC [85]. In continuation with this study, another author used adenovirus-mediated gene transfer of AXIN1 and initiated HCC cell apoptosis [86]. AXIN was found to be an inhibitor of tankyrase 1 and 2 through XAV 939 and hence can be used as a novel therapeutic target within Wnt signalling [87].

7.2 Telomere Maintenance

7.2.1 TERT

TERT promoter mutations were known to be associated with the pathogenesis of HCC. Previous studies reported that the TERT promoter showed mutation at the upstream of ATG translation start site at positions −124 (G > A) and −146 (G > A) [88, 89]. Mutations in TERT promoter sequences produce a de novo consensus binding region for the ETS (E-twenty-six) transcription factor that further triggers the increased production of TERT proteins that attenuate the telomerase activity and length [90, 91]. A recent study reported mutation of the TERT promoter in HCC patients at −297 (C > T) upstream of the ATG translational region, generating an AP2 consensus sequence [92]. It has been found that the protein expressed by the RB/E2F gene regulates the activity of the TERT promoter and contributes to liver cancer [93]. A past study also showed that TERT gene activation was triggered by the binding of RNA-binding fox-1 homolog 3 (RBFOX3) with AP2β, which activates telomerase and promotes HCC [94]. Another study found that SP1 and YAP1 activate the TERT gene expression in the HepG2 cell line [95].

7.3 Cell Cycle Regulation

7.3.1 TP53

Nearly 13–48% of patients with liver cancer exhibit TP53 mutations [74, 75]. TP53 gene suppresses the tumours by arresting the growth and apoptosis of cancerous cells [74]. A previously published study from West demonstrated that mutations in the TP53 gene, especially in patients with HCC, are associated with poor health outcomes and prognosis [75]. Another study found that non-inflamed tumors exhibit T-cell exclusion mediated either through TP53 gene mutations or an intermediate class [96, 97]. Authors of another study concluded that TP53 mutations, especially hot spot mutations at R249S and V157F, were associated with poor outcomes and prognosis of patients with HCC [98].

7.3.2 Tumour Necrosis Factor-Related Apoptosis-Inducing Ligand (TRAIL)/DR4/DR5.

TRAIL receptor 2/DR5 is a member of the TNF receptor family and is associated with chromosome 8p21-22. The study has reported mutations of TRAIL-R2 in cancer [99]. A similar study detected single point mutation in the DR5 domain among 1% of HCC patients, suggesting its importance in the pathogenesis of carcinogenesis [99]. TRAIL and IER3 proteins were found to trigger the inhibition of Wnt/β-catenin signalling [100]. The study suggested that the TRAIL/IER3/β-catenin axis plays an important role in HCC and can be explored as a checkpoint or therapeutic target against HCC [100]. DR4/5 clustering and oligomerization mediated by TRAIL protein recruit several adaptor factors to generate a death-inducing signalling complex (DISC) that, in turn, further activates the caspases-8 and 10 within this complex along with TRADD and RIP kinases [101–105].

7.3.3 CDKN2A, CCND1, FGF3, FGF4 or FGF19

Nearly 8% of the HCC cases seem to exhibit CDKN2A deletion mutation [78]. It was known that CDKN2A is also a tumour suppressor gene that triggers the arresting of the cell cycle at the G1 and G2 phases and can act as a potential checkpoint for HCC therapy. Moreover, it also inhibits the expression of CDK4/6 and MDM2, which are responsible for oncogenic action [106]. A previous study reported that loss of CDKN2A in patients with HCC attenuates the rise of CDK4/6 inhibitors in the advanced stage [107]. In liver cancer, it was found that nearly 5–7 and 4–6% of the patients exhibit mutations of CCND1 and FGF3, FGF4, or FGF19, respectively [79]. It was studied that augmentation of CCND1, FGF3, FGF4, or FGF19 in patients with resected HCC is associated with poor prognosis and outcome [57]. A plausible study showed suppression of 11q13.3 amplicon by anti-FGF19 antibody along with anti-sense RNA mediated knockdown of FGF19 or CCND1 [108].

8 Oxidative Stress

Hepatocytes exhibit numerous fatty acids that trigger oxidative stress along with endoplasmic reticulum (ER) stress. Furthermore, these stresses cause cellular damage and inflammation [109]. One of the animal studies found that ER stress can cause NASH-triggered HCC due to the activation of several pathways, including NF- κB and TNF [110]. One of the previously published studies suggested that ER stress is mediated through the activation of hepatosteatosis and secondly due to the promotion of SREBP1 activation and increasing the process of lipogenesis [90]. In association with steatosis, ER stress generates ROS in hepatocytes, which are the primary cause of oxidative stress and oncogenic mutations. These ROS trigger the lipotoxic death of hepatocytes and thereby activate the macrophages. Further release of TNF-α also triggers the activation of chemokines and growth factors that attenuate the inflammatory microenvironment of hepatocytes [110]. In addition, ROS production induces DNA damage due to mitochondrial dysfunction and hence contributes to the pathophysiology of HCC in humans [111]. The previously published study found that mTORC2 activations within the hepatocytes trigger the concentration of sphingolipid glucosylceramide, thereby increasing ROS generation and leading to HCC [111]. Impaired cholesterol metabolism also triggers the pathophysiology of HCC [94]. A clinical study demonstrated the trend of HCC in patients and found that NASH posed a higher risk for HCC pathogenesis than NAFLD [111].

9 Conclusion

Despite several drugs in the pharmaceutical market, HCC is a highly uncontrollable cancer that tends to metastasize to distant organs, including the lungs and stomach. Moreover, the gap between the etiology and genetic mutations contributes to poor treatment outcomes. The current boom in nanotechnology can provide new hope for the early intervention and treatment of HCC without any associated side effects, as in the case of drugs. Nanotechnology offers alterations to several nanoparticles, including 2D nanomaterial that have been widely implicated in biomedical research related to cancer therapeutics. 2D nanomaterial improves the accessibility of drugs to human cells and increases their metabolic tendency along with delayed and prolonged therapeutic action. Its modified surface area offers greater drug loading and mitigates the side effects of drugs. Enhanced penetration and retention mechanisms and active targeting exhibit highly specific targeted anticancer therapeutics. Due to their low or negligible toxicity, enhanced biocompatibility, and biodegradability, anticancer 2D nanomaterial have also been the research focus.

References

1. Global Burden of Disease Cancer Collaboration., Fitzmaurice, C., Allen, C., et al.: Global, regional, and national cancer incidence, mortality, years of life lost, years lived with disability, and disability-adjusted life-years for 32 cancer groups, 1990 to 2015: a systematic analysis for the global burden of disease study. JAMA Oncol. **3**(4):524–548 (2017)
2. Llovet, J.M., Kelley, R.K., Villanueva, A., et al.: Hepatocellular carcinoma. Nat. Rev. Dis. Primers. **7**(1), 6 (2021)

3. Yang, J., Pan, G., Guan, L., et al.: The burden of primary liver cancer caused by specific etiologies from 1990 to 2019 at the global, regional, and national levels. Cancer Med. **5**, 1357–1370 (2022)
4. Global Burden of Disease Liver Cancer Collaboration., Akinyemiju, T., Abera, S., et al.: The burden of primary liver cancer and underlying etiologies from 1990 to 2015 at the global, regional, and national level: results from the global burden of disease study 2015. JAMA Oncol. **3**(12), 1683–1691 (2017)
5. Kanwal, F., Kramer, J., Asch, S.M., et al.: Risk of hepatocellular cancer in HCV patients treated with direct-acting antiviral agents. Gastroenterology **153**(4), 996-1005.e1 (2017)
6. Estes, C., Razavi, H., Loomba, R., et al.: Modeling the epidemic of nonalcoholic fatty liver disease demonstrates an exponential increase in burden of disease. Hepatology **67**(1), 123–133 (2018)
7. Nishida, N.: Metabolic disease as a risk of hepatocellular carcinoma. Clin. Mol. Hepatol. **27**(1), 87–90 (2021)
8. Ascha, M.S., Hanouneh, I.A., Lopez, R., et al.: The incidence and risk factors of hepatocellular carcinoma in patients with nonalcoholic steatohepatitis. Hepatology **51**(6), 1972–1978 (2010)
9. Zhao, L.: The gut microbiota and obesity: from correlation to causality. Nat. Rev. Microbiol. **11**(9), 639–647 (2013)
10. WHO. data visualization tools for exploring the global cancer burden in 2020. https://gco.iarc.fr/today/. (2020). Accessed 20 May 2023
11. Kulik, L., El-Serag, H.B.: Epidemiology and management of hepatocellular carcinoma. Gastroenterology **156**(2), 477-491.e1 (2015)
12. Petrick, J.L., Florio, A.A., Znaor, A., et al.: International trends in hepatocellular carcinoma incidence, 1978–2012. Int. J. Cancer **147**(2), 317–330 (2020)
13. Zhang, C.H., Cheng, Y., Zhang, S., et al.: Changing epidemiology of hepatocellular carcinoma in Asia. Liver Int. **42**, 2029–2041 (2022)
14. Sung, H., Ferlay, J., Siegel, R.L., et al.: Global cancer statistics 2020: GLOBOCAN estimates of incidence and mortality worldwide for 36 cancers in 185 countries. CA Cancer J. Clin. **71**, 209–249 (2021)
15. Ferlay, J., Colombet, M., Soerjomataram, I., et al.: Cancer statistics for the year 2020: an overview. Int. J. Cancer **149**, 778–789 (2021)
16. Tellapuri, S., Sutphin, P.D., Beg, M.S., et al.: Staging systems of hepatocellular carcinoma: a review. Indian J. Gastroenterol. **37**(6), 481–491 (2018)
17. Liu, L., Chen, H., Wang, M., et al.: Combination therapy of sorafenib and TACE for unresectable HCC: a systematic review and meta-analysis. PLoS ONE **9**(3), e91124 (2014)
18. Chidambaranathan-Reghupaty, S., Fisher, P.B., Sarkar, D.: Hepatocellular carcinoma (HCC): epidemiology, etiology and molecular classification. Adv. Cancer Res. **149**, 1–61 (2021)
19. Huang, A., Yang, X.R., Chung, W.Y., et al.: Targeted therapy for hepatocellular carcinoma. Signal Transduct. Target. Ther. **5**(1), 146 (2020)
20. Li, M., Zhang, W., Wang, B., et al.: Ligand-based targeted therapy: a novel strategy for hepatocellular carcinoma. Int. J. Nanomed. **11**, 5645–5669 (2016)
21. Zeng, M., Xiao, Y., Liu, J., et al.: Exploring two-dimensional materials toward the next-generation circuits: from monomer design to assembly control. Chem. Rev. **118**(13), 6236–6296 (2018)
22. Tao, W., Zhu, X., Yu, X., et al.: Black phosphorus nanosheets as a robust delivery platform for cancer theranostics. Adv Mater. (1) (2017). https://doi.org/10.1002/adma.201603276
23. Lin, H., Chen, Y., Shi, J.: Insights into 2D MXenes for versatile biomedical applications: current advances and challenges ahead. Adv. Sci. (Weinh). **5**(10) (2018)
24. Global Burden of Disease Cancer Collaboration; Fitzmaurice, C., Allen, C., et al.: Global, regional, and national cancer incidence, mortality, years of life lost, years lived with disability, and disability-adjusted life-years for 32 cancer groups, 1990 to 2015: a systematic analysis for the global burden of disease study. JAMA Oncol. **3**(4), 524–548 (2017)

25. International Agency for Research on Cancer. GLOBOCAN 2018. IARC. https://gco.iarc.fr/today/online-analysis-map?v=2020&mode=population&mode_population=continents&population=900&populations=900&key=asr&sex=0&cancer=11&type=0&statistic=5&prevalence=0&population_groupearth&color_palette=default&map_scale=quantile&map_nb_colors=5&continent=0&rotate=%255B10%252C0%255D. Accessed 03 March 2023
26. Global Burden of Disease Liver Cancer Collaboration, Akinyemiju, T., Abera, S., et al.: The burden of primary liver cancer and underlying etiologies from 1990 to 2015 at the global, regional, and national level: results from the global burden of disease study 2015. JAMA Oncol. **3**(12), 1683–1691 (2017)
27. Ascha, M.S., Hanouneh, I.A., Lopez, R., et al.: The incidence and risk factors of hepatocellular carcinoma in patients with nonalcoholic steatohepatitis. Hepatology. **51**(6), 1972–8 (2010)
28. WHO. Data Visualization Tools for Exploring the Global Cancer Burden in 2020. https://gco.iarc.fr/today/ (2020). Accessed 20 Aug 2023
29. Kulik, L., El-Serag, H.B.: Epidemiology and management of hepatocellular carcinoma. Gastroenterology **156**(2), 477–491 (2019)
30. Sung, H., Ferlay, J., Siegel, R.L., et al.: Global Cancer Statistics 2020: GLOBOCAN estimates of incidence and mortality worldwide for 36 cancers in 185 countries. CA Cancer J. Clin. **71**, 209–49 (2021)
31. Liu, L., Chen, H., Wang, M., et al.: Combination therapy of sorafenib and TACE for unresectable HCC: a systematic review and meta-analysis. PLoS One **20**, 9(3) (2014)
32. Llovet, J.M., Kelley, R.K., Villanueva, A., et al.:. Hepatocellular carcinoma. Nat. Rev. Dis. Primers. **7**(1), 6 (2021). https://doi.org/10.1038/s41572-020-00240-3
33. Sirohi, B., Shrikhande, S.V., Gaikwad, V., et al.: Indian Council of Medical Research consensus document on hepatocellular carcinoma. Indian J. Med. Res. **152**, 468–474 (2020)
34. Kulkarni, A.V., Fatima, S., Sharma, M., et al.: Lenvatinib for unresectable hepatocellular carcinoma: the frst Indian experience. GastroHep. **3**, 407–408 (2021)
35. Kumar, M., Kumar, R., Hissar, S.S., et al.: Risk factors analysis for hepatocellular carcinoma in patients with and without cirrhosis: a case-control study of 213 hepatocellular carcinoma patients from India. J. Gastroenterol. Hepatol. **22**, 1104–1111 (2007)
36. Paul, S.B., Chalamalasetty, S.B., Vishnubhatla, S., et al.: Clinical profle, etiology and therapeutic outcome in 324 hepatocellular carcinoma patients at a tertiary care center in India. Oncology **77**, 162–171 (2009)
37. Sood, A., Midha, V., Goyal, O., et al.: Profle of hepatocellular carcinoma in a tertiary care hospital in Punjab in Northern India. Indian J. Gastroenterol. **33**, 35–40 (2014)
38. David, D., Raghavendran, A., Goel, A., et al.: Risk factors for nonalcoholic fatty liver disease are common in patients with non-B non-C hepatocellular carcinoma in India. Indian J. Gastroenterol. **36**, 373–379 (2017)
39. Kumar, A., Acharya, S.K., Singh, S.P., et al.: Update of Indian national association for study of the liver consensus on prevention, diagnosis, and management of hepatocellular carcinoma in India: the Puri Ii recommendations. J. Clin. Exp. Hepatol. **2020**, 43–80 (2019)
40. Koshy, A., Devadas, K., Panackel, C., et al.: Multi-center prospective survey of hepatocellular carcinoma in Kerala: More than 1,200 cases. Indian J. Gastroenterol. **42** (2023)
41. Krishan, S., Dhiman, R.K., Kalra, N., et al.: Joint consensus statement of the Indian National association for study of the liver and Indian radiological and imaging association for the diagnosis and imaging of hepatocellular carcinoma incorporating liver imaging reporting and data system. J. Clin. Exp. Hepatol. **9**, 625–651 (2019)
42. Tohra, S., Duseja, A., Taneja, S., et al.: Experience with changing etiology and nontransplant curative treatment modalities for hepatocellular carcinoma in a real-life setting-a retrospective descriptive analysis. J. Clin. Exp. Hepatol. **11**, 682–690 (2021)
43. Li, M., Zhang, W., Wang, B., et al.: Ligand-based targeted therapy: a novel strategy for hepatocellular carcinoma. Int. J. Nanomed. **11**, 5645–5669 (2016). https://doi.org/10.2147/IJN.S115727
44. Llovet, J.M., Castet, F., Heikenwalder, M., et al.: Immunotherapies for hepatocellular carcinoma. Nat. Rev. Clin. Oncol. **19**(3), 151–172 (2022). https://doi.org/10.1038/s41571-021-00573-2

45. Sia, D., Villanueva, A., Friedman, SL., et al.: Liver cancer cell of origin, molecular class, and effects on patient prognosis. Gastroenterology **152**(4), 745–761 (2017). https://doi.org/10.1053/j.gastro.2016.11.048
46. Zhu, W., Li, H., Luo, P.: Emerging 2D nanomaterials for multimodel theranostics of cancer. Front Bioeng Biotechnol. **9**, 769178 (2021)
47. Nafiujjaman, M., Nurunnabi, M., Lee, Y.K.: Photolumiscence high quantum yield graphene quantum dots for imaging guided photodynamic therapy. Drug Metab. Pharmacokinet. **1**(32), S46 (2017)
48. Nurunnabi, M., Nafiujjaman, M., Lee, S.J., et al.: Preparation of ultra-thin hexagonal boron nitride nanoplates for cancer cell imaging and neurotransmitter sensing. Chem. Commun. (Camb.) **52**(36), 6146–6149 (2016)
49. Khatun, Z., Nurunnabi, M., Nafiujjaman, M., et al.: A hyaluronic acid nanogel for photo-chemo theranostics of lung cancer with simultaneous light-responsive controlled release of doxorubicin. Nanoscale **7**(24), 10680–10689 (2015)
50. Ashfaq, M., Talreja, N, Chauhan, D., et al.: Two-dimensional (2D) hybrid nanomaterials for diagnosis and treatment of Cancer. J. Drug Deliv. Sci. Technol. **70**, 103268 (2022)
51. Wang, S.B., Ma, Y.Y., Chen, X.Y., et al.: Ceramide-graphene oxide nanoparticles enhance cytotoxicity and decrease HCC xenograft development: a novel approach for targeted cancer therapy. Front. Pharmacol. **10**, 69 (2019)
52. Sosnowska, M., Kutwin, M., Strojny, B., et al.: Graphene oxide nanofilm and chicken embryo extract decrease the invasiveness of HepG2 liver cancer cells. Cancer Nano **12**, 2 (2021)
53. Wang, K., Ruan, J., Song, H., et al.: Biocompatibility of graphene oxide. Nanoscale Res. Lett. **6**(8), 1–8 (2011)
54. Wang, Y., Shenouda, S., Baranwal, S., et al.: Integrin subunits alpha5 and alpha6 regulate cell cycle by modulating the chk1 and Rb/E2F pathways to affect breast cancer metastasis. Mol. Cancer **10**(1), 1–11 (2011)
55. Ouyang, R., Cao, P., Jia, P., et al.: Bistratal Au@Bi_2S_3 nanobones for excellent NIR-triggered/multimodal imaging-guided synergistic therapy for liver cancer. Bioact Mater. **6**(2), 386–403 (2020)
56. Zhao, W., Li, J., Zhong, C., et al.: Green synthesis of gold nanoparticles from Dendrobium officinale and its anticancer effect on liver cancer. Drug Deliv. **28**(1), 985–994 (2021)
57. Li, B., Fu, Y., Xie, M., et al.: Gold-based nanoparticles realize photothermal and photodynamic synergistic treatment of liver cancer and improve the anaerobic tumor microenvironment under near-infrared light. Front Bioeng Biotechnol. **10**, 957349 (2022)
58. Yadong, Z., Kandasamy, V., Fafu, D., et al.: Hyper-branched multifunctional carbon nanotubes carrier for targeted liver cancer therapy. Arab. J. Chem. **15**(3), 103649 (2022)
59. Iancu, C., Mocan, L., Bele, C., et al.: Enhanced laser thermal ablation for the in vitro treatment of liver cancer by specific delivery of multiwalled carbon nanotubes functionalized with human serum albumin. Int. J. Nanomedicine **6**, 129–141 (2011)
60. Das, T., Chakraborty, S., Sarma, H.D., et al.: 166Ho-labeled hydroxyapatite particles: a possible agent for liver cancer therapy. Cancer Biother. Radiopharm. **24**(1), 7–14 (2009)
61. Awwad, N.S., Saleh, K.A., Abbas, H.A., et al.: Induction apoptosis in liver cancer cells by altering natural hydroxyapatite to scavenge excess sodium without deactivate sodium-potassium pump. Mater. Res. Express **6**, 055403 (2019)
62. Ying, X., Yihan, K., Jiawen, X., et al.: Doxorubicin intercalated copper diethyldithiocarbamate functionalized layered double hydroxide hybrid nanoparticles for targeted therapy of hepatocellular carcinoma. Biomater. Sci. **8**, 897–911 (2020)
63. Lu, Y.F., Zhou, J.P., Zhou, Q.M., et al.: Ultra-thin layered double hydroxide-mediated photothermal therapy combine with asynchronous blockade of PD-L1 and NR2F6 inhibit hepatocellular carcinoma. J. Nanobiotechnol. **20**, 351 (2022)
64. Liao, K.H., Lin, Y.S., Macosko, C.W., Haynes, C.L.: Cytotoxicity of graphene oxide and graphene in human erythrocytes and skin fibroblasts. ACS Appl. Mater. Interfaces **3**(7), 2607–2615 (2011). https://doi.org/10.1021/am200428v

65. Jaworski, S., Sawosz, E., Kutwin, M., Wierzbicki, M., Hinzmann, M., Grodzik, M., Winnicka, A., Lipińska, L., Włodyga, K., Chwalibog, A.: In vitro and in vivo effects of graphene oxide and reduced graphene oxide on glioblastoma. Int. J. Nanomed. **25**(10), 1585–1596 (2015). https://doi.org/10.2147/IJN.S77591
66. Zhang, Y., Ali, S.F., Dervishi, E., Xu, Y., Li, Z., Casciano, D., Biris, A.S.: Cytotoxicity effects of graphene and single-wall carbon nanotubes in neural phaeochromocytoma-derived PC12 cells. ACS Nano **4**(6), 3181–3186 (2010). https://doi.org/10.1021/nn1007176
67. Teo, W.Z., Chng, E.L., Sofer, Z., Pumera, M.: Cytotoxicity of exfoliated transition-metal dichalcogenides (MoS2, WS2, and WSe2) is lower than that of graphene and its analogues. Chemistry **20**(31), 9627–9632 (2014). https://doi.org/10.1002/chem.201402680
68. Shafizadeh, N., Kakar, S.: Diagnosis of well-differentiated hepatocellular lesions: role of immunohistochemistry and other ancillary techniques. Adv Anat Pathol **18**(6), 438–445 (2011). https://doi.org/10.1097/PAP.0b013e318234abb4
69. Michalopoulos, G.K., Bhushan, B.: Liver regeneration: biological and pathological mechanisms and implications. Nat. Rev. Gastroenterol. Hepatol. **18**, 40–55 (2021)
70. Bruno, S., Herrera Sanchez, M.B., Chiabotto, G., Fonsato, V., Navarro-Tableros, V., Pasquino, C., Tapparo, M., Camussi, G.: Human liver stem cells: a liver-derived mesenchymal stromal cell-like population with pro-regenerative properties. Front Cell Dev Biol. **26**(9), 644088 (2021). https://doi.org/10.3389/fcell.2021.644088
71. Bayard, Q., Meunier, L., Peneau, C., Renault, V., Shinde, J., Nault, J.C., Mami, I., Couchy, G., Amaddeo, G., Tubacher, E., Bacq, D., Meyer, V., La Bella, T., Debaillon-Vesque, A., Bioulac-Sage, P., Seror, O., Blanc, J.F., Calderaro, J., Deleuze, J.F., Imbeaud, S., Zucman-Rossi, J., Letouzé, E.: Cyclin A2/E1 activation defines a hepatocellular carcinoma subclass with a rearrangement signature of replication stress. Nat. Commun. **9**(1), 5235 (2018). https://doi.org/10.1038/s41467-018-07552-9
72. Pfister, D., Núñez, N.G., Pinyol, R., Govaere, O., Pinter, M., Szydlowska, M., Gupta, R., Qiu, M., Deczkowska, A., Weiner, A., Müller, F., Sinha, A., Friebel, E., Engleitner, T., Lenggenhager, D., Moncsek, A., Heide, D., Stirm, K., Kosla, J., Kotsiliti, E., Leone, V., Dudek, M., Yousuf, S., Inverso, D., Singh, I., Teijeiro, A., Castet, F., Montironi, C., Haber, P.K., Tiniakos, D., Bedossa, P., Cockell, S., Younes, R., Vacca, M., Marra, F., Schattenberg, J.M., Allison, M., Bugianesi, E., Ratziu, V., Pressiani, T., D'Alessio, A., Personeni, N., Rimassa, L., Daly, A.K., Scheiner, B., Pomej, K., Kirstein, M.M., Vogel, A., Peck-Radosavljevic, M., Hucke, F., Finkelmeier, F., Waidmann, O., Trojan, J., Schulze, K., Wege, H., Koch, S., Weinmann, A., Bueter, M., Rössler, F., Siebenhüner, A., De Dosso, S., Mallm, J.P., Umansky, V., Jugold, M., Luedde, T., Schietinger, A., Schirmacher, P., Emu, B., Augustin, H.G., Billeter, A., Müller-Stich, B., Kikuchi, H., Duda, D.G., Kütting, F., Waldschmidt. D.T., Ebert, M.P., Rahbari, N., Mei, H.E., Schulz, A.R., Ringelhan, M., Malek, N., Spahn, S., Bitzer, M., Ruiz de Galarreta, M., Lujambio, A., Dufour, J.F., Marron, T.U., Kaseb, A., Kudo, M., Huang, Y.H., Djouder, N., Wolter, K., Zender, L., Marche, P.N., Decaens, T., Pinato, D.J., Rad, R., Mertens, J.C., Weber, A., Unger, K., Meissner, F., Roth, S., Jilkova, Z.M., Claassen, M., Anstee, Q.M., Amit, I., Knolle, P., Becher, B., Llovet, J.M., Heikenwalder, M.: NASH limits anti-tumour surveillance in immunotherapy-treated HCC. Nature **592**(7854), 450–456 2021. https://doi.org/10.1038/s41586-021-03362-0
73. Gellert-Kristensen, H., Richardson, T.G., Davey Smith, G., Nordestgaard, B.G., Tybjaerg-Hansen, A., Stender, S.: Combined effect of PNPLA3, TM6SF2, and HSD17B13 variants on risk of cirrhosis and hepatocellular carcinoma in the general population. Hepatology **72**(3), 845–856 (2020). https://doi.org/10.1002/hep.31238
74. Yang, J.D., Hainaut, P., Gores, G.J., Amadou, A., Plymoth, A., Roberts, L.R.: A global view of hepatocellular carcinoma: trends, risk, prevention and management. Nat. Rev. Gastroenterol. Hepatol. **16**(10), 589–604 (2019). https://doi.org/10.1038/s41575-019-0186-y
75. Nault, J.C., Letouzé, E.: Mutational processes in hepatocellular carcinoma: the story of aristolochic acid. Semin. Liver Dis. **39**(3), 334–340 (2019). https://doi.org/10.1055/s-0039-1685516

76. Guichard, C., Amaddeo, G., Imbeaud, S., Ladeiro, Y., Pelletier, L., Maad, I.B., Calderaro, J., Bioulac-Sage, P., Letexier, M., Degos, F., Clément, B., Balabaud, C., Chevet, E., Laurent, A., Couchy, G., Letouzé, E., Calvo, F., Zucman-Rossi, J.: Integrated analysis of somatic mutations and focal copy-number changes identifies key genes and pathways in hepatocellular carcinoma. Nat. Genet. **44**(6), 694–698 (2012). https://doi.org/10.1038/ng.2256
77. Takai, A., Dang, H.T., Wang, X.W.: Identification of drivers from cancer genome diversity in hepatocellular carcinoma. Int. J. Mol. Sci. **15**(6), 11142–11160 (2014). https://doi.org/10.3390/ijms150611142
78. Ahn, S.M., Jang, S.J., Shim, J.H., Kim, D., Hong, S.M., Sung, C.O., Baek, D., Haq, F., Ansari, A.A., Lee, S.Y., Chun, S.M., Choi, S., Choi, H.J., Kim, J., Kim, S., Hwang, S., Lee, Y.J., Lee, J.E., Jung, W.R., Jang, H.Y., Yang, E., Sung, W.K., Lee, N.P., Mao, M., Lee, C., Zucman-Rossi, J., Yu, E., Lee, H.C., Kong, G.: Genomic portrait of resectable hepatocellular carcinomas: implications of RB1 and FGF19 aberrations for patient stratification. Hepatology **60**(6), 1972–1982 (2014). https://doi.org/10.1002/hep.27198
79. Belinky, F., Nativ, N., Stelzer, G., Zimmerman, S., Iny Stein, T., Safran, M., Lancet, D.: PathCards: multi-source consolidation of human biological pathways. Database (Oxford). (2015) bav006
80. Schulze, K., Imbeaud, S., Letouzé, E., Alexandrov, L.B., Calderaro, J., Rebouissou, S., Couchy, G., Meiller, C., Shinde, J., Soysouvanh, F., Calatayud, A.L., Pinyol, R., Pelletier, L., Balabaud, C., Laurent, A., Blanc, J.F., Mazzaferro, V., Calvo, F., Villanueva, A., Nault, J.C., Bioulac-Sage, P., Stratton, M.R., Llovet, J.M., Zucman-Rossi, J.: Exome sequencing of hepatocellular carcinomas identifies new mutational signatures and potential therapeutic targets. Nat. Genet. **47**(5), 505–511 (2015). https://doi.org/10.1038/ng.3252
81. National Cancer Institute. The Cancer Genome Atlas. http://cancergenome.nih.gov/. Accessed 04 March 20
82. Tao, J., Xu, E., Zhao, Y., Singh, S., Li, X., Couchy, G., Chen, X., Zucman-Rossi, J., Chikina, M., Monga, S.P.: Modeling a human hepatocellular carcinoma subset in mice through coexpression of met and point-mutant β-catenin. Hepatology **64**(5), 1587–1605 (2016). https://doi.org/10.1002/hep.28601
83. Ruiz de Galarreta, M., Bresnahan, E., Molina-Sánchez, P., Lindblad, KE., Maier, B., Sia, D., Puigvehi, M., Miguela, V., Casanova-Acebes, M., Dhainaut, M., Villacorta-Martin, C., Singhi, A.D., Moghe, A., von Felden, J., Tal Grinspan, L., Wang, S., Kamphorst, A.O., Monga, S.P., Brown, B.D., Villanueva, A., Llovet, J.M., Merad, M., Lujambio, A.: β-Catenin activation promotes immune escape and resistance to anti-PD-1 therapy in hepatocellular carcinoma. Cancer Discov. **9**(8), 1124–1141 (2019). https://doi.org/10.1158/2159-8290.CD-19-0074
84. Tao, J., Krutsenko, Y., Moghe, A., Singh, S., Poddar, M., Bell, A., Oertel, M., Singhi, A.D., Geller, D., Chen, X., Lujambio, A., Tan Liu, S., Monga, S.P.: Nuclear factor erythroid 2-related factor 2 and β-Catenin coactivation in hepatocellular cancer: biological and therapeutic implications. Hepatology **74**(2), 741–759 (2021). https://doi.org/10.1002/hep.31730
85. Tan, X., Behari, J., Cieply, B., Michalopoulos, G.K., Monga, S.P.: Conditional deletion of beta-catenin reveals its role in liver growth and regeneration. Gastroenterology **131**(5), 1561–1572 (2006). https://doi.org/10.1053/j.gastro.2006.08.042
86. Miyoshi, Y., Nagase, H., Ando, H., Horii, A., Ichii, S., Nakatsuru, S., Aoki, T., Miki, Y., Mori, T., Nakamura, Y.: Somatic mutations of the APC gene in colorectal tumors: mutation cluster region in the APC gene. Hum. Mol. Genet. **1**(4), 229–233 (1992). https://doi.org/10.1093/hmg/1.4.229
87. Li, J., Quan, H., Liu, Q., Si, Z., He, Z., Qi, H.: Alterations of axis inhibition protein 1 (AXIN1) in hepatitis B virus-related hepatocellular carcinoma and overexpression of AXIN1 induces apoptosis in hepatocellular cancer cells. Oncol. Res. **20**(7), 281–288 (2013). https://doi.org/10.3727/096504013x13639794277608
88. Satoh, S., Daigo, Y., Furukawa, Y., Kato, T., Miwa, N., Nishiwaki, T., Kawasoe, T., Ishiguro, H., Fujita, M., Tokino, T., Sasaki, Y., Imaoka, S., Murata, M., Shimano, T., Yamaoka, Y., Nakamura, Y.: AXIN1 mutations in hepatocellular carcinomas, and growth suppression in

cancer cells by virus-mediated transfer of AXIN1. Nat. Genet. **24**(3), 245–250 (2000). https://doi.org/10.1038/73448. PMID: 10700176
89. Huang, S.M., Mishina, Y.M., Liu, S., Cheung, A., Stegmeier, F., Michaud, G.A., Charlat, O., Wiellette, E., Zhang, Y., Wiessner, S., Hild, M., Shi, X., Wilson, C.J., Mickanin, C., Myer, V., Fazal, A., Tomlinson, R., Serluca, F., Shao, W., Cheng, H., Shultz, M., Rau, C., Schirle, M., Schlegl, J., Ghidelli, S., Fawell, S., Lu, C., Curtis, D., Kirschner, M.W., Lengauer, C., Finan, P.M., Tallarico, J.A., Bouwmeester, T., Porter, J.A., Bauer, A., Cong, F.: Tankyrase inhibition stabilizes axin and antagonizes Wnt signalling. Nature **461**(7264), 614–620 (2009). https://doi.org/10.1038/nature08356
90. Totoki, Y., Tatsuno, K., Covington, K.R., Ueda, H., Creighton, C.J., Kato, M., Tsuji, S., Donehower, L.A., Slagle, B.L., Nakamura, H., Yamamoto, S., Shinbrot, E., Hama, N., Lehmkuhl, M., Hosoda, F., Arai, Y., Walker, K., Dahdouli, M., Gotoh, K., Nagae, G., Gingras, M.C., Muzny, D.M., Ojima, H., Shimada, K., Midorikawa, Y., Goss, J.A., Cotton, R., Hayashi, A., Shibahara, J., Ishikawa, S., Guiteau, J., Tanaka, M., Urushidate, T., Ohashi, S., Okada, N., Doddapaneni, H., Wang, M., Zhu, Y., Dinh, H., Okusaka, T., Kokudo, N., Kosuge, T., Takayama, T., Fukayama, M., Gibbs, R.A., Wheeler, D.A., Aburatani, H., Shibata, T.: Trans-ancestry mutational landscape of hepatocellular carcinoma genomes. Nat. Genet. **46**(12), 1267–1273 (2014). https://doi.org/10.1038/ng.3126
91. Pezzuto, F., Izzo, F., Buonaguro, L., Annunziata, C., Tatangelo, F., Botti, G., Buonaguro, F.M., Tornesello, M.L.: Tumor specific mutations in TERT promoter and CTNNB1 gene in hepatitis B and hepatitis C related hepatocellular carcinoma. Oncotarget **7**(34), 54253–54262 (2016). https://doi.org/10.18632/oncotarget.9801
92. Huang, F.W., Hodis, E., Xu, M.J., Kryukov, G.V., Chin, L., Garraway, L.A.: Highly recurrent TERT promoter mutations in human melanoma. Science **339**(6122), 957–959 (2013). https://doi.org/10.1126/science.1229259
93. Horn, S., Figl, A., Rachakonda, P.S., Fischer, C., Sucker, A., Gast, A., Kadel, S., Moll, I., Nagore, E., Hemminki, K., Schadendorf, D., Kumar, R.: TERT promoter mutations in familial and sporadic melanoma. Science **339**(6122), 959–961 (2013). https://doi.org/10.1126/science.1230062
94. Borah, S., Xi, L., Zaug, A.J., Powell, N.M., Dancik, G.M., Cohen, S.B., Costello, J.C., Theodorescu, D., Cech, T.R.: Cancer. TERT promoter mutations and telomerase reactivation in urothelial Cancer. Science. **347**(6225), 1006–10010 (2015). https://doi.org/10.1126/science.1260200
95. Lombardo, D., Saitta, C., Giosa, D., Di Tocco, F.C., Musolino, C., Caminiti, G., Chines, V., Franzè, M.S., Alibrandi, A., Navarra, G., Raimondo, G., Pollicino, T.: Frequency of somatic mutations in TERT promoter, TP53 and CTNNB1 genes in patients with hepatocellular carcinoma from Southern Italy. Oncol. Lett. **19**(3), 2368–2374 (2020). https://doi.org/10.3892/ol.2020.11332
96. Sirma, H., Kumar, M., Meena, J.K., Witt, B., Weise, J.M., Lechel, A., Ande, S., Sakk, V., Guguen-Guillouzo, C., Zender, L., Rudolph, K.L., Günes, C.: The promoter of human telomerase reverse transcriptase is activated during liver regeneration and hepatocyte proliferation. Gastroenterology. **141**(1), 326–37, 337.e1-3 2011. https://doi.org/10.1053/j.gastro.2011.03.047
97. Liu, T., Li, W., Lu, W., Chen, M., Luo, M., Zhang, C., Li, Y., Qin, G., Shi, D., Xiao, B., Qiu, H., Yu, W., Kang, L., Kang, T., Huang, W., Yu, X., Wu, X., Deng, W.: RBFOX3 promotes tumor growth and progression via hTERT signaling and predicts a poor prognosis in hepatocellular carcinoma. Theranostics. **7**(12), 3138–3154 (2017). https://doi.org/10.7150/thno.19506
98. Esopi, D., Graham, M.K., Brosnan-Cashman, J.A., Meyers, J., Vaghasia, A., Gupta, A., Kumar, B., Haffner, M.C., Heaphy, C.M., De Marzo, A.M., Meeker, A.K., Nelson, W.G., Wheelan, S.J., Yegnasubramanian, S.: Pervasive promoter hypermethylation of silenced TERT alleles in human cancers. Cell. Oncol. (Dordr.) **43**(5), 847–861 (2020). https://doi.org/10.1007/s13402-020-00531-7
99. Montironi, C., Castet, F., Haber, P.K., Pinyol, R., Torres-Martin, M., Torrens, L., Mesropian, A., Wang, H., Puigvehi, M., Maeda, M., Leow, W.Q., Harrod, E., Taik, P., Chinburen, J.,

Taivanbaatar, E., Chinbold, E., Solé Arqués, M., Donovan, M., Thung, S., Neely, J., Mazzaferro, V., Anderson, J., Roayaie, S., Schwartz, M., Villanueva, A., Friedman, S.L., Uzilov, A., Sia, D., Llovet, J.M.: Inflamed and non-inflamed classes of HCC: a revised immunogenomic classification. Gut **72**(1), 129–140 (2023). https://doi.org/10.1136/gutjnl-2021-325918
100. Villanueva, A., Hoshida, Y.: Depicting the role of TP53 in hepatocellular carcinoma progression. J. Hepatol. **55**(3), 724–725 (2011). https://doi.org/10.1016/j.jhep.2011.03.018
101. Jeng, Y.M., Hsu, H.C.: Mutation of the DR5/TRAIL receptor 2 gene is infrequent in hepatocellular carcinoma. Cancer Lett. **181**(2), 205–208 (2002). https://doi.org/10.1016/s0304-3835(02)00051-4
102. Liu, S., Qiu, J., He, G., et al.: TRAIL promotes hepatocellular carcinoma apoptosis and inhibits proliferation and migration via interacting with IER3. Cancer Cell Int. **21**, 63 (2021). https://doi.org/10.1186/s12935-020-01724-8
103. Salmena, L., Lemmers, B., Hakem, A., Matysiak-Zablocki, E., Murakami, K., Au, P.Y., Berry, D.M., Tamblyn, L., Shehabeldin, A., Migon, E., Wakeham, A., Bouchard, D., Yeh, W.C., McGlade, J.C., Ohashi, P.S., Hakem, R.: Essential role for caspase 8 in T-cell homeostasis and T-cell-mediated immunity. Genes Dev. **17**(7), 883–895 (2003). https://doi.org/10.1101/gad.1063703
104. Kim, I.K., Chung, C.W., Woo, H.N., Hong, G.S., Nagata, S., Jung, Y.K.: Reconstitution of caspase-8 sensitizes JB6 cells to TRAIL. Biochem. Biophys. Res. Commun. **277**(2), 311–316 (2000). https://doi.org/10.1006/bbrc.2000.3673
105. Kischkel, F.C., Lawrence, D.A., Chuntharapai, A., Schow, P., Kim, K.J., Ashkenazi, A.: Apo2L/TRAIL-dependent recruitment of endogenous FADD and caspase-8 to death receptors 4 and 5. Immunity **12**(6), 611–620 (2000). https://doi.org/10.1016/s1074-7613(00)80212-5
106. Lin, Y., Devin, A., Cook, A., Keane, M.M., Kelliher, M., Lipkowitz, S., Liu, Z.G.: The death domain kinase RIP is essential for TRAIL (Apo2L)-induced activation of IkappaB kinase and c-Jun N-terminal kinase. Mol. Cell Biol. **20**(18), 6638–6645 (2000). https://doi.org/10.1128/MCB.20.18.6638-6645.2000
107. Schneider, P., Thome, M., Burns, K., Bodmer, J.L., Hofmann, K., Kataoka, T., Holler, N., Tschopp, J.: TRAIL receptors 1 (DR4) and 2 (DR5) signal FADD-dependent apoptosis and activate NF-kappaB. Immunity **7**(6), 831–836 (1997). https://doi.org/10.1016/s1074-7613(00)80401-x
108. Knudsen, E.S., Gopal, P., Singal, A.G.: The changing landscape of hepatocellular carcinoma: etiology, genetics, and therapy. Am. J. Pathol. **184**(3), 574–583 (2014). https://doi.org/10.1016/j.ajpath.2013.10.028
109. Llovet, J.M., Villanueva, A., Lachenmayer A, Finn, R.S.: Advances in targeted therapies for hepatocellular carcinoma in the genomic era. Nat. Rev. Clin. Oncol. **12**(8), 436 (2015). https://doi.org/10.1038/nrclinonc.2015.121. Epub 2015 Jun 23. Erratum for: Nat. Rev. Clin. Oncol. (2015) **12**(7), 408 24
110. Sawey, E.T., Chanrion, M., Cai, C., Wu, G., Zhang, J., Zender, L., Zhao, A., Busuttil, R.W., Yee, H., Stein, L., French, D.M., Finn, R.S., Lowe, S.W., Powers, S.: Identification of a therapeutic strategy targeting amplified FGF19 in liver cancer by Oncogenomic screening. Cancer Cell **19**(3), 347–358 (2011). https://doi.org/10.1016/j.ccr.2011.01.040
111. Anstee, Q.M., Reeves, H.L., Kotsiliti, E., Govaere, O., Heikenwalder, M.: From NASH to HCC: current concepts and future challenges. Nat. Rev. Gastroenterol. Hepatol. **16**(7), 411–428 (2019). https://doi.org/10.1038/s41575-019-0145-7

Potential of 2D Materials: Novel Insights and Applications in Colorectal Cancer Research

Mohd Younis, Sanjeev Kumar Digra, Ashma Gupta, Vijaya Anand, and Parvinder Kumar

Abstract Colorectal cancer (CRC) is a prevalent and potentially fatal disease worldwide. Despite advances in conventional cancer research, there is a need for innovative approaches to improve early detection, targeted therapy, and patient outcomes. Two-dimensional (2D) materials have become a potential area in cancer research due to their distinctive features and wide range of uses. This comprehensive article explores the insights and applications of 2D materials in CRC research. We discuss the properties and synthesis methods of prominent 2D materials, together with graphene, transition metal dichalcogenides (TMDs), and black phosphorus (BP). Furthermore, we explore the use of 2D materials in tissue engineering, drug administration, imaging, and CRC biosensing. Specifically, we examine their potential in detecting CRC biomarkers, delivering therapeutics with enhanced precision, improving imaging modalities for accurate diagnosis and monitoring, and facilitating tissue regeneration for effective treatment. We also address the challenges and future directions in the integration of 2D materials in CRC research, including biocompatibility, toxicity concerns, and translational hurdles. This article highlights the promising applications of 2D materials in CRC research and emphasizes their potential to revolutionize

M. Younis (✉) · P. Kumar (✉)
Institution of Human Genetics, University of Jammu, Jammu, Jammu & Kashmir, India
e-mail: younisgenetic@gmail.com

P. Kumar
e-mail: parvinderkb2003@gmail.com

S. Kumar Digra
Department of Pediatrics, SMGS, Hospital, Govt. Medical College, Jammu, Jammu & Kashmir, India

A. Gupta
Department of Zoology, Govt. College for Women, Gandhi Nagar, Jammu, Jammu & Kashmir, India

M. Younis · V. Anand
Department of Human Genetics and Molecular Biology, Bharathiar University, Chennai, Tamil Nadu, India

M. Younis · P. Kumar
Department of Zoology, University of Jammu, Jammu, Jammu & Kashmir, India

N. Talreja et al. (eds.), *Two-dimensional Hybrid Composites*, Engineering Materials,
https://doi.org/10.1007/978-981-99-8010-9_13

cancer diagnostics, therapeutics, and personalized medicine, ultimately improving outcomes for CRC patients.

Keywords 2D materials · Colorectal cancer · Biosensing · Drug delivery · Imaging · Tissue engineering

1 Background on Colorectal Cancer

In terms of cancer-related morbidity and mortality, colorectal cancer (CRC) is a serious worldwide health concern [1]. It is brought on by the unchecked expansion of aberrant cells in the colon or rectum and is impacted by a complex combination of genetic, environmental, and lifestyle factors [2, 3]. Despite improvements in screening programmes and treatment choices, CRC is a primary cause of cancer-related mortality, highlighting the need for ongoing research and innovation in this sector. Recent years have seen a substantial increase in interest in the developing field of two-dimensional (2D) materials, particularly in the fields of biomedicine and cancer research [4]. 2D materials, which are atomically thin sheets, have a high surface-to-volume ratio, excellent mechanical strength, and exceptional electrical and optical properties [5]. These characteristics make them intriguing candidates for applications in CRC research, offering novel insights and potential solutions to address the challenges associated with early detection, targeted therapy, and personalized medicine.

The significance of research on 2D materials in CRC lies in its potential to transform the landscape of CRC diagnostics and treatment [6]. By harnessing the distinctive properties of 2D materials, researchers can develop innovative approaches that improve the sensitivity and specificity of CRC detection, enhance targeted drug delivery, and advance imaging modalities for accurate diagnosis and monitoring [7]. Furthermore, 2D materials have the potential for tissue engineering applications due to their capacity to generate biomimetic circumstances that aid in investigating CRC evolution and creating personalized treatment modalities [8]. The objective of this article is to comprehensively review the current state of research on 2D materials in the context of CRC. Our aim is to shed light on the characteristics and fabrication processes of well-known 2D materials, including graphene, transition metal dichalcogenides (TMDs), and black phosphorus (BP). Additionally, we will examine their uses in biosensing, drug administration, imaging, and tissue engineering in CRC research.

This article's scope thoroughly examines the possibilities of 2D materials in CRC research, emphasizing the fresh perspectives and uses they provide. We will go over the difficulties and restrictions that come with using 2D materials in CRC research, including issues with biocompatibility, toxicity, and translational barriers. Moreover, we will address the future directions and opportunities for further exploration in this rapidly evolving field. This article intends to stimulate additional study and collaboration by highlighting the potential of 2D materials in CRC research, ultimately opening the door for the creation of novel strategies and personalized interventions that can enhance patient outcomes in the management of CRC.

2 Overview of 2D Materials

2D materials are a class of nanomaterials that have thicknesses in the nanometer range while possessing an extensive lateral dimension, often extending to micrometres or more. Unlike traditional bulk materials with three dimensions, 2D materials are atomically thin and exhibit unique properties due to their reduced dimensionality [9]. Some of the most prominent 2D materials include graphene, TMDs, hexagonal boron nitride, and BP. They are made up of one or more layers of atoms organized in a two-dimensional lattice, and they display special qualities that set them apart from their bulk counterparts [10]. The medicinal, electronics, and energy applications are just a few of the scientific and technological disciplines that have taken notice of the extraordinary capabilities of 2D materials.

2.1 Properties of 2D Materials:

Their atomic thickness determines the length and width of 2D materials, and they are macroscopic in size [11]. They have a high surface-to-volume ratio because of their reduced dimensionality, which results in excellent physical and chemical properties [12]. Some key properties of 2D materials include: Mechanical Strength: 2D materials possess exceptional mechanical strength, making them highly resistant to deformation and breakage [13]. For instance, graphene, a well-known 2D material, is one of the strongest materials ever discovered. Electrical Conductivity: Many 2D materials exhibit high electrical conductivity, enabling efficient charge transport [14]. Graphene, for example, is an excellent conductor of electricity due to its unique electronic band structure. Optical Properties: 2D materials exhibit intriguing optical properties, including transparency, light absorption, and emission [12]. Their unique band structures allow for efficient light-matter interactions, making them useful for optical applications Fig. 1. Chemical Reactivity: The exposed surfaces of 2D materials provide a large number of reactive sites, enabling various chemical reactions and functionalization [10]. This reactivity can be harnessed for tailored applications and surface modifications.

2.2 Prominent 2D Materials

Several prominent 2D materials have been extensively studied, each with its distinct properties and potential applications. Some notable examples include: Graphene:

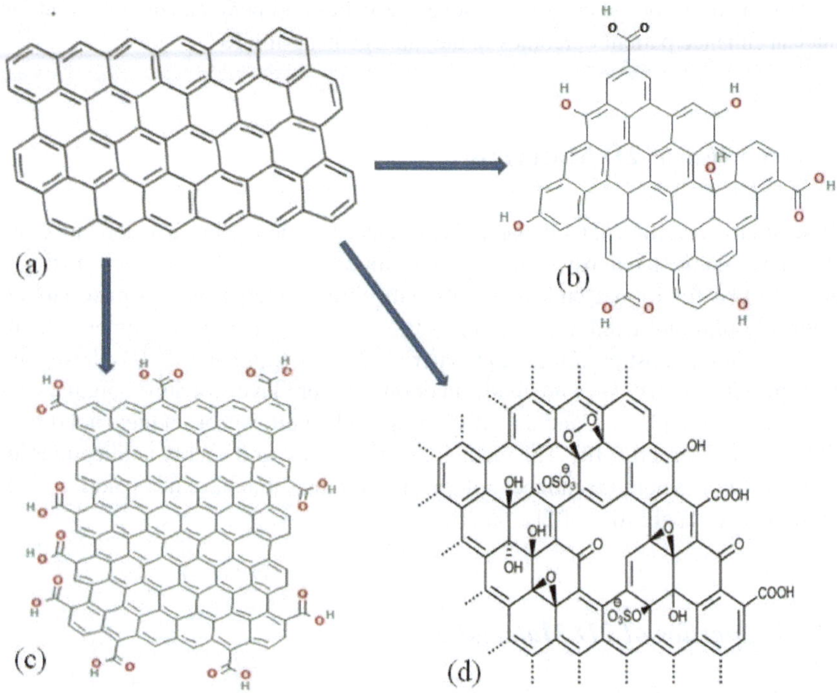

Fig. 1. Chemical structures of **a** graphene and its different forms **b** graphene quantum dot **c** reduced graphene oxide **d** graphene oxide and their processes (Pubchem)

Graphene is renowned for its high mechanical strength, electrical conductivity, and thermal conductivity. It is made up of a single sheet of carbon atoms organized in a hexagonal lattice [14, 15]. It has immense potential in electronics, energy storage, and biomedical applications. Transition Metal Dichalcogenides: the layered compounds called TMDs, such as tungsten diselenide, tungsten sulfide, and molybdenum disulfide, have unusual electrical characteristics [16]. In their monolayer form, they show a transition from an indirect to a direct band gap, making them excellent for photovoltaics and optoelectronics (Fig. 2a–c). Black Phosphorus: BP is another intriguing 2D material with a layered structure known as BP (Fig. 2d). It is desirable for applications in electronics, optoelectronics, and energy storage due to its configurable band gap, high carrier mobility, and outstanding electrochemical characteristics [17, 18].

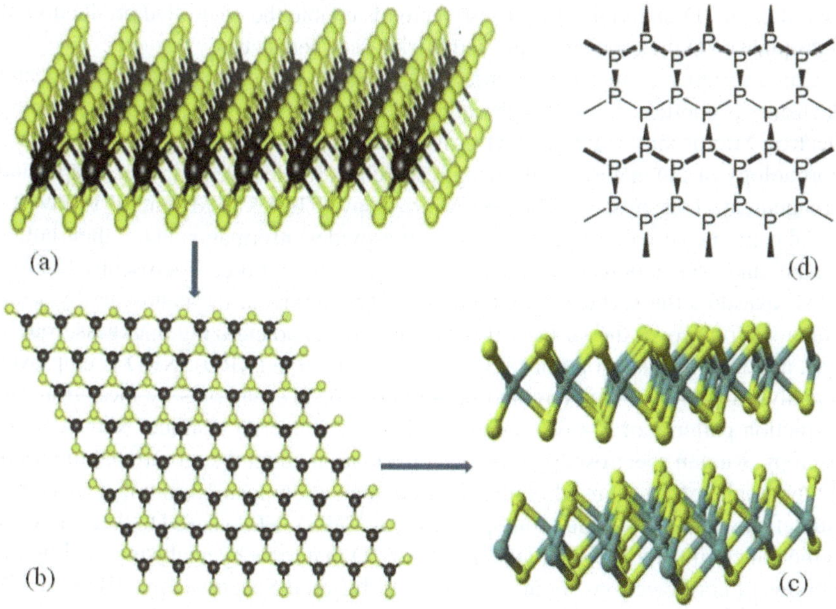

Fig. 2 Structures of **a** TMD Monolayer, black are transition metal atoms and yellow are the chalcogen atoms **b** hexagonal TMD monolayer (wikipedia.org/wiki/Transition_metal dichalcogenide_monolayers), **c** crystal structures of TMD ball and stick model (The American Mineralogist Crystal Structure Database) and **d** chemical structure of BP (wikipedia.org/wiki/Allotropes_of_phosphorus)

3 Synthesis and Characterization Techniques

There are several methods for synthesizing 2D materials, each tailored to specific material properties and applications. Some common synthesis techniques include: Mechanical Exfoliation: This method involves peeling individual layers from a bulk material using adhesive tape or other mechanical means. It was famously used to isolate graphene for the first time. While simple and effective, this technique is limited to small-scale production. Chemical Vapor Deposition (CVD): CVD involves the growth of 2D materials on a substrate by reacting precursor gases at high temperatures. This technique allows for large-scale production of high-quality 2D materials like graphene and TMDs. Liquid Exfoliation: In this approach, bulk materials are dispersed in a solvent, and ultrasonic or shear forces are applied to exfoliate the layers into 2D flakes. Liquid exfoliation is versatile and enables the production of various 2D materials. Sol–Gel Method: This technique involves the hydrolysis and condensation of metal-containing precursors to produce 2D metal oxides or hydroxides. Subsequent thermal treatment can lead to the formation of 2D materials. Electrochemical Exfoliation: Here, an electrical potential is applied to a bulk material to exfoliate it into 2D layers. This method is relatively simple and can produce large quantities of 2D materials [19]. These methods enable the controlled production of high-quality 2D materials with desired thicknesses and sizes.

Characterizing 2D materials is crucial to understand their structural, chemical, and electronic properties. Several techniques are employed for this purpose: Scanning Electron Microscopy (SEM): SEM provides high-resolution images of the surface morphology of 2D materials, allowing researchers to observe their size, shape, and arrangement. Transmission Electron Microscopy (TEM): TEM is used to investigate the atomic structure of 2D materials. It provides information about their lattice spacing and crystal defects at the atomic level. Atomic Force Microscopy (AFM): AFM measures the surface topography of 2D materials at the nanoscale by scanning a sharp tip over the surface. It is valuable for characterizing thickness, roughness, and mechanical properties [20]. X-ray Diffraction (XRD): XRD is employed to analyse the crystal structure and phase purity of 2D materials by measuring the diffraction patterns of X-rays interacting with the material's lattice. Raman Spectroscopy: Raman spectroscopy provides information about the vibrational modes of 2D materials, offering insights into their chemical composition, number of layers, and defects. X-ray Photoelectron Spectroscopy (XPS): XPS is used to determine the elemental composition and chemical state of 2D materials by analysing the binding energies of core electrons. Fourier Transform Infrared Spectroscopy (FTIR): FTIR is employed to study the vibrational modes of 2D materials, providing information about their chemical bonding and functional groups. Electrical and Optical Characterization: Techniques like Hall effect measurements, electrical conductivity, and optical absorption spectroscopy are used to assess the electrical and optical properties of 2D materials [21].

The combination of synthesis and characterization techniques allows researchers to produce high-quality 2D materials and gain a comprehensive understanding of their structure and properties. This knowledge is crucial for optimizing their use in various applications, including drug delivery systems, biosensing, imaging, and tissue engineering in colorectal cancer and other fields.

4 Biosensing Applications

4.1 *Functionalization Strategies of 2D Materials for Biosensing*

2D materials are intriguing candidates for biosensing applications, such as the detection of cancer biomarkers, due to their special characteristics. To enhance their sensing capabilities, 2D materials can be functionalized through various strategies. In order to enable the selective and sensitive detection of target analytes, functionalization entails altering the surface of 2D materials with certain molecules, such as antibodies, aptamers or DNA probes [22]. Functionalization can be achieved through covalent or non-covalent methods. Covalent functionalization is the process of chemically directly attaching functional molecules to the surface of 2D materials [23]. Van der Waals forces, stacking, electrostatic interactions, and other weak interactions are used in non-covalent functionalization to immobilize functional molecules on surfaces [22, 24].

4.2 Detection of CRC Biomarkers Using 2D Materials

CRC biomarkers are essential for early disease detection, prognosis, and monitoring. 2D materials can be utilized to detect specific CRC biomarkers, offering improved sensitivity and selectivity compared to traditional detection methods [25]. For instance, the functionalization of 2D materials with specific antibodies or aptamers enables the recognition and binding of target biomarkers present in biological samples [26]. Several CRC indicators, including DNA mutations, carbohydrate antigen 19–9 (CA19-9), and carcinoembryonic antigen (CEA), can be found utilizing 2D materials-based biosensors [27]. The interaction between the biomarker and the functionalized 2D material surface can induce changes in electrical conductivity, optical properties, or other measurable signals, allowing for the sensitive detection and quantification of the biomarker [25].

4.3 Early Diagnosis and Monitoring of CRC Using 2D Materials

Early diagnosis and effective monitoring of CRC are crucial for improving patient outcomes and reducing mortality rates. 2D materials have shown great potential in enhancing early diagnosis and monitoring of CRC due to their unique properties, which enable highly sensitive and specific detection of biomarkers associated with the disease. Early detection is possible with 2D materials-based biosensors since they make it possible to identify CRC biomarkers in a sensitive and specific manner [27]. Here are some ways 2D materials are being utilized for early diagnosis and monitoring of CRC: Biosensing of CRC Biomarkers: 2D materials, such as graphene and TMDs, can be functionalized with specific receptors or antibodies that recognize CRC biomarkers [27]. These biomarkers may include proteins, DNA mutations, or other molecules that are indicative of CRC. When the biomarkers bind to the functionalized 2D material surface, changes in the material's electrical, optical, or mechanical properties occur. These changes can be measured and quantified, enabling the detection and quantification of CRC biomarkers with high sensitivity and specificity.

Liquid Biopsy: Liquid biopsy is a non-invasive method that analyses tumour-derived material in body fluids, such as blood or stool. 2D materials can be used as biosensors in liquid biopsy platforms to detect circulating tumour cells, cell-free

DNA, or exosomes shed by CRC tumours. The ability to detect and analyse these biomarkers minimally invasively can facilitate early detection and monitoring of CRC progression. 2D materials' large surface area and superior electrical conductivity enable effective signal transduction and amplification, resulting in increased detection sensitivity [28]. Point-of-Care Diagnostics: The unique properties of 2D materials allow for the development of portable and rapid diagnostic devices for CRC. These point-of-care devices can be used in clinical settings or even remote areas to provide real-time results, allowing for timely diagnosis and monitoring of CRC patients. 2D materials can accelerate the creation of point-of-care tools for practical and quick CRC biomarker screening [27]. These portable biosensors can be used for on-site testing, enabling early diagnosis in resource-limited settings or remote areas.

Imaging Enhancements: 2D materials can act as contrast agents in various imaging modalities, such as MRI and optical imaging. When functionalized with targeting ligands, these materials can accumulate at the site of CRC tumours, enhancing the visualization and precise localization of cancerous lesions. This can aid in the early detection of CRC and help monitor treatment responses over time. Real-Time Monitoring of Treatment Response: Monitoring treatment response is crucial for tailoring therapeutic strategies for individual CRC patients. By using 2D material-based biosensors, changes in biomarker levels or other disease-related molecules can be monitored over the course of treatment. This real-time monitoring can provide valuable insights into the efficacy of the treatment and the need for adjustments in the therapeutic regimen. Personalized Medicine: The use of 2D materials in early diagnosis and monitoring can contribute to personalized medicine approaches in CRC. By analysing specific biomarkers or disease-related changes in individual patients, treatment plans can be tailored to optimize therapeutic outcomes for each patient.

In addition to early diagnosis, 2D materials-based biosensors hold promise for continuous monitoring of CRC. By incorporating these biosensors into wearable or implantable devices, real-time monitoring of biomarkers can be achieved, allowing for personalized treatment adjustments and disease progression tracking. The utilization of 2D materials in biosensing applications for CRC biomarkers offers exciting possibilities for improving early diagnosis, prognosis, and monitoring of the disease [26]. The construction of extremely specialized and trustworthy biosensors is made possible by the functionalization techniques and sensitive detecting capabilities of 2D materials. Continued research in this field may help create novel diagnostic instruments and personalized treatment plans for CRC. The integration of 2D materials in early diagnosis and monitoring of CRC holds tremendous potential to improve patient outcomes. Their unique properties enable sensitive and specific detection of CRC biomarkers, facilitating early diagnosis and real-time monitoring of treatment responses. The development of 2D material based biosensors and imaging enhancements offers promising avenues for advancing personalized medicine strategies in CRC management. The sustained research in this area will likely lead to further advancements in CRC diagnostic and monitoring, ultimately improving patient survival rates and quality of life.

5 Drug Delivery Systems

5.1 Utilizing 2D Materials as Drug Carriers

As drug carriers in drug delivery systems, 2D materials have special benefits. They are desirable candidates for encapsulating and delivering medicinal compounds due to their huge surface area, high loading capacity, and adjustable characteristics [29]. Numerous 2D materials, including graphene, TMDs, and BP, have demonstrated promise for use in drug delivery systems [4]. The capacity of 2D materials to shield the medications from deterioration and maintain their stability during storage and transit is one of the main benefits of employing them as drug carriers. The exterior of 2D materials can be modified to provide a controlled release of drugs, allowing for sustained and targeted delivery to the desired site. Below are the key aspects of utilizing 2D materials as drug carriers in drug delivery systems: Dressing capacity: 2D materials possess a high surface area-to-volume ratio, allowing for efficient drug loading. The large surface area allows drug molecules to be adsorbed, absorbed, or chemically bonded to the 2D material surface. This high drug loading capacity is particularly advantageous for hydrophobic or poorly water-soluble drugs that have limited solubility in biological fluids. Controlled drug release: The controlled and sustained release of drugs is essential for optimizing therapeutic outcomes and minimizing adverse effects. 2D materials can be engineered to achieve controlled drug release profiles, ensuring that drugs are released at a controlled rate over an extended period [29]. The release rate can be modulated by altering the surface properties, functionalization, or stacking of 2D materials, allowing for tailored drug delivery to meet specific therapeutic needs.

Enhanced stability: Drugs encapsulated or bound to 2D materials are protected from degradation by environmental factors such as light, temperature, and pH. This enhanced stability is crucial for preserving the drug's efficacy during storage, transportation, and delivery to the target site. The surface of 2D materials can be functionalized with targeting ligands, such as antibodies, peptides, or aptamers, that recognize and bind to specific receptors overexpressed on the surface of target cells, including cancer cells in CRC. By attaching these targeting ligands, 2D materials can act as vehicles to deliver drugs selectively to the desired site, increasing the drug's concentration at the target site while minimizing exposure to healthy tissues, thus reducing systemic side effects [6]. Synergistic therapy: 2D materials can be used to deliver multiple drugs simultaneously, allowing for combination therapy. This approach is particularly beneficial in CRC treatment, involving multiple pathways and targets. The co-delivery of multiple drugs can enhance therapeutic efficacy by simultaneously addressing various aspects of the disease, such as inhibiting tumour growth, promoting apoptosis, or suppressing angiogenesis. Response to stimuli: Some 2D materials possess stimuli-responsive behaviour, meaning that they can release drugs in response to specific external triggers, such as changes in pH, temperature, or light. This property can be harnessed to develop smart drug delivery systems, where drugs are released only at the tumour site or in response to other specific conditions within the body [30, 31].

5.2 *Targeted Delivery to CRC Cells*

To improve therapy efficacy and reduce adverse effects in the context of CRC, targeted delivery of therapeutic drugs to cancer cells is crucial [6]. Targeting ligands that precisely recognize and bind to receptors overexpressed on CRC cells, such as antibodies, peptides, or aptamers, can be added to 2D materials [30]. By adding targeting ligands to 2D materials, CRC cells may selectively take them up. This increases the concentration of treatment medicines at the tumour site while minimizing off-target effects [6]. This targeted delivery approach improves the therapeutic index and reduces the dosage required for effective treatment, minimizing toxicity to healthy tissues [31, 32]. Utilizing 2D materials as drug carriers in drug delivery systems offers exciting opportunities to enhance therapeutic agents' efficacy, specificity, and controlled release. The properties of 2D materials, such as their high surface area, tunable properties, and ability to be functionalized, make them attractive candidates for targeted and personalized drug delivery in CRC and other diseases. Continued research in this field holds the potential to revolutionize drug delivery strategies, leading to improved treatment outcomes and reduced side effects in CRC and other medical applications.

5.3 *Enhanced Therapeutic Efficacy and Reduced Side Effects*

By increasing the solubility, stability, and controlled release of anticancer medicines, 2D materials can improve their therapeutic efficacy [29]. Because 2D materials have a wide surface area, they have a higher drug-loading capacity, allowing more therapeutic agents to be transported to the tumour site [33, 34]. Additionally, the special qualities of 2D materials, such as their capacity to penetrate cell membranes, can make it easier for medications to be delivered intracellularly, improving their therapeutic efficacy. The controlled release capabilities of 2D materials can be utilized to achieve sustained drug release, ensuring a continuous therapeutic effect over an extended period. Photothermal therapy (PTT) is a non-invasive therapeutic approach that utilizes the photothermal effect of 2D materials to selectively ablate cancer cells. When exposed to near-infrared (NIR) light, 2D materials efficiently convert light into heat, leading to localized hyperthermia and cancer cell destruction. In CRC, PTT has shown promising results as an alternative or complementary treatment modality. Photodynamic therapy (PDT), 2D materials can also serve as photosensitizers in PDT, a therapeutic approach that utilizes light-induced reactive oxygen species (ROS) to destroy cancer cells. When combined with light, 2D materials can generate ROS, inducing cell death in CRC cells. PDT holds the potential for localized and targeted treatment of CRC.

By specifically targeting CRC cells and delivering therapeutic agents with improved efficiency, 2D materials-based drug delivery systems have the potential to enhance treatment outcomes while minimizing negative effects [33, 35]. This approach can contribute to personalized and precision medicine strategies for CRC, improving patient quality of life and overall therapeutic efficacy. There is significant potential for the targeted and effective delivery of therapeutic drugs to CRC cells through the use of 2D materials as drug carriers in drug delivery systems (Fig. 3). The unique properties of 2D materials allow for controlled release, enhanced drug stability, and increased drug loading capacity. By harnessing the potential of 2D materials, we can develop innovative drug delivery systems that improve treatment outcomes in CRC while minimizing adverse effects on healthy tissues.

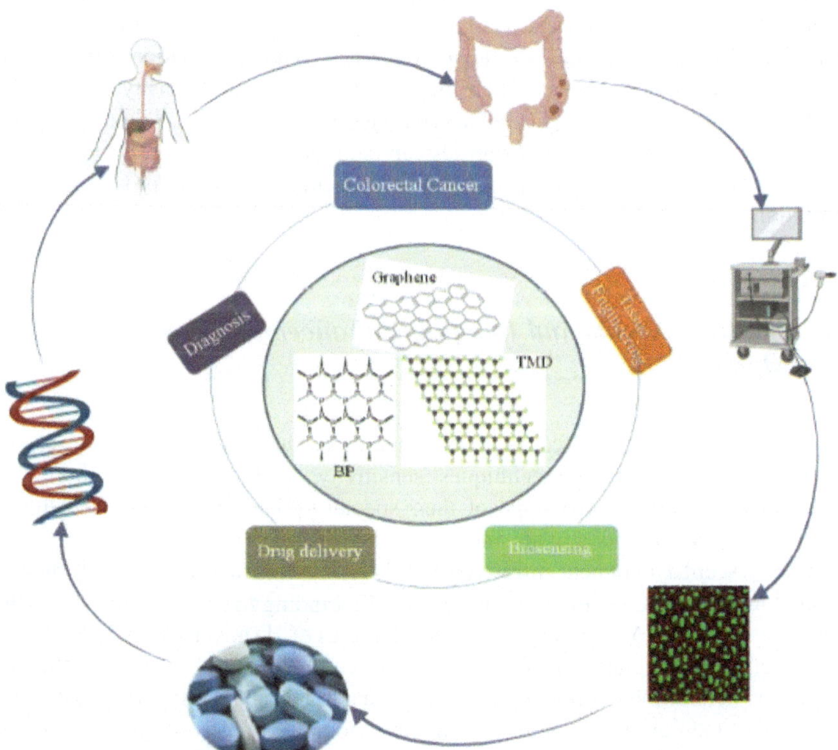

Fig. 3 Schematic diagram represents the utilization of grapheme, TMD, and BP in diagnosing, imaging, biosensing and transporting drugs, genetic materials to CRC cells and biomedical advancements

6 Imaging and Diagnostics

6.1 Role of 2D Materials in Improving Imaging Techniques

Magnetic resonance imaging (MRI), computed tomography (CT), and photoacoustic imaging (PA) are only a few of the imaging methods utilized in CRC diagnosis that have demonstrated significant improvement with the use of 2D materials [36–38]. High surface area, biocompatibility, and tunable optical and magnetic properties are just a few of the special qualities of 2D materials that make them useful instruments for improving imaging modalities. [5, 8]. In MRI, 2D materials can be utilized as contrast agents to improve the signal contrast between normal and cancerous tissues. Their magnetic properties can enhance the relaxation rates of water protons, resulting in increased signal intensity and improved image contrast [39, 40]. Additionally, the enormous surface area of 2D materials makes it possible to load contrast agents efficiently, allowing for focused and accurate imaging of CRC tumours. 2D materials can be used as contrast agents in CT imaging to increase X-ray attenuation and enhance the visibility of tumour tissues. Their high atomic number and density enable efficient X-ray absorption, leading to enhanced image contrast [41, 42]. Additionally, the surface functionalization of 2D materials can facilitate their selective accumulation in tumour tissues, allowing for precise tumour localization and improved imaging resolution.

6.2 Contrast Agents and Imaging Enhancements Using 2D Materials

2D materials can serve as effective contrast agents and imaging enhancers in various imaging modalities. Imaging techniques' sensitivity, resolution, and specificity can be enhanced by taking advantage of their special optical and magnetic features. 2D materials like graphene oxide and BP can be functionalized for optical imaging using fluorescent dyes or near-infrared probes [43, 44]. This functionalization enables targeted imaging of CRC tumours by specifically binding to cancer cells or specific biomarkers [45, 46]. Additionally, the robust features of 2D materials for light-matter interaction enable higher optical signal production, resulting in improved imaging sensitivity. In addition to optical imaging, 2D materials can be utilized as agents for multimodal imaging, where multiple imaging modalities are combined to provide complementary information. By incorporating 2D materials with both optical and magnetic properties, multimodal imaging techniques such as MRI/optical imaging or CT/optical imaging can be employed for accurate tumour localization, precise characterization, and improved diagnostic accuracy [8, 47, 48].

6.3 *Precise Tumour Localization and Characterization with 2D Materials*

Precision tumour localization and characterization in CRC are made possible by the use of 2D materials in imaging and diagnostics. Specific binding to CRC cells or tumour-associated biomarkers can be accomplished by functionalizing 2D materials with targeted ligands, such as antibodies or peptides [49–51]. This targeted approach allows for accurate tumour localization, distinguishing cancerous tissues from healthy tissues. A thorough evaluation of CRC tumours is also made possible by the special qualities of 2D materials, such as their high surface area and controllable optical and magnetic properties [48, 52]. For instance, the functionalization of 2D materials with specific probes can provide information about molecular markers, gene expression patterns, or cellular metabolism within the tumour microenvironment [53]. This characterization aids in understanding tumour heterogeneity, predicting treatment response, and guiding personalized treatment strategies.

By leveraging the capabilities of 2D materials in imaging and diagnostics, precise tumour localization and comprehensive characterization of CRC tumours can be achieved [54, 55]. This advancement can contribute to improved diagnostic accuracy, personalized treatment planning, and better patient outcomes in CRC management. In summary, 2D materials play a vital role in improving imaging techniques, acting as contrast agents and imaging enhancers in MRI, CT, and optical imaging. They enable precise tumour localization and characterization, providing valuable information about CRC tumours. The integration of 2D materials in imaging modalities offers great potential for enhancing diagnostics, guiding treatment decisions, and monitoring the therapeutic response in CRC.

7 Tissue Engineering and Regenerative Medicine

7.1 *Integration of 2D Materials in Scaffolds for CRC Cell Growth*

Regenerative medicine and tissue engineering have the potential to create novel methods for analysing the course of CRC and assessing prospective treatments [56]. Integrating 2D materials into scaffolds offers unique opportunities to mimic the tumour microenvironment and facilitate CRC cell growth in a controlled *in-vitro* setting [57]. Scaffolds provide a three-dimensional framework supporting cell adhesion, proliferation, and differentiation, resembling the native tissue environment [58]. The outstanding qualities of 2D materials, such as their high surface area, electrical conductivity, and mechanical strength, can improve the scaffolds' performance and usefulness [8, 59]. A suitable platform for cell development and tissue regeneration can be created by the scaffolds' 2D materials, which can encourage CRC cell adhesion and proliferation [60]. The surface properties of 2D materials can be modified to promote cell–matrix interactions, enhance cell adhesion receptors, and facilitate cell signalling processes, thus influencing CRC cell behaviour and function [61, 62]. Furthermore, to encourage desired cell behaviours and direct tissue regeneration in the setting of CRC, 2D materials can be functionalized with certain molecules, such as growth factors, peptides, or extracellular matrix components. A deeper knowledge of cancer progression and therapeutic responses is made possible by the controlled release of bioactive chemicals from the 2D material-functionalized scaffolds, which can replicate the signalling cues present in the tumour microenvironment.

7.2 Impact on Understanding Cancer Progression and Evaluating Therapies

Integrating 2D materials in scaffolds for CRC cell growth provides valuable insights into cancer progression and enables the evaluation of potential therapies [57]. Researchers can explore the interactions between CRC cells, surrounding stromal cells, and the extracellular matrix in a controlled and repeatable manner by using biomimetic systems replicating the tumour microenvironment [63–65]. These 2D material-based scaffolds can be used to investigate various aspects of CRC, including tumour invasion, angiogenesis, and the formation of metastases [66]. Systematic investigation of the impact of particular parameters on cancer growth, such as matrix stiffness, topographical cues, or the presence of inflammatory mediators, is made possible by the scaffolds' ability to be precisely controlled in terms of composition and qualities [58, 67].

Additionally, integrating 2D materials in the scaffolds enables the evaluation of potential therapeutic approaches for CRC. Researchers can test the efficacy of anticancer drugs, nanoparticles, or gene therapy vectors within these biomimetic systems [67, 68]. The use of 2D materials in the scaffolds can improve therapeutic targeting, increase drug delivery, and shed light on how CRC cells react to various treatments [69, 70]. Overall, the integration of 2D materials in scaffolds for CRC cell growth offers a powerful platform for understanding cancer progression and evaluating potential therapies. These biomimetic systems provide a controllable and tunable environment that closely resembles the tumour microenvironment, enabling researchers to study CRC cell behaviour, tumour-stroma interactions, and therapeutic responses. The insights collected from these studies can help create more efficient CRC treatment plans and personalized medicine techniques.

8 Challenges and Future Directions

8.1 Toxicity and Biocompatibility Considerations

The potential applications of 2D materials, such as graphene, TMDs, and BP, in various fields, including biomedicine, highlight the need to understand their toxicity and biocompatibility profiles thoroughly. As these materials come into contact with living organisms, including cells and tissues, their effects on biological systems must be carefully evaluated. Assuring the safety and biocompatibility of 2D materials used in CRC research is one of the main issues. While 2D materials offer unique properties and functionalities, their potential toxicity must be carefully evaluated to avoid adverse effects on healthy tissues and organs. It is crucial to thoroughly investigate the biocompatibility of 2D materials by assessing their interactions with cells, tissues, and biological systems [71]. Understanding the potential toxicity mechanisms, such as oxidative stress or inflammation, is essential for designing safe and biocompatible applications. Additionally, proper surface functionalization and modifications can enhance the biocompatibility of 2D materials, reducing potential cytotoxicity and improving their overall safety profile [72].

Here are some key considerations regarding the toxicity and biocompatibility of 2D materials: Physicochemical Properties: The physicochemical properties of 2D materials, such as size, shape, surface charge, and surface functionalization, play a critical role in determining their interactions with biological systems. For example, the size of 2D materials can influence their cellular uptake and biodistribution, while surface functionalization can affect their biocompatibility and potential for targeted drug delivery. Clearance and Biodegradation: The biocompatibility of 2D materials is influenced by their clearance and biodegradation rates. Ideally, the materials should be cleared from the body over time without causing excessive accumulation or long-term retention in vital organs. The biodegradability of 2D materials is particularly important for materials intended for temporary use or as drug carriers [71]. Immune Response: The interaction of 2D materials with the immune system is critical for assessing their biocompatibility. A robust immune response can lead to inflammation and potential adverse effects. Modulating the surface chemistry or functionalization of 2D materials can help minimize immunogenicity and promote biocompatibility.

Potential Toxicity Mechanisms: Understanding the potential toxicity mechanisms of 2D materials is essential for developing safe applications. For instance, generating reactive oxygen species (ROS) due to the interaction of 2D materials with biological environments can lead to oxidative stress and cellular damage. Studying these mechanisms aids in designing strategies to mitigate toxicity. Surface Functionalization: Surface functionalization of 2D materials can significantly influence their biocompatibility [72]. By modifying the surface with biocompatible molecules or coatings, researchers can tailor the interaction of 2D materials with biological systems, reducing potential cytotoxicity and improving their overall safety profile. Assessing the toxicity and biocompatibility of 2D materials is crucial for their safe application in biomedicine and other fields. A thorough understanding of their interactions with biological systems and careful surface functionalization and modification will pave the way for developing safe and effective 2D material-based biomedical applications.

8.2 Scalability and Clinical Translation of 2D Materials

Scalability is crucial for the clinical translation of 2D materials in CRC research. The synthesis and production methods of 2D materials need to be optimized to ensure large-scale production without compromising their properties and quality. Incorporating 2D materials into several applications, including drug delivery systems, biosensing platforms, and tissue engineering scaffolds, will be made easier by developing scalable production procedures [73, 74]. Furthermore, clinical translation requires rigorous testing and validation of the safety and efficacy of 2D materials-based technologies. Preclinical studies, including animal models and ex vivo experiments, are essential to demonstrate the effectiveness and safety of these materials in CRC treatment and diagnosis. Before in vivo investigations, researchers conducted in vitro studies to assess the cytotoxicity and biocompatibility of 2D materials using cell culture models. These studies evaluate the impact of 2D materials on cell viability, proliferation, apoptosis, and cellular functions [73]. It is essential to use relevant cell types to ensure that the results reflect the potential biological response of specific tissues or organs. In vivo, studies involving animal models are conducted to evaluate the biocompatibility of 2D materials in a complex biological environment. These studies assess systemic toxicity, biodistribution, tissue accumulation, and potential immune responses. In-depth investigation of the long-term effects of 2D materials is essential to ensure their safety for biomedical applications [74]. Long-term biocompatibility studies, biodistribution analyses, and thorough toxicity assessments are necessary to ensure the clinical viability of 2D materials.

8.3 Emerging Trends and Potential Synergies

As the field of 2D materials in CRC research continues to advance, emerging trends and potential synergies with other therapies are being explored. Combination therapies involving 2D materials, such as their integration with conventional chemotherapy drugs, immunotherapies, or targeted therapies, hold promise for enhanced treatment outcomes. The unique possessions of 2D materials, such as their capability to deliver therapeutic agents, act as imaging contrast agents, or modulate the tumour microenvironment, can be synergistically combined with other treatment modalities to improve therapeutic efficacy. For instance, the targeted delivery of chemotherapy to CRC cells can be improved by using 2D materials as drug carriers, and the immunomodulatory effects of these materials can be used to increase the effectiveness of immunotherapies [45, 75]. Furthermore, combining 2D materials with emerging technologies, such as nanomedicine, gene editing, or microfluidics, can open up new avenues for

CRC research and therapy [76, 77]. These synergistic approaches can potentially revolutionize CRC treatment by overcoming drug resistance, enhancing early diagnosis, and enabling personalized treatment strategies [78]. The key benefits and drawbacks of graphene, reduced-graphene oxide, graphene oxide, graphene quantum dot, TMD, and BP for biomedical applications (Table 1). The integration of 2D materials in CRC research faces challenges related to toxicity and biocompatibility, scalability for clinical translation, and ensuring long-term safety. However, by addressing these challenges and leveraging emerging trends and potential synergies, 2D materials are promising for advancing CRC diagnosis, treatment, and personalized medicine. Continued research and association among scientists, engineers, and clinicians will drive the field forward, benefiting CRC patients and improving clinical outcomes.

9 Conclusion

The promise of 2D materials in CRC research has been underlined in this thorough work. The incorporation of 2D materials into a number of CRC-related fields, including biosensing, drug delivery systems, imaging and diagnostics, and tissue engineering, offers fresh perspectives and potential uses that have the potential to have a big impact on the industry. The main conclusions of this article show that the unique qualities of 2D materials, like high surface area, controllable optical and magnetic properties, and biocompatibility, make them useful instruments in CRC research. The functionalization strategies of 2D materials enable the biosensing of CRC biomarkers, facilitating early diagnosis and monitoring of the disease. Additionally, the targeted delivery of drugs to CRC cells made possible by their usage as drug carriers improves therapy effectiveness while minimizing negative effects. Integration of 2D materials in imaging techniques improves imaging modalities, enabling precise tumour localization and characterization. In tissue engineering, 2D materials integrated into scaffolds provide platforms for studying CRC progression and evaluating therapies. These outcomes have significant ramifications for upcoming CRC research. Utilization of 2D materials can advance our understanding of CRC biology, tumour microenvironment, and therapeutic response. It can lead to the development of innovative diagnostic tools, targeted therapies, and regenerative medicine approaches for CRC treatment. Moreover, the integration of 2D materials in clinical settings can contribute to personalized medicine, improving patient outcomes and enhancing treatment efficacy.

Future research should concentrate on tackling the issues related to 2D materials, such as toxicity and biocompatibility considerations, in order to develop the subject. Additional in-depth in vitro and in vivo research is required to fully assess the security and long-term impacts of these compounds. The scalability of synthesis and manufacturing processes should be optimized to enable the translation of 2D materials into clinical practice. Additionally, the integration of 2D materials with other emerging technologies and therapies should be explored to harness their synergistic effects. Combination therapies that combine 2D materials with conventional

Table 1 The key benefits and drawbacks of grapheme, TMD, BP, and related materials for biomedical applications

S.No	Properties	Benefits	Drawbacks	Ref
1	Graphene	1. High electrical and thermal conductivities 2. High control on functionalization	1. Hydrophobicity 2. High cost 3. Difficult workability 4. Small production	[79, 80]
2	Reduced graphene oxide	1. High electrical and thermal conductivities 2. Good control of functionalization 3. Less expensive than neat graphene	1. Hydrophobicity 2. Difficult workability 3. Properties related to production methodology used	[80, 81]
3	Graphene oxide	1. Water dispersibility 2. Polar functionalization 3. Low cost 4. Easy workability	1. Lower electrical and thermal conductivity 2. Surface random functionalization 3. Poor control of post-preparation functionalization	[79, 80]
4	Graphene quantum dot	1. Increased effective viewing angles 2. High capability to absorb light 3. Have a lifetime of upto 14 years 4. Compatible with chip technology	1. Hard to control the size of the particles 2. Overall conversion efficiency is lower 3. Lower temperature operation	[82]
5	TMD	1. High yield, facile preparation , and low cost 2. Applied to preoperative CT, MR and PA imaging	1. Maintaining their structural and chemical stability 2. Dependent upon additional composite materials	[83, 84]
6	BP	1. Rapid synthesis process 2. Uses in biosensing bioimaging, drug delivery, and cancer therapy 3. Good biocompatibility and low cytotoxicity	1. Low yield 2. Limited ability to control size, shape, and thickness 3. Presence of chemical traces from the adhesive tapes 4. Poor air stability	[85, 86]

treatments, immunotherapies, or emerging technologies can potentially enhance therapeutic outcomes and overcome treatment resistance. The potential of 2D materials in CRC research is vast, with implications for diagnostics, therapeutics, and tissue engineering. Further investigation and integration of 2D materials in clinical settings hold promise for improving CRC management and patient outcomes. The collaboration between researchers, clinicians, and engineers will be crucial in realizing the full potential of 2D materials in CRC research and clinical practice.

10 Contributions

MY drafted the original manuscript, SKG, VA, AG, and PK made critical suggestions available. The final manuscript was read and approved by all writers.

11 Acceptance of Participation and Ethical Clearance

Not relevant.

12 Permission to Publish

Not relevant.

13 Conflicts of Interest

The writers say they have no conflicting agendas.

References

1. Hossain, M.S., Karuniawati, H., Jairoun, A.A., Urbi, Z., Ooi, D.J., John, A., Lim, Y.C., Kibria, K.K., Mohiuddin, A.K., Ming, L.C., Goh, K.W.: Colorectal cancer: a review of carcinogenesis, global epidemiology, current challenges, risk factors, preventive and treatment strategies. Cancer **29**, 732–741 (2022)
2. Sawicki, T., Ruszkowska, M., Danielewicz, A., Niedźwiedzka, E., Arłukowicz, T., Przybyłowicz, K.E.: A review of colorectal cancer in terms of epidemiology, risk factors, development, symptoms and diagnosis. Cancer **22**, 25–32 (2021)
3. Mohd, Y., Kumar, P., Kuchi Bhotla, H., Meyyazhagan, A., Balasubramanian, B., Ramesh Kumar, M.K., Pappusamy, M., Alagamuthu, K.K., Orlacchio, A., Keshavarao, S.,

Sampathkumar, P.: Transmission jeopardy of adenomatosis polyposis coli and methylenetetrahydrofolate reductase in colorectal cancer. JRAAS **10**, 21–29 (2021)
4. Derakhshi, M., Daemi, S., Shahini, P., Habibzadeh, A., Mostafavi, E., Ashkarran, A.A.: Two-dimensional nanomaterials beyond graphene for biomedical applications. J. Funct. Biomater. **9**, 27–36 (2022)
5. Baig, N.: Two-dimensional nanomaterials: A critical review of recent progress, properties, applications, and future directions. Compos. A Appl. Sci. Manuf. **9**, 62–74 (2022)
6. Krasteva, N., Georgieva, M.: Promising therapeutic strategies for colorectal cancer treatment based on nanomaterials. Pharm. **7**, 213–225 (2022)
7. Alrushaid, N., Khan, F.A., Al-Suhaimi, E.A., Elaissari, A.: Nanotechnology in cancer diagnosis and treatment. Pharmaceutics **22**, 25-36 (2023)
8. Ranasinghe, J.C., Jain, A., Wu, W., Zhang, K., Wang, Z., Huang, S.: Engineered 2D materials for optical bioimaging and path toward therapy and tissue engineering. J. Mater. Sci. **28**, 689–713 (2022)
9. Zavabeti, A., Jannat, A., Zhong, L., Haidry, A.A., Yao, Z., Ou, J.Z.: Two-dimensional materials in large-areas: synthesis, properties and applications. Nano-Micro Lett. **12**, 1–34 (2020)
10. Alam, S., Chowdhury, M.A., Shahid, A., Alam, R., Rahim, A.: Synthesis of emerging two-dimensional (2D) materials–advances, challenges and prospects. FlatChem **1**, 305–315 (2021)
11. Velicky, M., Toth, P.S.: From two-dimensional materials to their heterostructures: an electrochemist's perspective. Appl. Mater. Today. **1**, 68–103 (2017)
12. Wang, G., Hou, H., Yan, Y., Jagatramka, R., Shirsalimian, A., Wang, Y., Li, B., Daly, M., Cao, C.: Recent advances in the mechanics of 2D materials. Int. J. Extreme Manuf. **5**, 17–29 (2023)
13. Xiong, Z., Zhong, L., Wang, H., Li, X.: Structural defects, mechanical behaviors, and properties of two-dimensional materials. Materials **3**, 192–199 (2021)
14. Majumder, M., Santosh, M.S., Viswanatha, R., Thakur, A.K., Dubal, D.P., Jayaramulu, K.: Two-dimensional conducting metal-organic frameworks enabled energy storage devices. Energy Stor. Mater. 1, 396–416 (2021)
15. Dethan, J.F.: Mechanical properties and thermal conductivity of newly introduced graphene-like borophanes: a reactive molecular dynamics study. Phys. Chem. Chem. Phys. **23**, 09–17 (2021)
16. Chen, E., Xu, W., Chen, J., Warner, J.H.: 2D layered noble metal dichalcogenides (Pt, Pd, Se, S) for electronics and energy applications. Mater. Today Adv. **1**, 76–85 (2020)
17. Shaikh, J.S., Shaikh, N.S., Sabale, S., Parveen, N., Patil, S.P., Mishra, Y.K., Kanjanaboos, P., Praserthdam, S., Lokhande, C.D.: A phosphorus integrated strategy for supercapacitor: 2D black phosphorus–doped and phosphorus-doped materials. Mater. Today Chem. **1**, 480–491 (2021)
18. Zeng, Y., Guo, Z.: Synthesis and stabilization of black phosphorus and phosphorene: Recent progress and perspectives. Iscience **22**, 116–127 (2021)
19. Shanmugam, V., Mensah, R.A., Babu, K., Gawusu, S., Chanda, A., Tu, Y., Neisiany, R.E., Försth, M., Sas, G., Das, O.: A review of the synthesis, properties, and applications of 2D materials. Part Part Syst Charact. **39**, 31–44 (2022)
20. Stylianou, A., Kontomaris, S.V., Grant, C., Alexandratou, E.: Atomic force microscopy on biological materials related to pathological conditions. Scanning **12**, 19–28 (2019)
21. Li, S., Liu, M., Qiu, X.: Scanning probe microscopy of topological structure induced electronic states of graphene. Small Methods **4**, 683–695 (2020)
22. Sekhon, S.S., Kaur, P., Kim, Y.H., Sekhon, S.S.: 2D graphene oxide–aptamer conjugate materials for cancer diagnosis. NPJ 2D Mater. Appl. **17**, 21–36 (2021)
23. Thurakkal, S., Zhang, X.: Recent advances in chemical functionalization of 2D black phosphorous nanosheets. Adv. Sci. **7**, 359–367 (2020)
24. Georgakilas, V., Tiwari, J.N., Kemp, K.C., Perman, J.A., Bourlinos, A.B., Kim, K.S., Zboril, R.: Noncovalent functionalization of graphene and graphene oxide for energy materials, biosensing, catalytic, and biomedical applications. Chem. Rev. **11**, 464–519 (2016)
25. Zhang, X., Tan, X., Wang, P., Qin, J.: Application of polypyrrole-based electrochemical biosensor for the early diagnosis of colorectal cancer. Nanomaterials **9**, 674–688 (2023)

26. Macovei, D.G., Irimes, M.B., Hosu, O., Cristea, C., Tertis, M.: Point-of-care electrochemical testing of biomarkers involved in inflammatory and inflammatory-associated medical conditions. Anal. Bioanal. Chem. **415**, 33–63 (2023)
27. Rao Bommi, J., Kummari, S., Lakavath, K., Sukumaran, R.A., Panicker, L.R., Marty, J.L., Yugender Goud, K.: Recent trends in biosensing and diagnostic methods for novel cancer biomarkers. Biosensors **18**, 398–412 (2023)
28. Sulleiro, M.V., Dominguez-Alfaro, A., Alegret, N., Silvestri, A., Gómez, I.J.: 2D materials towards sensing technology: from fundamentals to applications. Sens. Bio-Sens. **25**, 540–555 (2022)
29. Davis Jr., R., Urbanowski Jr., R.A., Gaharwar, A.K.: 2D layered nanomaterials for therapeutics delivery. Curr. Opin. Biomed. **1**, 319–328 (2021)
30. Zhou, G., Wilson, G., Hebbard, L., Duan, W., Liddle, C., George, J., Qiao, L.: Aptamers: a promising chemical antibody for cancer therapy. Oncotarget **3**, 446–458 (2016)
31. Younis, M., Arun, M., Iqbal, J., Kumar, M., Arun, S., Cacabelos, R., Sasikala, K., Anand, V.: Analysis of genetic alterations in colorectal cancer (CRC) patients in south Indian population. Int. J. Hum. Genet. **1**, 210–218 (2018)
32. Kong, X., Liu, Y., Huang, X., Huang, S., Gao, F., Rong, P., Zhang, S., Zhang, K., Zeng, W.: Cancer therapy based on smart drug delivery with advanced nanoparticles. Anticancer Agents Med Chem. **1**, 720–730 (2019)
33. Wei, Y., Zhou, F., Zhang, D., Chen, Q., Xing, D.: A graphene oxide based smart drug delivery system for tumor mitochondria-targeting photodynamic therapy. Nanoscale **8**, 530–538 (2016)
34. Chen, W., Ouyang, J., Liu, H., Chen, M., Zeng, K., Sheng, J., Liu, Z., Han, Y., Wang, L., Li, J., Deng, L.: Black phosphorus nanosheet-based drug delivery system for synergistic photodynamic/photothermal/chemotherapy of cancer. Adv. Mater. **29**, 864–877 (2017)
35. Wang, K., Shen, R., Meng, T., Hu, F., Yuan, H.: Nano-drug delivery systems based on different targeting mechanisms in the targeted therapy of colorectal cancer. Molecules **6**, 981–993 (2022)
36. Stuker, F., Ripoll, J., Rudin, M.: Fluorescence molecular tomography: principles and potential for pharmaceutical research. Pharmaceutics **26**, 229–274 (2011)
37. Khadir, S., Bon, P., Vignaud, D., Galopin, E., McEvoy, N., McCloskey, D., Monneret, S., Baffou, G.: Optical imaging and characterization of graphene and other 2D materials using quantitative phase microscopy. ACS Photonics **20**, 130–139 (2017)
38. García-Figueiras, R., Baleato-González, S., Padhani, A.R., Luna-Alcalá, A., Marhuenda, A., Vilanova, J.C., Osorio-Vázquez, I., Martínez-de-Alegría, A., Gómez-Caamaño, A.: Advanced imaging techniques in evaluation of colorectal cancer. Radiographics **38**, 740–765 (2018)
39. Caspani, S., Magalhães, R., Araújo, J.P., Sousa, C.T.: Magnetic nanomaterials as contrast agents for MRI. Materials 5, 586–599 (2020)
40. Chang, Z.X., Li, C.H., Chang, Y.C., Huang, C.Y., Chan, M.H., Hsiao, M.: Novel monodisperse FePt nanocomposites for T2-weighted magnetic resonance imaging: biomedical theranostics applications. Nanoscale Adv. **4**, 377–386 (2022)
41. Rajiah, P., Parakh, A., Kay, F., Baruah, D., Kambadakone, A.R., Leng, S.: Update on multienergy CT: physics, principles, and applications. Radiographics **40**, 284–308 (2020)
42. Jiang, Z., Zhang, M., Li, P., Wang, Y., Fu, Q.: Nanomaterial-based CT contrast agents and their applications in image-guided therapy. Theranostics **13**, 483–497 (2023)
43. Zhang, J., Chen, H., Zhao, M., Liu, G., Wu, J.: 2D nanomaterials for tissue engineering application. Nano Res. **13**, 19–34 (2020)
44. Malik, M., Iqbal, M.A., Choi, J.R., Pham, P.V.: 2D materials for efficient photodetection: Overview, mechanisms, performance and UV-ir range applications. Front. Chem. **5**, 10–17 (2022)
45. Fusco, L., Gazzi, A., Peng, G., Shin, Y., Vranic, S., Bedognetti, D., Vitale, F., Yilmazer, A., Feng, X., Fadeel, B., Casiraghi, C.: Graphene and other 2D materials: a multidisciplinary analysis to uncover the hidden potential as cancer theranostics. Theranostics **10**, 435–449 (2020)
46. Lei, Z.L., Guo, B.: 2D material-based optical biosensor: status and prospect. Adv. Sci. **9**, 924–942 (2022)

47. Eom, S., Choi, G., Nakamura, H., Choy, J.H.: 2-Dimensional nanomaterials with imaging and diagnostic functions for nanomedicine; a review. Bull. Chem. Soc. Jpn. **15**, 1–2 (2020)
48. Murar, M., Albertazzi, L., Pujals, S.: Advanced optical imaging-guided nanotheranostics towards personalized cancer drug delivery. Nanomaterials **26**, 399–412 (2022)
49. Feng, L., Dong, Z., Tao, D., Zhang, Y., Liu, Z.: The acidic tumor microenvironment: a target for smart cancer nano-theranostics. Natl. Sci. Rev. **1**, 269–286 (2018)
50. Soni, G.K., Manhas, P., Sharma, R.K.: Peptide-based optical biosensors: a promising approach for early-stage cancer detection. Biosens. Bioelectron. **6**, 259–273 (2022)
51. Djermane, R., Nieto, C., Vega, M.A., Del Valle, E.M.: Antibody-loaded nanoplatforms for colorectal cancer diagnosis and treatment: an update. Pharmaceutics **17**, 514–531 (2023)
52. Younis, M., Sujeetha, P., Sangeetha, T., Kaviya, M., Bharathi, K., Anand, A.: Molecular heterogeneity in colorectal cancer: insights from genetic profiling. J. Adv. Appl. Sci. Res. **15**, 6–15 (2021)
53. Xiao, Z., Dai, Z., Locasale, J,W.: Metabolic landscape of the tumor microenvironment at single cell resolution. Nat. Commun. **21**, 763–779 (2019)
54. Tamandl, D., Mang, T., Ba-Ssalamah, A.: Imaging of colorectal cancer–the clue to individualized treatment. Innov. Surg. Sci. **1**, 3–15 (2018)
55. Lin, J.R., Wang, S., Coy, S., Chen, Y.A., Yapp, C., Tyler, M., Nariya, M.K., Heiser, C.N., Lau, K.S., Santagata, S., Sorger, P.K.: Multiplexed 3D atlas of state transitions and immune interaction in colorectal cancer. Cell **19**, 363–381 (2023)
56. Mansouri, V., Beheshtizadeh, N., Gharibshahian, M., Sabouri, L., Varzandeh, M., Rezaei, N.: Recent advances in regenerative medicine strategies for cancer treatment. Biomed. Pharmacother.Pharmacother. **1**, 875–892 (2021)
57. Bregenzer, M.E., Horst, E.N., Mehta, P., Novak, C.M., Raghavan, S., Snyder, C.S., Mehta, G.: Integrated cancer tissue engineering models for precision medicine. PLoS One. **10**, 564–575 (2019)
58. Nikolova, M.P., Chavali, M.S.: Recent advances in biomaterials for 3D scaffolds: a review. Bioact. Mater. **1**, 271–292 (2019)
59. Lee, J.H., Park, S.J., Choi, J.W.: Electrical property of graphene and its application to electrochemical biosensing. Nanomaterials **20**, 297–312 (2019)
60. Ouyang, J., Rao, S., Liu, R., Wang, L., Chen, W., Tao, W., Kong, N.: 2D materials-based nanomedicine: from discovery to applications. Adv. Drug Deliv. Rev. **8**, 268–281 (2022)
61. Joyce, K., Fabra, G.T., Bozkurt, Y., Pandit, A.: Bioactive potential of natural biomaterials: Identification, retention and assessment of biological properties. Curr. Signal Transduct. Ther.. Signal Transduct. Ther. **19**, 122–136 (2021)
62. Santos, J., Moschetta, M., Rodrigues, J., Alpuim, P., Capasso, A.: Interactions between 2D materials and living matter: a review on graphene and hexagonal boron nitride coatings. Front. Bioeng. Biotechnol. **27**, 669–685 (2021)
63. Baghban, R., Roshangar, L., Jahanban-Esfahlan, R., Seidi, K., Ebrahimi-Kalan, A., Jaymand, M., Kolahian, S., Javaheri, T., Zare, P.: Tumor microenvironment complexity and therapeutic implications at a glance. Cell Commun. SignalCommun. Signal **18**, 1–9 (2020)
64. Mohd, Y., Balasubramanian, B., Meyyazhagan, A., Kuchi Bhotla, H., Shanmugam, S.K., Ramesh Kumar, M.K., Pappusamy, M., Alagamuthu, K.K., Keshavarao, S., Arumugam, V.A.: Extricating the association between the prognostic factors of colorectal cancer. J. Gastrointest. Cancer. **52**, 1022–1028 (2021)
65. Kim, J., Jang, J., Cho, D.W.: Recapitulating the cancer microenvironment using Bioprinting technology for precision medicine. Micromachines **17**, 122–143 (2021)
66. Metzcar, J., Wang, Y., Heiland, R., Macklin, P.: A review of cell-based computational modeling in cancer biology. JCO Clin. Cancer Inform. **2**, 1–3 (2019)
67. Park, J.H., Jo, S.B., Lee, J.H., Lee, H.H., Knowles, J.C., Kim, H.W.: Materials and extracellular matrix rigidity highlighted in tissue damages and diseases: implication for biomaterials design and therapeutic targets. Bioact. Mater. **1**, 381–403 (2023)
68. Roshni, K., Younis, M., Ilakkiyapavai, D., Basavaraju, P., Puthamohan, V.M.: Anticancer activity of biosynthesized silver nanoparticles using Murraya koenigii leaf extract against HT-29 colon cancer cell line. J. Cancer Sci. Ther. **10**, 72–75 (2018)

69. Meli, V.S., Veerasubramanian, P.K., Atcha, H., Reitz, Z., Downing, T.L., Liu, W.F.: Biophysical regulation of macrophages in health and disease. J. Leukoc. Biol. **106**, 283–299 (2019)
70. Cheng, Z., Li, M., Dey, R., Chen, Y.: Nanomaterials for cancer therapy: current progress and perspectives. J. Hematol. Oncol. Pharm. **14**, 1–27 (2021)
71. Bauleth-Ramos, T., Feijão, T., Gonçalves, A., Shahbazi, M.A., Liu, Z., Barrias, C., Oliveira, M.J., Granja, P., Santos, H.A., Sarmento, B.: Colorectal cancer triple co-culture spheroid model to assess the biocompatibility and anticancer properties of polymeric nanoparticles. J. Control Release. **10**, 398–411 (2020)
72. Martín, C., Kostarelos, K., Prato, M., Bianco, A.: Biocompatibility and biodegradability of 2D materials: graphene and beyond. Chem Comm. **55**, 540–546 (2019)
73. Sarkhosh-Inanlou, R., Shafiei-Irannejad, V., Azizi, S., Jouyban, A., Dolatabadi, J.E., Mobed, A., Adel, B., Soleymani, J., Hamblin, M.R.: Applications of scaffold-based advanced materials in biomedical sensing. TrAC Trends Anal. Chem. **1**, 342–361 (2021)
74. Liu, L., Ma, Q., Cao, J., Gao, Y., Han, S., Liang, Y., Zhang, T., Song, Y., Sun, Y.: Recent progress of graphene oxide-based multifunctional nanomaterials for cancer treatment. Cancer Nanotechnol. **12**, 1–31 (2021)
75. Veselov, V.V., Nosyrev, A.E., Jicsinszky, L., Alyautdin, R.N., Cravotto, G.: Targeted delivery methods for anticancer drugs. Cancer. 26, 622–641 (2022)
76. Kim, J., Kim, J., Jeong, C., Kim, W,J.: Synergistic nanomedicine by combined gene and photothermal therapy. Adv. Drug Deliv. Rev. **1**, 99–112 (2016)
77. Preetam, S., Nahak, B.K., Patra, S., Toncu, D.C., Park, S., Syväjärvi, M., Orive, G., Tiwari, A.: Emergence of microfluidics for next generation biomedical devices. Biosens Bioelectron. **8**, 106–123 (2022)
78. Haider, M., Zaki, K.Z., El Hamshary, M.R., Hussain, Z., Orive, G., Ibrahim, H.O.: Polymeric nanocarriers: a promising tool for early diagnosis and efficient treatment of colorectal cancer. J. Adv. Res. **1**, 237–255 (2022)
79. Castelletto, S., Boretti, A.: Advantages, limitations, and future suggestions in studying graphene-based desalination membranes. RSC Adv. **11**, 7981–8002 (2021)
80. Catania, F., Marras, E., Giorcelli, M., Jagdale, P., Lavagna, L., Tagliaferro, A., Bartoli, M.: A review on recent advancements of graphene and graphene-related materials in biological applications. Appl. Sci. **10**, 614–625 (2021)
81. Razaq, A., Bibi, F., Zheng, X., Papadakis, R., Jafri, S,H., Li, H.: Review on graphene, graphene oxide, reduced graphene oxide-based flexible composites: from fabrication to applications. Materials **28**, 1012–1031 (2022)
82. Tian, P., Tang, L., Teng, K.S., Lau, S.P.: Graphene quantum dots from chemistry to applications. Mater. Today Chem. **1**, 221–258 (2018)
83. Nafiujjaman, M., Nurunnabi, M.: Graphene and 2d materials for phototherapy. In: Biomedical Applications of Graphene and 2D Nanomaterials, vol. 1, pp. 105–117. Elsevier (2019)
84. Hasan, M.N., Nafiujjaman, M., Lee, Y.K.: 2D nanomaterials for gene delivery. In: Biomedical Applications of Graphene and 2D Nanomaterials, , vol. 1, pp. 87–104. Elsevier (2019)
85. Illarionov, Y.Y., Waltl, M., Rzepa, G., Knobloch, T., Kim, J.S., Akinwande, D., Grasser, T.: Highly-stable black phosphorus field-effect transistors with low density of oxide traps. npj 2D Mater. Appl. **17**, 23–42 (2017)
86. Choi, J.R., Yong, K.W., Choi, J.Y., Nilghaz, A., Lin, Y., Xu, J., Lu, X.: Black phosphorus and its biomedical applications. Theranostics **8**, 1005–1024 (2018)

[illegible]

An Overview of Two-Dimensional Materials and Their Applications in Dentistry

Mohmed Isaqali Karobari and Ali A. Assiry

Abstract Due to their multitudinous properties, including physiochemical, antibacterial, mechanical, and stem cell differentiating and bone regenerative properties, research on the applications of two-dimensional (2D) materials in dentistry has recently garnered increased attention. Although their functionality has been extensively studied in a variety of medical applications, the biocompatibility of 2D materials in dentistry has received scant attention. In addition, the buccal environment differs significantly from that of the body, which must be taken into account when evaluating biocompatibility requirements for dental applications. This chapter provides a comprehensive analysis of the 2D material properties and their applications in various dental disciplines, as well as a discussion of their future perspectives that will guide future research.

Keywords Two-dimensional materials · Graphene · Bioactive glass · Endodontics · Implants · Restorative dentistry

1 Introduction

Since graphene's exfoliation in 2004, two-dimensional (2D) materials have attracted the interest of researchers towards nanomaterials [1]. The 2D materials are nanomaterials that exhibit a freestanding sheet-like structure [2]. The thickness of 2D materials is usually a few angstroms to a few nanometers, whereas the lateral size may range

M. Isaqali Karobari (✉)
Dental Research Unit, Centre for Global Health Research, Saveetha Institute of Medical and Technical Sciences, Chennai, Tamil Nadu 600077, India
e-mail: dr.isaq@gmail.com

Centre for Dental Health Research, Saveetha Dental College & Hospitals, Saveetha Institute of Medical and Technical Sciences University, Chennai, Tamil Nadu 600077, India

A. A. Assiry
Preventive Dental Science Department, Faculty of Dentistry, Najran University, Najran, Kingdom of Saudi Arabia

N. Talreja et al. (eds.), *Two-dimensional Hybrid Composites*, Engineering Materials,
https://doi.org/10.1007/978-981-99-8010-9_14

from tens of nanometers to micrometres or greater [3]. This makes their thickness-to-lateral size ratio high [4]. To date, multitudinous 2D materials have been identified, such as graphene, silicene, germanene, and their derivatives. In contrast to three-dimensional (3D) bulk materials, 2D materials possess a number of unique properties, such as an exceptionally high specific surface area, carrier mobility, optical properties, and thermal conductivity, allowing them to be utilized in a broad range of disciplines like optics, sensors, phototherapy, and storage of energy [5]. Few 2D materials display excellent biodegradability and biocompatibility, proving their applicability in biomedicine. Since the surface of 2D materials does not have dangling bonds, enabling a new technique for enhancing the biocompatibility, biodegradability, or stability of particular 2D materials and expanding their biomedical applications can be used with photonic materials [6].

In a dental setting, materials used in the oral cavity are in constant contact with saliva and are susceptible to mastication, abrasion, and extreme temperatures, which can lead to mechanical failures and replacement over time. In addition, they remain in close contact with oral tissues for an extended period; consequently, they must be highly biocompatible and noncytotoxic to function effectively while interacting with the host tissues. Thus, recent trends show increased attention towards developing dental materials with enhancing properties.

To date, there have been no comprehensive studies describing the function of various 2D materials in dentistry; however, numerous studies evaluating the properties of individual dental materials have been reported. The majority of them concentrated on using 2D materials to enhance mechanical properties, biocompatibility, cell growth, differentiation, and antibacterial properties. Various disciplines of dentistry have reported the functions and applications of 2D materials such as graphene and its derivatives, silica-based compounds, molybdenum, tungsten disulphide, and boron [7–9]. Although research on 2D materials in dentistry is still in its infancy, their unique properties and potential to function as biomaterials, either alone or in conjunction with other materials, present numerous clinical application opportunities. This chapter provides a comprehensive overview of the functions and applications of various 2D materials in various dental specialities. Figure 1 depicts different types of 2D materials that are involved in dentistry.

2 Two-Dimensional Materials, Synthesis, and Properties

2.1 Graphene

Graphene is the strongest, thinnest, and most promising 2D carbon-based material. Andre Geim and Konstantin Novoselov effectively isolated graphene in 2004 and demonstrated its exceptional mechanical, physiochemical, and optical properties [10]. To date, there are nearly 3,000 studies investigating the function of graphene-based materials in various dental applications. Graphene and its derivatives have been

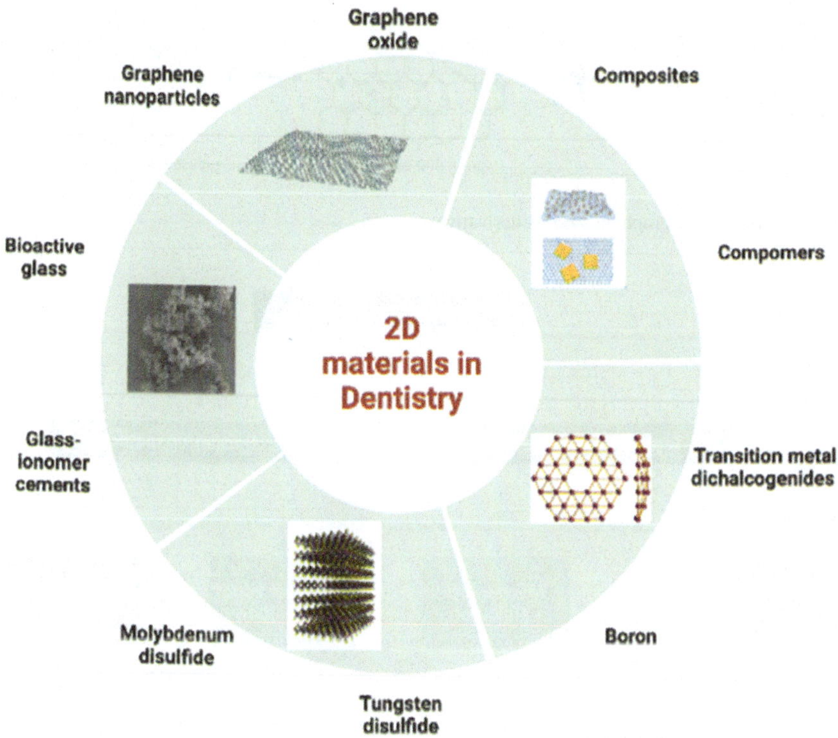

Fig. 1 Different 2D materials involved in various aspects of dentistry

investigated for applications in dentistry, bone cement, dental implants, and scaffold engineering [11] due to their exceptional biocompatibility and conductivity.

Graphene, with its one-atom-thick layer, is currently the strongest and thinnest material. Its structure entails six-membered rings stacked parallelly, consisting of one sp2-hybridized carbon atom layer organized in a hexagonal setting (Fig. 2) [12]. Graphene's strong mechanical strength, wide surface area, superior conductivity, etc., piqued curiosity [13]. Graphene and graphene-based materials like reduced graphene oxide (rGO) and graphene oxide (GO), with identical structures but distinct functional groups, have more flexible physical and chemical properties. There are two graphene synthesis methods: top-down and bottom-up [14]. The top-down method includes chemical exfoliation, mechanical exfoliation, and liquid-phase exfoliation, while the bottom-up methods include chemical vapour deposition (CVD) and solid-phase deposition methods. Figure 3 illustrates different isolation methods of graphene.

In 2004, Novoselov et al. isolated graphene using mechanical exfoliation methods with adhesive tape; graphene was subsequently desorbed and obtained [10]. However, the yield was extremely low despite the absence of chemical groups and high purity. During the process of liquid-phase exfoliation, Van der Waals forces between graphite layers are weakened by graphite suspension forces in a solvent that is organic. Then,

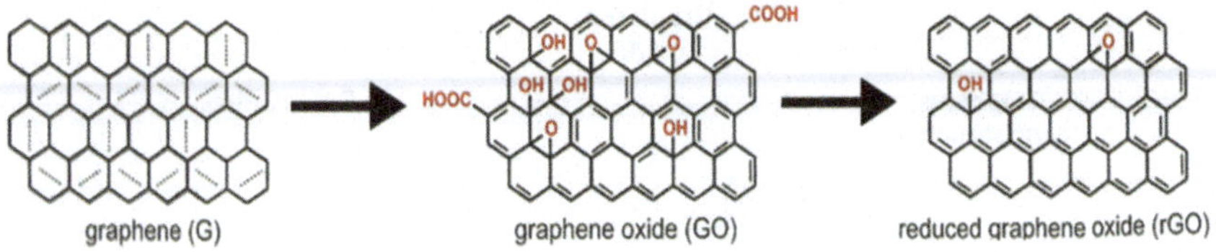

Fig. 2 Structure of graphene and its derivatives

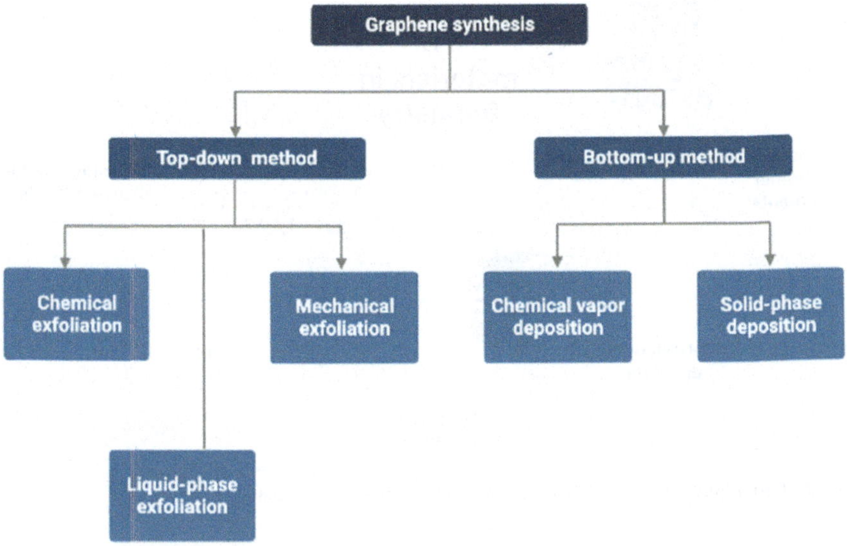

Fig. 3 Different isolation methods of graphene

ultrasonic waves at a voltage were used to convert graphite into graphene sheets [15]. This method is advantageous for small-scale graphene synthesis, even though the yield is still low. The chemical exfoliation method can produce graphene synthesis on a large scale. GO is produced by stirring or ultrasonic reaction of graphite with sodium nitrate, sulfuric acid, and potassium permanganate in water. GO exfoliation requires high temperatures of 1,000 °C. Subsequently, reducers transform GO into rGO. In the end, heat or chemical treatments transform rGO into graphene. Moreover, bottom-up methods, such as CVD exfoliation, are the most successful methods, as their economical and abundant synthesis of improved-quality graphene monolayer on metal surfaces is widely accepted. From methane, ethane, or propane, high-temperature pyrolysis forms carbon atoms on metal foils like Ru, Pt, Cu, Fe, Ni, and Fe. Thereby, graphene was produced from free carbon atoms [16]. In order to develop graphene-based dental materials, it is necessary to assess their bioactive, biocompatibility, cytotoxicity, and antibacterial properties. Few studies have reported that GO's toxicity to fibroblasts is dose-dependent [17], and as GO concentration increased, so did in vivo toxicity. In addition, certain graphene-based 2D materials induced tissue inflammation in and around the dental implants' soft and bone tissue [18]. The ability to stimulate cell differentiation is another characteristic of this type. Multiple in vitro experiments have revealed that graphene and graphene-based products can induce dental pulp regeneration and osteogenic differentiation [19–21]. While researching the ability of dental pulp stem cells (DPSCs) to undergo osteogenic differentiation, osteogenic genes, and proteins were discovered to be upregulated on graphene. Graphene-based nanomaterials have also been shown to possess neurogenic differentiation capability [22]. In addition, the GO-modified scaffold promoted periodontal tissue regeneration in vivo [23]. Last but not least, the antibacterial property of biomaterials is crucial for their application, and numerous studies have confirmed the antibacterial effect of graphene-based materials [24].

2.2 *Silicate-Based Compounds*

Silicon acts as a half-metallic compound that, at high temperatures, oxidizes to form silicon dioxide (SiO_2). Silicates are esters and salts of mono-silicic acid (H_4SiO_4). Silicate and silicate compounds can often be used as filling materials in various dental fillers. Some of the silicate-based compounds include composites, compomers, glass-ionomer cement (GICs), and adhesive systems. They exhibit various physiochemical properties, which can be regulated depending on their application.

Wilson and Kent first introduced GICs as dental fillers in the late 1960s [25]. Glass-ionomer cement (GICs) were constructed from silicate and carboxylate cement containing crystalline and liquid components. Combining strontium-fluoro-silicate and calcium-aluminium–silicate glasses with polycarbonic and polyacrylic acids produces precipitates of these substances. The subsequent acid–base response produces a silica gel matrix [26]. The fillers are created by simultaneously melting the various components between 1200 and 1550 degrees Celsius. GICs may bind to hard dental tissues via carboxylate groups of chain-like acid molecules. Dental composite resins are complex substances composed of an organic matrix, such as bisphenol A-glycidyl methacrylate, and an inorganic filler, such as silica. Typically, silane is employed as a bridge between these two organic and inorganic phases [27]. The filler material frequently determines the properties of the resin, such as wear resistance, mechanical strength, and translucency.

In addition, silica glass containing only SiO2 molecules contains very few metallic impurities. They are mechanically weaker but exhibit superior strength when etched with resin bonding and can be used as resin-bonding cement. Silica-based ceramic crowns demonstrate superior optical properties due to their higher translucent nature than crowns based on zirconia or alumina. Due to greater translucency, silica/glass-based systems demonstrate superior esthetics compared to crowns based on zirconia or alumina when used on tooth preparations with favourable tooth colour [28].

2.3 Molybdenum and Tungsten Disulfide

Similar to graphene, tungsten disulfide (WS_2) and molybdenum disulfide (MoS_2) are layered transition metal dichalcogenides (TMDCs). These composites are generated using materials having a multilayered arrangement, which includes an even surface of hexagonal (H) crystals of respective metals, either Mo or W, incorporated within two layers of sulfur atoms with anionic charge, which are 2H-WS2 and 2H-MoS2, respectively. In specific circumstances, these potentially precarious structures can bend and close in on themselves, connecting the edge atoms to form graphite-like 20–200 nm particles [29]. The unique orientation of WS_2 and MoS_2 nanoparticles makes them outstanding dense lubricants, enhancing wear and friction characteristics under dry and wet conditions on various load levels [30]. WS_2 nanoparticles penetrating into the interaction between rubbed surfaces may cause this. Nanoparticles alter and exfoliate as the load between bodies increases. Thus, Coating the interface asperities permits low-shear force movement among the two interacting bodies.

Graphene's exfoliation from bulk graphite allows the "Scotch tape method" to generate more graphene-like 2D materials [31]. Mechanical exfoliation, like graphene, is suitable only for low-yield production and has disadvantages in modulating layer number and sheet size. Chemical exfoliation increases production more than mechanical exfoliation. However, sonication throughout the method leads to 2D lattice arrangement defects and lowers flake size to a minimum of a few 1000 nm, restricting 2D nanosheet applications in big-scale integrated circuits as well as electronic devices [32]. Currently, regulated exfoliation of 2D TMDCs with large-area homogeneity continues to be challenging. CVD could produce wafer-scale 2D TMDCs as a continuous single film of a specific thickness.

2.4 Boron

Numerous elements and mixtures have been capable of forming 2D materials since the discovery of graphene. Boron's orientation on the periodic table is more mysterious than that of its peers. It possesses various bulk stages, and even more diverse are its nanostructures, which include planar and cage-like clusters, nanowires, 1D nanotubes, nanofilms, and 2D sheets. It is complementary to graphene, hexagonal boron nitride (h-BN), and TMDCs due to the 2D boron's metallic nature. It is an isomorph of graphene with an identical layered structure. CVD remains the preferred technique for synthesizing 2D boron [33]. 2D-hBN is a more recent 2D material that attracted significant interest due to its different characteristics like high-thermal stability, corrosion resistance, electric insulation, easy synthesis, and chemical inertness. It can be easily incorporated with other 2D materials like graphene and TMDCs to be used in multiple electronic and optical devices. Despite its insulating properties, 2D-hBN's properties and functionalities can be enhanced using several techniques such as hybridization, substitution, and doping, allowing it to be an ideal material type for different applications. In addition, various other boron-containing compounds were reported to have superior strength and anti-inflammatory, anti fungal, and cariostatic activity. Combined with biomaterials, they may play a role in osteogenic development or differentiation [34].

3 Applications in Dentistry

Owing to their improved methods of synthesis, expanded types and derivatives, and enhanced properties, various 2D materials have been shown to have applications across different fields of dentistry. Figure 4 illustrates the different applications of 2D dental materials. Table 1 indicates the various properties and functions of graphene-based materials.

Fig. 4 Different applications of 2D materials across various fields of dentistry

Table 1 Various applications, properties, and functions of various graphene-based materials

Function	Graphene-based materials	Properties	Application type
Periodontal tissue regeneration	GO	Osteogenic differentiation	Scaffolds
	GO	Cementoblast differentiation	Scaffolds
Dental implants	Monolayer graphene	Osteogenic differentiation	Coatings
	GO/HA	Osteogenic differentiation	Coatings
	rGO	Osteogenic differentiation	Coatings
	GO	Antibacterial property	Coatings
Collagen member	GO	Roughness and stiffness	Coatings
	Graphene	Anti-inflammatory properties	Coatings
Bone tissue engineering	Graphene/HA	Biomimetic mineralization	Scaffolds
	rGO	Osteogenic differentiation	Coatings
	rGO/HA	Mineralization	Scaffolds
	GO/chitosan	Osteogenic differentiation	Coatings
Dental pulp regeneration	Graphene dispersion	Neural differentiation	Scaffolds
	rGO	Neural differentiation	Scaffolds
	GO	Osteogenic differentiation	Scaffolds

3.1 Endodontics

Root canal treatment is the most common procedure where an affected root canal is cleaned to treat pathogens, disinfect the remaining infested tissue, and evade additional infection, and a dental filling is used to seal the space. With multiple contributing factors such as persisting infection and choice of dental filling material, endodontic file separation has recently become a recurring issue affecting root canal treatment success rates [35]. In a recent photodynamic therapy, indocyanine green (ICG), a nontoxic photosensor material used for effective root canal disinfection, has concerns regarding its stability and aggregation. However, when used in conjugation with GO, it reduced bacterial activity remarkably and enhanced the ICG's bioavailability and stability [36]. Similarly, incorporating graphene into silver nanoparticles increased sodium hypochlorite's efficacy, biocompatibility, and antibacterial properties, a common intracanal irrigant [37].

The most popular bioactive cements, such as Biodentine and Endocem-Zr, have limitations such as a long setting time and a high pull-out bond strength. According to one study, incorporating graphene nanosheets reduced both types of cement' setting time [38]. Another recently developed cement, mineral trioxide aggregate (MTA), has obtained acceptance in endodontic treatments because of its potential to promote odontoblast differentiation and calcium deposition [39]. Nonetheless, its limited physical properties, such as poor wear resistance and brittleness, limit its application. The reinforcement of graphene nanomaterials (GFN) has been shown to improve grain size refinement, brittle index, and fracture durability. Furthermore, GFN incorporation promoted the proliferation of human osteoblastic cells (hFOB) [40]. A small amount of GO (0.03%) added to Portland cement can boost its tensile and compressive strength by 40% [39]. Overall, the incorporation of graphene into dental cement improves pore refinement, thereby strengthening and inhibiting bacterial invasion. In addition, silicate-based compounds are widely used as fillers in dental filling materials like composites, calcium silicate cements, and adhesive systems. These fillers in these materials enhance their mechanical properties like thermal expansion coefficient and physical resistance. A boron-rich diet improved the mineral composition of teeth. Studies have shown that boron exerts a cariogenic effect, as it was determined that boron levels in healthy non-carious teeth were shown to be greater compared to carious teeth [41]. The combination of boron in dental materials has been further shown to regulate the development of dental caries development, exhibiting significant antibacterial effects [42].

3.2 *Periodontics*

Barrier membranes are key biomaterials in periodontal bone defect treatment via guide tissue and bone regeneration. They separate soft connective tissue from regenerating bone to allow bone growth and quicker mesenchymal cell differentiation into odontoblast/osteoblast. Studies have aimed to modulate barrier membranes to enhance their biocompatibility. Radunovic et al. demonstrated the effects of collagen membrane coated with GO on metabolic activity and viability of DPSCs [23]. Results revealed that GO coating regulates inflammation and promotes DPSCs differentiation, owing to the large surface area of GO. Furthermore, an in vivo study demonstrated the wound-healing ability of GO in dogs. The GO scaffold successfully regenerated cementum-like tissue formation prior to alveolar bone formation [23].

Recently, irrigation with of boronic acid revealed possible adjuvant in treating mild to moderate gingivitis and thus serves as a prognostic and treatment strategy in periodontal disease [43–45]. AN0128, a B-containing synthetic boronic acid, has been extensively explored. In vitro, this demonstrated anti-inflammatory and antibacterial properties towards periodontal disease caused by bacteria *Treponema denticola, Prevotella intermedia, Eubacterium nodatum,* and *P. gingivalis* [46]. Silica-based bioactive glass (BG) is a non-crystalline ceramic that binds to living tissues and induces fresh tissue development whilst disintegrating gradually; consequently, BG is regarded as a desirable target for tissue engineering applications [47]. Incorporating a variety of polymers with BG granules in composite resins helps overcome the limitations of BG crystallization. Modified BG, BGMS100 was incorporated with polyethylene glycol (PEG) and collagen to form BGMS/C compound for dental applications, showing improved cell proliferation. BG-containing products like PerioGlas induce periodontal bone regeneration when packed inside the periodontal defect site [48]. Furthermore, during endodontic treatment, PerioGlas promotes effective apical bone structure regeneration [49]. Another BG product, ERMI (endosseous ridge maintenance implant), is introduced in periodontics to aid in maintaining alveolar ridge height from resorption. Moreover, BG revealed remarkable improvement in gingivitis-related signs such as gingival bleeding and plaque formation [50]. Topical application of BG reduced inflammation of gingival tissues in patients with gingivitis [51].

3.3 Implantology

Osseointegration requires implant material with osteogenic characteristics, whereas a tight epithelial sealing inhibiting bacterial invasion is necessary at the soft tissue contact. Infection by bacteria and colonization from seal leaks at any interface may impede osteogenesis and cause bone loss. Among several alternatives, Ti and its composites remain the vastly popular biomaterial of choice for dental implants [52]. Recent studies have demonstrated that altering the properties of titanium, such as surface topography, composition, and geometry, can impact the efficiency of osseointegration, making it an excellent coating material [53]. Coating titanium substrate with graphene, the surface hydrophobic property of graphene reduced the surface adhesion of bacteria such as S. mutans and S. Sanguinis [54]. Another study revealed that GO-Ti substrate facilitated PDLSC adhesion, differentiation, and proliferation compared to the of Na-Ti substrate [55]. GO/chitosan/hydroxyapatite-Ti composite exhibited superior bioactivity in vivo by enhancing the bone marrow stem cell attachment, differentiation, and proliferation, and superior osseointegration in vitro [56]. GO incorporation on Ti surface enabled SP and BMP-2's dual delivery, with most of the fresh bone regeneration in the murine calvarium occurring on Ti implants. BGs pose a higher prospect of bonding and integrating with human tissue in comparison to metal implants, owing to their excellent bioactivity and biocompatibility. Surface coating with BGs offers various advantages like restoring injured tissue as well as injured bone, integrating effectively to the host body's environment, enabling tissue restoration, and degradation at a rate identical to tissue restoration. Novel glass compositions have been suggested for coatings and dental fillings [57]. This configuration flexibility enables researchers to add functions like Sr2 + -enhanced bone growth and Cu2-enhanced angiogenesis. For example, to address the thermal expansion differential between Bioglass and titanium, the Na2O and CaO in the Bioglass system are changed with K2O and MgO, respectively [58]. BG-coated implants revealed increased bone development and corrected thermal expansion coefficients using precise composition proportions when compared to non-coated implants. Moreover, research has demonstrated that upon releasing ions such as boron, dental material exhibits antimicrobial, acid-resistant, and remineralization properties [59]. In addition, the impact of incorporating boron on the biocompatibility and strength of sintered implants was examined.

3.4 Restorative Dentistry

Because of its excellent linear thermal expansion coefficient and dynamic fluoride release, glass ionomers (GI) have recently been used in various clinical functionalities. Recent efforts are being made to combine nanomaterial obtained from graphene into swiftly accessible GI for strengthening. Graphene-based materials incorporated with GI improved their physiochemical characteristics only to an extent [60]. However, when mixed with glass ionomer, fluoride graphene generates GICs/FG composite, significantly enhancing the mechanical and antimicrobial properties of glass ionomer [61]. Reinforcing resin matrices made of polymers containing graphene gold nanoparticles as fillers improved conversion and surface attributes, improving dental nanocomposite physicochemical properties [62]. Moreover, GFNs are proposed as a potential antibiofilm as well as antimicrobial filler in dental adhesives because they significantly inhibit the attachment and development of *S. mutans*.

Burke et al. arguably suggested silica/glass-based ceramic crowns as ideal for restorations [63]. A study has suggested that applying ceramic crowns made entirely of silica/glass in conjunction with different resin cements presents the potential to give a more aesthetically pleasing alternative over standard ceramic–metal crowns [64]. In restoration dentistry, MTA is suggested as a promising material for direct capping and preserving pulp vitality, owing to its excellent biocompatibility and bioactive properties [65]. Another silicate composite, Biodentine, is suggested in both direct and indirect pulp capping. It has also been used as enamel and dentine restorative material [66]. Biodentine is an effective choice for deep dental caries treatment, including reversible pulpal inflammation cases. BiodentineTM's bioactive characteristics may seal pulp-dentin, improving pulp response and tooth decay tissues.

The remineralization of hard dental tissues (enamel and dentin) is a complex issue that dentists worldwide face on a regular basis. BGs can release huge quantities of phosphate and calcium ions when treated with an acidic medium, remineralizing hard dental tissues based on previous studies [67–70]. BGs derivatives like Bioglass and fluoride bioactive glass have demonstrated promising remineralizing results in enamel and dentine tissues. Furthermore, bio-silicate (P2O5–Na2O–CaO–SiO2), a wholly crystalline bioactive glass–ceramic, is proposed for the DH treatment by hydroxyl carbonate apatite incorporation in exposed dentinal tubules [71].

3.5 Tissue Regeneration

In designing biomaterials for bone tissue regeneration applications in dentistry, osteointegration, and stem cell differentiation are key factors to be taken into account. Graphene is the most extensively studied 2D material for its role in tissue engineering applications. Figure 5 illustrates the applications of graphene-based materials in dental tissue regeneration. It is shown to facilitate cell processes such as cell attachment, cell differentiation, and proliferation in dental tissues. Unique properties like high surface area, functionalization feasibility, and mechanical strength are ideal for tissue engineering. Several methods to synthesize graphene-based materials for bone tissue engineering include CVD, electrospinning, solvent casting, and freeze-dying methods [56, 72–74]. CVD may produce tissue engineering scaffolds and graphene films. Using CVD, dental pulp stem cells (DPSCs) are utilized to evaluate graphene-incorporated dental materials' biocompatibility, proliferation, and cell differentiation. DPSCs entail a muti potency, allowing them to differentiate into odontoblasts, which aid in bone matrix formation and dentine regeneration in tooth [75].

GO-coated substrates were shown to improve the cell attachment and proliferation of DPSCs and further upregulated the expression of RUNX2 proteins, which play a role in the differentiation of osteoblasts [76]. CVD-based graphene films showed better osteogenic functionalities for bone tissue regeneration, and their scaffolds

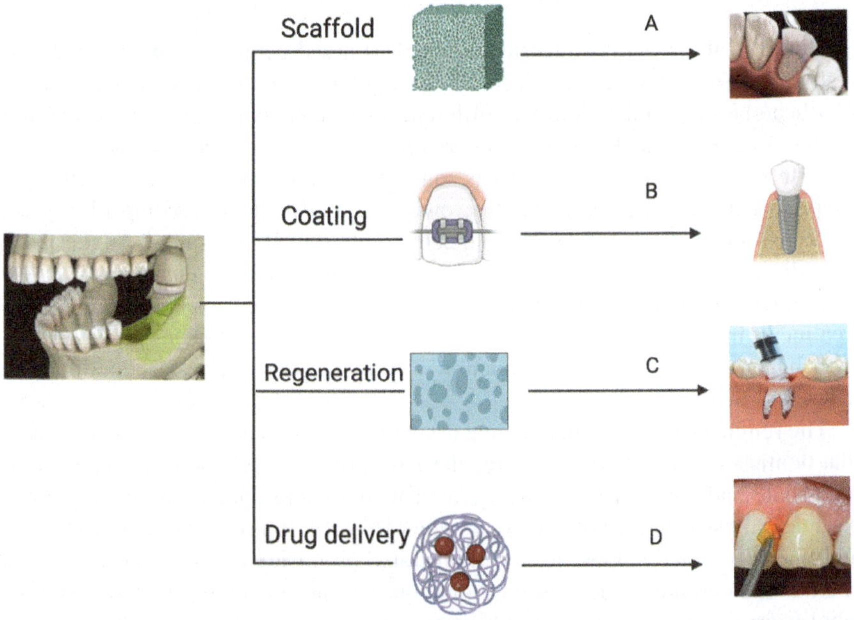

Fig. 14.5 Graphene-based materials incorporated in scaffold materials for tissue regeneration (A), dental coatings for osteointegration (B), directed bone membrane regeneration (C), and drug delivery systems (D)

facilitate osteogenesis in vivo [21]. Moreover, combining GO on polyether ether ketone (PEEK)-based materials enhanced antibacterial effects against oral bacteria (*Porphyromonas gingivalis*) and proliferation in embryonic osteoblast MC3T3-E1 [77]. Furthermore, graphene coating on titanium implants has been shown to enhance antibacterial properties and cell proliferation [78]. GO and silver composite coatings can be incorporated with NiTi alloy implants to facilitate dental pulp fibroblasts' biocompatibility and lower inflammatory responses. It demonstrated anti-corrosive properties and upregulated expression levels of a few anti-inflammatory mediators like IL-6 and IL-8 [79].

Overall, biocompatible substrates with graphene-based materials facilitate cell adhesion, viability, and proliferation. In addition, silica-based biomaterials like calcium silicate (CS) are recently been studied for their excellent antimicrobial, biocompatibility, osteogenesis, and osteoinduction properties. CS bone cement was proposed for marrow capping, where it was found that light-cured CS bone cement as an indirect pulp capping agent can be a reliable and simple strategy for the restoration of severe caries teeth [80]. Due to their unique physical, biological, and mechanical properties, molybdenum disulfide-based materials find widespread use in biomedicine. The application of bioactive glass scaffolds containing MoS2 nanoparticles in bone tissue engineering was evaluated. It was discovered that the polymer coating layer and the presence of MoS2 nanoparticles in the polymer matrix enhance the mechanical properties of the scaffolds, which also demonstrate a high radiation-shielding property [81].

4 Conclusions and Future Perspectives

Evidently, though there are many applications and advantages of 2D materials in dentistry, the research is still in its infancy with most studies in in vivo and in vitro stages. These findings are yet to be validated in humans. Over the years, many challenges have been resolved through multitudinous studies. Almost all materials exhibited excellent physiochemical properties such as anti-inflammatory, antimicrobial, biocompatibility, and bioactive properties. The research on 2D materials mainly focussed on two aspects: first, to construct novel dental materials using individual materials and second, to alter the properties and integrate the common dental materials. Among all the 2D-based materials, graphene was the most extensively studied material for its application in dentistry. GFNs enhanced biomaterials' physical, chemical, and mechanical properties, bearing immense potential for developing novel dental therapeutic approaches. Silicate-based compounds like calcium-silicate cement and Bioactive glass are one of the extensively studied dental biomaterials. BG composites have been created to improve BGs' bioactive characteristics. Boron-containing compounds are still making their way into applications as dental materials with their excellent antimicrobial and anti-inflammatory properties. Molybdenum disulfide is the least explored 2D material in dentistry, with few studies iterating its role in orthodontic stainless archwires with strong mechanical properties. Molybdenum disulfide and titanium disulfide are often incorporated as a coating to enhance the properties of other biomaterials. This chapter provides an overview of different 2D materials, their synthesis, and their role in different disciplines of dentistry.

Emphasis should be put on the safety and toxicity of 2D materials in an oral setting to ensure harmless application. Moreover, recent research in regenerative medicine has accumulated so many findings, making it more promising in the near future. GFNs and BGs have been widely studied in vitro and in vivo across various human body tissues; however, no human clinical trials have been conducted, and detailed studies are still lacking in oral setup. Therefore, additional studies are required into their in-depth interactive mechanisms within oral tissues, such as antimicrobial properties, cell signaling, osteogenesis, and metabolic pathways.

References

1. Wei, Y., Tang, X., Shang, J., Ju, L., Kou, L.: Two-dimensional functional materials: from properties to potential applications. Int. J. Smart Nano Mater. **11**(3), 247–264 (2020)
2. Gray, E.P., Browning, C.L., Wang, M., Gion, K.D., Chao, E.Y., Koski, K.J., Hurt, R.H., et al.: Biodissolution and cellular response to MoO_3 nanoribbons and a new framework for early hazard screening for 2D materials. Environ. Sci. Nano **5**(11), 2545–2559 (2018)
3. Cheng, L., Wang, X., Gong, F., Liu, T., Liu, Z.: 2D nanomaterials for cancer theranostic applications. Adv. Mater. **32**(13), 1902333 (2020)
4. Machnicki, C.E., Fu, F., Jing, L., Chen, P.Y., Wong, I.Y.: Mechanochemical engineering of 2D materials for multiscale biointerfaces. J. Mater. Chem. B **7**(41), 6293–6309 (2019)
5. Wang, X., Han, X., Li, C., Chen, Z., Huang, H., Chen, J., Zhang, H., et al.: 2D materials for bone therapy. Adv. Drug Deliv. Rev. (2021)
6. Zheng, W., Zheng, B., Yan, C., Liu, Y., Sun, X., Qi, Z., Pan, A., et al.: Direct vapor growth of 2D vertical heterostructures with tunable band alignments and interfacial charge transfer behaviors. Adv. Sci. **6**(7), 1802204 (2019)
7. Ge, Z., Yang, L., Xiao, F., Wu, Y., Yu, T., Chen, J., Zhang, Y., et al.: Graphene family nanomaterials: properties and potential applications in dentistry. Int. J. Biomater. (2018)
8. Lührs, A.K., Geurtsen, W.: The application of silicon and silicates in dentistry: a review. Biosilica Evol. Morphogenesis, Nanobiotechnol. Case Study Lake Baikal, 359–380 (2009)
9. Corral-Núñez, C., Fernández-Godoy, E., Casielles, J.M., Estay, J., Bersezio-Miranda, C., Cisternas-Pinto, P., Batista-de Oliveira, O.: The current state of calcium silicate cements in restorative dentistry: a review. Revista Facultad de Odontología Universidad de Antioquia **27**(2), 425–441 (2016)
10. Novoselov, K.S., Geim, A.K., Morozov, S.V., Jiang, D.E., Zhang, Y., Dubonos, S.V., Firsov, A.A., et al.: Electric field effect in atomically thin carbon films. Science **306**(5696), 666–669 (2004)
11. Li, X., Liang, X., Wang, Y., Wang, D., Teng, M., Xu, H., Han, L., et al.: Graphene-based nanomaterials for dental applications: principles, current advances, and future outlook. Front. Bioeng. Biotechnol. **10** (2022)
12. Bei, H.P., Yang, Y., Zhang, Q., Tian, Y., Luo, X., Yang, M., Zhao, X.: Graphene-based nanocomposites for neural tissue engineering. Molecules **24**(4), 658 (2019)
13. Du, Z., Wang, C., Zhang, R., Wang, X., Li, X.: Applications of graphene and its derivatives in bone repair: advantages for promoting bone formation and providing real-time detection, challenges and future prospects. Int. J. Nanomed., 7523–7551 (2020)

14. Liao, C., Li, Y., Tjong, S.C.: Graphene nanomaterials: synthesis, biocompatibility, and cytotoxicity. Int. J. Mol. Sci. **19**(11), 3564 (2018)
15. Ghuge, A.D., Shirode, A.R., Kadam, V.J.: Graphene: a comprehensive review. Curr. Drug Targets **18**(6), 724–733 (2017)
16. Gu, M., Liu, Y., Chen, T., Du, F., Zhao, X., Xiong, C., Zhou, Y.: Is graphene a promising nanomaterial for promoting surface modification of implants or scaffold materials in bone tissue engineering? Tissue Eng. Part B Rev. **20**(5), 477–491 (2014)
17. Wang, K., Ruan, J., Song, H., Zhang, J., Wo, Y., Guo, S., Cui, D.: Biocompatibility of graphene oxide. Nanoscale Res. Lett. **6**, 1–8 (2011)
18. Bressan, E., Ferroni, L., Gardin, C., Bellin, G., Sbricoli, L., Sivolella, S., Zavan, B., et al.: Metal nanoparticles released from dental implant surfaces: potential contribution to chronic inflammation and peri-implant bone loss. Materials **12**(12), 2036 (2019)
19. Lee, J.H., Shin, Y.C., Jin, O.S., Kang, S.H., Hwang, Y.S., Park, J.C., Han, D.W., et al.: Reduced graphene oxide-coated hydroxyapatite composites stimulate spontaneous osteogenic differentiation of human mesenchymal stem cells. Nanoscale **7**(27), 11642–11651 (2015)
20. Norahan, M.H., Amroon, M., Ghahremanzadeh, R., Rabiee, N., Baheiraei, N.: Reduced graphene oxide: osteogenic potential for bone tissue engineering. IET Nanobiotechnol.Nanobiotechnol. **13**(7), 720–725 (2019)
21. Xie, H., Chua, M., Islam, I., Bentini, R., Cao, T., Viana-Gomes, J.C., Rosa, V., et al.: CVD-grown monolayer graphene induces osteogenic but not odontoblastic differentiation of dental pulp stem cells. Dent. Mater. **33**(1), e13–e21 (2017)
22. Seonwoo, H., Jang, K.J., Lee, D., Park, S., Lee, M., Park, S., Chung, J.H., et al.: Neurogenic differentiation of human dental pulp stem cells on graphene-polycaprolactone hybrid nanofibers. Nanomaterials **8**(7), 554 (2018)
23. Kawamoto, K., Miyaji, H., Nishida, E., Miyata, S., Kato, A., Tateyama, A., Sugaya, T., et al.: Characterization and evaluation of graphene oxide scaffold for periodontal wound healing of class II furcation defects in dog. Int. J. Nanomed.Nanomed. **13**, 2365 (2018)
24. Gholibegloo, E., Karbasi, A., Pourhajibagher, M., Chiniforush, N., Ramazani, A., Akbari, T., Khoobi, M., et al.: Carnosine-graphene oxide conjugates decorated with hydroxyapatite as promising nanocarrier for ICG loading with enhanced antibacterial effects in photodynamic therapy against Streptococcus mutans. J. Photochem. Photobiol. B.Photochem. Photobiol. B. **181**, 14–22 (2018)
25. Kent, B.E.: The properties of a glass ionomer cement. Br. Dent. J. **135**, 322–326 (1973)
26. Nicholson, J.W., Brookman, P.J., Lacy, O.M., Wilson, A.D.: Fourier transform infrared spectroscopic study of the role of tartaric acid in glass-ionomer dental cements. J. Dent. Res. **67**(12), 1451–1454 (1988)
27. Becker, J.: Kariologie und Füllungstherapie, vol. 4. In: Heidemann, D. (ed.). Urban & Schwarzenberg, München (1999)
28. Raigrodski, A.J.: All-ceramic full-coverage restorations: concepts and guidelines for material selection. Pract. Proced. Aesthetic Dent. PPAD **17**(4), 249–256 (2005)
29. Tenne, R., Margulis, L., Genut, M.E., Hodes, G.: Polyhedral and cylindrical structures of tungsten disulphide. Nature **360**(6403), 444–446 (1992)
30. Rapoport, L., Bilik, Y., Feldman, Y., Homyonfer, M., Cohen, S.R., Tenne, R.: Hollow nanoparticles of WS_2 as potential solid-state lubricants. Nature **387**(6635), 791–793 (1997)
31. Li, H., Wu, J., Yin, Z., Zhang, H.: Preparation and applications of mechanically exfoliated single-layer and multilayer MoS_2 and WSe_2 nanosheets. Acc. Chem. Res. **47**(4), 1067–1075 (2014)
32. Li, X., Zhu, H.: Two-dimensional MoS_2: properties, preparation, and applications. J. Materiomics **1**(1), 33–44 (2015)
33. Zhang, Z., Penev, E.S., Yakobson, B.I.: Two-dimensional boron: structures, properties and applications. Chem. Soc. Rev. **46**(22), 6746–6763 (2017)
34. Mitruţ, I., Scorei, I. R., Manolea, H. O., Biţă, A., Mogoantă, L., Neamţu, J., Mogoşanu, G.D.: Boron-containing compounds in dentistry: a narrative review. Rom. J. Morphol. Embryol. Revue Roumaine de Morphologie et Embryologie **63**(3), 477–483 (2022)

35. Yamaguchi, M., Noiri, Y., Itoh, Y., Komichi, S., Yagi, K., Uemura, R., Ebisu, S., et al.: Factors that cause endodontic failures in general practices in Japan. BMC Oral Health **18**, 1–5 (2018)
36. Akbari, T., Pourhajibagher, M., Hosseini, F., Chiniforush, N., Gholibegloo, E., Khoobi, M., Bahador, A., et al.: The effect of indocyanine green loaded on a novel nano-graphene oxide for high performance of photodynamic therapy against Enterococcus faecalis. Photodiagn. Photodyn. Ther.. Photodyn. Ther. **20**, 148–153 (2017)
37. Sharma, D.K., Bhat, M., Kumar, V., Mazumder, D., Singh, S.V., Bansal, M.: Evaluation of antimicrobial efficacy of graphene silver composite nanoparticles against E. faecalis as root canal irrigant: an ex-vivo study. Int. J. Pharm. Med. Res **3**(5), 267–272 (2015)
38. Dubey, N., Rajan, S.S., Bello, Y.D., Min, K.S., Rosa, V.: Graphene nanosheets to improve physico-mechanical properties of bioactive calcium silicate cements. Materials **10**(6), 606 (2017)
39. Gong, K., Pan, Z., Korayem, A.H., Qiu, L., Li, D., Collins, F., Duan, W.H., et al.: Reinforcing effects of graphene oxide on portland cement paste. J. Mater. Civ. Eng. **27**(2), A4014010 (2015)
40. Mehrali, M., Moghaddam, E., Seyed Shirazi, S.F., Baradaran, S., Mehrali, M., Latibari, S.T., Osman, N.A.A., et al.: Mechanical and in vitro biological performance of graphene nanoplatelets reinforced calcium silicate composite. PLoS ONE **9**(9), e106802 (2014)
41. Rüya, K., Balan, G., Yilmaz, Ş., Taşli, P.N., Akyüz, S., Yarat, A., & Şahin, F. : The level of two trace elements in carious, non-carious, primary, and permanent teeth. Eur. Oral Res. **54**(2), 77–80 (2020)
42. Demirci, S., Kaya, M.S., Doğan, A., Kalay, Ş., Altin, N. Ö., Yarat, A., Şahin, F., et al.: Antibacterial and cytotoxic properties of boron-containing dental composite. Turk. J. Biol. 39(3), 417–426 (2015)
43. Hernandez, V., Crépin, T., Palencia, A., Cusack, S., Akama, T., Baker, S.J., Plattner, J.J., et al.: Discovery of a novel class of boron-based antibacterials with activity against gram-negative bacteria. Antimicrob. Agents Chemother.. Agents Chemother. **57**(3), 1394–1403 (2013)
44. Sağlam, M., Arslan, U., Buket Bozkurt, Ş, Hakki, S.S.: Boric acid irrigation as an adjunct to mechanical periodontal therapy in patients with chronic periodontitis: a randomized clinical trial. J. Periodontol.Periodontol. **84**(9), 1297–1308 (2013)
45. Kilic, A., Beyazsakal, L., Findik, B.T., Incebay, H.: Synthesis and electrochemical investigation of chiral amine bis (phenolate)-boron complexes: In vitro antibacterial activity screening of boron compounds. Inorg. Chim. Acta. Chim. Acta **510**, 119777 (2020)
46. Luan, Q., Desta, T., Chehab, L., Sanders, V.J., Plattner, J., Graves, D.T.: Inhibition of experimental periodontitis by a topical boron-based antimicrobial. J. Dent. Res. **87**(2), 148–152 (2008)
47. Bellucci, D., Salvatori, R., Giannatiempo, J., Anesi, A., Bortolini, S., Cannillo, V.: A new bioactive glass/collagen hybrid composite for applications in dentistry. Materials **12**(13), 2079 (2019)
48. Lovelace, T.B., Mellonig, J.T., Meffert, R.M., Jones, A.A., Nummikoski, P.V., Cochran, D.L.: Clinical evaluation of bioactive glass in the treatment of periodontal osseous defects in humans. J. Periodontol.Periodontol. **69**(9), 1027–1035 (1998)
49. Pantchev, A., Nohlert, E., Tegelberg, Å.: Endodontic surgery with and without inserts of bioactive glass PerioGlas®—a clinical and radiographic follow-up. Oral Maxillofac. Surg.Maxillofac. Surg. **13**, 21–26 (2009)
50. Tai, B.J., Bian, Z., Jiang, H., Greenspan, D.C., Zhong, J., Clark, A.E., Du, M.Q.: Anti-gingivitis effect of a dentifrice containing bioactive glass (NovaMin®) particulate. J. Clin. Periodontol.Periodontol. **33**(2), 86–91 (2006)
51. Eberhard, J., Reimers, N., Dommisch, H., Hacker, J., Freitag, S., Acil, Y., Jepsen, S., et al.: The effect of the topical administration of bioactive glass on inflammatory markers of human experimental gingivitis. Biomaterials **26**(13), 1545–1551 (2005)
52. Torii, D., Tsutsui, T.W., Watanabe, N., Konishi, K.: Bone morphogenetic protein 7 induces cementogenic differentiation of human periodontal ligament-derived mesenchymal stem cells. Odontology **104**, 1–9 (2016)

53. Rasouli, R., Barhoum, A., Uludag, H.: A review of nanostructured surfaces and materials for dental implants: surface coating, patterning and functionalization for improved performance. Biomater. Sci. **6**(6), 1312–1338 (2018)
54. Dubey, N., Ellepola, K., Decroix, F.E., Morin, J.L., Castro Neto, A.H., Seneviratne, C.J., Rosa, V.: Graphene onto medical grade titanium: an atom-thick multimodal coating that promotes osteoblast maturation and inhibits biofilm formation from distinct species. Nanotoxicology **12**(4), 274–289 (2018)
55. Zhou, Q., Yang, P., Li, X., Liu, H., Ge, S.: Bioactivity of periodontal ligament stem cells on sodium titanate coated with graphene oxide. Sci. Rep. **6**(1), 19343 (2016)
56. Suo, L., Jiang, N., Wang, Y., Wang, P., Chen, J., Pei, X., Wan, Q., et al.: The enhancement of osseointegration using a graphene oxide/chitosan/hydroxyapatite composite coating on titanium fabricated by electrophoretic deposition. J. Biomed. Mater. Res. B Appl. Biomater.Biomater. **107**(3), 635–645 (2019)
57. Joy-anne, N.O., Su, Y., Lu, X., Kuo, P.H., Du, J., Zhu, D.: Bioactive glass coatings on metallic implants for biomedical applications. Bioact. Mater. **4**, 261–270 (2019)
58. Moritz, N., Rossi, S., Vedel, E., Tirri, T., Ylänen, H., Aro, H., Närhi, T.: Implants coated with bioactive glass by CO_2-laser, an in vivo study. J. Mater. Sci. Mater. Med. **15**, 795–802 (2004)
59. Kargozar, S., Montazerian, M., Fiume, E., Baino, F.: Multiple and promising applications of strontium (Sr)-containing bioactive glasses in bone tissue engineering. Front. Bioeng. Biotechnol. **7**, 161 (2019)
60. Malik, S., Ruddock, F.M., Dowling, A.H., Byrne, K., Schmitt, W., Khalakhan, I., Hill, J.P., et al.: Graphene composites with dental and biomedical applicability. Beilstein J. Nanotechnol.Nanotechnol. **9**(1), 801–808 (2018)
61. Sun, L., Yan, Z., Duan, Y., Zhang, J., Liu, B.: Improvement of the mechanical, tribological and antibacterial properties of glass ionomer cements by fluorinated graphene. Dent. Mater. **34**(6), e115–e127 (2018)
62. Sarosi, C., Biris, A.R., Antoniac, A., Boboia, S., Alb, C., Antoniac, I., Moldovan, M.: The nanofiller effect on properties of experimental graphene dental nanocomposites. J. Adhes. Sci. Technol.Adhes. Sci. Technol. **30**(16), 1779–1794 (2016)
63. Burke, F.J., Qualtrough, A.J., Hale, R.W.: The dentine-bonded ceramic crown: an ideal restoration? Br. Dent. J. **179**(2), 58–63 (1995)
64. Mizrahi, B.: All-ceramic silica/glass-based crowns–clinical protocols. Br. Dent. J. **211**(6), 257–262 (2011)
65. Mente, J., Geletneky, B., Ohle, M., Koch, M.J., Ding, P.G.F., Wolff, D., Pfefferle, T., et al.: Mineral trioxide aggregate or calcium hydroxide direct pulp capping: an analysis of the clinical treatment outcome. J. Endod.Endod. **36**(5), 806–813 (2010)
66. Koubi, G., Colon, P., Franquin, J.C., Hartmann, A., Richard, G., Faure, M.O., Lambert, G.: Clinical evaluation of the performance and safety of a new dentine substitute, Biodentine, in the restoration of posterior teeth—a prospective study. Clin. Oral Invest. **17**, 243–249 (2013)
67. Bakhsh, T.A., Bakry, A.S., Mandurah, M.M., Abbassy, M.A.: Novel evaluation and treatment techniques for white spot lesions: an in vitro study. Orthod. Craniofac. Res. **20**(3), 170–176 (2017)
68. Bakry, A.S., Abbassy, M.A., Alharkan, H.F., Basuhail, S., Al-Ghamdi, K., Hill, R.: A novel fluoride containing bioactive glass paste is capable of remineralizing early caries lesions. Materials **11**(9), 1636 (2018)
69. Bakry, A.S., Marghalani, H.Y., Amin, O.A., Tagami, J.: The effect of a bioglass paste on enamel exposed to erosive challenge. J. Dent. **42**(11), 1458–1463 (2014)
70. Bakry, A.S., Takahashi, H., Otsuki, M., Sadr, A., Yamashita, K., Tagami, J.: CO_2 laser improves 45S5 bioglass interaction with dentin. J. Dent. Res. **90**(2), 246–250 (2011)
71. Tirapelli, C., Panzeri, H., Soares, R.G., Peitl, O., Zanotto, E.D.: A novel bioactive glass-ceramic for treating dentin hypersensitivity. Braz. Oral Res. **24**, 381–387 (2010)
72. Rodriguez, C.L., Kessler, F., Dubey, N., Rosa, V., Fechine, G.J.: CVD graphene transfer procedure to the surface of stainless steel for stem cell proliferation. Surf. Coat. Technol. **311**, 10–18 (2017)

73. Mahdavi, R., Belgheisi, G., Haghbin-Nazarpak, M., Omidi, M., Khojasteh, A., Solati-Hashjin, M.: Bone tissue engineering gelatin–hydroxyapatite/graphene oxide scaffolds with the ability to release vitamin D: Fabrication, characterization, and in vitro study. J. Mater. Sci. Mater. Med. **31**, 1–14 (2020)
74. Wang, Q., Chu, Y., He, J., Shao, W., Zhou, Y., Qi, K., Cui, S., et al.: A graded graphene oxide-hydroxyapatite/silk fibroin biomimetic scaffold for bone tissue engineering. Mater. Sci. Eng. C **80**, 232–242 (2017)
75. Sabbagh, J., Ghassibe-Sabbagh, M., Fayyad-Kazan, M., Al-Nemer, F., Fahed, J.C., Berberi, A., Badran, B.: Differences in osteogenic and odontogenic differentiation potential of DPSCs and SHED. J. Dent. **101**, 103413 (2020)
76. Rosa, V., Xie, H., Dubey, N., Madanagopal, T.T., Rajan, S.S., Morin, J.L.P., Neto, A.H.C., et al.: Graphene oxide-based substrate: physical and surface characterization, cytocompatibility and differentiation potential of dental pulp stem cells. Dent. Mater. **32**(8), 1019–1025 (2016)
77. Guo, C., Lu, R., Wang, X., Chen, S.: Graphene oxide-modified polyetheretherketone with excellent antibacterial properties and biocompatibility for implant abutment. Macromol. Res.. Res. **29**(5), 351–359 (2021)
78. Qiu, J., Geng, H., Wang, D., Qian, S., Zhu, H., Qiao, Y., Liu, X., et al.: Layer-number dependent antibacterial and osteogenic behaviors of graphene oxide electrophoretic deposited on titanium. ACS Appl. Mater. Interfaces **9**(14), 12253–12263 (2017)
79. Srimaneepong, V., Rokaya, D., Thunyakitpisal, P., Qin, J., Saengkiettiyut, K.: Corrosion resistance of graphene oxide/silver coatings on Ni–Ti alloy and expression of IL-6 and IL-8 in human oral fibroblasts. Sci. Rep. **10**(1), 1–12 (2020)
80. García-Mota, L.F., Hardan, L., Bourgi, R., Zamarripa-Calderón, J.E., Rivera-Gonzaga, J.A., Hernández-Cabanillas, J.C., Cuevas-Suárez, C.E.: Light-cured calcium silicate based-cements as pulp therapeutic agents: a meta-analysis of clinical studies. J. Evid. Based Dental Pract. 101776 (2022)
81. Deliormanli, A.M., Ensoylu, M., ALMisned, G., Tekin, H.O.: Two-dimensional molybdenum disulfide/polymer-coated bioactive glass scaffolds for tissue engineering: fabrication, structural, mechanical, bioactivity, and radiation interaction properties. Ceram. Int. (2023)

Two-Dimensional-Based Hybrid Materials for Agriculture System

Smruti Ranjan Padhan, Pratikshya Mishra, Kaushik Kumar Panigrahi, and Ayesha Mohanty

Abstract The integration of two-dimensional (2D) materials in agriculture holds tremendous potential for revolutionizing farming practices and promoting sustainable and efficient agricultural systems. The potential benefits of these materials in agriculture are diverse and impactful. These materials offer advantages such as precise nutrient delivery, improved crop growth and productivity, soil remediation and pollution control, enhanced pesticide and fertilizer efficiency, water management, and irrigation. By leveraging the unique properties and functionalities of 2D materials and combining them with other materials, we can address key agricultural challenges and endorse resource efficiency. Continued research and development in the field of agriculture are crucial to unlocking their full potential. Ongoing efforts should focus on improving material synthesis and fabrication techniques, optimizing performance, scalability, and cost-effectiveness, and assessing long-term efficacy and safety. Collaboration and interdisciplinary research play a vital role in fostering innovation, knowledge exchange, and practical application in real-world agricultural settings. The insinuations of these materials for sustainable and efficient agricultural practices are momentous. These materials contribute to food security, water conservation, and climate resilience. They align with sustainable development goals, including zero hunger, clean water and sanitation, sustainable cities and communities, responsible consumption and production, and climate action. Addressing key agricultural challenges can pave the way for a more sustainable, resilient, and food-secure future. Future directions and potential breakthroughs in the field of hybrid materials in agriculture include the development of multifunctional materials, integration with the Internet of Things (IoT) technology, collaboration, and interdisciplinary research, and the establishment of supportive policies and regulatory frameworks. These improvements will further enhance these materials' performance, functionality, and practical implementation in agriculture. Hence, the integration of these hybrid materials in agriculture offers immense opportunities for transforming farming practices and

S. R. Padhan
Indian Agricultural Research Institute (IARI), New Delhi, India

P. Mishra · K. K. Panigrahi (✉) · A. Mohanty
Odisha University of Agriculture and Technology (OUAT), Bhubaneswar, India
e-mail: kkpanigrahi@ouat.ac.in

N. Talreja et al. (eds.), *Two-dimensional Hybrid Composites*, Engineering Materials,
https://doi.org/10.1007/978-981-99-8010-9_15

achieving sustainable and efficient agricultural systems. The recap of potential benefits, the importance of continued research and development, and the implications for sustainable agriculture highlight the significance of them in addressing key agricultural challenges and promoting resource efficiency. With ongoing advancements, 2D-based hybrid materials have the potential to revolutionize the agricultural sector and contribute to a more sustainable and food-secure future.

Keywords Two-dimensional materials · Hybrid materials · Graphene · Nutrient delivery · Soil remediation · Pesticide efficiency

1 Introduction

Agriculture is vital in ensuring food security and sustainable development, nourishing a growing global population. However, traditional agricultural practices face numerous challenges, including diminishing arable land, environmental degradation, and increasing demands for food production [1]. Innovative approaches incorporating advanced materials have emerged to address these challenges and enhance the efficiency of agricultural systems. Among these materials, two-dimensional (2D) based hybrid materials have shown great promise in revolutionizing the agriculture industry. Two-dimensional materials refer to thin sheets of materials with only one or a few atoms in thickness [2]. These materials possess unique properties due to their ultrathin nature, high surface-to-volume ratio, and exceptional mechanical, electrical, and chemical properties [3, 4]. They have garnered significant attention in various scientific disciplines, including materials science, electronics, energy storage, and agriculture [2]. The utilization of 2D-based hybrid materials in agriculture offers several advantages over conventional practices [5]. Firstly, their large surface area allows for increased absorption and retention of nutrients and water, leading to improved plant growth and development. Secondly, their unique electrical and optical properties enable the development of sensors and actuators that can monitor and regulate various environmental factors, optimizing crop growth conditions [6]. Additionally, the incorporation of 2D materials into hybrid structures can enhance mechanical strength, stability, and durability, ensuring long-term performance in agricultural applications.

Graphene, a single layer of carbon atoms arranged in a honeycomb lattice, is one of the most extensively studied 2D materials [7, 8]. Its exceptional electrical conductivity, thermal stability, and mechanical strength make it an ideal candidate for various agricultural applications [9, 10]. For instance, graphene-based sensors can detect soil moisture, temperature, and nutrient levels in real-time, providing precise data for efficient irrigation and fertilization. Graphene-based nanocomposites have also demonstrated the ability to improve soil water-holding capacity and nutrient retention, reducing water and fertilizer consumption while enhancing crop yields [11, 12]. Transition metal dichalcogenides (TMDs) are another class of 2D materials that hold tremendous potential in agriculture. These materials exhibit unique optical and electronic properties, making them suitable for applications such as photocatalysis and plant disease detection. TMD-based hybrid materials have shown promise in the removal of pollutants from soil and water, contributing to soil remediation and environmental sustainability [13, 14]. Moreover, the integration of TMDs with conventional materials can enhance the efficiency of pesticide delivery systems, reducing chemical usage and minimizing environmental risks [15].

The synthesis and fabrication of 2D-based hybrid materials for agriculture require precise control and manipulation of the material's structure and properties. Various techniques such as chemical vapor deposition (CVD) [16, 17], physical vapor deposition (PVD) [18], and electrochemical deposition have been employed to synthesize 2D materials with desired properties. The combination of these techniques with conventional fabrication methods allows for the development of hybrid materials with tailored characteristics suitable for specific agricultural applications. The potential applications of 2D-based hybrid materials in agriculture are vast and diverse. Improved nutrient delivery systems can be achieved through the incorporation of 2D materials in fertilizers, enabling controlled release and efficient utilization of nutrients by plants. Enhanced crop growth and productivity can be realized by optimizing environmental conditions using 2D-based sensors and actuators. Additionally, the remediation of contaminated soils and water bodies can be facilitated by the use of 2D-based adsorbents and catalysts [14].

Despite the significant potential of 2D-based hybrid materials, several challenges need to be addressed for their widespread implementation in agricultural systems. Environmental and health concerns associated with the use of these advanced materials require thorough investigation to ensure their safe and sustainable use [19]. Furthermore, economic feasibility and scalability need to be considered to ensure the practicality of these materials for small-scale and large-scale farming operations. Continued research and development efforts are required to overcome these challenges and further unlock the full potential of 2D-based hybrid materials in agriculture. The future perspectives of 2D-based hybrid materials in agriculture are highly promising. Ongoing research aims to explore new combinations of 2D materials with other components to enhance their functionality and performance. For instance, the integration of these materials with biodegradable polymers can lead to the development of environmentally friendly and sustainable agricultural products [20]. Additionally, nanotechnology and material engineering advancements will enable the production of 2D-based hybrid materials with tailored properties for specific agricultural needs.

One area of future exploration is the development of smart nanomaterials that can respond to environmental stimuli and regulate plant growth processes [21]. These materials can potentially release growth-promoting substances, such as hormones or signaling molecules, in response to specific conditions, optimizing plant development and increasing crop yields. Furthermore, the integration of nano sensors into 2D-based hybrid materials can facilitate real-time monitoring of crop health, pest infestation, and disease outbreaks, enabling early detection and targeted intervention [22]. The utilization of 2D-based hybrid materials in precision agriculture is another area of great interest. Precision agriculture aims to optimize farming practices by using technology and data-driven decision-making. By incorporating 2D materials into precision farming systems, farmers can gather accurate and timely information about soil conditions, nutrient levels, and plant health, leading to efficient resource management and reduced environmental impact. Collaboration between scientists, engineers, agronomists, and farmers is crucial for the successful implementation of 2D-based hybrid materials in agriculture. The interdisciplinary nature of this field requires a holistic approach, considering not only the technical aspects but also the social, economic, and environmental implications [23]. Regulatory frameworks and policies need to be developed to ensure these materials' safe and responsible use, addressing concerns related to human health, ecosystem impact, and ethical considerations.

Therefore, two-dimensional-based hybrid materials hold tremendous potential to revolutionize the agriculture system. Their unique properties and functionalities offer new opportunities to address the challenges faced by traditional farming practices.

2 Overview of Various Two-Dimensional Materials in Agriculture

Two-dimensional (2D) materials are a class of materials that hold a layered structure with a thickness of only a few atomic layers or a single layer. These materials are composed of atoms or molecules arranged in a two-dimensional plane, unveiling unique properties due to their ultrathin nature [4]. The confinement of electrons in the plane leads to significant changes in their electronic, optical, thermal, and mechanical properties compared to their bulk counterparts. There are numerous 2D materials available with befitting features for agriculture systems (Table 1). Some of them are given as follows:

2.1 Graphene

Graphene is the most well-known and extensively studied 2D material and is composed of a single layer of carbon atoms arranged in a hexagonal lattice, exhibiting exceptional electrical conductivity, high mechanical strength, and excellent thermal properties. These characteristics make it suitable for various agricultural applications, including sensors, nanocomposites for soil improvement, and water filtration systems [7, 8].

Table 1 Varied types of two-dimensional (2-D) materials in agriculture along with their functions

S. no	Type of 2D material	Important functions in agriculture
i.	Graphene [8]	Augments nutrient absorption and transport in plants
		Improves soil moisture retention and water-use efficiency
		Acts as a barrier against pests and diseases
ii.	Transition Metal Dichalcogenides (TMDs) [13]	Enhances photosynthesis and plant growth
		Improves crop yield and stress tolerance
		Provides antimicrobial properties for disease control
iii.	Boron Nitride [25]	Serves as a protective coating for seeds and fertilizers
		Facilitates controlled release of nutrients in the soil
iv.	Molybdenum Disulfide (MoS_2) [34]	Acts as a biosensor for detecting plant diseases
		Augments water filtration and purification
v.	Silicene [31]	Enhances plant root development and nutrient uptake
		Improves soil structure and water retention
vi.	MXene [30]	Improves water and nutrient uptake efficiency
		Acts as a protective barrier against soil erosion
vii.	Phosphorene [35]	Boosts plant growth and photosynthesis
		Improves nitrogen fixation in leguminous crops
		Increases nutrient efficiency in the soil
viii.	Black Phosphorus [32]	Acts as a nanocarrier for targeted delivery of agrochemicals
		Expands nutrient absorption in plant roots

2.2 *Transition Metal Dichalcogenides (TMDs)*

TMDs are a class of 2D materials composed of transition metal atoms (e.g., molybdenum, tungsten) bonded to chalcogen atoms (e.g., sulphur, selenium). They exhibit unique electronic and optical properties due to quantum confinement effects. TMDs have shown potential in agricultural applications such as photocatalysis for water splitting and degradation of pollutants, as well as in the advance of nano sensors for detecting plant diseases and monitoring environmental parameters [13, 14].

2.3 Boron Nitride (BN)

BN is a 2D material composed of boron and nitrogen atoms settled in a hexagonal lattice, similar to graphene. However, unlike graphene, BN is an insulator with exceptional thermal stability and high chemical resistance [24]. BN has potential applications in agriculture as a protective coating for crop seeds, letting for controlled release of nutrients and protection against pests and diseases [25].

2.4 Phosphorene

Phosphorene is a 2D material composed of phosphorus atoms agreed in a puckered lattice structure. It displays high carrier mobility and excellent semiconducting properties [26]. Phosphorene has potential applications in agriculture, such as in the development of high-performance sensors for monitoring soil conditions, optimizing irrigation, and detecting nutrient deficiencies in plants [27].

2.5 Metal Oxides

Metal oxides, such as titanium dioxide (TiO_2) and zinc oxide (ZnO), can be fabricated into 2D forms [28]. These materials have photocatalytic properties and can be used for water purification, eliminating organic pollutants and harmful microorganisms from agricultural water sources.

2.6 MXenes

These are a family of 2D materials formed by selectively etching layers from layered transition metal carbides, nitrides, or carbonitrides known as MAX phases [29]. MXenes exhibit high electrical conductivity and excellent mechanical properties. They have potential applications in agricultural sensors, energy storage devices, and water treatment systems [30].

2.7 Silicene

It is a 2D material composed of a single layer of silicon atoms arranged in a honeycomb lattice structure, similar to graphene. It shows unique electronic properties and has potential applications in electronics, optoelectronics, and energy storage devices. Silicene has garnered noteworthy attention due to its compatibility with existing silicon-based technology [31].

2.8 Black Phosphorus

Black phosphorus, also known as phosphorene, is a 2D material composed of phosphorus atoms arranged in a layered structure. It exhibits a high carrier mobility, making it suitable for electronics and photonics applications. Black phosphorus has shown promise in areas such as transistors, photodetectors, and energy storage devices [32, 33].

2.9 Hexagonal Boron Nitride (h-BN)

Hexagonal boron nitride, often referred to as "white graphene", is a 2D material composed of alternating boron and nitrogen atoms arranged in a hexagonal lattice structure [36]. It is an insulator with excellent thermal and chemical stability. h-BN is commonly used as a dielectric material in electronics, as a lubricant due to its low friction properties, and as a protective coating for various applications.

2.10 Metal Dichalcogenides (MoS_2, WS_2 etc.)

These are a class of 2D materials composed of a transition metal atom (such as molybdenum or tungsten) bonded with chalcogen atoms (such as sulphur or selenium). They exhibit a layered structure with excellent semiconducting properties [37]. Metal dichalcogenides have gained attention for their potential applications in electronics, optoelectronics, catalysis, and energy storage [34].

2.11 Metal–Organic Frameworks (MOFs)

Metal–organic frameworks are a class of 2D materials consisting of metal ions or clusters coordinated with organic ligands. They possess high porosity and large surface areas, making them suitable for applications such as gas storage and separation, drug delivery, and catalysis [38]. MOFs have the potential to improve soil fertility and nutrient management in agriculture.

2.12 Phosphorene Oxide

Phosphorene oxide is a 2D material derived from black phosphorus by introducing oxygen functionalities. It exhibits unique properties and can be used for applications such as energy storage, sensing, and catalysis [39]. Phosphorene oxide has shown potential for use in agriculture, particularly in the development of efficient electrochemical sensors [35] for soil analysis and plant health monitoring.

2.13 Carbon Nitride

Carbon nitride is a 2D material composed of carbon and nitrogen atoms. It exhibits a range of electronic and optical properties, depending on the composition and structure. Carbon nitride materials have potential applications in photocatalysis, water splitting, and energy conversion devices [40, 41]. They can contribute to sustainable agriculture by harnessing solar energy for various processes, such as water purification and crop growth enhancement.

It is important to note that the field of 2D materials is continuously expanding, and new materials with unique properties are being discovered and synthesized. Each material enjoys unique properties that can be tailored for specific applications in agriculture, electronics, energy, and other fields. The versatility and tunability of 2D materials make them promising candidates for addressing the challenges and advancing innovation in innumerable industries. Therefore, 2D materials hold great promise for transforming agricultural systems with their unique properties, such as large surface area, high conductivity, and adsorption capabilities, enabling enhanced nutrient delivery, water management, soil remediation, disease and pest management, and upgraded crop growth and quality. Graphene, TMDs, BN, phosphorene, metal oxides, and MXenes are just a few specimens of the diverse range of 2D materials that can be utilized in agriculture. By harnessing the potential of these materials and fitting them into agricultural practices, we can pave the way for a more sustainable, efficient, and resilient agricultural industry [42]. Continued research and development efforts in this field are crucial to further explore the capabilities of 2D materials and realize their full potential in addressing the challenges of modern agriculture.

3 Synthesis and Fabrication Techniques of Hybrid Materials in Agriculture

Hybrid materials, which combine two or more different materials with complementary properties, have gained significant attention in the field of agriculture. These materials offer unique functionalities and enhanced performance that can revolutionize agricultural practices and address various challenges. To harness the potential of hybrid materials, it is crucial to understand the synthesis and fabrication techniques involved in their production. Some of the techniques are explained as below (Fig. 1).

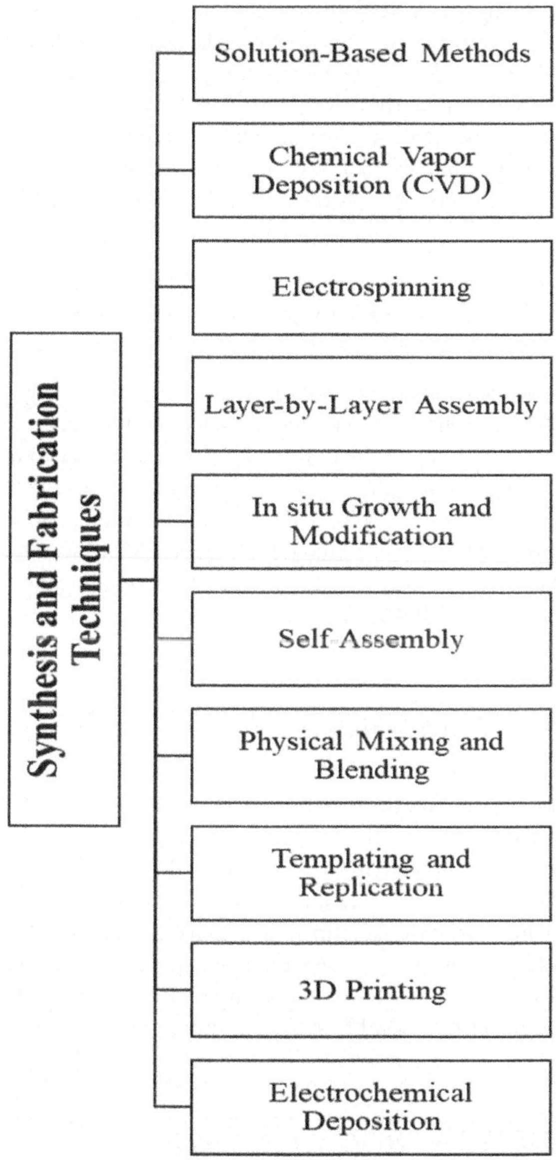

Fig. 1 Different synthesis and fabrication techniques of 2D-hybrid materials

3.1 Solution-Based Methods

Solution-based methods are extensively used for synthesizing hybrid materials in agriculture. One common technique is the *sol–gel method*, where precursor solutions are mixed and undergo hydrolysis and condensation reactions to form a gel-like structure followed by subsequent drying and heat treatment result in the formation of hybrid materials [43]. This procedure allows for the incorporation of different materials, such as 2D materials, polymers, nanoparticles, and organic compounds, into the final product. Solution-based methods offer versatility and control over the composition, morphology, and properties of these materials [44].

3.2 Chemical Vapour Deposition (CVD)

Chemical vapour deposition is a technique frequently employed for the synthesis of 2D materials and their integration into hybrid materials. In CVD, precursor gases are introduced into a reaction chamber, where they decompose and deposit on a substrate to form a thin film. By controlling the reaction parameters, such as temperature, pressure, and gas composition, hybrid materials can be fabricated with precise switch over the composition and morphology. CVD allows for the growth of large-area films and the integration of multiple materials, enabling the development of complex hybrid structures [16, 17].

3.3 Electrospinning

Electrospinning is a versatile technique used for fabricating hybrid materials in the form of nanofibers. In this course, a high voltage is applied to a polymer solution or melt, causing it to form a charged jet that is drawn towards a grounded collector [45]. By incorporating 2D materials, nanoparticles, or other additives into the polymer solution, hybrid nanofibers with enhanced properties can be shaped. Electrospinning offers control over fibre diameter, porosity, and surface area, making it suitable for applications like controlled-release systems, filtration membranes, and tissue engineering scaffolds in agriculture.

3.4 Layer-By-Layer Assembly

Layer-by-layer (LbL) assembly is a technique used for fabricating thin films with precise control over the layer thickness and composition. This method involves the sequential adsorption of oppositely charged materials onto a substrate [46].

By alternating between different materials, such as polymers, 2D materials, and nanoparticles, hybrid thin films can be constructed with tailored properties. LbL assembly allows for the incorporation of multiple components, enabling the design of functional coatings, sensors, and controlled-release systems for agricultural applications.

3.5 *In Situ Growth and Modification*

In situ growth and modification techniques involve the synthesis or incorporation of materials directly within the agricultural system. For example, 2D materials can be grown on substrates or incorporated into hydrogels or porous structures through direct synthesis or modification techniques [47]. These in situ techniques enable the integration of hybrid materials into specific agricultural environments, such as soil amendments or plant growth substrates, facilitating their direct interaction with crops and enhancing their performance in agricultural systems.

3.6 *Self-Assembly*

Self-assembly is a technique where materials organize themselves into ordered structures through molecular interactions. By designing the molecular structure and properties of the components, self-assembly can be used to fabricate hybrid materials with specific functionalities. For example, amphiphilic molecules can self-assemble into micelles or vesicles that encapsulate 2D materials or nanoparticles, creating hybrid structures with controlled release properties or enhanced stability [46].

3.7 *Physical Mixing and Blending*

Physical mixing and blending involve the simple mixing of different materials to create hybrid composites. This technique is straightforward and cost-effective, making it suitable for large-scale production [48]. By combining 2D materials with polymers, nanoparticles, or other additives, hybrid composites can be formed with improved mechanical, thermal, or electrical properties. Physical mixing and blending offer flexibility in controlling the composition and concentration of the materials, allowing for customized hybrid materials for specific agricultural applications.

3.8 Templating and Replication

Templating and replication techniques involve using templates or molds to create hybrid materials with desired structures. For example, 2D materials can be used as templates to fabricate hierarchical hybrid structures by filling their interlayer spaces with other materials [49]. Alternatively, replica molding can be employed to transfer the surface texture and properties of one material onto another, creating hybrid materials with unique surface functionalities.

3.9 3D Printing

3D printing, also known as additive manufacturing, is a technique that allows for the fabrication of complex three-dimensional structures layer by layer. By incorporating 2D materials, polymers, or nanoparticles into the printing materials, hybrid structures with tailored compositions and functionalities can be created. 3D printing offers precise control over the geometry and internal architecture of the fabricated objects, enabling the development of customized agricultural tools, sensors, and structures [50].

3.10 Electrochemical Deposition

Electrochemical deposition involves the electrodeposition of materials onto a conductive substrate using an electrolyte solution. By controlling the deposition parameters, such as current density and deposition time, hybrid materials can be synthesized with controlled composition, morphology, and thickness. Electrochemical deposition is a versatile technique that can be used to integrate 2D materials, nanoparticles, or other materials onto electrodes, enabling the development of hybrid electrochemical sensors or energy storage devices for agricultural applications.

The field of hybrid materials in agriculture continues to advance with the development of new synthesis and fabrication techniques. Each method offers unique advantages in terms of control over composition, structure, and properties. Solution-based methods, chemical vapour deposition, electrospinning, layer-by-layer assembly, in situ growth and modification, self-assembly, physical mixing and blending, templating and replication, 3D printing, and electrochemical deposition are just some of the techniques employed to create hybrid materials with enhanced functionalities for agricultural systems [44]. By leveraging these techniques, researchers and engineers can tailor hybrid materials to address specific challenges in crop yield improvement, soil quality enhancement, and sustainable agriculture [48].

4 Applications of 2D-Based Hybrid Materials in Agriculture

The 2D materials are going to be the backbone of next generation agriculture to produce more by expensing less. Some of its potential applications are outlined and summarized below (Fig. 2).

4.1 Improved Nutrient Delivery Systems

Efficient nutrient delivery is essential for promoting healthy plant growth and maximizing agricultural productivity. Traditional methods of nutrient application, such as soil fertilization and foliar spraying, often suffer from issues such as nutrient leaching, low absorption rates, and uneven distribution [51]. In recent years, the integration of two-dimensional (2D) hybrid materials in agriculture has shown great promise in addressing these challenges. By combining the unique properties of 2D materials with other materials, hybrid systems offer innovative solutions for improving nutrient delivery and enhancing plant nutrition. This document aims to explore the concept of improved nutrient delivery systems through 2D-hybrid materials in agriculture and highlight their potential benefits and applications.

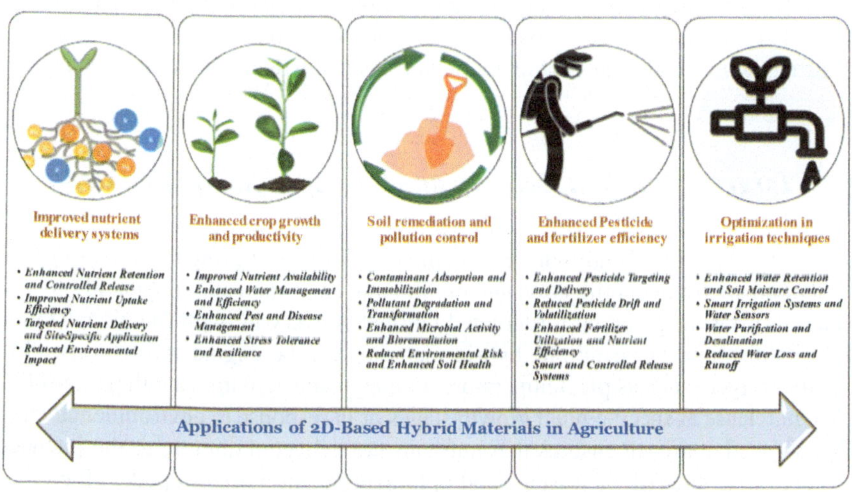

Fig. 2 Various applications of 2D-based hybrid materials in Agriculture

4.2 Enhanced Nutrient Retention and Controlled Release

One of the key advantages of using 2D-hybrid materials in nutrient delivery systems is their ability to enhance nutrient retention and facilitate controlled release [52]. By incorporating 2D materials, such as graphene or transition metal dichalcogenides (TMDs), into nutrient carriers or matrices, hybrid materials can improve the adsorption and retention of nutrients. These materials possess large surface areas and high adsorption capacities, allowing them to effectively bind and retain essential nutrients, preventing their loss through leaching. Additionally, the controlled release properties of hybrid materials can be tailored by adjusting factors such as composition, morphology, and interlayer spacing, enabling nutrients to be released gradually over time, matching the plant's uptake requirements.

4.3 Improved Nutrient Uptake Efficiency

2D-hybrid materials can also enhance nutrient uptake efficiency by promoting root growth and improving nutrient absorption. Incorporating 2D materials into soil amendments or hydrogels can enhance soil structure, porosity, and water-holding capacity, thereby creating a favourable environment for root development. The increased surface area of 2D materials facilitates the adsorption of nutrients in the root zone, making them readily available for plant uptake [53]. Furthermore, the use of hybrid materials can enhance the solubility and bioavailability of nutrients, ensuring their effective absorption by plant roots and reducing nutrient wastage.

4.4 Targeted Nutrient Delivery and Site-Specific Application

The precise targeting of nutrients to specific plant organs or growth stages is critical for optimizing nutrient utilization and minimizing environmental impact. 2D-hybrid materials offer the potential for targeted nutrient delivery through functionalization and modification. For example, hybrid materials can be engineered to respond to specific triggers such as pH, temperature, or enzymatic activity, enabling controlled nutrient release at specific plant growth stages or in response to environmental cues. This targeted approach ensures that nutrients are delivered directly to the intended sites, reducing the risk of wastage and optimizing plant nutrient uptake efficiency) [52].

4.5 *Reduced Environmental Impact*

Improved nutrient delivery systems through 2D-hybrid materials have the potential to reduce the environmental impact associated with conventional nutrient application methods. The controlled-release nature of hybrid materials minimizes nutrient leaching, decreasing the risk of water contamination and eutrophication [54]. Additionally, the site-specific nutrient delivery enabled by hybrid materials ensures that nutrients are delivered where they are most needed, reducing excess nutrient application and subsequent environmental pollution [55]. These sustainable nutrient delivery systems contribute to environmentally responsible agricultural practices and promote the conservation of natural resources [56].

5 Enhanced Crop Growth and Productivity

The global population is steadily increasing, putting significant pressure on agricultural systems to meet the rising demand for food. To address this challenge, innovative approaches are required to enhance crop growth and productivity [52, 57]. One such approach is the integration of two-dimensional (2D) hybrid materials in agriculture. These materials, combined with other components, offer unique properties and functionalities that can positively impact crop performance [58].

5.1 *Improved Nutrient Availability*

Optimal nutrient availability is essential for promoting plant growth and maximizing crop productivity. 2D-hybrid materials play a significant role in improving nutrient availability and uptake efficiency. By incorporating 2D materials, such as graphene or transition metal dichalcogenides (TMDs), into fertilizers or soil amendments, hybrid systems can enhance nutrient retention and release characteristics [53]. The high surface area and adsorption capacity of 2D materials enable them to bind and retain nutrients, preventing their loss through leaching. Moreover, these materials can improve the solubility and bioavailability of nutrients, making them more accessible to plant roots and increasing their uptake efficiency.

5.2 *Enhanced Water Management and Efficiency*

Water scarcity is a major challenge in agriculture, particularly in regions with limited water resources. 2D-hybrid materials offer the potential to improve water management and efficiency in crop production. By incorporating hydrophilic 2D materials into soil amendments or mulches, hybrid systems can enhance water retention in the root zone, reducing water loss through evaporation and improving the availability of water to plants [59]. Additionally, the use of 2D materials with water-responsive properties allows for controlled water release, ensuring that plants receive water when needed. This improved water management can lead to better crop growth, especially in drought-prone areas, and contribute to sustainable water usage in agriculture [58].

5.3 *Enhanced Pest and Disease Management*

Pests and diseases pose significant threats to crop health and productivity. Integrating 2D-hybrid materials in pest and disease management strategies can provide novel solutions to combat these challenges. For instance, incorporating antimicrobial 2D materials into crop protection formulations can enhance the efficacy of pesticides, reducing the risk of disease outbreaks and improving crop health. Similarly, the use of 2D materials with insect-repellent properties can help deter pests and minimize crop damage [60]. The antimicrobial and insect-repellent functionalities of 2D-hybrid materials offer environmentally friendly alternatives to conventional pesticide use, promoting sustainable pest and disease management practices [61].

5.4 *Enhanced Stress Tolerance and Resilience*

Crop plants often face various environmental stresses, such as heat, drought, salinity, and heavy metal contamination, which can negatively impact their growth and productivity. 2D-hybrid materials have shown potential in enhancing stress tolerance and resilience in crops [62]. By incorporating stress-responsive 2D materials into plant growth substrates or seed coatings, hybrid systems can provide plants with additional protection and support under challenging environmental conditions [63]. These materials can help regulate plant responses to stress, mitigate oxidative damage, and promote the production of stress-related signalling molecules, ultimately enhancing crop resilience and productivity in adverse environments.

6 Soil Remediation and Pollution Control

Soil pollution is a significant environmental issue that affects agricultural productivity and poses risks to human health. Contamination of soil by heavy metals, pesticides, industrial pollutants, and other toxic substances can have detrimental effects on plant growth, soil fertility, and ecosystem health. In recent years, the integration of two-dimensional (2D) hybrid materials in agriculture has emerged as a promising approach for soil remediation and pollution control [64]. These materials, combined with other components, offer unique properties and functionalities that can effectively remove, immobilize, or degrade pollutants in the soil.

6.1 Contaminant Adsorption and Immobilization

One of the key advantages of using 2D-hybrid materials in soil remediation is their high adsorption capacity and affinity for contaminants. Certain 2D materials, such as graphene oxide and layered double hydroxides, possess large surface areas and abundant functional groups that can effectively adsorb a wide range of pollutants and heavy materials [65]. By incorporating these materials into soil amendments or barriers, hybrid systems can enhance the removal and immobilization of contaminants. The adsorption mechanisms involve electrostatic interactions, ion exchange, and surface complexation, allowing 2D-hybrid materials to selectively capture and retain pollutants, preventing their movement through the soil profile and reducing the risk of groundwater contamination.

6.2 Pollutant Degradation and Transformation

In addition to adsorption and immobilization, some 2D-hybrid materials possess inherent catalytic properties that can facilitate pollutant degradation and transformation. For example, certain transition metal dichalcogenides (TMDs) and metal–organic frameworks (MOFs) can serve as catalysts for the degradation of organic pollutants through processes like photocatalysis or advanced oxidation [55]. These materials can generate reactive oxygen species or promote redox reactions, breaking down pollutants into less toxic compounds or mineralizing them into harmless substances. The integration of such materials in soil remediation strategies offers the potential for in situ pollutant degradation and the restoration of soil quality.

6.3 Enhanced Microbial Activity and Bioremediation

Microorganisms play a crucial role in the degradation and detoxification of pollutants in the soil. 2D-hybrid materials can enhance microbial activity and facilitate bioremediation processes. By incorporating materials with antimicrobial properties or those that can promote microbial growth and activity, hybrid systems create a favourable environment for pollutant-degrading microorganisms [66]. These materials can serve as carriers for beneficial microorganisms, providing protection, nutrients, and a suitable microenvironment for their growth and activity. This synergy between 2D materials and microorganisms can accelerate the natural biodegradation processes and enhance the effectiveness of bioremediation strategies.

6.4 *Reduced Environmental Risk and Enhanced Soil Health*

The use of 2D-hybrid materials in soil remediation and pollution control offers several benefits in terms of reducing environmental risks and enhancing soil health. By effectively adsorbing or degrading pollutants, these materials can mitigate the mobility and bioavailability of contaminants, reducing their potential for environmental exposure and subsequent risks to ecosystems and human health [67]. Moreover, the immobilization or transformation of pollutants can lead to the restoration of soil fertility and the enhancement of soil structure and microbial activity. This, in turn, promotes the recovery of soil health, nutrient cycling, and plant growth, contributing to sustainable agricultural practices [68].

7 Enhanced Pesticide and Fertilizer Efficiency

Pesticides and fertilizers play a crucial role in modern agriculture by protecting crops from pests and providing essential nutrients for optimal growth. However, their excessive and inefficient use can have detrimental effects on the environment, human health, and agricultural sustainability. To address these challenges, the integration of two-dimensional (2D) hybrid materials in agriculture has emerged as a promising approach to enhance the efficiency and effectiveness of pesticides and fertilizers [69]. These materials, combined with other components, offer unique properties and functionalities that can improve pesticide targeting, reduce environmental pollution, and enhance fertilizer utilization [70].

7.1 *Enhanced Pesticide Targeting and Delivery*

One of the key advantages of using 2D-hybrid materials in agriculture is their ability to improve pesticide targeting and delivery. By incorporating these materials into pesticide formulations or coatings, hybrid systems can enhance the adhesion and absorption of pesticides to plant surfaces, improving their efficacy and reducing the amount of pesticide required [71]. The high surface area and functionalization of 2D materials allow for better interaction with pests and enhance the penetration of active ingredients into plant tissues. This targeted and controlled delivery of pesticides reduces off-target effects, minimizes environmental pollution, and ensures maximum pest control efficiency [72].

7.2 Reduced Pesticide Drift and Volatilization

Pesticide drift and volatilization are significant concerns in agricultural practices, as they can lead to unintended contamination of neighbouring crops, water bodies, and ecosystems. 2D-hybrid materials offer the potential to reduce pesticide drift and volatilization by incorporating them into drift-reducing formulations or as adjuvants [73]. These materials can improve the viscosity and surface tension of spray solutions, enhancing droplet size and reducing drift. Additionally, the use of 2D materials with barrier properties can limit the release of volatile pesticides into the atmosphere, minimizing the potential for air pollution and human exposure.

7.3 Enhanced Fertilizer Utilization and Nutrient Efficiency

Fertilizer efficiency is a critical aspect of sustainable agriculture, as excessive fertilizer use not only wastes resources but also leads to nutrient runoff, water pollution, and soil degradation. 2D-hybrid materials offer solutions to enhance fertilizer utilization and nutrient efficiency. By incorporating these materials into fertilizer formulations or soil amendments, hybrid systems can improve nutrient retention, slow-release properties, and nutrient availability to plants [72]. The high adsorption capacity of 2D materials allows them to bind and hold nutrients, preventing their leaching and loss from the root zone. This ensures that plants have a continuous and balanced supply of nutrients, leading to improved nutrient uptake, reduced fertilizer application, and minimized environmental impact.

8 Smart and Controlled Release Systems

The controlled release of pesticides and fertilizers can optimize their effectiveness and minimize waste. 2D-hybrid materials enable the development of smart and controlled release systems for agricultural inputs. These materials can be functionalized with stimuli-responsive properties, such as pH, temperature, or moisture sensitivity, allowing for the release of active ingredients based on specific environmental conditions or plant needs [74]. This precise control over the release of pesticides and fertilizers ensures targeted application, minimizes losses, and maximizes their efficacy, resulting in improved crop protection and nutrient management [75].

The integration of 2D-hybrid materials in agriculture offers significant potential to enhance the efficiency and effectiveness of pesticides and fertilizers. Through improved pesticide targeting and delivery, reduced drift and volatilization, enhanced fertilizer utilization, and smart release systems, these materials contribute to sustainable agricultural practices.

9 Optimization in Irrigation and Water Management Techniques

Water scarcity and the efficient management of water resources are critical challenges in agriculture. With growing global population and changing climatic conditions, sustainable water management practices are essential for ensuring food security and optimizing agricultural productivity. The integration of two-dimensional (2D) hybrid materials in water management and irrigation systems has emerged as a promising approach to enhance water use efficiency, improve soil moisture retention, and mitigate water-related stresses in agriculture which are discussed in detail as follows.

9.1 Enhanced Water Retention and Soil Moisture Control

One of the key advantages of using 2D-hybrid materials in agriculture is their ability to improve water retention and control soil moisture. These materials, such as graphene oxide and layered double hydroxides, possess high water-holding capacity due to their large surface areas and unique interlayer structures [76]. By incorporating these materials into soil amendments or hydrogels, hybrid systems can enhance the water-holding capacity of soils, reducing water loss through evaporation and promoting efficient water uptake by plant roots [77]. This ensures better soil moisture management, particularly in arid and semi-arid regions, leading to improved plant growth, reduced irrigation requirements, and increased water use efficiency.

9.2 Smart Irrigation Systems and Water Sensors

2D-hybrid materials also enable the development of smart irrigation systems and water sensors. By integrating these materials into sensors or membranes, hybrid systems can detect and regulate soil moisture levels in real time [78]. These sensors can provide accurate information on soil moisture content, enabling farmers to optimize irrigation scheduling and prevent overwatering or underwatering of crops. Additionally, the use of 2D materials in irrigation membranes can help control the release of water, ensuring a steady and controlled water supply to plants [79]. This targeted irrigation approach minimizes water waste and optimizes water distribution, contributing to water conservation and sustainable irrigation practices.

9.3 Water Purification and Desalination

In regions where water quality is a challenge, 2D-hybrid materials offer potential solutions for water purification and desalination [80]. Certain 2D materials, such as graphene-based membranes, exhibit excellent filtration properties, allowing them to selectively remove contaminants, impurities, and salts from water sources [81]. These materials can be used in water treatment systems to improve water quality for irrigation purposes. Additionally, the integration of 2D materials in desalination processes, such as reverse osmosis, can enhance the efficiency of saltwater conversion into freshwater, expanding the availability of irrigation water in coastal regions.

9.4 Reduced Water Loss and Runoff

Water loss and runoff are significant concerns in agriculture, as they result in inefficient water use and potential environmental pollution [82]. The use of 2D-hybrid materials can help address these issues by reducing water loss and runoff. By incorporating these materials into mulching films, irrigation tapes, or hydrophilic coatings, hybrid systems can minimize water evaporation from the soil surface and prevent excessive runoff [69]. This allows for better utilization of irrigation water, reduces the need for frequent irrigation, and minimizes the risk of nutrient leaching and soil erosion.

Moreover, the adoption of 2D-hybrid materials in water management requires awareness and education among farmers and stakeholders. It is essential to promote knowledge transfer and provide training on the benefits, applications, and proper utilization of these materials. Additionally, supportive policies, incentives, and financial mechanisms can facilitate the adoption of innovative water management technologies, including 2D-hybrid materials, in agriculture.

10 Current Challenges in the Development and Implementation of 2D-Based Hybrid Materials in Agriculture

The integration of two-dimensional (2D) materials in agriculture has opened up new possibilities for enhancing crop productivity, improving resource efficiency, and promoting sustainable agricultural practices [83]. The unique properties of 2D materials, such as graphene and transition metal dichalcogenides, make them attractive for a range of applications in agriculture. However, despite their potential, there are several challenges that need to be addressed for the successful development and implementation of 2D-based hybrid materials in agriculture. Some of the major challenges are explained as follows (Fig. 3).

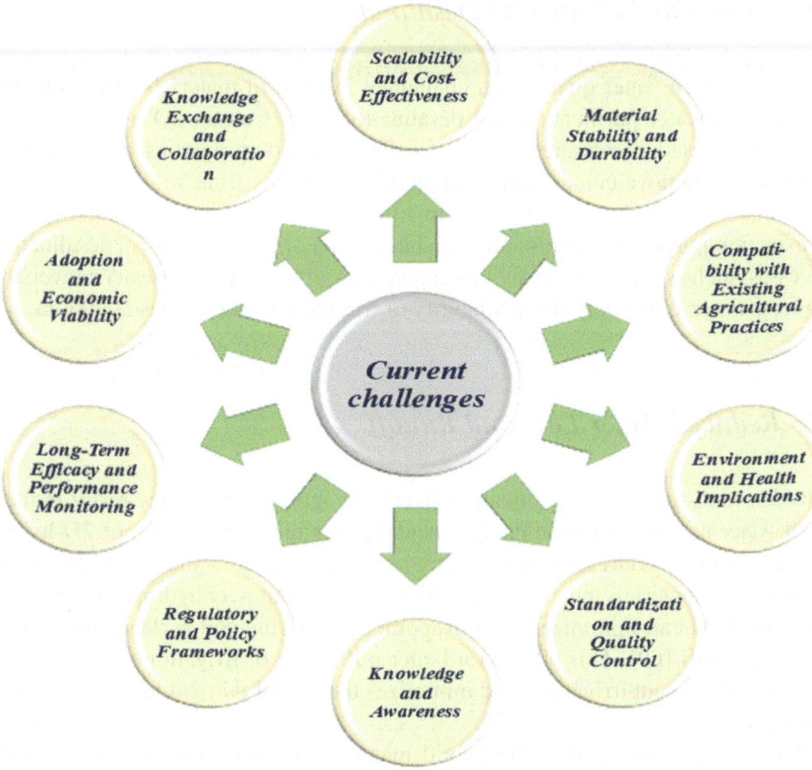

Fig. 3 Current challenges in the development and implementation of 2D-based hybrid materials in agriculture

10.1 Scalability and Cost-Effectiveness

One of the primary challenges in the development and implementation of 2D-based hybrid materials in agriculture is achieving scalability and cost-effectiveness. Currently, the production of high-quality 2D materials is mostly limited to the laboratory scale, which poses challenges in meeting the demand for large-scale agricultural applications [84]. Developing scalable synthesis methods and production techniques is crucial to make 2D-based hybrid materials economically viable and accessible to farmers. Additionally, efforts should be made to optimize the production process and reduce the cost of manufacturing these materials, ensuring their affordability for widespread use in agriculture [85].

10.2 Material Stability and Durability

Another challenge is ensuring the stability and durability of 2D-based hybrid materials in agricultural environments. Agricultural systems are exposed to various harsh conditions, including temperature fluctuations, moisture, and chemical interactions [83]. It is essential to assess the long-term stability and performance of 2D materials under these conditions to ensure their effectiveness and reliability. Strategies such as functionalization, encapsulation, and composite formulations can be explored to enhance the stability and durability of 2D materials, enabling their sustained performance in agricultural applications.

10.3 Compatibility with Existing Agricultural Practices

The compatibility of 2D-based hybrid materials with existing agricultural practices is crucial for their successful implementation. It is essential to consider the integration of these materials into conventional agricultural systems without requiring significant modifications or disruptions to existing practices [52]. Compatibility includes factors such as the formulation of 2D materials into existing products, their interaction with soil and crops, and their potential impact on the environment. Collaboration between scientists, engineers, and farmers is essential to ensure that the use of 2D-based hybrid materials aligns with established agricultural practices and facilitates their adoption in the field.

10.4 Environmental and Health Implications

Understanding the potential environmental and health implications of 2D-based hybrid materials is critical for their safe and sustainable use in agriculture. While 2D materials offer numerous benefits, their interaction with the environment and living organisms must be carefully assessed [66]. Studies should be conducted to evaluate the long-term effects of these materials on soil health, crop growth, microbial communities, and ecosystem dynamics. Additionally, the potential release of nanoparticles or toxic components from 2D materials should be investigated to minimize any adverse effects on human health and the environment. Regulations and guidelines should be developed to ensure the responsible and sustainable use of 2D-based hybrid materials in agriculture.

11 Knowledge and Awareness

Another challenge lies in raising awareness and disseminating knowledge about 2D-based hybrid materials among farmers, agricultural practitioners, and policymakers. Many stakeholders may have limited familiarity with these emerging technologies and their potential applications in agriculture. Efforts should be made to bridge the knowledge gap through education, training programs, workshops, and information-sharing platforms [86]. Collaborative initiatives between researchers, industry, and agricultural organizations can facilitate knowledge transfer and technology adoption, promoting the successful integration of 2D-based hybrid materials in agricultural systems.

12 Standardization and Quality Control

Establishing standardized protocols and quality control measures for the production, characterization, and application of 2D-based hybrid materials is essential [87]. Consistency in material properties, performance, and functionality is crucial for ensuring reproducibility and reliability across different agricultural settings. Developing standardized testing methods and quality assurance guidelines will enable farmers and agricultural practitioners to make informed decisions about the selection and utilization of 2D-based hybrid materials.

13 Regulatory and Policy Frameworks

The development and implementation of regulatory and policy frameworks specific to 2D-based hybrid materials in agriculture are important [52]. These frameworks should address aspects such as safety assessments, labelling requirements, and environmental risk assessments. Clear guidelines will provide regulatory certainty and foster responsible use, ensuring that the potential benefits of 2D-based hybrid materials are realized while minimizing any potential risks to human health and the environment.

14 Long-Term Efficacy and Performance Monitoring

Long-term efficacy and performance monitoring of 2D-based hybrid materials in agricultural systems are crucial to assess their sustained effectiveness. Monitoring studies should evaluate factors such as material degradation, nutrient release rates, pest and disease control, and overall crop performance over extended periods [88].

These studies will provide valuable insights into the long-term benefits and limitations of using 2D-based hybrid materials, allowing for continuous improvement and optimization of their agricultural applications.

14.1 Adoption and Economic Viability

Encouraging the adoption of 2D-based hybrid materials in agriculture requires demonstrating their economic viability and return on investment for farmers. Cost–benefit analyses and case studies that showcase the potential economic advantages of using these materials, such as increased crop yield, reduced inputs, and improved resource efficiency, can incentivize farmers to incorporate them into their agricultural practices [83]. Additionally, exploring funding mechanisms, subsidies, or incentives for farmers to adopt 2D-based hybrid materials can help overcome initial financial barriers.

Therefore, while the integration of 2D-based hybrid materials in agriculture presents immense potential, several challenges need to be addressed to ensure their successful development and implementation. Overcoming scalability and cost-effectiveness barriers, ensuring material stability and compatibility with existing practices, assessing environmental and health implications, and promoting knowledge exchange and Collaboration are crucial steps in realizing the full benefits of 2D-based hybrid materials for sustainable agriculture. By addressing these challenges, we can unlock the transformative potential of these materials and pave the way for a more efficient, resilient, and sustainable agricultural future.

15 Future Directions and Potential Breakthroughs in the Field

The integration of two-dimensional (2D) materials in agriculture has demonstrated immense potential for enhancing crop productivity, resource efficiency, and sustainability. As researchers continue to explore the applications of 2D-based hybrid materials in agriculture, several future directions and potential breakthroughs are emerging which are summarized in this section below.

15.1 Development of Novel 2D Materials

One of the key future directions is the development of novel 2D materials tailored specifically for agricultural applications. Researchers are exploring the synthesis and characterization of new materials with enhanced properties, such as improved stability, durability, and selectivity. The discovery of new 2D materials or the modification of existing ones can lead to breakthroughs in areas such as controlled release of nutrients, targeted delivery of agrochemicals, and improved stress tolerance in crops [69]. The exploration of materials beyond graphene and transition metal dichalcogenides opens up new possibilities for addressing specific agricultural challenges.

15.2 Smart and Responsive Systems

Future breakthroughs in 2D-based hybrid materials will involve the development of smart and responsive systems for precision agriculture. Integration of sensors, actuators, and stimuli-responsive materials with 2D hybrids can enable real-time monitoring and precise control of agricultural processes. These systems can detect and respond to environmental changes, nutrient deficiencies, or pest infestations, triggering targeted actions for optimized resource utilization and crop management [89]. Such smart systems have the potential to revolutionize the way we monitor and manage agricultural systems, leading to increased efficiency and reduced environmental impact.

15.3 Nanotechnology and Nanoscale Delivery Systems

Nanotechnology holds great promise for the future of 2D-based hybrid materials in agriculture. Researchers are exploring the use of nanoscale delivery systems to improve the targeted delivery and controlled release of nutrients, pesticides, and growth-promoting substances [90]. Nanoencapsulation and nanocarrier technologies can protect active compounds, increase their stability, and enhance their bioavailability [75]. The combination of 2D materials with nanotechnology offers opportunities for precise and efficient delivery of agrochemicals, minimizing their environmental impact and improving their efficacy.

15.4 Bioinspired and Biocompatible Approaches

Future breakthroughs in 2D-based hybrid materials will likely involve bioinspired and biocompatible approaches. Researchers are looking to nature for inspiration, studying the principles and mechanisms employed by plants and microorganisms to overcome environmental challenges [91]. By mimicking these natural processes, it is possible to develop bioinspired materials that exhibit enhanced functionality and adaptability. Biocompatible materials can integrate seamlessly with biological systems, promoting symbiotic interactions and minimizing any potential harm to the environment or organisms. These approaches have the potential to revolutionize pest and disease management, soil remediation, and crop enhancement strategies.

15.5 *Data-Driven Agriculture and Machine Learning*

The integration of 2D-based hybrid materials with data-driven agriculture and machine learning techniques represents a promising direction for future breakthroughs [92]. By harnessing the power of big data, advanced analytics, and machine learning algorithms, it is possible to develop predictive models for crop growth, disease outbreaks, and resource requirements. The integration of sensor data, satellite imagery, and real-time monitoring with 2D-based hybrid materials can enable proactive decision-making, precise resource allocation, and optimized crop management [93]. This data-driven approach has the potential to revolutionize the efficiency and sustainability of agricultural systems.

15.6 *Integration of Renewable Energy Sources*

The integration of renewable energy sources with 2D-based hybrid materials offers exciting possibilities for the future of agriculture. Harnessing solar energy through the use of photovoltaic materials integrated with 2D-based hybrid materials can provide a sustainable and efficient energy source for agricultural systems [94]. The integration of solar panels made from 2D materials, such as graphene, can capture solar energy and convert it into electricity to power various agricultural processes. This integration can reduce reliance on conventional energy sources and promote greener and more sustainable farming practices.

15.7 *Multifunctional Materials*

Future breakthroughs in 2D-based hybrid materials will involve the development of multifunctional materials that can perform multiple tasks simultaneously. Researchers are exploring the incorporation of different functionalities into a single material, such as nutrient delivery, water retention, and pest control [95]. This integration can lead to the development of advanced materials that offer comprehensive solutions for multiple agricultural challenges. Multifunctional materials can streamline agricultural processes, reduce the need for multiple applications, and enhance overall efficiency [79].

15.8 Integration of Internet of Things (IoT)

The integration of 2D-based hybrid materials with the Internet of Things (IoT) technology presents a significant opportunity for the future of agriculture [78]. By connecting sensors, devices, and agricultural equipment through wireless networks, farmers can monitor and control various aspects of their operations remotely. This integration can facilitate real-time data collection, analysis, and decision-making, leading to optimized resource management, reduced waste, and improved productivity [79]. The combination of 2D-based hybrid materials and IoT can enable precision agriculture on a large scale.

15.9 Collaboration and Interdisciplinary Research

To unlock the full potential of 2D-based hybrid materials in agriculture, collaboration, and interdisciplinary research are essential. Collaboration between material scientists, agronomists, engineers, and farmers can facilitate knowledge exchange, innovation, and the practical implementation of these materials in agricultural systems [52]. Interdisciplinary research efforts can address complex agricultural challenges from multiple perspectives, leading to holistic solutions that consider the social, economic, and environmental aspects of agriculture.

15.10 Policy and Regulatory Support

The future success of 2D-based hybrid materials in agriculture relies on the development of supportive policies and regulatory frameworks. Governments and regulatory bodies need to keep pace with advancements in the field and establish guidelines for the safe and responsible use of these materials [52]. Supporting research funding, promoting technology transfer, and providing incentives for the adoption of 2D-based hybrid materials can accelerate their development and implementation in agriculture.

The future of 2D-based hybrid materials in agriculture holds tremendous potential for transforming farming practices, enhancing resource efficiency, and promoting sustainability. With ongoing advancements, we can anticipate a future where agriculture is more sustainable, efficient, and resilient.

16 Conclusion

The integration of two-dimensional (2D) materials in agriculture has the potential to revolutionize farming practices and pave the way for sustainable and efficient agricultural systems. The utilization of 2D-based hybrid materials in agriculture offers a wide range of potential benefits. These materials have shown promise in enhancing nutrient delivery systems, improving crop growth and productivity, remediating soils, increasing pesticide and fertilizer efficiency, managing water resources, and mitigating environmental pollution. By leveraging the unique properties of 2D materials and combining them with other materials, we can achieve enhanced functionality and performance, resulting in more sustainable and efficient agricultural practices.

Continued research and development in the field of 2D-based hybrid materials in agriculture are crucial for unlocking their full potential. There is still much to discover and understand about these materials, including their synthesis and fabrication techniques, performance under different environmental conditions, long-term efficacy, and safety aspects. Furthermore, exploring novel 2D materials, integrating them with emerging technologies such as nanotechnology and IoT, and investigating their interactions with plants, soil microorganisms, and ecosystems will further advance our knowledge and capabilities. Research and development efforts should focus on optimizing the properties and functionalities of 2D-based hybrid materials, improving their scalability, cost-effectiveness, and environmental compatibility, and addressing any potential risks or limitations associated with their use. Additionally, collaborative research endeavours that bring together scientists, engineers, agronomists, policymakers, and farmers will facilitate the exchange of knowledge, foster innovation, and promote the practical application of these materials in real-world agricultural settings.

The implementation of 2D-based hybrid materials in agriculture aligns with the broader goals of sustainable development, including the United Nations' Sustainable Development Goals (SDGs). These materials have the potential to contribute to multiple SDGs, such as zero hunger, clean water and sanitation, sustainable cities and communities, responsible consumption and production, and climate action. Therefore, by addressing key agricultural challenges and promoting resource-efficient practices, 2D-based hybrid materials can play a vital role in achieving a more sustainable, resilient, and food-secure future.

References

1. Singh, P.K., Naresh, R.K., Bhatt, R., Tiwari, H., Rajput, A., Sampath, L., Padhan, S.R., Goyal, A.M.: Effect of climate-smart agriculture practices on energy, greenhouse gas mitigation and resource use efficiency of rice-wheat cropping system in North West IGP: a review. Pharma Innov. **12**, 1976–1987 (2023)

2. Balandin, A.A.: Thermal properties of graphene and nanostructured carbon materials. Nat. Mater. **10**, 569–581 (2011). https://doi.org/10.1038/nmat3064
3. Quintana, M.: 2D materials and van der Waals heterostructures platforms for advanced sensing of COVID-19. In: Biomedical Innovations to Combat COVID-19, pp. 241–252. Elsevier (2022)
4. Varghese, S., Varghese, S., Swaminathan, S., Singh, K., Mittal, V.: Two-dimensional materials for sensing: graphene and beyond. Electronics (Basel) **4**, 651–687 (2015). https://doi.org/10.3390/electronics4030651
5. Avouris, P.: Graphene: electronic and photonic properties and devices. Nano Lett. **10**, 4285–4294 (2010). https://doi.org/10.1021/nl102824h
6. Novoselov, K.S., Geim, A.K., Morozov, S.V., Jiang, D., Zhang, Y., Dubonos, S.V., Grigorieva, I.V., Firsov, A.A.: Electric field effect in atomically thin carbon films. Science **306**, 666–669 (2004). https://doi.org/10.1126/science.1102896
7. Geim, A.K., Novoselov, K.S.: The rise of graphene. Nat. Mater. **6**, 183–191 (2007). https://doi.org/10.1038/nmat1849
8. Lee, C., Wei, X., Kysar, J.W., Hone, J.: Measurement of the elastic properties and intrinsic strength of monolayer graphene. Science **321**, 385–388 (2008). https://doi.org/10.1126/science.1157996
9. Kumar, S., Kaushik, S., Pratap, R., Raghavan, S.: Graphene on paper: a simple, low-cost chemical sensing platform. ACS Appl. Mater. Interfaces **7**, 2189–2194 (2015). https://doi.org/10.1021/am5084122
10. Schwierz, F.: Graphene transistors. Nat. Nanotechnol.Nanotechnol. **5**, 487–496 (2010). https://doi.org/10.1038/nnano.2010.89
11. Rumyantsev, S., Liu, G., Shur, M.S., Potyrailo, R.A., Balandin, A.A.: Selective gas sensing with a single pristine graphene transistor. Nano Lett. **12**, 2294–2298 (2012). https://doi.org/10.1021/nl3001293
12. Wehling, T.O., Novoselov, K.S., Morozov, S.V., Vdovin, E.E., Katsnelson, M.I., Geim, A.K., Lichtenstein, A.I.: Molecular doping of graphene. Nano Lett. **8**, 173–177 (2008). https://doi.org/10.1021/nl072364w
13. Blanco, E., Rocha, L., Pozo, M.D., Vázquez, L., Petit-Domínguez, M.D., Casero, E., Quintana, C.: A supramolecular hybrid sensor based on cucurbit [8] uril, 2D-molibdenum disulphide and diamond nanoparticles towards methyl viologen analysis. Anal. Chim. Acta **1182**, 338940 (2021). https://doi.org/10.1016/j.aca.2021.338940
14. Ozcelikay, G., Karadurmus, L., Bilge, S., Sınağ, A., Ozkan, S.A.: New analytical strategies amplified with 2D carbon nanomaterials for electrochemical sensing of food pollutants in water and soils sources. Chemosphere **296**, 133974 (2022). https://doi.org/10.1016/j.chemosphere.2022.133974
15. Selamneni, V., Sahatiya, P.: Mixed dimensional transition metal dichalcogenides (TMDs) vdW Heterostructure based photodetectors: a review. Microelectron. Eng. **269**, 111926 (2023). https://doi.org/10.1016/j.mee.2022.111926
16. Drogowska-Horná, K., Frank, O., Kalbac, M.: Chemical vapor deposition (CVD) growth of graphene films. In: Graphene, pp. 199–222. Elsevier (2021)
17. Zhang, J., Wang, F., Shenoy, V.B., Tang, M., Lou, J.: Towards controlled synthesis of 2D crystals by chemical vapor deposition (CVD). Mater. Today (Kidlington) **40**, 132–139 (2020). https://doi.org/10.1016/j.mattod.2020.06.012
18. Abdullah, M.Z.B., Ahmad, M.A., Abdullah, A.N., Othman, M.H., Hussain, P., Zainuddin, A.: Metal release of multilayer coatings by physical vapour deposition (PVD). Procedia Eng. **148**, 254–260 (2016). https://doi.org/10.1016/j.proeng.2016.06.612
19. Lv, Y., Zhang, L., Wei, X., Qiu, B., Zhang, W., Qin, Q., Jia, D., He, X., Liu, Z., Wei, F.: The emerging of zinc-ion hybrid supercapacitors: advances, challenges, and future perspectives. Sustain. Mater. Technol. **35**, e00536 (2023). https://doi.org/10.1016/j.susmat.2022.e00536
20. Glavin, N.R., Nam, S.: 2D layered materials and heterostructures: past, present, and a bright future. Matter **6**, 4–6 (2023). https://doi.org/10.1016/j.matt.2022.11.030
21. Nautiyal, H., Kumari, S., Tyagi, R., Rao, U.S., Khatri, O.P.: Evaluation of tribological performance of copper-based composites containing nano-structural 2D materials and their hybrid. Tribol. Int.. Int. **153**, 106645 (2021). https://doi.org/10.1016/j.triboint.2020.106645

22. Wen, F., Zhu, C., Li, L., Zhou, B., Zhang, L., Han, C., Li, W., Yue, Z., Wu, W., Wang, G., Zhang, S.: Enhanced energy storage performance of polymer nanocomposites using hybrid 2D ZnO@MoS2 semiconductive nano-fillers. Chem. Eng. J. **430**, 132676 (2022). https://doi.org/10.1016/j.cej.2021.132676
23. Elloh, V.W., Boadu, E.O., Gebreyesus, G., Mishra, A.K., Dodoo-Arhin, D., Yaya, A.: Design of novel hybrid 2D nanomaterials for optical, optoelectronic and micro-electro-mechanical systems applications. Hybrid Adv. **3**, 100045 (2023). https://doi.org/10.1016/j.hybadv.2023.100045
24. Yu, J., Xiong, S., Wang, B., Wang, R., He, B., Jin, J., Wang, H., Gong, Y.: Constructing boron-doped graphitic carbon nitride with 2D/1D porous hierarchical architecture and efficient N2 photofixation. Colloids Surf A Physicochem Eng AspA Physicochem Eng Asp **656**, 130481 (2023). https://doi.org/10.1016/j.colsurfa.2022.130481
25. Daneshnia, S., Shams, A., Daraei, D., Abdouss, M., Daneshmayeh, M.: Novel thin film nanocomposite membrane modified with Boron Nitride Nanosheets for water treatment. Colloids Surf A Physicochem Eng AspPhysicochem Eng Asp **651**, 129768 (2022). https://doi.org/10.1016/j.colsurfa.2022.129768
26. Irshad, R., Tahir, K., Li, B., Sher, Z., Ali, J., Nazir, S.: A revival of 2D materials, phosphorene: its application as sensors. J. Ind. Eng. Chem. **64**, 60–69 (2018). https://doi.org/10.1016/j.jiec.2018.03.010
27. Zuluaga-Hernandez, E.A., Flórez, E., Dorkis, L., Mora-Ramos, M.E., Correa, J.D.: Small molecule gas adsorption onto blue phosphorene oxide layers. Appl. Surf. Sci. **530**, 147039 (2020). https://doi.org/10.1016/j.apsusc.2020.147039
28. Liu, J., Xu, Y., Yi, F.: Fabrication of the ZnO nanowires/TiO2 nanowires/Si micropillars structures for the gas sensor application. Sens. Actuators A Phys. **345**, 113665 (2022). https://doi.org/10.1016/j.sna.2022.113665
29. Kailasa, S.K., Joshi, D.J., Koduru, J.R., Malek, N.I.: Review on MXenes-based nanomaterials for sustainable opportunities in energy storage, sensing and electrocatalytic reactions. J. Mol. Liq. **342**, 117524 (2021). https://doi.org/10.1016/j.molliq.2021.117524
30. Huang, H., Dong, D., Li, W., Zhang, X., Zhang, L., Chen, Y., Sheng, X., Lu, X.: Synergistic effect of MXene on the flame retardancy and thermal degradation of intumescent flame retardant biodegradable poly (lactic acid) composites. Chin. J. Chem. Eng. **28**, 1981–1993 (2020). https://doi.org/10.1016/j.cjche.2020.04.014
31. Dávila, M.E., Le Lay, G.: Silicene: Genesis, remarkable discoveries, and legacy. Mater. Today Adv. **16**, 100312 (2022). https://doi.org/10.1016/j.mtadv.2022.100312
32. Lin, S., Li, Y., Qian, J., Lau, S.P.: Emerging opportunities for black phosphorus in energy applications. Mater. Today Energy. **12**, 1–25 (2019). https://doi.org/10.1016/j.mtener.2018.12.004
33. Zeng, Y., Guo, Z.: Synthesis and stabilization of black phosphorus and phosphorene: recent progress and perspectives. iScience **24**, 103116 (2021). https://doi.org/10.1016/j.isci.2021.103116
34. Sahoo, D., Senapati, S., Naik, R.: Progress and prospects of 2D VS2 transition metal dichalcogenides. FlatChem. **36**, 100455 (2022). https://doi.org/10.1016/j.flatc.2022.100455
35. Tapia, M.A., Gusmão, R., Serrano, N., Sofer, Z., Ariño, C., Díaz-Cruz, J.M., Esteban, M.: Phosphorene and other layered pnictogens as a new source of 2D materials for electrochemical sensors. Trends Analyt. Chem. **139**, 116249 (2021). https://doi.org/10.1016/j.trac.2021.116249
36. Park, M., Chun, Y., Kim, S., Sohn, K.Y., Jeon, M.: Effects of hexagonal boron nitride insulating layers on the driving performance of ionic electroactive polymer actuators for light-weight artificial muscles. Int. J. Mol. Sci. **23**, 4981 (2022). https://doi.org/10.3390/ijms23094981
37. Joseph, S., Mohan, J., Lakshmy, S., Thomas, S., Chakraborty, B., Thomas, S., Kalarikkal, N.: A review of the synthesis, properties, and applications of 2D transition metal dichalcogenides and their heterostructures. Mater. Chem. Phys. **297**, 127332 (2023). https://doi.org/10.1016/j.matchemphys.2023.127332
38. Mohan, B., Dhiman, D., Sun, Q., Jan, M., Singh, G., Raghav, N.: Metal-organic frameworks (MOFs) structural properties and electrochemical detection capability for cancer biomarkers. Microchem. J. 108956 (2023). https://doi.org/10.1016/j.microc.2023.108956

39. Venkateshalu, S., Subashini, G., Bhardwaj, P., Jacob, G., Sellappan, R., Raghavan, V., Jain, S., Pandiaraj, S., Natarajan, V., Al Alwan, B.A.M., Al Mesfer, M.K.M., Alodhayb, A., Khalid, M., Grace, A.N.: Phosphorene, antimonene, silicene and siloxene based novel 2D electrode materials for supercapacitors-a brief review. J. Energy Storage. **48**, 104027 (2022). https://doi.org/10.1016/j.est.2022.104027
40. Sudhaik, A., Raizada, P., Shandilya, P., Jeong, D.-Y., Lim, J.-H., Singh, P.: Review on fabrication of graphitic carbon nitride based efficient nanocomposites for photodegradation of aqueous phase organic pollutants. J. Ind. Eng. Chem. **67**, 28–51 (2018). https://doi.org/10.1016/j.jiec.2018.07.007
41. Yu, S., Wang, X., Pang, H., Zhang, R., Song, W., Fu, D., Hayat, T., Wang, X.: Boron nitride-based materials for the removal of pollutants from aqueous solutions: a review. Chem. Eng. J. **333**, 343–360 (2018). https://doi.org/10.1016/j.cej.2017.09.163
42. Wen, W., Song, Y., Yan, X., Zhu, C., Du, D., Wang, S., Asiri, A.M., Lin, Y.: Recent advances in emerging 2D nanomaterials for biosensing and bioimaging applications. Mater. Today (Kidlington) **21**, 164–177 (2018). https://doi.org/10.1016/j.mattod.2017.09.001
43. Benrezgua, E., Deghfel, B., Zoukel, A., Basirun, W.J., Amari, R., Boukhari, A., Yaakob, M.K., Kheawhom, S., Mohamad, A.A.: Synthesis and properties of copper doped zinc oxide thin films by sol-gel, spin coating and dipping: a characterization review. J. Mol. Struct. **1267**, 133639 (2022). https://doi.org/10.1016/j.molstruc.2022.133639
44. Deepa, C., Rajeshkumar, L., Ramesh, M.: Preparation, synthesis, properties and characterization of graphene-based 2D nanomaterials for biosensors and bioelectronics. J. Mater. Res. Technol. **19**, 2657–2694 (2022). https://doi.org/10.1016/j.jmrt.2022.06.023
45. Yadav, A., Kumar, H., Shrama, R., Kumari, R.: Synthesis, processing, and applications of 2D (nano)materials: a sustainable approach. Surf. Interfaces. **39**, 102925 (2023). https://doi.org/10.1016/j.surfin.2023.102925
46. Wang, C., Park, M.J., Yu, H., Matsuyama, H., Drioli, E., Shon, H.K.: Recent advances of nanocomposite membranes using layer-by-layer assembly. J. Memb. Sci. **661**, 120926 (2022). https://doi.org/10.1016/j.memsci.2022.120926
47. Gao, Q., Li, J., Liu, B., Liu, C.: In-situ synthesis of direct Z-scheme 2D/2D ZnIn2S4@CeO2 heterostructure toward enhanced photodegradation and Cr(VI) reduction. J. Alloys Compd. **931**, 167430 (2023). https://doi.org/10.1016/j.jallcom.2022.167430
48. Naikoo, G.A., Arshad, F., Almas, M., Hassan, I.U., Pedram, M.Z., Aljabali, A.A.A., Mishra, V., Serrano-Aroca, Á., Birkett, M., Charbe, N.B., Goyal, R., Negi, P., El-Tanani, M., Tambuwala, M.M.: 2D materials, synthesis, characterization and toxicity: a critical review. Chem. Biol. Interact. **365**, 110081 (2022). https://doi.org/10.1016/j.cbi.2022.110081
49. Pal, N., Bhaumik, A.: Soft templating strategies for the synthesis of mesoporous materials: inorganic, organic-inorganic hybrid and purely organic solids. Adv. Colloid Interface Sci. **189–190**, 21–41 (2013). https://doi.org/10.1016/j.cis.2012.12.002
50. Chakraborty, P.K.: 3D printing of 2D nano-inks for multifarious applications. In: Smart Multifunctional Nano-inks, pp. 91–124. Elsevier (2023)
51. Mahana, S., Padhan, S.R., Padhan, S.R.: An insight into GreenSeeker technology: a vital tool for precision nutrient management. Biotica Res. Today. **4**, 026–028 (2022)
52. Singhal, J., Verma, S., Kumar, S.: The physio-chemical properties and applications of 2D nanomaterials in agricultural and environmental sustainability. Sci. Total. Environ. **837**, 155669 (2022). https://doi.org/10.1016/j.scitotenv.2022.155669
53. Wang, S., Liu, Y., Wang, X., Xiang, H., Kong, D., Wei, N., Guo, W., Sun, H.: Effects of concentration-dependent graphene on maize seedling development and soil nutrients. Sci. Rep. **13**, 2650 (2023). https://doi.org/10.1038/s41598-023-29725-3
54. Lee, C.W., Suh, J.M., Jang, H.W.: Chemical sensors based on two-dimensional (2D) materials for selective detection of ions and molecules in liquid. Front. Chem. **7**, 708 (2019). https://doi.org/10.3389/fchem.2019.00708
55. Chen, X., Li, R., Li, Y., Wang, Y., Zhang, F., Zhang, M.: Research of the multifunctional rGO/MoS2 material in the sensing field: human breathing and Hg (II) pollution detection. Mater. Sci. Semicond. Process. **138**, 106268 (2022). https://doi.org/10.1016/j.mssp.2021.106268

56. Shyam, C.S., Rathore, S.S., Shekhawat, K., Singh, R.K., Padhan, S.R., Singh, V.K.: Precision nutrient management in maize (Zea mays) for higher productivity and profitability. Indian J. Agric. Sci. **91** (2021). https://doi.org/10.56093/ijas.v91i6.114303
57. Padhan, S.R., Rathore, S.S., Prasad, S.M., Shekhawat, K., Singh, V.K.: Influence of precision nutrient and weed management on growth and productivity of direct-seeded upland rice (Oryza sativa) under Eastern Plateau and Hills Region. Indian J. Agron. **66**, 366–369 (2021)
58. Kallenbach, C.M., Conant, R.T., Calderón, F., Wallenstein, M.D.: A novel soil amendment for enhancing soil moisture retention and soil carbon in drought-prone soils. Geoderma **337**, 256–265 (2019). https://doi.org/10.1016/j.geoderma.2018.09.027
59. Malik, S., Chaudhary, K., Malik, A., Punia, H., Sewhag, M., Berkesia, N., Nagora, M., Kalia, S., Malik, K., Kumar, D., Kumar, P., Kamboj, E., Ahlawat, V., Kumar, A., Boora, K.: Superabsorbent polymers as a soil amendment for increasing agriculture production with reducing water losses under water stress condition. Polymers (Basel) **15**, 161 (2022). https://doi.org/10.3390/polym15010161
60. Kumar, R., Kumar, N., Rajput, V.D., Mandzhieva, S., Minkina, T., Saharan, B.S., Kumar, D., Sadh, P.K., Duhan, J.S.: Advances in biopolymeric nanopesticides: a new eco-friendly/eco-protective perspective in precision agriculture. Nanomaterials (Basel). **12** (2022). https://doi.org/10.3390/nano12223964
61. Malik, S., Singh, S., Pooja, Meenu, Radheshyam, Padhan, S.R.: Growth behavior of parthenium hysterophorus as influenced by environmental factors. Int. J. Environ. Climate Change **13**, 1158–1168 (2023). https://doi.org/10.9734/ijecc/2023/v13i82055
62. Seleiman, M.F., Al-Suhaibani, N., Ali, N., Akmal, M., Alotaibi, M., Refay, Y., Dindaroglu, T., Abdul-Wajid, H.H., Battaglia, M.L.: Drought stress impacts on plants and different approaches to alleviate its adverse effects. Plants **10**, 259 (2021). https://doi.org/10.3390/plants10020259
63. Le Cointe, R., Simon, T.E., Delarue, P., Hervé, M., Leclerc, M., Poggi, S.: Reducing the use of pesticides with site-specific application: The chemical control of Rhizoctonia solani as a case of study for the management of soil-borne diseases. PLoS ONE **11**, e0163221 (2016). https://doi.org/10.1371/journal.pone.0163221
64. Akhtar, M., Tahir, A., Zulfiqar, S., Hanif, F., Warsi, M.F., Agboola, P.O., Shakir, I.: Ternary hybrid of polyaniline-alanine-reduced graphene oxide for electrochemical sensing of heavy metal ions. Synth. Met. **265**, 116410 (2020). https://doi.org/10.1016/j.synthmet.2020.116410
65. Wang, J., Yu, P., Kan, K., Lv, H., Liu, Z., Sun, B., Bai, X., Chen, J., Zhang, Y., Shi, K.: Efficient ultra-trace electrochemical detection of Cd2+, Pb2+ and Hg2+ based on hierarchical porous S-doped C3N4 tube bundles/graphene nanosheets composite. Chem. Eng. J. **420**, 130317 (2021). https://doi.org/10.1016/j.cej.2021.130317
66. Shakeel, A., Rizwan, K., Farooq, U., Iqbal, S., Iqbal, T., Awwad, N.S., Ibrahium, H.A.: Polymer based nanocomposites: a strategic tool for detection of toxic pollutants in environmental matrices. Chemosphere **303**, 134923 (2022). https://doi.org/10.1016/j.chemosphere.2022.134923
67. Vasseghian, Y., Dragoi, E.-N., Almomani, F., Le, V.T.: Graphene-based materials for metronidazole degradation: a comprehensive review. Chemosphere **286**, 131727 (2022). https://doi.org/10.1016/j.chemosphere.2021.131727
68. Padhan, S.R., Rathore, S.S., Prasad, S.M., Singh, R.K., Shekhawat, K.: Influence of nutrient and weed management on weed dynamics and productivity of upland rice (Oryza sativa). Indian J. Agric. Sci. **91** (2021). https://doi.org/10.56093/ijas.v91i7.115141
69. Ali, S.S., Al-Tohamy, R., Koutra, E., Moawad, M.S., Kornaros, M., Mustafa, A.M., Mahmoud, Y.A.-G., Badr, A., Osman, M.E.H., Elsamahy, T., Jiao, H., Sun, J.: Nanobiotechnological advancements in agriculture and food industry: applications, nanotoxicity, and future perspectives. Sci. Total. Environ. **792**, 148359 (2021). https://doi.org/10.1016/j.scitotenv.2021.148359
70. Panigrahi, K.K., Mohanty, A., Padhan, S.R., Guru, R.K.S.: Genotoxicity and DNA damage induced by herbicides and toxins in plants. In: Induced Genotoxicity and Oxidative Stress in Plants, pp. 29–63. Springer Singapore, Singapore (2021)

71. Zobir, S.A.M., Ali, A., Adzmi, F., Sulaiman, M.R., Ahmad, K.: A review on nanopesticides for plant protection synthesized using the supramolecular chemistry of layered hydroxide hosts. Biology (Basel). **10**, 1077 (2021). https://doi.org/10.3390/biology10111077
72. Mustafa, I.F., Hussein, M.Z., Saifullah, B., Idris, A.S., Hilmi, N.H.Z., Fakurazi, S.: Synthesis of (hexaconazole-zinc/aluminum-layered double hydroxide nanocomposite) fungicide nanodelivery system for controlling Ganoderma disease in oil palm. J. Agric. Food Chem. **66**, 806–813 (2018). https://doi.org/10.1021/acs.jafc.7b04222
73. Saratale, R.G., Saratale, G.D., Shin, H.S., Jacob, J.M., Pugazhendhi, A., Bhaisare, M., Kumar, G.: New insights on the green synthesis of metallic nanoparticles using plant and waste biomaterials: current knowledge, their agricultural and environmental applications. Environ. Sci. Pollut. Res. Int. **25**, 10164–10183 (2018). https://doi.org/10.1007/s11356-017-9912-6
74. Godínez-García, F.J., Guerrero-Rivera, R., Martínez-Rivera, J.A., Gamero-Inda, E., Ortiz-Medina, J.: Advances in two-dimensional engineered nanomaterials applications for the agro- and food-industries. J. Sci. Food Agric. (2023). https://doi.org/10.1002/jsfa.12556
75. Fan, G., Li, F., Evans, D.G., Duan, X.: Catalytic applications of layered double hydroxides: recent advances and perspectives. Chem. Soc. Rev. **43**, 7040–7066 (2014). https://doi.org/10.1039/c4cs00160e
76. Rajanna, G.A., Manna, S., Singh, A., Babu, S., Singh, V.K., Dass, A., Chakraborty, D., Patanjali, N., Chopra, I., Banerjee, T., Kumar, A., Khandelwal, A., Parmar, B.S.: Biopolymeric superabsorbent hydrogels enhance crop and water productivity of soybean-wheat system in Indo-Gangetic plains of India. Sci. Rep. **12**, 11955 (2022). https://doi.org/10.1038/s41598-022-16049-x
77. Womack, N.C., Piccoli, I., Camarotto, C., Squartini, A., Guerrini, G., Gross, S., Maggini, M., Cabrera, M.L., Morari, F.: Hydrogel application for improving soil pore network in agroecosystems: preliminary results on three different soils. Catena **208**, 105759 (2022). https://doi.org/10.1016/j.catena.2021.105759
78. Premkumar, S., Sigappi, A.N.: IoT-enabled edge computing model for smart irrigation system. J. Intell. Syst.Intell. Syst. **31**, 632–650 (2022). https://doi.org/10.1515/jisys-2022-0046
79. García, L., Parra, L., Jimenez, J.M., Lloret, J., Lorenz, P.: IoT-based smart irrigation systems: an overview on the recent trends on sensors and IoT systems for irrigation in precision agriculture. Sensors (Basel). **20**, 1042 (2020). https://doi.org/10.3390/s20041042
80. Fatima, J., Shah, A.N., Tahir, M.B., Mehmood, T., Shah, A.A., Tanveer, M., Nazir, R., Jan, B.L., Alansi, S.: Tunable 2D nanomaterials; their key roles and mechanisms in water purification and monitoring. Front. Environ. Sci. **10** (2022). https://doi.org/10.3389/fenvs.2022.766743
81. Dai, Y., Liu, M., Li, J., Kang, N., Ahmed, A., Zong, Y., Tu, J., Chen, Y., Zhang, P., Liu, X.: Graphene-based membranes for water desalination: a literature review and content analysis. Polymers (Basel). **14**, 4246 (2022). https://doi.org/10.3390/polym14194246
82. Boretti, A., Al-Zubaidy, S., Vaclavikova, M., Al-Abri, M., Castelletto, S., Mikhalovsky, S.: Outlook for graphene-based desalination membranes. Npj Clean Water. **1** (2018). https://doi.org/10.1038/s41545-018-0004-z
83. Kulandaivalu, S., Azman, N.H.N., Sulaiman, Y.: Advances in layered double hydroxide/carbon nanocomposites containing Ni2+ and Co2+/3+ for supercapacitors. Front. Mater. **7** (2020). https://doi.org/10.3389/fmats.2020.00147
84. Rafiei-Sarmazdeh, Z., Morteza Zahedi-Dizaji, S., Kafi Kang, A.: Two-dimensional nanomaterials. In: Nanostructures. IntechOpen (2020)
85. Ismail, S.O., Akpan, E., Dhakal, H.N.: Review on natural plant fibres and their hybrid composites for structural applications: recent trends and future perspectives. Compos. Part C Open Access. **9**, 100322 (2022). https://doi.org/10.1016/j.jcomc.2022.100322
86. Saini, S., Mallick, S., Padhan, S.R.: Participatory extension approach: empowering farmers. Biotica Res. Today **5**, 326–328 (2023)
87. Sinha, S., Kim, H., Robertson, A.W.: Preparation and application of 0D–2D nanomaterial hybrid heterostructures for energy applications. Mater. Today Adv. **12**, 100169 (2021). https://doi.org/10.1016/j.mtadv.2021.100169

88. Ashfaq, M., Talreja, N., Singh, N., Chauhan, D.: 2D-nanolayer (2D-NL)-based hybrid materials: a next-generation material for dye-sensitized solar cells. Electronics (Basel) **12**, 570 (2023). https://doi.org/10.3390/electronics12030570
89. Guleria, G., Thakur, S., Shandilya, M., Sharma, S., Thakur, S., Kalia, S.: Nanotechnology for sustainable agro-food systems: the need and role of nanoparticles in protecting plants and improving crop productivity. Plant Physiol. Biochem.Biochem. **194**, 533–549 (2023). https://doi.org/10.1016/j.plaphy.2022.12.004
90. Li, F., Zhang, L., Evans, D.G., Forano, C., Duan, X.: Structure and thermal evolution of Mg–Al layered double hydroxide containing interlayer organic glyphosate anions. Thermochim. Acta. Acta **424**, 15–23 (2004). https://doi.org/10.1016/j.tca.2004.05.007
91. Anjum, S., Vyas, A., Sofi, T.: Fungi-mediated synthesis of nanoparticles: characterization process and agricultural applications. J. Sci. Food Agric. **103**, 4727–4741 (2023). https://doi.org/10.1002/jsfa.12496
92. Shrestha, Y.R., Krishna, V., von Krogh, G.: Augmenting organizational decision-making with deep learning algorithms: Principles, promises, and challenges. J. Bus. Res. **123**, 588–603 (2021). https://doi.org/10.1016/j.jbusres.2020.09.068
93. Bourechak, A., Zedadra, O., Kouahla, M.N., Guerrieri, A., Seridi, H., Fortino, G.: At the confluence of artificial intelligence and edge computing in IoT-based applications: a review and new perspectives. Sensors (Basel) **23** (2023). https://doi.org/10.3390/s23031639
94. Mabindisa, R., Tambwe, K., Mciteka, L., Ross, N.: Organic nanostructured materials for sustainable application in next generation solar cells. Appl. Sci. (Basel). **11**, 11324 (2021). https://doi.org/10.3390/app112311324
95. Singh, H., Sharma, A., Bhardwaj, S.K., Arya, S.K., Bhardwaj, N., Khatri, M.: Recent advances in the applications of nano-agrochemicals for sustainable agricultural development. Environ Sci Process Impacts **23**, 213–239 (2021). https://doi.org/10.1039/d0em00404a

[illegible]

SPRINGER NATURE

GPSR Compliance

The European Union's (EU) General Product Safety Regulation (GPSR) is a set of rules that requires consumer products to be safe and our obligations to ensure this.

If you have any concerns about our products, you can contact us on ProductSafety@springernature.com

In case Publisher is established outside the EU, the EU authorized representative is:

Springer Nature Customer Service Center GmbH
Europaplatz 3
69115 Heidelberg, Germany

The manufacturer's authorised representative in the EU is Springer Nature Customer Service Centre GmbH, Europaplatz 3, 69115 Heidelberg, Germany. If you have any concerns regarding our products, please contact ProductSafety@springernature.com

Printed and bound by CPI Group (UK) Ltd, Croydon, CR0 4YY
30/09/2026
02225324-0016